Antiaging:
Physiology to Formulation

Antiaging: Physiology to Formulation

ISBN 978-1-932633-16-0

Copyright 2006, by Allured Publishing Corporation. All Rights Reserved.

Neither this book nor any part may be reproduced or transmitted in any form by any means, electronic or mechanical, including photocopying, microfilming and recording, or by any information storage retrieval system without permission in writing from the publisher.

Allured Publishing Corporation
362 South Schmale Road, Carol Stream, IL 60188-2787 USA
Tel: 630/653-2155   Fax: 630/653-2192
E-mail: books@allured.com

Table of Contents

INTRODUCTION 1

CHAPTER 1 – PHYSIOLOGY

1. Wrinkles from UVA Exposure, *Motoyoshi et al.* 3
2. Participation of Metalloproteinases in Photoaging, *Rieger* 17
3. Pigmentation, Swelling and Wrinkling in the Eye Area, *Draelos* 25
4. Cell Adhesion: A New Approach to Tissue Protection, *Moreau et al.* 35
5. Measured Dermal Effect of Applying Retin-A,
   Andrassy and Maibach 45

CHAPTER 2 – INTERVENTION AND INGREDIENTS
Botanicals

1. Oat Fractions, *Hart et al.* 49
2. New Active Ingredient for Aging Prevention, *Jay et al.* 61
3. Galactomannan and Xyloglucan: Bio-Active Polysaccharides,
   Pauly et al. 73
4. Take Tea and See, *Brewster* 93
5. Sweet WhiteLupine Extract as a Skin Restructuring Agent,
   Closs and Pautique 95
6. *Centella asiatica* and Skin Care, *Loiseau and Mercier* 103
7. Tinged Autumnal Leaves of Maple and Cherry Trees as Potent
   Antioxidant Sources, *Lee et al.* 111
8. β-(1,6)-Branched β-(1,3)-Glucan in Skin Care, *Kim et al.* 123
9. Artemia Extract: Toward More Extensive Sun Protection,
   Domloge et al. 133
10. Topically Applied Soy Isoflavones Increase Skin Thickness,
    Schmid and Zülli 147

11. Whitening Complex with *Waltheria indica* Extract and
    Ferulic Acid, *Maeyama* 155
12. A Botanical Anti-Sagging/Firming Blend, *Gillis et al.* 161
13. Anti-Wrinkle Activity of Hydrolyzed Ginseng Saponins, *Yeom et al.* ... 167
14. Phytoestrogens: Applications of Soy Isoflavones in Skin Care,
    Kawai 173
15. Multifunctional Ingredients: The Novel Face of Natural,
    Prakash et al. 181
16. Innovative Natural Active Ingredient with Anti-Inflammatory
    Properties, *Segond et al.* 189
17. Low Molecular Weight Tannins of *Phyllanthus emblica*: Antiaging
    Effects, *Chaudhuri et al.* 197
18. A New Active from Germinated Seeds Boosts HSP Expression
    in Skin's Natural Defenses, *Jeanmarie et al.* 213

Hyaluronans

1. Hyaluronan: History and Biochemistry, *Neudecker et al.* 221
2. Hyaluronan: Biology, Pathology and Pharmacology, *Neudecker et al.* ... 229
3. Hyaluronan: Metabolism and Modulation of Hyaluronan
   Levels in Skin, *Neudecker et al.* 245

Lighteners and Whiteners

1. Skin-Whitening Agents, *Zhai and Maibach* 255

Metalloproteinase Inhibitors

1. The Inhibition of Metalloproteinase by *Macrocystis pyrifera* Extract,
   Ansaldi and Bosmann 263
2. Controlling MMPs During Skin Whitening, *Thibodeau* 267

Vitamins

1. Ascorbic Acid and Its Derivatives in Cosmetic Formulations,
   Campos and Silva 271

2. The Synergistic Antioxidative Effect of Ascorbyl 2-Phosphate
   and α-Tocopheryl Acetate, *Jentzsch et al.* 277
3. Skin Antioxidants, *Zhia and Maibach* 287

Water (seaweed, etc.)

1. Efficacy Testing of a Brown Seaweed Extract, *Bennett and Vallee* 293
2. Active Substances From The Sea, *Briand* 301
3. Phytoplankton: The New Frontier for Stress-Relieving Cosmetic
   Ingredients, *Andre et al.* 313

Permeability

1. Penetration of Vitamin A Palmitate into the Skin,
   Campos and Eccleston. 323
2. Defending Against Photoaging: A New Perspective for Retinol,
   Jentzsch et al. 331
3. Moisturizing Potential of d-α-tocopherol, *Tamburic et al.* 339
4. Polyethoxylated Retinamide as an Anti-Wrinkle Agent, *Song et al.* 349

Other

1. A Collagen III Amplifier System, *Chaudhuri et al.* 359
2. Effects of Gelatin-Glycine on Oxidative Stress, *Morganti and Fabrizi.* ... 373
3. Potassium Azeloyl Digycinate: A Multifunctional Skin Lightener,
   Maramaldi and Esposito 383
4. Dietary/Nutritional Supplements: The New Ally to Topical Cosmetic
   Formulations?, *Thibodeau and Lauzier* 391
5. New Laminin Peptide for Innovative Skin Care Cosmetics,
   Bauza et al.. 401
6. Copperceuticals and the Skin, *Pickart* 411
7. Targeting the Cutaneous Nervous Network, *Nicolay and Imbert* 417
8. Reflecting on Soft Focus, *Brewster* 427
9. Building a Better Barrier for the Inside Out, *Yarosh and Brown* 433

10. Applications of Essential Fatty Acid in Skin Care, Cosmetics and Cosmeceuticals, *Brenner* 441

CHAPTER 3 – CLAIM SUBSTANTIATION & TESTING
1. Scratching the Skin Surface, *Nole et al.* 449
2. Photoaging and Photodocumentation, *Pagnoni* 455

CHAPTER 4 – FORMULATION CONSIDERATIONS
1. Influence of a Formulation's pH on Cutaneous Absorption of Ascorbic Acid, *Silva and Campos* 467
2. O/w Emulsions Enriched with Vitamin E, *Djordjevic et al.* 471
3. A Triply Stabilized System to Improve Retinol Stability, *Ji and Choi* 481
4. Delivering Antiaging Actives, *Brewster* 491

INDEX..497

Introduction

The market for anti-aging products remains the largest global segment of skin care with conservative expectations to continue to grow.

This rapidly growing market is driven by the Baby Boomers yet; consumers in their 20s also comprise part of this segment as they seek ways to stave off the aging process.

In addition, consumers desire a feeling to follow a path of wellness, inside and out. Although cosmetic surgeries and medi-spas are also on the rise, consumers prefer noninvasive alternatives to surgery. And those who do prefer longer-lasting forms of improvement, highly effective skin care lines are necessary.

From physiology to finished product, this book is crucial to formulators working interested in growing as rapidly as this sector.

In this compilation, you will find the same expert authors who contribute to *Cosmetics & Toiletries* magazine. Chapters include: ingredients, intervention, testing, claim substantiation and physiology. Topics covered run the gamut, such as: naturals, new methodologies, skin lighteners, novel ingredients and age-effect inhibitors.

For those who are interested in formulating either mass-market products or prestige lines, this book is beneficial in keeping one step ahead of the game.

Enjoy.

The articles have been edited from their original publication for style consistency, but otherwise remain mostly unchanged from their original publication in *C&T* magazine.

Wrinkles from UVA Exposure

Keywords: elastase, anti-inflammatory agents, wrinkles, honeysuckle flower, Engelhardtia chrysolepis, UVA

Research on mechanisms of wrinkle formation and methods of prevention

Recently, the increase in UV irradiation due to progressive depletion of the Earth's ozone layer has become a serious human health concern. In Japan, the frequency of skin cancer of persons living in sunny regions has showed a clear increasing trend from the 1970s to the 1990s. Solar keratosis, a precancerous condition, is also increasing rapidly.

Many UV protective products are now available on the market. However, regardless of the amount of the product used, completely protecting against UV radiation under the intensity of the midsummer sun is difficult. In autumn, winter and spring, UV protective cosmetics are rarely used, making adequate protection impossible.

UV radiation reaching the Earth's surface may increase substantially in the future. What is needed for protection may not only be sunscreen products with high SPF, but also anti-photoaging products, which can be used irrespective of region, season, lifestyle, gender and age.

Of the effects of UV irradiation on the Earth's surface, skin damage caused by UVB irradiation has been extensively investigated. However, the mechanisms of skin damage and photoaging caused by long-term exposure to UVA irradiation have not yet been sufficiently elucidated.

This study investigated the mechanism of skin sagging caused by long-term exposure to UVA radiation, in hope that better understanding of the mechanism will assist the development of new concepts in skin care and new products for the prevention of photoaging.

Chronic UVA and Sagging

We used hairless mice SKh:HR-1[a] as our animal model for chronic UVA damage to skin, modifying the method of Bissett et al.[1] At the start of experiments, the mice were nine to 10 weeks old.

We produced UVA radiation by passing output from eight lamps[b] through a 5 mm thick glass filter to cut out wavelengths below 320 nm. Animals housed individually in glass boxes with walls 2 mm thick were irradiated five times weekly for six months with

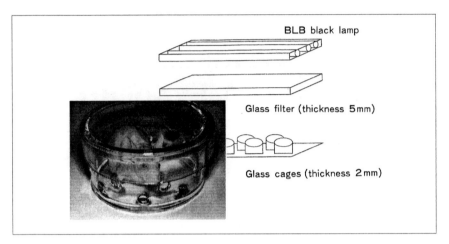

Figure 1. Experimental apparatus

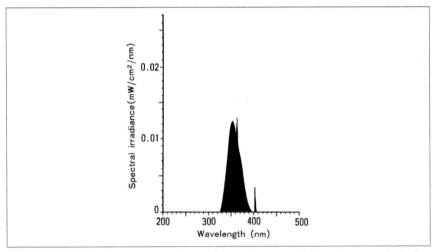

Figure 2. Spectrum of UVA light from the Toshiba BLB lamp passed through a 5 mm thick glass filter

27 J/cm^2 UVA radiation per daily exposure. The irradiation energy was determined using a radiometer.[c] The radiation spectrum through the glass boxes was determined using an spectroradiometer[d] (Figures 1 and 2).

Visual examination: The degree of sagging was graded on a scale of 0-3 (Table 1), according to the method of Bissett et al.[1]

After long-term exposure (24 weeks) to UVA radiation, hairless mice exhibited large, loose folds (sagging)[1] on the dorsal and lateral skin. At three months after the first exposure, the dorsal skin became blanched and the skin texture became rough. After 6 months, there was complete blanching of the skin, nodular texture

[a]Charles River Laboratories, Boston MA, USA
[b]Toshiba-BLB, Toshiba, Tokyo, Japan

Table 1. Scale for grading sagging, 0 to 3

0 = Pale pink color, fine striation (head to tail), no loose folds
1 = Slight blanching, slight fine striation, no loose folds
2 = Complete blanching, no fine striation, slight loose folds
3 = Complete blanching, no fine striation, large loose folds

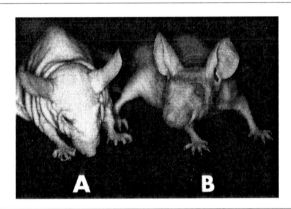

Figure 3. Visible changes at 24 weeks
    A = UVA irradiated hairless mouse
    B = nonirradiated hairless mouse

and loose folds (Figures 3, 4). Figure 5 shows the time-course for development of sagging.

Histological examination: On the last day of the experimental period, biopsies of the dorsal skin were performed on five animals per group. We fixed tissue specimens in formalin, embedded them in paraffin and sectioned to 5 µm in thickness. The sections were stained with H-E, van Gieson's and Luna stains, then examined for several parameters. One of the tissue specimens was reserved for transmission electron microscopy (TEM).

Histological and TEM findings after exposure to UVA for 24 weeks were difficult to analyze. Marked thickening of the epidermis and dermis was observed. In the dermis, an increase in mast cells and definite hypertrophic cysts were observed. In addition, partial infiltration into the dermis of inflammatory cells, including polymorphonuclear leukocytes (PMNs), was observed. In the upper layer of the dermis, there was a marked decrease in collagen fibers. As for elastin fibers, aggregations were intermingled with areas of partial absence (Figures 6 through 14).

Determination of elastase activity: We removed the subcutaneous tissue from circular, dorsal skin tissue specimens (2.1 cm in diameter) not used in the histological

[c]Topcon-UV radiometer 305/365 DII, Topcon, Tokyo, Japan
[d]Ushio spectroradiometer USR-20B, Ushio Denki, Yokohama, Japan

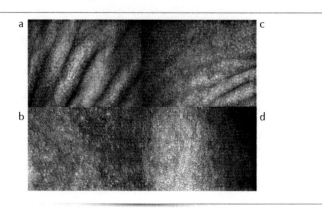

Figure 4. Visible changes at 24 weeks
    a,b = UVA irradiated hairless mouse skin
    c,d = nonirradiated hairless mouse skin
    a,c = Dorsal (back) skin
    b,d = Abdominal skin

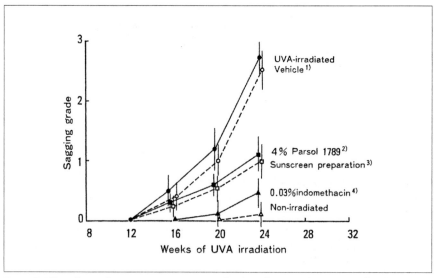

Figure 5. Effect of topical sunscreen preparation and indomethacin on the sagging of hairless mouse skin caused by UVA irradiation
    1) Control vehicle:1/1 w/w propylene glycol and ethanol
    2) Butyl methoxydibenzoylmethane in carbitol
    3) 5% Butyl methoxydibenzoylmethane and 3% octyl methoxycinnamate
    4) Nonirradiated control, treated with control vehicle

examinations. These were homogenized in a glass homogenizer with 2 mL of buffer (0.1 M N-2-hydroxyethyl-piperazine-N'-2-ethanesulfonic acid and 0.5 M sodium chloride, pH 7.5). We centrifuged the homogenate at 15,000 rpm for 10 min. We assayed enzyme activity using 0.1 mL of supernatant and 1.4 mL of enzyme substrate

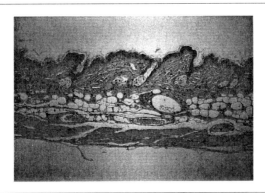

Figure 6. Intact (nonirradiated) skin; H-E stain, 400x

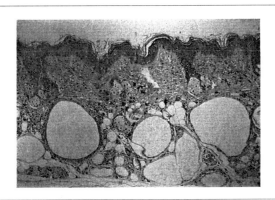

Figure 7. Thickening and hyperplasia of epidermis and dermis after 24 weeks UVA exposure; dermal mast cells increase in number while dermal cysts increase in both number and size. H-E stain, 400x

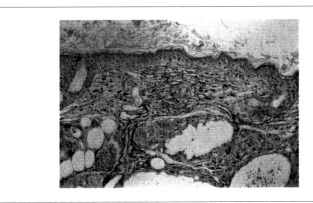

Figure 8. Increase of elastin fibers in focal area of UVA damage after 24 weeks UVA exposure; Luna stain, 400x

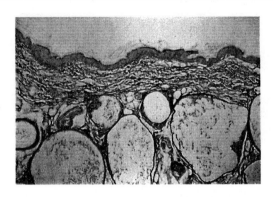

Figure 9. Damaged collagen fibers in the upper dermis after 24 weeks UVA exposure; Van Gieson stain, 400x

solution at 37°C for 1 h. The amount of p-nitroaniline released was determined spectrophotometrically at 405 nm using a spectrophotometer[e] at 25°C.

The assay for elastase activity showed a significant increase in the activity in UVA irradiated skin compared with nonirradiated skin (Figure 15).

Chronic UVA Exposure

Peroxidation of membrane lipids in the epidermis and dermis may be promoted by active oxygen and other free radicals produced by long-term exposure to UVA in hairless mice.[9] Skin damage and inflammation may be caused by the resultant peroxylipids.[10] Considering our histological and biochemical results, the infiltration into the dermal layer of cells that caused inflammation observed in UVA irradiated hairless mice supports the idea that mild and persistent inflammation occurs in UVA irradiated skin. Among the inflammatory cells infiltrated into the dermal layer, PMNs in particular release large amounts of elastase. The released elastase may attack the proteins of connective tissue, resulting in damage of elastin and collagen fibers, and finally causing sagging (Figure 16).

Histological and TEM examination of irradiated tissue specimens show partial absence and aggregation of elastin fibers in the dermis, supporting the above stated hypothesis. Normally elastin fibers are not well developed in the skin of hairless mice, and a partial increase in elastin fibers occurred by a feed back mechanism after degradation of elastin fibers by elastase, which collapsed to form aggregates.

Screening for Photoaging Inhibition

For mice receiving topical treatment of drugs, 50 μL of the test solution was applied to the dorsal skin surface 1 h prior to each exposure to UVA radiation. The test solution was applied to the skin using an Eppendorf pipette and spread evenly over the entire dorsal skin surface using the flat part of the pipette tip.

[e]Hitachi U-3210, Hitachi, Tokyo, Japan

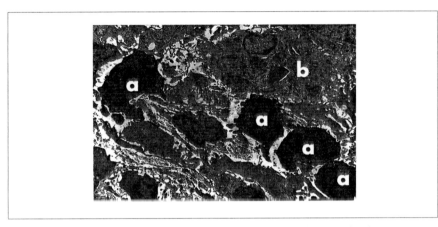

Figure 10. PMNs (a) and a macrophage (b) are seen in the dermis after 24 weeks of UVA exposure; TEM, scale bar = 1 μm

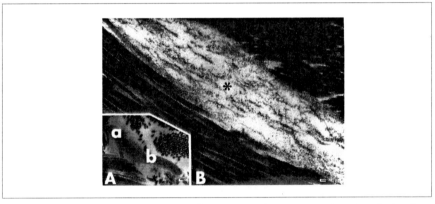

Figure 11. TEM of elastin fibers; A = normal fibers in intact skin, tannic acid positive elastic microfilaments (a) and tannic acid negative elastic microfilaments (b). B = damage after 24 week UVA exposure, tannic acid nagative indistinct elastic microfilaments (*). Tannic acid positive elastic microfilaments disappeared. Tannic acid stain. Scale bar = 0.1 μm

For sunscreen products, the test solution was applied to the skin using a disposable syringe and spread evenly over the skin with a glass rod to achieve 2 mg/cm^2 coverage.

To investigate our hypothesis on the photoaging mechanism, it was necessary to determine which mechanism, antioxidation or anti-inflammation, might inhibit UVA induced skin damage. We therefore included various antioxidants and anti-inflammatory agents in our photoaging inhibition study on hairless mice. In addition, we screened several Chinese homeopathic preparations for both antioxidative and anti-inflammatory activities

UV induced erythema: We used male Hartley guinea pigs weighing 700 to 900 g, with six animals per group. Wide adhesive tape punctured with six 1.5 cm x 1.5 cm holes (two rows each with three holes) was applied on the shaved dorsal skin,

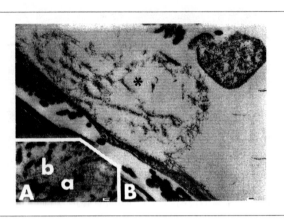

Figure 12. TEM of elastin fibers; A = normal fibers in intact skin, tannic acid positive elastic microfilaments (a) and tannic acid negative elastic microfilaments (b). B = damage after 24 weeks UVA exposure, tannic acid nagetive indistinct elastic microfilaments (*). Tannic acid positive elastic microfilaments disappeared. Tannic acid stain. Scale bar = 0.1 μm

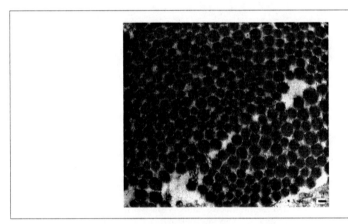

Figure 13. Intact (nonirradiated) skin; TEM, scale bar = 1 μm

and UV radiation was irradiated over the tape.[f] The amount of irradiation energy was controlled at 768 mJ/cm². After 24 h, a uniform erythematous reaction was observed at the six exposed sites. We administered 5 μl/site of the test drugs five times, at 1 h intervals beginning immediately after UV exposure. We evaluated the erythematous reaction at 24 h after exposure according to Table 2.

We calculated percent inhibition in erythema caused by each drug using the following equation:

$$\% \text{ inhibition in erythema} = \{(Ev_c - Ev_d)/Ev_c\} \times 100$$

Ev_c = Mean erythema value at the vehicle-treated site
Ev_d = Mean erythema value at the drug-treated site

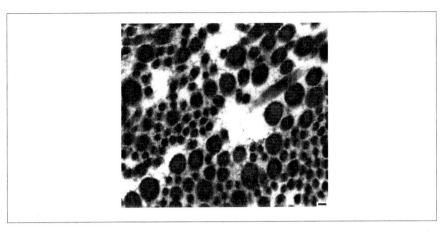

Figure 14. Skin after 24 weeks UVA irradiation, showing huge and small collagen fibers in the upper dermis, TEM, scale bar = 1 μm

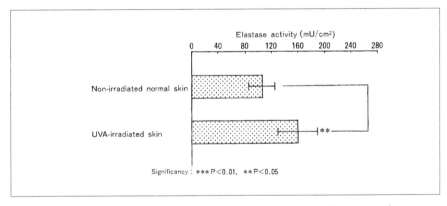

Figure 15. Effects of 24 weeks UVA irradiation on elastase activity in hairless mouse skin

Sunscreen preparations: Although a 4% solution of butyl methoxydibenzoylmethane[g] in carbitol and a sunscreen mixture containing 5% butyl methoyxydibenzoylmethane and 3% octyl methoxycinnamate absorb radiation, they only moderately inhibited the photoaging response (skin sagging) caused by long-term UVA irradiation in hairless mice (Figure 5).

Antioxidants: Ascorbic acid (0.3%) and α-tocopherol (0.3%) did not inhibit the photoaging response in hairless *mice, as shown in Figure 17.*

Anti-inflammatory agents: Dipotassium glycyrrhetic acid at 0.3% and 0.3% ε-aminocaproic acid each weakly inhibited the photoaging response in hairless mice, but there was no significant difference from the control group receiving the vehicle. The potent anti-inflammatory agents indomethacin (0.03%) and ibuprofen piconol (0.3%) markedly inhibited the photoaging response in hairless mice as shown in Figure 17. The skin of hairless mice treated with these potent anti-inflammatory

[g]Toshiba FL30-SE lamp, Toshiba, Tokyo, Japan

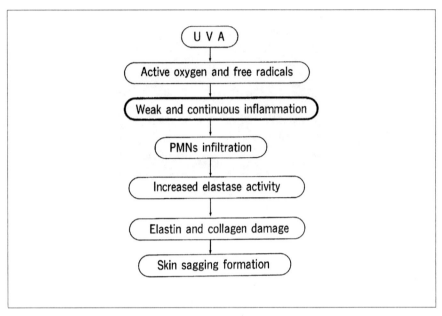

Figure 16. Hypothetical mechanism for development of sagging skin in chronically UVA irradiated hairless mouse skin

Table 2. Erythema scale, 0 to 3
0 = No erythematous reaction is observed
1 = Slight erythema or erythema with vague border
2 = Moderate erythema with well-defined border
3 = Intense erythema with well-defined border (occasionally associated with edema)

agents showed that inflammatory cell infiltration into the dermis was markedly inhibited. The inhibitory effect of these anti-inflammatory agents on the photoaging response in hairless mice was proportional to the anti-inflammatory potency of these agents (Figure 18).

New Anti-Inflammatory Substances

Because potent anti-inflammatory agents such as indomethacin have been reported to produce many adverse reactions,[8] they cannot be added to cosmetics for reasons of safety. We therefore screened various Chinese homeopathic preparations that were considered to have less tendency to cause adverse reactions. Among them, we selected the extracts of honeysuckle flower and *Engelhardtia chrysolepis* Hance, which showed moderate anti-inflammatory effects. These two Chinese homeopathic preparations markedly inhibited photoaging in hairless mice. They inhibited

---
9Parsol 1789, Hoffmann LaRoche, Basel, Switzerland

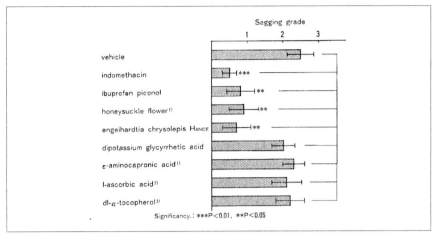

Figure 17. Inhibitory effects of various drugs on UVA induced skin sagging in hairless mice; indomethacin was used at 0.03%, all other drugs at 0.3%. ***P < 0.01, **P < 0.05. Vehicle was propylene glycol and ethanol (1/1 w/w) except as marked:
1) Distilled water/ethanol 1/1
2) Distilled water/ethanol 3/2
3) Ethanol

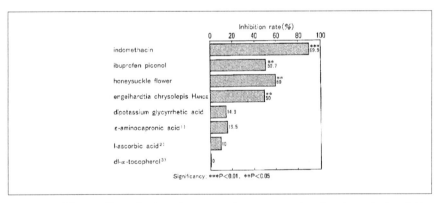

Figure 18. Inhibitory effects of various drugs on erythematous reactions in guinea pig dorsal skin; indomethacin was used at 0.03%, all other drugs at 0.3%. ***P < 0.01, **P < 0.05. Vehicle was propylene glycol and ethanol (1/1 w/w) except as marked:
1) Distilled water/ethanol 1/1
2) Distilled water/ethanol 3/2
3) Ethanol

both infiltration of PMNs into the dermis and elastase activity in the skin of UVA irradiated hairless mice as can be seen in Figures 19, 20 and 21. The capacity for UV absorbance of these two preparations was extremely low as compared with butyl methoxydibenzoylmethane, as shown in Figure 22.

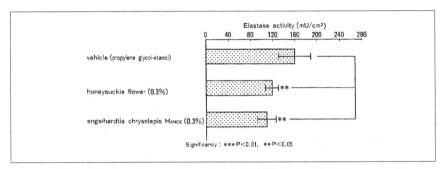

Figure 19. Effects of various drugs on elastase activity in UVA irradiated hairless mouse skin

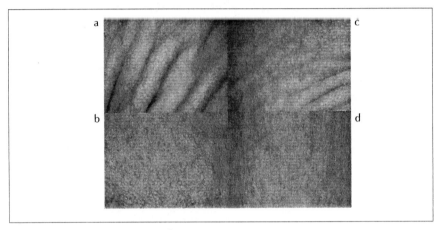

Figure 20. Visible changes at 24 weeks
    a,b = Vehicle-treated UVA irradiated hairless mouse skin
    c,d = UVA irradiated hairless mouse skin treated
    a,c = Dorsal (back) skin
    b,d = Abdominal skin

The usefulness of sunscreen preparations to prevent or reverse photoaging in hairless mice has been well documented.[5] Repair of collagen degradation, glycosaminoglycan deposition and elastin aggregation was observed with administration of high SPF sunscreen preparations to mice.[6] However, whether or not the sunscreen preparations may be truly useful and whether the usefulness may be altered by the vehicle remains unclear.

Recently, significant inhibition of elastin aggregation was reported in humans treated with UVA/UVB sunscreen preparations as compared with the vehicle. However, no significant differences were reported in inflammatory cell infiltration into the dermis, epidermal thickening and atrophy of keratinocytes between sunscreen-treated or vehicle-treated groups.[3]

In this study, the sunscreen preparation containing UVA absorbers moderately inhibited the photoaging response caused by long-term UVA irradiation. However, the skin of hairless mice treated with UVA absorbers showed some epidermal thickening, inflammatory cell infiltration into the dermis and partial elastin aggregation.

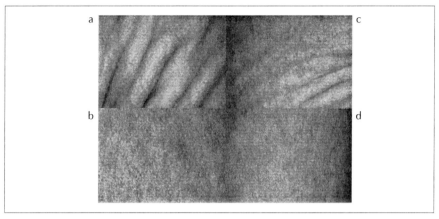

Figure 21. Visible changes at 24 weeks
    a,b = Vehicle-treated UVA irradiated hairless mouse skin
    c,d = UVA irradiated hairless mouse skin treated with
    a,c = Dorsal (back) skin
    b,d = Abdominal skin

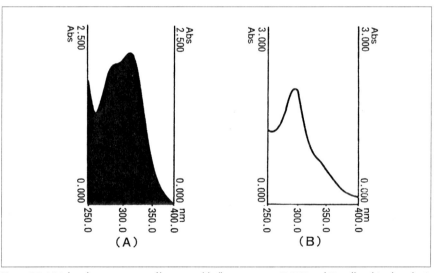

Figure 22. UV absorbance spectra of honeysuckle flower extract (0.1%) and *Engelhardtia chysolepis* (0.01%); K = UV absorbance/sample conc. in g/l

	K	λ peak (nm)
Honeysuckle flower	2.1	323
Engelhardtia chrysolepis	19.2	290
Butyl methoxydibnzoylmethane	111.0	357

Based on these results, we concluded that UVA absorbing sunscreen preparations are somewhat, but not completely, effective in preventing photoaging.

Most commercially available sunscreen preparations that offer protection from UV radiation contain chemical absorbents and scattering agents. Our study results suggest that the use of sunscreen preparations alone may not be sufficient in preventing photoaging caused by UVA irradiation. Thus, the question arises: What should we use for the anti-photoaging cosmetics for the future?

We believe that anti-photoaging cosmetics should have a built-in anti-inflammatory effect and should be intended for daily use by all people, irrespective of geographical region, season, lifestyle, gender or age. It seems that skin care cosmetics for daily use containing the extracts of honeysuckle flower and *Engelhardtia chrysolepis* Hance may be useful for protecting the skin from photoaging.

Conclusions

1. The infiltration of PMNs into the dermis as observed in hairless mice submitted to long-term exposure to UVA radiation may play an important role in the photoaging response. The PMNs may release elastase which attack connective tissue proteins, resulting in damages of elastin and collagen fibers, and finally causing sagging.
2. Sunscreen preparations containing only UVA absorbents and scattering agents do not effectively inhibit the photoaging response to UVA irradiation.
3. The photoaging response was significantly inhibited by anti-inflammatory agents such as indomethacin, the inhibitory effect of which was proportional to the anti-inflammatory potency. However, anti-inflammatory agents such as indomethacin cannot be incorporated into cosmetics because of the possible adverse reactions. Therefore, we selected alternative anti-inflammatory agents—honeysuckle flower and *Engelhardtia chrysolepis* Hance extracts—because they showed the greatest anti-inflammatory effect among the Chinese homeopathic preparations tested. These two homeopathic preparations effectively inhibited the photoaging response and elastase activity.
4. The use of cosmetics having both anti-inflammatory effect and high safety level is a new anti-photoaging concept. Such cosmetics, when used on a daily basis, can prevent sagging skin caused by exposure to UVA radiation.

—**Katsuhiro Motoyoshi, PhD, Yutaka Ota, Yuko Takuma and Masanori Takenouchi,**
Pola R&D Laboratories, Pola Corporation, Yokohama, Japan

References

1. DL Bissett et al, *Photochem Photobiol* 46 367-378 (1987)
2. DL Bissett et al, *Photochem Photobiol* 50 763-769 (1989)
3. AS Boyd et al, *J Am Acad Dermatol* 33 941-946 (1995)
4. R Kasai et al, *Chem Pharm Bull* 36 (10) 4167-4170 (1988)
5. LH Kligman et al, *J Invest Dermatol* 78 181-189 (1982)
6. LH Kligman et al, *J Invest Dermatol* 81 98-102 (1983)
7. NJ Lowe et al, *J Invest Dermatol* 105 739-743 (1995)
8. Nippon Igaku Joho Center, Iryoyaku Nippon Iyakuhinshu, 10th ed, Yakugyo Jihosha, Tokyo (1986)
9. R Ogura et al, *J Invest Dermatol* 97 1044-1047 (1991)
10. H Ueda, *Active Oxygen & Free Radical* 3 291-297 (1992)

Participation of Metalloproteinases in Photoaging

Keywords: matrix, metalloproteinases, UVB, photoaging, minimal erythemal dose (MED)

The basic biology of metalloproteinases and a novel approach to our understanding of photoaging

Matrix Metalloproteinases

Extracellular matrix metalloproteinases (MMP) have been studied for many years. The best known of this ever-increasing number of proteolytic enzymes are the collagenases.[1]

The body's requirement for the presence of MMPs arises from the need for degradation of essentially insoluble polymeric ground substance components. Proteolysis of fibrillar macromolecules requires fragmentation (or partial solubilization) to allow the better known phagocytic pathways to complete the task of removing unwanted or damaged polypeptides. The role played by MMPs in dermal homeostasis is complex and not fully understood.

MMPs degrade extracellular matrices during wound repair and remove damaged tissue. They play a part in the remodeling processes during tissue repair. The number of identified MMPs is quite large (about 15), and many of them are capable of degrading the same substrate. This redundancy, shown in Table 1, is unexplained, and many unknowns remain. It is generally agreed[1,2] that plasmin converts proenzymes to active MMPs. Once formed, their proteolytic activity must be controlled. Removal of zinc atoms from a MMP by chelation, for example with ethylenediamine tetraacetic acid, destroys the activity. In vivo, macroglobulins and the so-called tissue inhibitors of MMPs (TIMPs) control the activity of the MMPs. TIMPs, especially, have attracted the interest of skin scientists. TIMPs form non-covalent bi-molecular complexes with active MMPs and inhibit their proteolytic activity. The amino acid sequences of some TIMPs are known, but the mechanism for regulating TIMP expression is not fully understood.

Proteolytic action of MMPs: Despite some uncertainties about the precise mechanism, the proteolytic action of MMPs in healthy skin is evidently necessary. In

Table 1. Partial Listing of Metalloproteinases

MMP-1	Matrix Collagenase (Fibroblast Collagenase)	Collagens I, II, III, VII, and X
MMP-8	Neutrophil Collagenase	Collagens I, II, and III, Link Protein, Aggrecan
MMP-13	Collagenase 3	Collagens I, II, and III
MMP-18	Collagenase 4	Collagen I
MMP-2	Gelatinase A (Mol. Wt. 72K)	Gelatins, Collagens I, IV, VII, and XI, Fibronectin, Laminin, Elastin
MMP-9	Gelatinase B (Mol. Wt. 92K)	Gelatins, Collagens IV, V, and XIV, Aggrecan, Elastin
MMP-3	Stromelysin 1	Aggrecan, Gelatin, Fibronection, Laminin, Collagens III, IV, IX, and X
MMP-10	Stromelysin 2	Aggrecan
MMP-14	(membrane type)	Collagens I, II, and III, Laminin
MMP-15	—— ?	
MMP-17	—— ?	
MMP-7	Matrilysin	Aggrecan, Fibronectin
MMP-11	Stromelysin 3	Fibronectin
MMP-12	Metalloelastase (Macrophage)	Elastin

fibrotic skin diseases, such as scleroderma, excessive TIMP-3 levels are suspected.[3] The expression of mRNA for TIMP-3 is enhanced by diverse mitogenic stimuli. High levels of TIMP-3 have been shown to be accompanied by increased amounts of collagen I in cultured fibroblasts.[4]

TIMP

A specific TIMP, TIMP-1, has also been implicated in the accumulation of collagen in scleroderma.[5] Additional reports on the inhibition of MMPs have appeared in recent years with an emphasis on wound healing.[6-8] In addition, TIMP-2 reportedly has a somewhat different beneficial effect because it inhibits tumor cell metastasis in basal cell carcinoma.[9] Growth stimulating effects on keratinocytes of TIMP have also been reported.[10] These observations are pertinent to this discussion because thickened epidermis has been associated with photo-adaptation of exposed skin.

Therefore, it is safe to assume that various TIMPs have different effects on skin and that the expression of TIMP genes plays an important role in skin health. By analogy, the activation of MMPs in the extracellular environment must also contribute to skin health. The precise balance between proteinases and their inhibitors in skin health is a topic of serious research throughout the world. For example, Boelsma et al.[11] reported that a proteinase inhibitor, Skin Derived Antileukoproteinase (SKALP), was elevated in skin exposed to SLS or oleic acid and that this skin-derived inhibitor is a marker for skin irritancy. The expression of SKALP was induced by serum, epidermal growth factor, and fibroblasts but was inhibited by 1,2-dihydroxyvitamin D_3 and retinoic acid in vitro; it was unaffected by tocopherol and ascorbic acid. There is no evidence at this time that SKALP also inhibits MMPs. In contrast, McIlrath et al.[12] reported that the MMPs in pigskin cultures could be modified differentially. MMP-9 was increased by retinoic acid, while MMP-2 was slightly decreased. Retinoic acid appears to modify dermal proteinase activity. On the other hand, these results suggest that the repair of photoaged skin by retinoic acid is not simply due to enhanced proteolytic activity of the extracellular matrix.

It is becoming increasingly apparent that MMP activity and its induction is significant in a variety of human skin conditions.[13-18] MMP activity is initiated by a plasminogen activator commonly expressed by basal keratinocytes. One unexpected phenomena is the observation that TIMP stimulates the growth of human keratinocytes in tissue culture but has no effect on fibroblasts.[10]

Proteolysis in Photoaging

Most readers are familiar with Kligman's work on photoaging and its reversal in the hairless mouse.[21-23] Her findings can be summarized by the statement that UVB irradiation can cause damage to elastic fibers, collagen and glycosaminoglycans (GAG) in the extracellular matrix (ECM) within the dermis. Two aspects of her reports are especially pertinent: the damage was attributed to overproduction of dermal constituents, and reversal was enhanced by retinoic acid. Such repair in the absence of UVB exposure is a relatively slow reconstruction process.

The generally accepted mechanisms for photoaging include direct excitation of DNA bases by UVB and dimeric photo additions, type II photoreaction (singlet oxygen), type I photoreactions (electron abstraction, for example, free radical formation), or superoxide generation (followed by a Fenton-type reaction). The end result of these reactions is attack on the bases in the DNA.

Data on photoaging of human skin under normal conditions are not easily available. Thus, Kligman relied on Skh-1 and Skh-2 mice. More recently, results on photoaging obtained with cultured human fibroblasts were reported from Japan.[24] The fibroblast culture was exposed to externally generated reactive oxygen species (ROS), and the end-point parameter was synthetic activity of the fibroblasts. Elastin and GAG were increased in this experiment, while collagen production was decreased. Interestingly, the investigator reports that the level of TIMP-1 was increased by ROS. These results support the suspicion that ROS may effect biological (metabolic) changes in the dermis.

The approach by Fisher et al. provides a different explanation for the mechanism of photoaging with participation by MMPs.[19] In their first paper, they report on their investigation of UVB radiation damage on human buttock skin. They noted that messenger ribonucleic acids (RNAs) for MMP and subsequent MMP activation occur very rapidly and at doses "well below those that cause skin reddening." Transcription factors of MMP genes are activated by very low doses of UVB (0.5 mW/cm^2) in subjects exhibiting minimal erythemal doses (MEDs) of 30-50 mJ/m^2. The reported observations are of considerable importance. Most research on the impact of UV light on skin depends on erythema for quantification. If Fisher's data are valid, the MED may provide a false sense of security with regard to photodamage.

The investigators specifically assayed for a 54K collagenase (MMP-1) and for a 92K gelatinase (MMP-9) after exposures below one MED and report a sharp and rapid increase in these two proteinases. The authors also noted that retinoic acid effects transrepression of the transcription factor. The striking feature of these biological events after irradiation is the rapidity with which the system responds to low doses of UVB. The investigations described are impressive and evidently have not been challenged.

No response to this paper, nor a follow-up by other scientists, appeared for some time. The publication presented an explanation for dermatoheliosis and photoaging that did not call for the involvement of reactive oxygen species or free radicals. The authors did not provide a mechanism for the induction of the MMP messenger RNA, and the terms free radical or ROS cannot be found in their papers. They described effects in the dermis after irradiation as repair processes requiring substantial destruction of damaged ECM components by specialized enzymes. The only contribution to this idea was provided by Tanaka,[24] who observed upregulation of TIMP-1.

Another feature that made the scientific community uncomfortable was the speed with which the repair process (MMP formation) was initiated, and that it occurred at levels below the MED. The latter must be distressing to those who assess skin damage and its potential prevention on the basis of erythema. Instead, Fisher's data suggest that insolation, at less than one MED, triggers photoactivation of MMPs in the dermis.

The second shoe fell about a year later in a second publication[20] by the Fisher group. In this case, too, the authors exposed buttock skin of Caucasians to low doses of UVB. Skin samples for analysis were obtained by biopsy or dermatotome from irradiated and non irradiated sites.

Collagenase (MMP-1, MMP-9, and MMP-3; cf Table 1) m-RNAs appeared to be almost absent in non-irradiated skin, but rapidly elevated after irradiation. The m-RNA for these enzymes were found predominantly in the epidermis. The enzymatic activity was found in the epidermis and dermis, but stromelysin (MMP-3) activity was located primarily in the dermis. Addition of TIMP reduced hydrolytic activity to normal. Continued (once a day) exposure to UV maintained the increased levels of MMPs for at least four to five days.

Pretreatment of the skin with all transretinoic acid before UV exposure reduced the induction of the three studied MMPs. A significant portion of the total MMPs is synthesized in the epidermis and then transported to the dermis. Retinoic acid appears to inhibit MMPs without elevating levels of TIMPs.

UV Induced Photoaging

These observations prompted the authors to provide a hypothetical model for UV induced photoaging.[20] Levels of UV light, causing no detectable sunburn, induce formation of MMPs in epidermal keratinocytes and dermal fibroblasts, with resulting degradation of the ECM. Keratinocytes also produce TIMPs, reducing the activity of MMPs. This process is followed by formation of imperfectly repaired collagen. During multiple intermittent UV exposure, this process, if repeated, results in the formation of more severely damaged collagen, for example, the solar scars visualized by Kligman.

In a third paper by Fisher et al., the authors began the arduous task of sorting out sequences of biological reactions which, after UV irradiation, trigger the induction of MMPs in human skin.[25] It is noteworthy that another group of investigators recently reported on the upregulation of MMPs and of structural protein formation by cytokines and integrins after UV irradiation of hairless mice.[26] It seems safe to predict that current research will clarify the relationship between UV light, MMP and photoaging.

The following needs to be done to clarify the mysteries of MMP and photoaging:

- Research for a sound reason for the poor repair process described in the Fisher model. One would expect that the repair process, or the formation of new collagen, should be "perfect"
- Research into agents that are totally unreactive with ROS or free radicals, but still may prevent photoaging
- Research leading to a decision as to whether inhibition of MMPs by diverse approaches would be appropriate because the enzymatic destruction of damaged ECM proteins or GAGs is likely to be required

Clearly, direct confirmation of the ideas presented by Fisher et al. at the University of Michigan would be welcome. In the meantime, cosmetic formulators might start the search for actives having the potential of controlling the repair process following the action of MMPs.

Protection Against Photoaging

Full recognition of the significance of the Fisher et al. findings has been slow. However, publicity in the May/June 1998 $R_x emedy$ is likely to alert professionals, as well as laymen, to the hazards of sunbathing. If the very short exposures to UVB studied by Fisher can trigger a rapid cascade of events in human skin, even a partial UV blocker is unlikely to afford full protection against skin photoaging.

Much current thinking about photoaging relies on the oncogenic activity of insolation. There is considerable debate on the merits of sunscreens and the optimal sun protective factor in the prevention of skin cancers. This debate obviously impacts the benefits of sunscreens in the prevention of photoaging. Despite regulatory uncertainties, and in the absence of sound clinical data, a rational argument can be made for the use of sunscreens for reducing the damage of sun (UV) exposure. Treatment options for photodamaged skins are currently limited to retinoids and facial peeling.

It would appear premature to abandon the use of agents acting as antioxidants or free radical scavengers. There is sufficient clinical evidence to support the use of compounds that can inactivate ROS and similar UV-created substances to reduce pathological consequences.

For future control of photoaging, it would appear wise to examine a battery of diverse ingredients that can inhibit the formation of this and other pathological skin conditions. Retinoids, as a group, seem to be able to interfere with photoaging, but the mechanism for their action remains obscure. Some recent patents (PCT Int. Appl. WO 98 113,017 and PCT Int. Appl. WO 98 113,018) extol the benefits of retinal and retinyl esters in skin care products. There is no reason to assume, however, that retinoids are the only chemicals able to alleviate photoaging. Even if the mechanism proposed by Fisher et al. is not fully confirmed, the findings of these investigators suggest that there may be other cosmetic ingredients that can modulate photoaging by controlling some of the biochemical or transcriptional events in the skin.

Summary

Investigations published in recent years relate the stigmata of photoaging to improperly repaired ECM proteins after low levels of insolation. At this time, it is not clear how the repair process, requiring proteolytic activity in the dermis as a first step, occurs and why it should fail to restore the ECM to its original state. Following the hydrolytic action of MMPs, the proteolytic potential of MMPs must be reduced to allow synthesis of replacement proteins. The solution to this conundrum can be expected to provide formulators and clinicians with unexpected tools for preventing or repairing photoaged skin.

—**Martin Rieger,** *M&A Rieger Associates, Morris Plains, New Jersey USA*

References

1. H Birkedal-Hansen, WGIMoore, MKBodden, LJ Windsor, B Birkedal-Hansen, A DeCarlo and JA Engler, Matrix metalloproteinases: A review, Critical Reviews, *Oral Biology and Medicine* 4 (2) 197-250 (1993)
2. G Murphy, R Ward, J Gavrilovic and S Atkinson, Physiological mechansim for metalloproteinase activitation, Matrix, Special Suppl. 1 224-230 (1992)
3. L Mattila, K Airola, M Ahonen, M Hietarinta, C Black, U Saarialho-Kere and VM Kähäri, Activation of tissue inhibitor of metalloproteinases-3 (TIMP-3) mRNA expression in scleroderma skin fibroblasts, *J Invest Dermatol* 110 416-421 (1998)
4. EC LeRoy, M McGuire and N Chen, Increased collagen synthesis by scleroderma skin fibroblasts in vitro, *J Clin Invest* 54 880-889 (1974)
5. K Kikuchi, T Kadomo, M Furue, and K Tamaki, Tissue inhibitor of metalloproteinase 1 (TIMP-1) may be an autocrine growth factor in scleroderma fibroblasts, *J Invest Dermatol,* 108 281-284 (1997)

6. F Grinnell and M Zhu, Fibronectin degradation in chronic wounds depends on the relative levels of elastase, 1-proteinase inhibitor, and macroglobulin, *J Invest Dermatol* 106 335-341 (1996)
7. PW Park, K Biedermann, L Mecham, DL Bissett and RP Mecham, Lysozyme binds to elastin and protects elastin from elastase-mediated degradation, *J Invest Dermatol* 106 1075-1080 (1996)
8. F Grinnell, M Zhu, and WC Parks, Collagenase-1 complexes with 2-macrogobulin in the acute and chronic would environments, *J Invest Dermatol* 110 771-776 (1998)
9. SN Wagner, HM Ockenfels, C Wagner, HP Soyer, and M Goos, Differential expression of tissue inhibitor of metalloproteinases-2 by cutaneous squamous and basal cell carcinomas, *J Invest Dermatol* 106 321-326 (1996)
10. B Bertaux, W Hornebeck, AZ Eisen, and L Dubertret, Growth stimulation of human keratinocytes by tissue inhibitor of metalloproteinases, *J Invest Dermatol* 97 679-685 (1991)
11. E Boelsma, S Gibbs and M Ponec, Expression of skin-derived antileukoproteinase (SKALP) in reconstructed human epidermis and its value as a marker for skin irritation, *Acta Derm Venereol (Stockh)* 78 107-113 (1998)
12. EM McIlrath, U Santhanam, and MR Greene, The effect of retinoic acid treatment on matrix metalloproteinase (MMP) and plasminogen activator (PA) production by pig skin organ culture, *Brit J Dermatol* 138 752 (1998)
13. R DeCastro, Y Zhang, H Guo, H Kataoka, MK Gordon, BP Toole and C Biswas, Human keratinocytes express EMMPRIN, an extracellular matrix metalloproteinase inducer, *J Invest Dermatol* 106 1260-1265 (1996)
14. M Weckroth, A Vaheri, J Lauharanta, T Sorsa and YT Konttinen, Matrix metalloproteinases, gelatinase and collagenase, in chronic leg ulcers, *J Invest Dermatol* 106 1119-1124 (1996)
15. A Oikarinen, M Kylmäniemi, H Autio-Harmainen, P Autio and T Salo, Demonstration of 72-kDa and 92-kDa forms of type IV collagenase in human skin: Variable expression in various blistering diseases, induction during re-epithelialization, and decrease by topical glucocorticoids, *J Invest Dermatol* 101 205-210 (1993)
16. DR Yager, L-Y Zhang, H-X Liang, RF Diegelmann and IK Cohen, Wound fluids from human pressure ulcers contain elevated matrix metalloproteinase levels and activity compared to surgical wound fluids, *J Invest Dermatol* 107 743-748 (1996)
17. K Airola, T Reunala, S Salo, and U Saarialho-Kere, Urokinase plasminogen activator is expressed by basal keratinocytes before interstitial collagenase, stromelysin-1, and laminin-5 in experimentally induced dermatitis herpetiformis lesions, *J Invest Dermatol* 108 7-11 (1997)
18. L Vaalamo, L Mattila, N Johansson, A-L Kariniemi, M-L Karjalainen-Lindsberg, V-M Kähäri, and U Saarialho-Kere, Distinct populations of stromal cells express collagenase-3 (MMP13) and collagenase-1 (MMP-1) in chronic ulcers but not in normally healing wounds, *J Invest Dermatol* 109 96-101(1997)
19. GJ Fisher, SC Datta, HS Talwar, Z-Q Wang, J.Varani, S Kang and J.J. Voorhees, Molecular basis of sun-induced premature skin aging and retinoid antagonism, *Nature* 379 335 (1996)
20. GJ Fisher, Z-Q Wang, SC Datta, HS Talwar, J. Varani, S Kang, and J.J. Voorhees, Pathophysiology of premature skin aging induced by ultraviolet light, *N Engl J Med* 337 1419-28 (1997)
21. L Kligman, Connective tissue photodamage in the hairless mouse is partially reversible, *J Invest Dermatol* 88 21s-17s (1987)
22. L Kligman, Skin changes in photoaging: characteristics, prevention, and repair, in *Aging and the Skin*, AK Balin and AM Kligman (eds), Raven Press, New York (1989)
23. L Kligman, The hairless mouse and photoaging, *Photochem Photobiol* 54 1109-18 (1991)
24. H Tanaka, Alterations in metabolism of elastin, collagen and glycosaminoglycan induced by reactive oxygen specis. Analysis of photoaging mechanisms using cultured human dermal

fibroblasts as an experimental model, *Kyoto-furitsu Ika Daigaku Zasshi* 107 (2) 207-222 (1998) through Chem Abstr 128 268632m (1998)
25. GF Fisher, HS Talwar, J Lin, P Lin, F McPhillips, Z Wang, X Li, Y Wan, S Kang and JJ Voorhees, Retinoic acid inhibits induction of c-Jun protein by ultraviolet radiation that occurs subsequent to activation of mitogen-activated protein kinase pathways in human skin in vivo, *J Clin Invest* 101(6), 1432-1440 (1998)
26. E Schwartz, AN Sapadin and LH Kligman, Ultraviolet B radiation increases steady-state mRNA levels for cytokines and integrins in hairless mouse skin; modulation by topical tretinoin, *Arch Dermatol Res* 290(3) 137-144 (1998)

Pigmentation, Swelling and Wrinkling in the Eye Area

Keywords: eye, pigmentation, swelling, wrinkles, anatomy

Product development for the eye area requires a basic understanding of the area's unique anatomy and skin physiology

The eyes are an extraordinarily important cosmetic and functional unit. They have been described as the mirror of the soul and represent the only source of visual input humans have with the external world. The eyes also provide a picture of the general health of the individual, appearing clear and sparkling during times of wellness, but also appearing tired and drawn in illness. Emotions are also most poignantly expressed by peering into the eyes. Tears represent sadness and anger, yet many persons also smile with their eyes. Given all of these important considerations, it is no surprise that the eyes are given considerable attention through the use of eye-care products and color cosmetics, as well as through the development of surgical procedures designed to restore a youthful appearance to the eye area.

Cosmetic considerations around the eye area include the presence of pigmentation or darkening, swelling presenting as under-eye bags, and wrinkling of either the upper and lower eyelid skin or the crow's-feet area lateral to the eye. There are a variety of lotions, creams and serums designed to improve the cosmetic appearance of these common eye-area concerns. Products attempt to lighten the pigmentation, firm the eye-area tissues and enhance moisturization to minimize the presence of fine lines and wrinkles. Unique delivery systems, such as gel patches and occlusive tapes, have been developed to enhance penetration of certain actives to perhaps improve their efficacy. Surgical procedures that remove excess upper and lower eyelid skin accompanied by the removal of periorbital fat are also performed attempting to physically remove wrinkles and tissue fullness.

Certainly the market for eye-area products is immense. Every man and woman alive would like to have youthful, healthy, vibrant eyes for a lifetime. Yet the current products on the market are somewhat limited in their ability to meet expectations. This situation creates an opportunity for the development of new technology aimed at cosmetically enhancing the appearance of the eye area.

Successful product development requires a thorough understanding of the problems that afflict the skin surrounding the eyes. This article presents a dermatologic perspective on how the anatomy and physiology of the eyes contributes to eye-area cosmetic problems.

Anatomic Considerations

Most of the cosmetic problems afflicting the eye area are in some way linked to the unique anatomy found only in this area of the body. The eyelid skin is transition skin between the dry facial skin, possessing a well-developed stratum corneum, and the moist conjunctival membrane covering the eyeball. This means that the eyelid tissues frequently become wet with tears, which can translate into compromised barrier function if the rewetting and drying is sustained.

Furthermore, the eyelid tissue also forms a transition between the relatively immobile facial skin of the cheeks and the tremendously mobile eyeball. This is due to the fact that the eyelid skin covers an extensive interlocking array of muscles encircling the eye allowing movement and expression. There are approximately 22 facial muscles that control movement around the eyes, with 14 muscles dedicated to opening and closing of the eye.

Skin thickness: The eyelid skin is the thinnest on the body, measuring approximately 0.5 mm. The thinness of the skin allows quick healing following surgery with minimal scarring, unlike surgery on the back where the thicker skin heals slowly and poorly. The difference in healing rates between the eye and back is apparent when considering suture removal time. Sutures are typically removed following eyelid surgery at 5-7 days since sufficient healing has occurred by then to allow the wound edges to remain approximated. However, sutures are usually left in the back for 14 days and wound strength is poor even then. Eyelid incisions generally heal to a fine, imperceptible white line while incisions on the back typically thicken and stretch, leaving an unsightly broad, lumpy, red scar.

Muscles, nerves and blood supply: The skin around the eyes covers a complex network of underlying structures. There are numerous capillaries and venules that supply the area with blood containing oxygen and nutrients for muscle movement and tissue repair. This extensive vascular supply is under sensitive nerve control allowing quick changes in blood supply and return. This is demonstrated by the rapidity with which the eyes can swell following ingestion of an allergic substance, such as a drug or a food. The blood vessels and nerves are intertwined among muscles that are quick to respond with fine movements to voluntary and involuntary central commands.

Oil and sweat glands: Abundant oil glands and specialized sweat glands reside in the skin around the eyeball. The oil glands, which become more apparent with advancing age and thinning skin, are seen as tiny yellow specks on the lower eyelid. The specialized sweat glands, known as apocrine glands or glands of Moll, secrete a clear watery liquid that produces a characteristic smell when mixed with other body secretions and bacteria. Both of these glands can enlarge and produce the appearance of pebbly skin around the eyes.

Bone and fat: Beneath the skin, the eyeball rests in a bony cradle known as the orbit. The bone above the eyeball is labeled the superior orbital ridge and the bone beneath the eyeball is similarly called the inferior orbital ridge. To protect the eyeball from trauma, there are a variety of periorbital fat pads located behind the eye and beneath the upper and lower eyelids. There are two fat pads in the upper eyelid and three fat pads in the lower eyelid. The fat is held in place by the orbital septum, which prevents the fat from pushing the overlying skin outward.

Eyelid margin: At the junction of the eyelid skin and the eyeball lies a fine row of eyelash hairs. The eyelashes contribute to the cosmetic value of the eye by forming an attractive frame. Their anatomic function is to prevent foreign substances from entering the eye by acting as enhanced touch receptors. The lid margin also possesses its own oil glands, known as Meibomian glands, that function to lubricate the movement of the eyelid over the eyeball by secreting a thin lubricating film that retards tear evaporation.

With this understanding of basic eye anatomy, we may now move on to insights into how cosmetic problems arise in the eye area and how the appearance of the skin around the eye can be improved.

Cosmetic Considerations

There are a variety of cosmetic problems important to both men and women involving the tissues around the eye. The problems can be summarized as pigmentation, swelling and wrinkling. These problems can occur individually or simultaneously, with some present only transiently while others are present for a lifetime.

Pigmentation

Perhaps the most common eye-area complaint worldwide is excess pigmentation or dark circles around the eyes. The darkness gives the appearance of sunken eyes that are tired and drawn. The pigmentation can affect only the lower eyelid or both the upper and lower eyelids simultaneously. It is important to recognize that the causes of this pigmentation are multifactorial, each of which must be carefully assessed to provide cosmetic improvement. The causes of eyelid pigmentation include deep set eyes, melanin production, hemosiderin deposition, increased blood flow, and edema.

The architecture of the eye is a common cause of dark circles. The appearance and color of the skin is determined by the amount of light reflected from the skin surface back to the observer's eye. Smooth skin around the eye reflects a large amount of light and therefore appears shiny and vibrant, while rough skin appears dull and tired. Also, recessed areas tend to appear dark due to the presence of shadows while protuberances appear lighter. This is the case with individuals possessing deep-set eyes.

If the eyes are deep within the orbital bone, a shadow is cast by the superior orbital ridge over the inferior orbital ridge. This creates the illusion of darkness around the eyes even though stretching the skin to eliminate the depression reveals skin of normal color and texture. Here, the "pigmentation" is merely an optical illusion caused by shadows on the skin surface. Increased light reflection would eliminate the appearance of pigmentation.

Pigmentation around the eye area can also be due to substances within the skin. The normal skin color is a combination of melanin, blood flow, iron pigments (chiefly hemosiderin) and other plasma constituents. Because the skin around the eyes is thin, any pigments present in or beneath the skin are more readily apparent than they would be elsewhere.

Melanin hyperpigmentation: The most common pigment accounting for darkening around the eyes is melanin. Melanin is a combination of brown pigments (eumelanins) and red pigments (pheomelanins) that are produced in response to genetic background, sun exposure and injury. Persons from India and the Middle East seem more likely to possess a familial predisposition for increased melanin production around the eyes. This darkening can be worsened by sun exposure and injury, and it is extraordinarily difficult to eliminate.

Pigmentation around the eyes due to sun or UV exposure is part of the pigment darkening response to injury. Melanin production or tanning occurs after UVA light exposure to a greater or lesser degree based upon the individual's ability to produce pigment. Fair-complected individuals are much less likely to have melanin pigmentation around the eyes than those who tan easily. Certain medications, such as the estrogen found in hormone-replacement therapy and oral contraceptives, can enhance pigmentation of the eye area, as well as on the lateral forehead and lateral jawline. Use of sun-protective measures such as sunblocks and sunglasses to prevent UVA light from reaching the periorbital skin can prevent further darkening. Such precautions may eventually result in skin lightening as well, once pigment production is no longer being stimulated.

Pigment can also be produced by the body in response to other insults besides ultraviolet radiation exposure. In short, anything that irritates the skin around the eyes can cause darkening. The causes of eyelid dermatitis are listed in Table 1.[1,2] Treatment of the eyelid dermatitis accompanied by topical anti-inflammatory agents, such as low-potency topical corticosteroids, is the first step to improving this post-inflammatory hyperpigmentation.

Hemosiderin: Other pigments, chiefly hemosiderin, may be deposited around the eye area. Hemosiderin is an iron-based breakdown product derived from the iron-rich hemoglobin found in the red blood cells. An injury to the eye area, such as a trauma or a surgical procedure, can result in bruising and the deposition of hemosiderin. In time, the hemosiderin is removed by the body through white blood cells and the lymphatic system.

There are certain rare disease states where hemosiderin deposition can occur as well. Hemosiderin deposition gives the eyelid skin a blue/purple color initially, which can turn to an orange/brown color with the passage of time.

The last two causes of pigment within the eyelid skin mentioned are increased blood flow and edema. These causes are always found with eyelid swelling and will be discussed next.

Table 1. Causes of eyelid irritation possibly resulting in pigmentation

Mechanical rubbing	Atopic dermatitis
Irritant contact dermatitis	Psoriasis
Allergic contact dermatitis	Collagen vascular disease
Infection	Conjunctivitis
Photoirritation	Seborrheic blepharitis
Contact urticaria	Idiopathic causes

Swelling

Swelling of the tissues around the eye lead to a puffy appearance that may be combined with discoloration, as previously discussed. Edema is the medical name for eyelid puffiness; it is caused by the presence of excess fluid in the skin and subcutaneous tissues around the eye. The presence of this excess fluid gives the skin a blue/purple appearance and contributes to upper eyelid redundancy and lower eyelid "bags."

Transient swelling after sleeping: Excess fluid around the eyes is generally present upon waking from sleep. During prolonged periods of reclining, the body redistributes the extracellular fluid evenly, since the effect of gravity is equivalent on all parts of the body. After standing or sitting for a period of time, the extracellular fluid is drawn from the eyelid tissues (usually to the lower body parts, such as the feet) due to gravity. The condition of the blood flow in the vessels and the amount of fluid accumulated determine how long it takes for the puffiness to resolve.

Transient swelling from vasodilation: Transient swelling of the tissues around the eye can be seen under conditions of blood vessel flux. For example, persons with inhaled allergies such as hay fever will experience eyelid puffiness. This is due to the release of histamine, which causes increased blood flow and enlargement of the blood vessels, also known as vasodilation, resulting in leaking vessel walls and escape of water into the extravascular space. Treatment of the allergies, usually accomplished with antihistamines, may reduce the puffiness. Ingestion of antihistamines will not only improve the swelling, but also improve the color of the tissues around the eye. Vasodilation leading to puffiness may also be induced by weeping.

Chronic fluid-related swelling: There are some individuals, however, in whom the fluid is not adequately reabsorbed and eyelid puffiness persists. This can be due to blood flow problems, because of which the fluid cannot be adequately removed from the tissues and placed back in the blood vessels. The fluid is actually water that is held in the extravascular space by mucopolysaccharides, primarily hyaluronic acid. The same mechanism of extracellular and extravascular fluid retention that is important in skin moisturization results in eyelid swelling. While plumpness of the skin is desirable on the cheeks to minimize the appearance of fine lines on an aging face, it is not desirable around the eyes.

Other conditions in which swelling occurs relate to an increase in body water, also labeled "fluid retention." Fluid retention refers to an expansion of the extravascular fluid volume and can often be seen in women just prior to menstruation and during pregnancy. Increased salt intake can also expand the extravascular fluid volume in susceptible individuals. Reduced salt intake and the use of diuretics can aid in reduction of the fluid retention and improve the appearance of eyelid swelling.

Other causes: Swelling of the eyelid tissues that is not transient may be due to other causes besides increased extracellular fluid. Herniation of the fat through the orbital septum may also give the appearance of eyelid puffiness, especially in mature individuals. This fat herniation is a surgical problem; it does not respond to changes in body water content or changes in periorbital blood flow.

Wrinkling

The more common problem affecting the skin on and around the eye is wrinkles. Wrinkles can take the form of fine lines or deep furrows around the eye; they may be present permanently or only transiently, during certain facial expressions such as smiling. Wrinkling around the eyes is due to three causes: laxity of the skin, photodamage and muscle insertions.

Laxity of the skin is seen in some individuals primarily on the upper eyelids and is medically termed blepharochalasis. This inherited condition accounts for dramatic redundancy of the upper eyelid skin that can interfere with vision. It is best remedied by surgical revision of the eyelids.

Photodamage, also termed dermatoheliosis, accounts for the fine permanent lines present around the eye area. Dermatoheliosis results from collagen and elastin degradation induced by cumulative UVA radiation exposure. This wrinkling is more pronounced in lighter-complected individuals and is sometimes casually termed crow's-feet.

Wrinkles that are present with facial expression are due to muscle insertions pulling on the skin surface. Treatments such as *Botulinum* toxin A injections, which prevent muscle contraction, can eliminate these movement-induced wrinkles.

The best treatment for eye-area wrinkling is sun protection in the form of topical sunscreens to prevent elastin and collagen degradation, and the use of sunglasses to prevent squinting and muscle contraction. Avoidance of cigarette smoking and second-hand cigarette smoke can also prevent premature wrinkling of the eye-area skin. There is no known medical prevention or treatment for hereditary relaxation of the eyelid tissues, thus blepharoplasty at the appropriate time is recommended.

Product Development Considerations

The market for products that improve the appearance of the skin around the eye is huge, since problems with eye-area pigmentation, swelling, and/or wrinkling eventually affect all men and women. This means that product development in this area is an important undertaking. What are some of the unique considerations when formulating for this body area?

Initially, raw materials must be carefully selected, adhering as closely as possible to the guidelines presented in Table 2. It is important to avoid substances that are known to cause allergic contact dermatitis because the eye area is particularly prone to problems due to enhanced penetration through the thin skin. Allergic contact dermatitis of the eyelid may cause dramatic swelling, so severe that the eyes cannot be opened. The most common cosmetic ingredients that have been reported to cause allergic contact dermatitis of the eyelids are presented in Table 3.

There are some women, however, who will experience eyelid problems even with the most carefully formulated skin care products and cosmetics. These eyelid problems will manifest as swelling, itching and burning, with eventual pigmentation. The recommendations that I provide to these women for eye-area color cosmetic selection are listed in Table 4.

Treatment Considerations

Minimizing pigmentation: Let's first address the issue of pigmentation. Possible cosmetic approaches to darkening around the eye include the use of light reflective materials or white pigments to minimize the appearance of the dark skin. This can be accomplished through the use of mica, fish scale, bismuth oxychloride, titanium dioxide or zinc oxide. Of these, the most suitable substance for eye-area use in sensitive skin consumers is zinc oxide of particle size large enough to lighten the skin, but not so large as to create an opaque white film over the skin surface. Zinc oxide is also an excellent choice because it can function as an effective physical

Table 2. Formulation considerations for eye-area products

1. Common allergens and irritants must be eliminated from the formulation or, if this is not possible, reduced in concentration.
2. High quality, pure materials without contaminants should be selected.[3]
3. Autoxidation products, which may be responsible for hypersensitivity reactions, should be prevented through the use of suitable antioxidants.[4,5]
4. Volatile vehicles and substances producing cutaneous stimulation should be eliminated.
5. Solvents that promote skin penetration (propylene glycol, ethanol) should be avoided.
6. Surfactants, used either for cleansing purposes or as emulsifiers, should be carefully selected.[6-8]
7. Preservatives with low sensitizing potential (parabens) should be selected over those with a higher sensitizing potential (formaldehyde and formaldehyde releasers).[9,10]

Table 3. Possible sources of allergic contact eyelid dermatitis

Preservatives[11]
  parabens
  phenyl mercuric acetate
  imidazolidinyl urea
  quaternium-15
  potassium sorbate
Antioxidants
  butylated hydroxyanisole[12]
  butylated hydroxytoluene[13]
  di-tert-butyl-hydroquinone[14]

Resins
  colophony[15]
Pearlescent Additives
  bismuth oxychloride[16]
Emollients
  lanolin[17]
  propylene glycol[18]
Fragrances[19]
Pigment Contaminants
  nickel[20]

UVA sunblock, especially if the particles are silicone coated. Thus, both the need for cosmetic lightening of the eye-area skin and the need for UVA protection to prevent further pigment darkening are met with one ingredient.

Careful selection of non-irritating ingredients can prevent additional pigmentation by avoiding post-inflammatory hyperpigmentation. In other words, eye-area products can help minimize undesirable periorbital pigmentation by preventing further darkening.

It is also possible to reduce post-inflammatory hyperpigmentation through the use of anti-inflammatory agents. Anti-inflammatories that may be of benefit in the over-the-counter realm include bisabolol and allantoin, but low-potency prescription topical corticosteroids are the most effective treatments currently available. Emollients that can smooth and soothe eyelid tissues (such as vitamin E, dimethicone and grape seed oil) may also be of value.

A variety of substances have been identified for cosmetic use that may decrease the presence of melanin in the skin. These pigment-lightening agents may also be incorporated into eye-area creams. The most effective lightening agent on the market in the United States is hydroquinone.

It is worth mentioning that hydroquinone is not allowed in formulations distributed in Japan and some other parts of the world. Hydroquinone is toxic to melanocytes and is a strong irritant; thus it is not appropriate for use on the upper eyelids.

Other over-the-counter pigment-lightening agents include azelaic acid, kojic acid, pulp mulberry extract and licorice extract. Unfortunately, the pigment-lightening abilities of these botanicals are slow at best. Vitamins C and A are also believed by some to induce pigment lightening, but few well-controlled scientific studies are

Table 4. Cosmetic selection criteria in patients with recurrent eyelid dermatitis

1. When possible, patients should use powder cosmetics rather than cream or lotion formulations.
2. All cosmetics applied should be easily removed by water; patients should use no waterproof cosmetics.
3. Patients should discard old cosmetics and purchase fresh products.
4. Patients should use black eyeliner and mascara.
5. Patients should use pencil forms of eyeliner and eyebrow cosmetics.
6. Patients should select eye shadows from the light earth tones, colors such as cream or tan.
7. Patients should use eye shadows with matte finishes, those without particulate light reflective materials.
8. Patients should use cosmetics without organic chemical sunscreen agents (such as oxybenzone, methoxycinnamates, etc.).
9. Patients should purchase cosmetic products with a low number of ingredients.

available for review. Exfoliation and other physical methods of lightening the skin, such as skin peeling, are generally not appropriate for the tender eyelid tissues in a consumer setting.

Reducing swelling: Swelling of the eyelid tissues is a slightly more difficult issue to tackle topically. For patients with allergies, oral over-the-counter and prescription antihistamines and vasoconstrictive eyedrops can be highly effective. Unfortunately, there are no good topical antihistamine ingredients available. It is possible, however, to induce vasoconstriction through the use of certain botanical extracts, such as witch hazel, horsetail and elder flower. Caffeine also has a vasoconstrictive effect. As the blood vessels shrink in size, they become less permeable, resulting in less water leaking into the tissues.

Minimizing appearance of wrinkles: Wrinkling of the tissue around the eyes can be improved in most patients, since the eyelid skin is frequently poorly moisturized due to a paucity of sebaceous glands in the eye area and (for many women) frequent cleansing to remove eye cosmetics. Most eye-area products that provide superior moisturization will be accepted with enthusiasm by the consumer. A good eye-area moisturizer should contain both occlusive ingredients (such as dimethicone) and humectant ingredients (such as glycerin) to maximize skin hydration. In general, it is not possible to worsen eyelid puffiness through the use of nonirritating topical moisturizers. Eyelid puffiness results from water residing in the dermis and subcutaneous tissues, while topical moisturizers enhance stratum corneum and epidermal hydration.

As mentioned previously, sun avoidance and the decision not to smoke are the two most important steps people can take to help prevent eye-area wrinkling. Yet it is possible to reverse *a very small amount* of the collagen and elastin fiber damage through the use of retinoids. The most effective retinoid is prescription tretinoin, yet there is excellent evidence that small amounts of retinol are converted in the skin to retinoic acid or tretinoin. Thus, topical stabilized retinol may be of value in the eye area to improve wrinkling when formulated in a moisturizing vehicle.

Conclusion

Pigmentation, swelling, and wrinkling of the eye area tissues detract from the facial appearance of health, youth and vitality desired by both men and women. Product development to meet the needs of the eye area requires a basic understanding of the unique anatomy and physiology of the skin. Once the possible etiologies of eye-area problems have been explained, efficacious formulations can be developed.

—**Zoe Diana Draelos, MD,** *Department of Dermatology, Wake Forest University School of Medicine, Winston-Salem, North Carolina USA*

References

1. HI Maibach and PG Engasser, Dermatitis due to cosmetics, in *Contact Dermatitis*, edition 3, AA Fisher, ed, Philadelphia: Lea & Febiger (1986) 378-379
2. HI Maibach, PG Engasser and B Ostler, Upper eyelid dermatitis syndrome, *Dermatologic Clinics* 10 549-554 (1992)

3. A Dooms-Goossens, Reducing sensitizing potential by pharmaceutical and cosmetic design, *J Am Acad Dermatol* 10 547-553 (1984)
4. EW Clark and GF Kitchen, Autoxidation and its inhibition in anhydrous lanolin, *J Pharm Pharmacol* 13 172-183 (1961)
5. EW Clark, A Blondeel, E Cronin et al, Lanolin of reduced sensitizing potential, *Cont Derm* 7 80-83 (1981)
6. M Rieger, Human epidermis responses to sodium lauryl sulfate exposure, *Cosmet Toil* 109(5) 65-74 (1994)
7. FR Bettley, The influence of detergents and surfactants on epidermal permeability, *Brit J Dermatol* 77 98-100 (1965)
8. BW Barry, *Dermatological Formulations*, New York: Marcel Dekker (1983) pp 170-172
9. AF Fransway and NA Schmitz, The problem of preservation in the 1990s: II. Formaldehyde and formaldehyde-releasing biocides, *Am J Cont Derm* 2(2) 78-88 (1991)
10. DC Steinberg, Cosmetic preservation: Current international trends, *Cosmet Toil* 107(9) 77-82 (1992)
11. JG Marks and VA DeLeo, Preservatives and vehicles, in *Contact and Occupational Dermatology*, St. Louis: CV Mosby (1992) pp 107-133
12. IR White, CR Lovell and E Cronin, Antioxidants in cosmetics, *Cont Derm* 11 265-267 (1984)
13. Ibid, CR Lovell and E Cronin, Antioxidants in cosmetics, *Cont Derm* 11 265-267 (1984)
14. CD Calnan, Ditertiary butylhydroquinone in eye shadow, *Cont Derm Newsletter* 14 402 (1973)
15. AA Fisher, Allergic contact dermatitis due to rosin (colophony) in eyeshadow and mascara, *Cutis* 42 505-508 (1988)
16. HJ Eiermann, W Larsen, HI Maibach and JS Taylor, Prospective study of cosmetic reactions: 1977-1980, *J Am Acad Dermatol* 6 909-917 (1982)
17. WF Schorr, Lip gloss and gloss-type cosmetics, *Cont Derm Newsletter* 14 408 (1973)
18. M Hannuksela, V Pirila and OP Salo, Skin reactions to propylene glycol, *Cont Derm* 1 112-116 (1975)
19. WG Larsen, Cosmetic dermatitis due to a perfume, *Cont Derm* 1 142-145 (1975)
20. CL Goh, SK Ng and SF Kwok, Allergic contact dermatitis from nickel in eyeshadow, *Cont Derm* 20 380-381 (1989)

Cell Adhesion: A New Approach to Tissue Protection

Keywords: skin firmness, UV, integrins

A study of the effect of UV on cell adhesion and expression of fibroblast integrins – two factors on which skin firmness depends

The interaction of cells with their environment is mediated by specific receptors on their surface. In the skin's extracellular matrix, a family of membrane receptors called integrins is believed to play an important role in intercellular adhesion and cell protein adhesion, and thereby in the organization and assembly of the extracellular matrix on which skin firmness depends.

A great deal has been published on the influence of age and UV on the expression of keratinocyte integrins, but there is little published data on the influence of UV on the expression of fibroblast integrins and cell adhesion.

In order to better understand the potential participation of integrins in the adhesion phenomenon and in particular the adhesion of skin fibroblasts, we studied the adhesion capacities and the expression of membrane integrins of cells in the presence or absence of exposure to UV radiation. Those results are reported here, as well as the possibility of an active ingredient to protect cells from the deleterious effects of UV.

Membrane Receptors

Membrane receptors are classified according to their structure into four families: selectins, cadherins, immunoglobulins and integrins.

The family of selectins or LEC-CAMs (lectin cell adhesion molecules) includes proteins involved in the adhesion of leukocytes to endothelial cells in the course of inflammation.[1]

Cadherins are calcium-dependent proteins that are the principal adhesion molecules of intercellular junctions.[1]

Immunoglobulins ensure recognition between cells.[1] The most well-known molecules of this family are those participating in the recognition of T lymphocyte antigen-presenting cells in the immune response.[1]

Table 1. The β integrins	
β integrin	Function or description
β–1	VLA (very late antigen)
β–2	LEU-CAMs, leukocyte adhesion molecules
β–3	expressed by endothelial cells
β–4	component of hemidesmosomes
β–5	vitronectin receptor
β–6	fibronectin receptor
β–7	adhesion molecules for B and T lymphocytes, immunoregulating molecules
β–8	glioma cell substratum attachments

Integrins are heterodimeric glycoproteins composed of an alpha sub-unit and a smaller beta sub-unit that are non-covalently bound. They participate in intercellular adhesion and cell protein adhesion in the extracellular matrix.[1] Integrins are placed in eight sub-families (Table 1) depending on the composition of their beta chain.[2-4] For example, β-1 integrins – described as very late antigen (VLA) because they were discovered two to three weeks after in vitro activation[1] – mediate dermal fibroblast attachment to type VII collagen[5] and are the major keratinocyte collagen binding receptors.[6]

The extracellular domains of the α and β sub-units are non-covalently bound to form a binding site for proteins of the extracellular matrix.

The Role and Behavior of Integrins

Skin firmness: The firmness of the skin depends on the status and the organization of the extracellular matrix.[7] The extracellular matrix (ECM) is made up of fibrous components (collagens, elastin), proteoglycans and structural glycoproteins (fibronectin). This complex system participates actively in the physical support of the dermis.

Integrins are believed to play a very important role in the organization and assembly of the matrix[2] by participating in cell anchoring.[1] Most proteins of the ECM, laminin, collagens and fibronectin are integrin ligands. Cell matrix interactions thus guarantee the integrity of connective tissue.

Tissue repair: Integrins also have an important function in various biological and physiological processes of the skin. Cell matrix interactions participate in embryonic development, in immune and nonimmune defense mechanisms and in the healing process.

As part of the healing process, some integrins participate in tissue repair. Among them are α–2 β–1 integrins that play an important role in the contraction of collagen I by human fibroblasts in the course of healing.[8] Alpha β–5, α–3 β–1 and α–5 β–1 integrins are the principal receptors involved in the migration of keratinocytes towards the site of the wound.[9,10] α–6 chains contribute to the migration of Langerhans cells from the epidermis to the lymph nodes, thereby participating in the initiation of the skin response to an allergen.[11]

Effect of age: Among the factors affecting the expression of integrins is age. Prior studies have shown that skin fibroblasts in the elderly contain fewer β–1 sub-units associated with α–2 sub-units in comparison to young subjects.[12]

Response to UV radiation: Integrins are also sensitive to UV radiation. UVA penetrates deep in the skin and is known to modify both cellular organization and that of the dermal matrix.[1] It can modify cell adhesion differently as a function of the dose used: a high dose (10 to 20 J/cm^2) leads to a dose-dependent inhibition of cell adhesion, whereas low doses (2 J/cm^2) lead to stimulation.

UVA causes a dose dependent decrease of alpha2 beta1 integrins, thus interfering with the flow of information between fibroblasts and their extracellular environment, delaying their intervention in healing processes.[13]

The majority of UVB (70%) is absorbed by the stratum corneum, but 10% reaches the superficial dermis. It also has a negative effect on the expression of integrins, thus causing a decrease in the quantities of α–2 and β–6 integrins.[6]

A Study of Fibroblast Integrins Exposed to UV

The influence of age and UV on the expression of keratinocyte integrins has been described in the literature, but not much has been written about the influence of UV on the expression of fibroblast integrins and cell adhesion. Therefore, we studied the adhesion capacities and the expression of membrane integrins of fibroblast cells in the presence or absence of exposure to UV radiation. We also observed the potential of an extract of *Oryza sativa* (rice) to limit the deleterious effects of UV on cells, in particular on membrane receptors.

Cell culture and treatments: A pool of human fibroblasts obtained from several donors was inoculated in complete MEM medium[a] supplemented with 10% fetal calf serum[b]. The cells were then incubated for two days in a 37°C incubator in a humid atmosphere containing 5% CO_2.

The products to test were diluted (1/1) in complete MEM medium and incubation was continued for 24 hours.

Oryza sativa extract: We obtained a particular *Oryza sativa* extract from a concentrated protein fraction of rice thanks to enzymatic engineering technology.[15] This extract is characterized by its richness in di- and tripeptides. The *Oryza sativa* extract is used directly in cell culture without specific treatment.

Ultraviolet irradiation of cells: Cells were irradiated[c] with UVA (2 to 40 J/cm^2) and UVB (0.2 to 0.6 J/cm^2).

Adhesion test: 96-Well plates[d] were coated with 18 µg/ml of type I collagen (solution at 2 µg/ml of rat tail collagen[e]) and incubated overnight at 4°C.

Inoculation was with 30,000 cells per well, followed by incubation for 1 h at 37°C. Adherent cells were fixed with a 1% solution of

---

[a] Invitrogen 41090-028, France
[b] Invitrogen 10270, France

glutaraldehyde[f] and stained with 0.1% crystal violet[g]. After 30 min at room temperature, the stain in cells was dissolved with 1% sodium dodecyl sulfate. Absorbance of each well was determined at 560 nm.

Immunolabeling and quantification: After 24 h of incubation with the products, immunolabeling of α–2 β–1 integrins was carried out. The cells were recovered and incubated for 45 min at 4°C with an anti-α–2 β–1 integrins antibody[h] and a second, FITC-conjugated antibody FITC[i], was then added.

Tubes containing immunolabeled cells were analyzed with a cell sorter[j]. Throughput was at a rate of 500 cells per second in front of an argon laser light source (emission 488 nm, absorption 530 nm). Fluorescence was quantified for 10,000 cells.

Statistical processing of data: Student's test for paired data was used for statistical analyses.

Results of the Study

Adhesive properties of fibroblasts after UV irradiation: Figures 1 and 2 present photomicrographs of stained fibroblasts in the adhesion test. The dark spots are fibroblasts stained with crystal violet. They adhere to the collagen I fixed on the bottom of petri dishes (shown in white or slightly blue). Increased adherence is shown by a greater number of crystal-violet-stained fibroblasts. Cells were irradiated with UVA (Figure 1) and UVB (Figure 2). Nonirradiated cells (controls) adhered to collagen I fixed on the support. Following irradiation by UVA (15 J/cm^2) or UVB (0.5 J/cm^2), the number of cells adhering to the support decreased. When the cells were irradiated with higher doses of UVA or UVB (UVA 30 J/cm^2, UVB 0.6 J/cm^2), adhesion decreased further.

Cell adhesion, recorded as absorbance (A) of adherent cells, was also quantified by the spectrophotometric determination of their number. Increasing doses of UVA or UVB were delivered to determine the possible dose effect of UV on the adhesion capacity of cells. The results with UVA and UVB are shown in Figures 3 and 4. The value of 0.77 corresponds to the absorbance value of the control (a well with nonirradiated cells, UVA = 0 J/cm^2).

The effect of UVA (Figure 3) reduces the adhesion of cells. This decrease is rapid up to 15 J/cm^2 and then decreases for higher UVA doses. The effect of UVB (Figure 4) also causes a decrease in cell adhesion as a function of dose administered. The slope of this decrease is rapid up to 0.3 J/cm^2 and then stabilizes. Cell adhesion is thus affected by UV irradiation. The curves of adhesion of human fibroblasts vs. irradiation dose shows the number of cells fixed to collagen I is dose dependent.

---

[c] BioSun UV irradiator, Vilbert-Lourmat, France
[d] BD-Falcon Microtest, VWR, France
[e] Jacques Boy, France
[f] G-5822, SIGMA, France
[g] C-6158, SIGMA, France
[h] Immunotech, France
[i] F0479, Dako, Denmark
[j] FACS Vantage cell sorter, Becton Dickson, USA

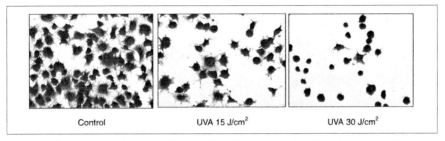

Figure 1. Photomicrographs (x10) of human fibroblasts exposed to UVA doses of 15 J/cm² (center), 30 J/cm² (right) and control (left), showing dose effect of UVA on cell adhesion

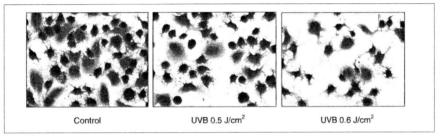

Figure 2. Photomicrographs (x10) of human fibroblasts exposed to UVB doses of 0.5 J/cm² (center), 0.6 J/cm² (right) and control (left), showing dose effect of UVB on cell adhesion

Expression of membrane integrins after UV irradiation: The quantities of α–2 β–1 integrins on the surface of cells were determined following irradiation with UVA (15 J/cm²) or UVB (0.5 J/cm²). These doses were chosen on the basis of the results of the adhesion test. Expression of α–2 β–1 integrins was reduced on the order of 18% by UVA and 15% by UVB.

In these first two evaluations, there was a reduction in both cell adhesion and the quantities of α–2 β–1 integrins on the surface of cells. Cell adhesion and the expression of α–2 β–1 integrins are thus sensitive to UV irradiation.

Preserving fibroblast adhesion after UV irradiation: The effects of the *Oryza sativa* extract at 2% on the adhesion of cells irradiated with UVA or UVB are shown in Figures 5 and 6.

Both UVA (15 J/cm²) and UVB (0.5 J/cm²) reduced the adhesion capacity of fibroblasts to collagen I by 33% and 25%, respectively. When cells were treated with the *Oryza sativa* extract and then irradiated with UVA, their adhesion capacity was reduced by only 14%, compared to 33% in the absence of pretreatment with the extract. Results were even more dramatic when the extract-treated cells were irradiated with UVB. In this case, irradiation led to a very small decrease in cell adhesion of 3%, compared to 25% for cells with no pretreatment. This 3% reduction in adhesion capacity was close to that of nontreated and nonirradiated controls. Pretreated cells clearly retain a higher adhesion capacity.

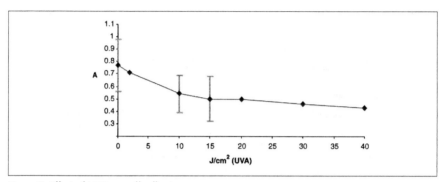

Figure 3. Effect of UVA on cell adhesion
x = UVA doses (J/cm²)
y = Absorbance

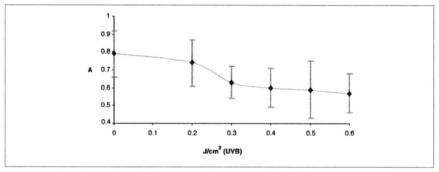

Figure 4. Effect of UVB on cell adhesion
x = UVA doses (J/cm²)
y = Absorbance

Limiting the reduction of integrins expression after UV: The effects of the *Oryza sativa* extract at 2% on the expression of α–2 β–1 integrins after irradiation with UVA and UVB are shown in Figures 7 and 8.

Following UVA irradiation, the quantity of integrins on the surface of fibroblasts pretreated with the 2% *Oryza sativa* extract decreased by 1%, whereas cells not pretreated with the extract showed an 18% decrease.

When irradiation was with UVB, the expression of α–2 β–1 integrins on cell surfaces decreased by 15%. When they were pretreated with the *Oryza sativa* extract before irradiation, there was no decrease but rather a 3% increase in expression. The *Oryza sativa* extract thus provides considerable cellular protection against UV induced damage.

Conclusion

It was initially shown in this study that UV irradiation has a deleterious effect on both cell adhesion and the expression of integrins on their surface. The adhesion

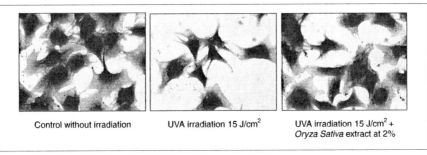

Control without irradiation | UVA irradiation 15 J/cm² | UVA irradiation 15 J/cm² + Oryza Sativa extract at 2%

Figure 5. Photomicrographs (x20) of human fibroblasts un-irradiated (left), after 15 J/cm² of UVA irradiation (center), and after pretreatment with Oryza sativa extract at 2% followed by 15 J/cm² of UVA irradiation (right), showing effect of Oryza sativa extract on the adhesion of fibroblasts irradiated with UVA

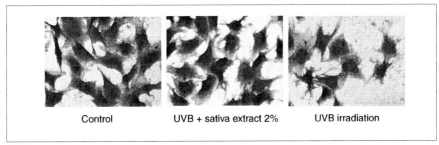

Control | UVB + sativa extract 2% | UVB irradiation

Figure 6. Photomicrographs (x20) of human fibroblasts un-irradiated (left), after 0.5 J/cm² of UVB irradiation (center), and after pretreatment with Oryza sativa extract at 2% followed by 0.5 J/cm² of UVB irradiation (right), showing effect of Oryza sativa extract on the adhesion of fibroblasts irradiated with UVB.

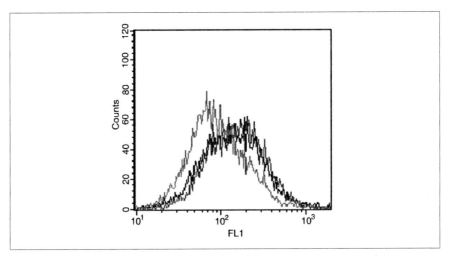

Figure 7. Flow cytometry profile showing the effect of the Oryza sativa extract on the expression of α–2 β–1 integrins after UVA irradiation.
x = arbitrary fluorescence units on a logarithmic scale
y = number of cells

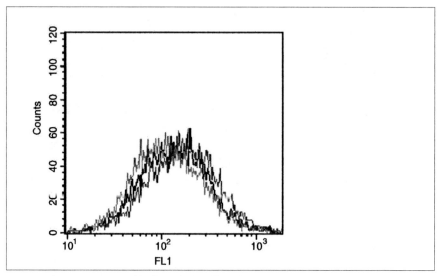

Figure 8. Flow cytometry profile showing the effect of the Oryza sativa extract on the expression of α–2 β–1 integrins after UVB irradiation.
x = arbitrary fluorescence units on a logarithmic scale
y = number of cells

of human fibroblasts to collagen I undergoes a dose-dependent modification when cells are irradiated with UV.

We also investigated the expression of integrins on the surface of cells following irradiation with UVA and UVB to better understand the potential participation of these proteins in the adhesion process. We were also able to demonstrate a reduction in the expression of α–2 β–1 integrins of the order of 18% when cells were irradiated with UVA and of 15% after UVB.

The harmful effects of UV radiation can ultimately lead to the disorganization of skin tissue. Cells, more precisely integrin membrane receptors, are sensitive to UV radiation. The decreased expression of the receptors limits the interactions of cells with their environment, because they no longer possess the same functional characteristics. Membrane receptors lose their reactivity and no longer ensure their mediator role.

We investigated the protective effects of an extract of *Oryza sativa* at 2% to limit the harmful effects of UV and those resulting from chronological aging on the expression of membrane integrins and cell adhesion. The *Oryza sativa* extract at 2% enables cells to retain their adhesion capacity and limits the reduction of integrin expression on the surface of fibroblasts when they are irradiated with UVA or UVB. Cells are thus better protected and cell matrix interactions are preserved.

Based on the results obtained in this work, it is seen that cell adhesion and the expression of integrins on the surface of fibroblasts have comparable behavior: When cell adhesion decreases, the expression of integrins decreases similarly.

According to published results,[14] the use of an antibody that can block either the α chain or the β chain of integrins leads to a reduced adhesion when cells are irradiated with UVA. In the present work, the use of an antibody against α–2 β–1 integrin supports the idea that there is a link between cell adhesion and membrane integrins. The anti-α–2 β–1 antibody prevents cells from adhering to collagen bound to the surface of the support. To extend this work and prove a correlation between these two factors, it would be of interest to use increasing quantities of the anti-α–2 β–1 antibody and determine a dose-dependent reduction of cell adhesion and the expression of integrins.

—Laetitia Moreau, Sylvie Bordes, Maud Jouandeaud and Brigitte Closs,
SILAB S.A., Brive, France

References

1. MJ Staquet, Intégrines et épiderme, INSERM Seminar 211 71-81 (1991)
2. J Labat-Robert, Les intégrines, Path Biol 40 9 883-888 (1992)
3. A van der Flier and A Sonnenberg, Function and interactions of integrins, Cell Tissue Res 305 285-298 (2001)
4. N Belot, S Rorive, I Doyen, F Lefranc, E Bruyneel, R Dedecker, S Micik, J Brotchi, C Decaestecker, I Salmon, R Kiss and I Camby, Molecular characterization of cell substratum attachments in human glial tumors relates to prognostic features, Glia 36(3) 375-390 (2001)
5. M Chen, EA O'Toole, YY Li and DT Woodley, Alpha2 beta1 integrin mediates dermal fibroblast attachment to type VII collagen via a 158 amino-acid segment of the NC1 domain, Exp Cell Res 249(2) 231-239 (1999)
6. BK Pilcher, M Wang, XJ Qin, WC Parks, RM Senior and HG Welgus, Role of matrix metalloproteinases and their inhibition in cutaneous wound healing and allergic contact hypersensitivity, Ann N Y Acad Sci 878 12-24 (1999)
7. I Katz, C Martin, C Chesne and AJ Brin, Actin and integrin mRNA levels in human skin fibroblast and keratinocyte cultures after UV irradiation, Proceedings of the 18th International IFSCC Congress (1994) 853-857
8. KC Eberhard, D Dressel, T Steinmayer, C Mauch, B Eckes, T Krieg, RB Bankert and L Weber, Integrin $\alpha_2\beta_1$ is upregulated in fibroblasts and highly aggressive melanoma cells in three-dimensional collagen lattices and mediates the reorganization of collagen I fibrils, J Cell Biol 115(5) 1427-1436 (1991)
9. A Cavani, G Zambruno, A Marconi, V Manca, M Marchetti and A Gianetti, Distinctive integrin expression in the newly forming epidermis during wound healing in humans, J Invest Dermatol 101(4) 600-604 (1993)
10. J Gailit and RA Clark, Studies in vitro on the role of alpha v and beta 1 integrins in the adhesion of human dermal fibroblasts to provisional matrix proteins fibronection, vitronection and fibrinogen, J Invest Dermatol 106(1) 102-108 (1996)
11. AA Price, M Cumberbatch, I Kimber and A Arger, α_6 Integrins are required for Langerhans cell migration from the epidermis, J Exp Med 186(10) 1725-1735 (1997)
12. B Le Varlet, C Chaudagne, A Saunois, P Barre, C Sauvage, B Berthouloux, A Meybeck, M Dumas and F Bonte, Age-related functional and structural changes in human dermo-epidermal junction components, J Invest Dermatol Symp Proc 3(2) 172-179 (1998)
13. K Scharffeter-Kochanek, M Wlaschek, G Heinen, H Behrendt, S Knaub, JC Van Der Leun, F De Gruijl, E Holzle and G Goerz, UVA-irradiation reduces collagen-lattice contraction by the specific downregulation of the $\alpha_2\beta_1$-integrin receptor, J Invest Dermatol 100(4) 439 (1993)

14. A Tupet, C Lebreton-De Coster, L Dubertret and B Coulomb, Low doses of ultraviolet A radiation stimulate adhesion of human dermal fibroblasts by integrins in a protein kinase C-dependent pathway, *J Photochem Photobiol* B 49 150-155 (1999)
15. Pat FR 01 08035, 19-Jun-2001

Measured Dermal Effect of Applying Retin-A

Keywords: Retin-A, collagen, hyaluronic acid

Effects of topical application of Retin-A on dermis of an in vivo mature rat model

Skin is organized horizontally into three distinct layers: epidermis, dermis and the subcutis. The epidermis is a layer of stratified squamous epithelium that is keratinized. The dermis consists of a complex fibrous matrix set into a mucopolysaccharide gel infiltrated with blood vessels, lymphatic channels, nerves, glandular ducts and hair follicles. The third layer, the subcutis, is populated primarily by adipocytes. The research reported here focused on the dermis because it is the site of the primary water retention molecule, hyaluronic acid (HA), in the skin.

The primary structural components of the dermis are collagen and a small amount of elastin, all set into a ground substance, the matrix, composed mainly of glycoproteins, glycosaminoglycans (GAGs) and water, both free and bound. Collagen has high tensile strength and resists shear forces, while elastin is an elastic protein, which maintains normal skin tensions.

The hydrated matrix (consisting primarily of HA) is closed inside an envelope of collagen fibers. This envelope resists compressional forces by pushing outward against the fibers, providing turgidity (smoothness) to the skin. HA is thought to be the principal water-regulating molecule in the matrix. It can bind up to 6,000 times its weight in water.[1]

As skin ages it develops a distinct phenotypic pattern. There is a change in the basic architectural components and elasticity of the skin, which causes obvious effects: wrinkling, thinning and dryness. These changes may be influenced largely in part by a decrease in hydration of the dermal compartment due to changes in HA. It is hypothesized that changes in the matrix, particularly in HA, cause decreased hydration of the skin, collapse of collagen, and subsequent wrinkling and thinning of the skin.[2]

The effects of topical Retin-A (tretinoin, RA) on the biology, biochemistry and phenotypic expression of skin are well known. It has been shown to stimulate overall metabolic activity in the skin, increasing collagen and GAG synthesis, both in vitro and in vivo.[3-5] In this article, we report quantitative assessments of the effects of topical application of RA on collagen, HA and water in the dermis of an in vivo mature rat model.

Materials and Methods

The effect of topical RA application on key dermal constituents was quantified by using a group of mature Sprague-Dawley hairless rats. The animals were identified with number tags, and a test substance (.5 mL of .1% RA in corn oil) was applied to a previously marked dorsal area in a group of four animals. Two control groups were established: four received corn oil only, and four were left untreated. The test substance and the corn oil were applied twice daily for a period of 13 days. After this time, the animals were anesthetized and punch biopsies of the treated areas were taken. All samples were 100 mg.

A modified HPLC procedure as described by Chun et al.[6] was utilized to determine total HA content.

Collagen was quantified by using the established hydroxyproline method that uses 2 to 10 mg of tissue hydrolyzed in 6N HCl (6 normal hydrochloric acid) for 24 h at 110°C. The HCl was removed in a vacuum desiccator, the sample redissolved and reacted with chloramine-T and Ehrlich's reagent in a multi-well ELISA dish, and then read with an ELISA reader at 550 nm. The hydroxyproline content represented approximately 15% of the dry weight of collagen present.

A programmable moisture-determining microwave[a] determined water content of a weighed sample of skin.

Results

Results showed that topical RA measurably stimulated the baseline level of HA, collagen and water in the system.

The RA treated animals showed consistently higher amounts of HA than either the vehicle controls or the untreated animals with a mean weight of .0795 mg HA per 100 mg dry weight versus .0615 mg HA per 100 mg dry weight of control and untreated skin (P=< .001).

Collagen increased to a mean value of 78 mg per 100 mg dry weight versus 71 mg per 100 mg dry weight of the control and untreated animals (P=<.001).

Water content was substantially increased in the treated animals to a mean value of 80.25 mg per 100 mg wet tissue versus 70.25 mg per 100 mg wet tissue (P=<.001).

Conclusions

This study showed a demonstrable increase in HA, collagen and water as a result of topical RA application. It proposes a mechanism for wrinkle effacement where the increase in water retention due to increased amounts of HA pushes outward against a rigid lattice of collagen to provide turgidity and wrinkle effacement.

As an adjunct to this study, it was noted that since HA, collagen and water in the skin were quantifiable, this assay might also find use as a method to substantiate putative skin stimulation (anti-wrinkle) products.

---

[a] CEM Labwave 9000, Manufactured by CEM, Matthews, North Carolina, USA

—**George Andrassy and Howard I. Maibach,** *Department of Dermatology, School of Medicine, University of California, San Francisco, California, USA*

References
1. JE Silbert, Mucopolysaccharides of ground substance, in *Dermatology in General Medicine*, TB Fitzpatrick and AT Eisen, eds, New York: McGraw Hill (1979) 189-199
2. AM Kligman and AK Balin, Aging of human skin, in *Aging and the Skin*, AK Balin and AM Kligman, eds, New York: Raven Press (1989)
3. AM Kligman et al, Topical tretinoin for photoaged skin, *J Am Acad Dermatol* 15 836-859 (1986)
4. J Varani et al, All trans retin-a stimulates growth and extracellular matrix production in growth inhibited cultured human fibroblasts, *J Invest Dermatol* 94 (5) 712-723 (May 1999)
5. R Tammi, JA Ripellino, RU Margolis and HI Maibach, RA stimulation of hyaluronate in skin organ culture, *J Invest Dermatol* 92 326-332 (1989)
6. LE Chun, TJ Koob and DR Eyre, Quantification of hyaluronic acid in tissues by ion-pair reverse phase high performance liquid chromatography of oligosachharide cleavage products, *Anal Biochem* 171(1) 197-206 (1988)

Oat Fractions

Keywords: UV, oat fractions, hair care

Their rejuvenating effects on skin and hair

Aging of the skin involves intrinsic and extrinsic factors. Intrinsic factors include genetics, immunological function, hormones and biological senescence. Included among extrinsic factors are diet, availability of health care and environmental conditions, one of the most important being exposure to solar radiation (photodamage). It is possible that at least 90% of age-associated cosmetic problems of the skin are caused by photodamage.[6]

The American Academy of Dermatology (AAD) concluded that cutaneous changes produced by UV damage include surface roughness (fine lines/wrinkles, skin texture), coarse wrinkles (deeper, permanent lines and furrows), coarseness (leathery/stiffness), mottled pigmentation (melasma, freckles, age spots), laxity, scaling/xerosis, sallowness, telangiectasia (red, finely branching skin capillaries), actinic lentigines (age spots) and actinic keratoses (pre cancerous lesions).[6,9]

Stringent efficacy studies, both clinical and in vitro, show that different oat fractions can have substantial benefits to the skin and hair in ameliorating the damage associated with UV radiation and aging. In particular, the fractions we studied were oat β-glucan, hydrolyzed oat protein and oat extract. Skin and hair care products formulated with these materials can reduce wrinkles and protect and repair damaged skin and hair.

Oat β-glucan on Skin and Hair

Purified β-glucan from oats is a linear, unbranched polysaccharide composed of 4-O-linked β-D-glucopyranosyl units (70%) and 3-O-linked β-D-glycopyranosyl units (30%).[1,2,7,11] Oat β-glucan (Figure 1) is a water-soluble, nonchemically-derived, high molecular weight polysaccharide.

The many uses of oat β-glucan extend to food additives, wound management, dermal care and personal care products. Oat β-glucan is the primary component responsible for the reduction of cholesterol and risk of heart disease through regular consumption of oats. On January 21, 1997, the U.S. Food and Drug Administration took a major step forward in helping consumers choose healthy diets by its final approval of the first food specific health claim: "Soluble fiber from oatmeal, as a part of a low saturated fat, low cholesterol diet may reduce the risk of heart disease."[3]

A β-glucan collagen matrix dressing available for the management of partial thickness burns, donor sites and shallow abrasion type wounds is currently on the market. Several trials have shown that β-glucan is a biological response modifier.

In keeping with the modern moist wound healing concept, this natural material can be a useful adjunct to the healing of difficult wounds.[5]

In relation to personal care, we will discuss at length the following efficacious attributes of oat β-glucan: it stimulates collagen synthesis, provides protection against UVA/UVB damage, acts as a natural moisture barrier and film forming agent, promotes cell turnover, reduces fine lines and wrinkles, improves skin viscoelasticity, decreases hyperpigmentation due to photodamage, increases hair strength and helps reduce damaging effects of bleaching.

Alleviate extrinsic signs of aging: An eight week photoaging clinical trial was sponsored to evaluate the ability of oat β-glucan to alleviate selected extrinsic signs of aging. Two products were used. One was an active containing carbopol gel with 0.1% oat β-glucan; the other was carbopol gel as a placebo. Because Colorado has a dry climate and relatively high exposure to ultraviolet radiation, that state was selected as the site for the study.

Using a half-face design and randomly assigned products, each of 27 subjects treated the left and right sides of the face with the appropriate product twice daily (once in the morning and once in the evening) for eight weeks. Subjects were instructed to apply the products as they would normally apply a facial moisturizer. They were further instructed to apply product to one side of the face, then thoroughly cleanse hands before applying the second product to the other side of the face. Subjects observed a three day conditioning period immediately prior to baseline measurements. Skin was evaluated for changes from baseline values of the following parameters:

- fine lines and wrinkles
- skin laxity
- skin elasticity
- sallowness
- roughness
- scaling
- hyperpigmentation

This clinical study included subjective and objective assessments. An expert grader evaluated skin elasticity by pinch recoil at the crow's-feet area of the eye. The grader evaluated each of the other parameters by placing a mark to the nearest tenth of a centimeter on a 10 cm analog scale. These assessments were recorded at baseline and again at two, four and eight weeks. Improvements were noted if the parameter received a higher "score" than it received in a prior evaluation. Subjects also answered questionnaires regarding these same parameters.

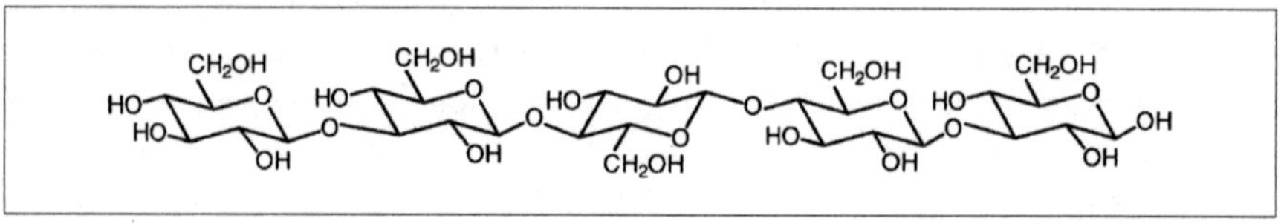

Figure 1. Physical structure of oat β-glucan showing mixed β-1,3 and β-1,4 linkages

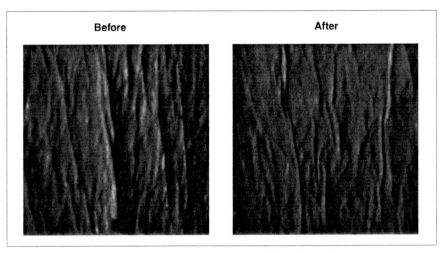

Figure 2. Images generated from silicone replicas of crow's feet before and after eight weeks of treatment with oat β-glucan

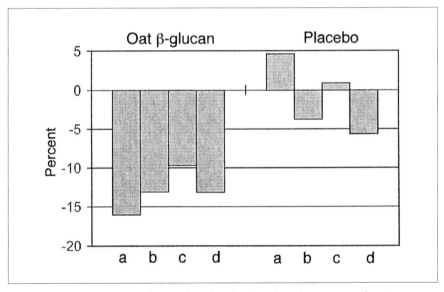

Figure 3. Average percentage changes from baseline in selected parameters, from image analysis of silicone replicas of crow's feet treated with oat β-glucan for eight weeks, compared to treatment with a placebo

a = depth of the deepest wrinkle
b = average depth of all wrinkles
c = average number of peaks
d = overall roughness

− = decrease from baseline
+ = increase from baseline

For the evaluation of fine lines and wrinkles, silicone replicas of the outer canthus (crow's-feet) area of the eye were subjected to digital image analysis. Additionally, macrophotography enabled further evaluation of changes in fine lines/wrinkles and changes in hyperpigmentation. Finally, a device[a] was used to measure changes in skin elasticity.

The resulting data from the expert grader's subjective assessments of the seven parameters indicated more incidents of improvement with oat β-glucan than with the placebo. In the subjects' own end point self assessment questionnaires, oat β-glucan received significantly higher positive responses to improvement in photodamage conditions than the control, and forced comparison results showed that it was the top choice among the tested products.

- Image analysis: Silicone replicas of the outer canthus of the left and right eye (crow's-feet) analyzed by digital image analysis demonstrated that all 13 parameters evaluated in this analysis indicated a smoothing of the cutaneous surface after eight weeks treatment with oat β-glucan. Figure 2 shows the "before" and "after" images generated from silicone replicas. Figure 3 shows data from four of the 13 parameters.
- Macrophotography: Macrophotography of the left and right sides of the face also demonstrated reduction in lines and wrinkles. Photographs were graded for change as follows: five points for improvement between measurements, five points deducted if conditions worsened and zero points if no change was observed. Sites treated with oat β-glucan had an average score of 6.07 compared to 3.46 for the control treated sites. In macrophotographic evaluations of hyperpigmentation, the average score for sites treated with oat β-glucan was 4.29 compared to 1.15 for control treated sites.
- Skin elasticity: Data for skin elasticity was obtained from measurements[a] on the area immediately adjacent and inferior to the outer canthus of the left and right eyes. Suction (vacuum of 200 mbar) was applied for 5 sec to a circular area of skin 8 mm in diameter. The vacuum was released and measurement continued for 10 sec. This suction and release constituted one cycle. Each elasticity measurement was performed over two cycles. Data indicated a statistically significant increase after 8 weeks in skin extensibility (softening effect) at oat β-glucan treated sites, while the control treated sites showed no such increase. Average cycle 1 and cycle 2 values (of the device's parameter EMAX) for skin elasticity at the oat β-glucan treated sites were 0.096 mm and 0.099 mm, respectively. Corresponding values were 0.03 mm and 0.025 mm at the control treated sites.

Protect against UV damage: This study used a multi-layered, highly differentiated model[b] of the human epidermis consisting of normal human derived keratinocytes. These tissues were treated with a 1.0% concentration of oat β-glucan powder and exposed to 1.5 MED/h/cm^2. The assay included positive (tissue irradiated but not treated with test material) and negative (nonirradiated tissue) controls.

[a] Cutometer SEM 575, Courage + Khazaka, Cologne, Germany
[b] MatTek Epiderm skin model, MatTek Corp., Ashland, Massachusetts, USA

After the tissues were incubated for 24 h, an MTT colorimetric assay assessed cell viability. 3-(4,5-Dimethylthiazol-2-yl)-2,5-diphenyltetrazolium bromide (MTT) is incorporated into living cells via the mitochondria. This results in the formation of insoluble purple formazin crystals that remain internal to the cells until extracted with isopropanol. The intensity of the purple color in the alcohol varies directly with the viability of the tissues (that is, the darker the purple color, the more cells are viable). Cell viability is inversely proportional to test agent toxicity. The less toxic a product, the more cells are viable.

Part A of the study was a test to determine whether or not the test material, oat β-glucan, is a cytotoxin. In this case, the control was octoxynol-9[c], a known cytotoxin. Data (Figure 4) indicated that oat β-glucan is not a cytotoxin.

In Part B of the study, the control was culture media. In this case, data (Figure 4) showed that oat β-glucan significantly ($p \leq 0.5$) reduced damage associated with UVA/UVB irradiation when compared to the irradiated (positive) control. Specifically, the average percentage of viable cells for the treated tissues was 48.65 after UV exposure, compared to 26.82 for the corresponding untreated tissues.

[c]Triton X-100

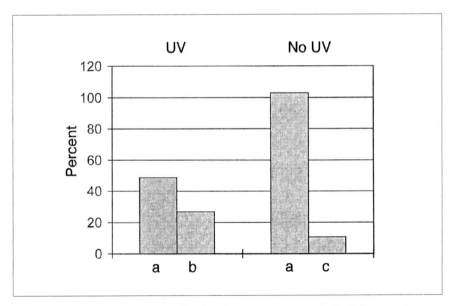

Figure 4. Average percentage changes from baseline in a two part study of cellular protection with oat β-glucan. The test for cytotoxicity (right) shows that oat β-glucan does not behave like a known cytotoxin. After exposure to UV (left), cells survive better with oat β-glucan in the culture medium than do untreated cells.

a = 1% oat β-glucan powder
b = no oat β-glucan positive control
c = octoxynol-9 positive control

Stimulate metabolic activity: Oat β-glucan was evaluated for its ability to stimulate an increase in the metabolic activity of human neonatal fibroblasts in nutrient poor media; that is, the focus was its ability to act as a cytostimulant. The culture media served as the negative control, while the culture media with 10% fetal bovine serum served as the positive control. Metabolic activity was determined by an MTT colorimetric assay after incubating the media for 48 h. The resulting data indicated that treatment with 1% oat β-glucan powder resulted in increased metabolic activity when compared to the positive control.

Activate collagen synthesis: Activation of collagen synthesis was determined indirectly by measuring procollagen using an enzyme linked immuno supressant assay (ELISA) kit. For every molecule of collagen produced, one molecule of procollagen is released.[10] Concentrations of 1.0%, 0.5% and 0.1% oat β-glucan were exposed to macrophage for macrophage activation. Macrophage, if activated by an effective material, will produce various biologically active intermediates. After the macrophage were removed, the macrophage intermediates were exposed to fibroblasts to determine the effects of these biologically active components on collagen formation.

The resulting data showed that 1% oat β-glucan powder stimulated an increase in procollagen (to 334.27 ng/mL) when compared to the positive control (100 μg/mL ascorbic acid, a known collagen stimulator, which produced 251.09 ng/mL of procollagen). Negative controls (culture media) produced less procollagen (221.66 ng/mL) than the 0.10% concentration (291.88 ng/mL) and slightly more than the 0.50% concentration (221.66 ng/mL).

Improve tensile strength of hair: To determine the effect of oat β-glucan on the tensile strength of hair, stress/strain evaluation was conducted on 100 individual hair fibers from a product treated tress and 100 fibers from a similar placebo treated tress. Product was 5% oat β-glucan solution (1% oat β-glucan powder in water) in a creme base. Placebo was the same creme formulation without oat β-glucan.

After five cycles of washing, rinsing, treating and drying, product treated and placebo treated hair fibers were placed in a stress/strain device[d] that stretched them at a rate of 20 mm/min until they broke. Relative humidity and temperature were maintained at $50\% \pm 2\%$ and 20°C, respectively.

The resulting data indicated the rate of premature failure was significantly higher for the placebo group ($p < 0.05$). The results also indicated a significant effect in reducing the premature failure of bleach damaged fibers. Finally, results showed a significant increase in plateau stress (yield region), break extension, break stress and total work done in the group treated with oat β-glucan.

Hydrolyzed Oat Protein

The superior nutritional value of oat protein has long been recognized. Although oat protein quality does not approach the quality of protein from animal sources, it surpasses that of other cereals such as wheat, rice and barley, as shown by feeding tests and by comparison of its amino acid balance with the FAO standard.[8]

---

[d] MT600 Autosampler and Fibre Dimensional Analysis System, both from Dia-Stron Ltd., Andover, UK

Due to its low molecular weight, hydrolyzed oat protein will penetrate the hair shaft, form a thin protective film on the hair and skin without buildup. It will also increase shine, dry combability and moisturization of the hair.

An in vitro study using the microfluorometric (fluorescence intensity) technique evaluated film forming and penetration characteristics of hydrolyzed oat protein. In order to be able to establish the presence of hydrolyzed oat protein on or within the hair fiber, a site-specific fluorochrome—dansyl chloride—was reacted with the basic sites (amine end groups) of the oat protein, which becomes highly fluorescent and visible only after the reaction.

Forms film on hair: After washing, some fibers of bleached human hair tresses were set aside as controls. Others were evaluated before treatment, after one treatment and after 10 treatments with a 5% solution of dansylated hydrolyzed oat protein. Three sets of scans were made along each of 20 hair fibers viewed longitudinally at a wavelength of 500 nm. Scans for untreated fibers were carried out at 450 nm.

Increasing fluorescence intensity indicated a significant adsorption of protein residues on the hair surface after a single treatment (Figure 5). There was no significant increase in fluorescence after 10 applications compared to a single treatment.

Penetrate to cuticle of hair shaft: To prepare hair fibers for cross-sectional viewing, they were embedded in resin[e], cured and microtomed at 25 µm thickness. Cross sections of the hair fibers were observed for evidence of penetration. The fibers treated once and 10 times with the dansylated hydrolyzed oat protein showed a highlighted cuticula (Figure 6). It was especially pronounced in fibers treated multiple times. The absence of greenish/yellow fluorescence in the fiber interior strongly suggests that the diffusion of the large molecule of oat protein is limited to the cuticular region.

Improve combability and feel: Conditioning properties of hydrolyzed oat protein were evaluated using a seven day clinical salon half head study of product (with 8% hydrolyzed oat protein) vs. placebo. The 24 subjects received treatment in the salon on days one, three, five and eight; the subjects did not treat their hair between salon visits. Treatment was randomly assigned to left/right side. Hair was evaluated wet and after drying by an expert grader and also by subject self assessment.

Data from the expert grader and the subjects showed dry combing was easier on the product treated side after the initial shampoo. After the seven day use period, product treated hair exhibited a significantly silkier feel ($p \leq 0.05$) compared to its feel before the treatment. The placebo treated hair showed no such improvement.

Oat Extract

Oat extract is a complex extract from oats. It consists mainly of proteins and sugars with small fractions of UV absorbers, phenolics, flavonoids and saponins. Oat extract

---

[e] Spurr's Low Viscosity Resin

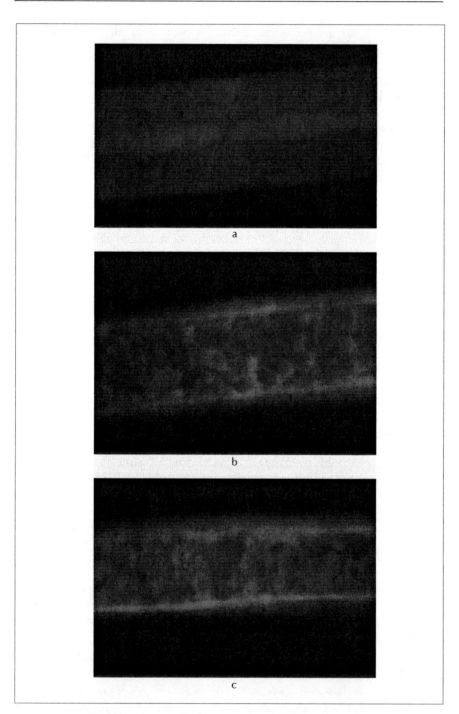

Figure 5. SEMs of hair fibers showing film forming action of hydrolyzed oat protein: a) before treatment; b) after a single application; c) after 10 applications

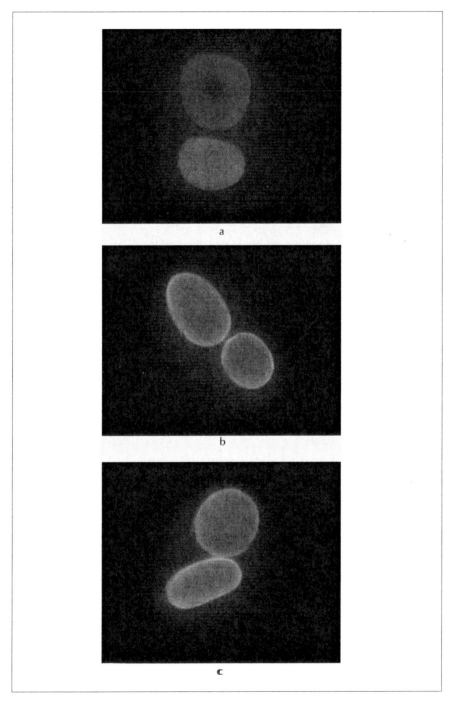

Figure 6. SEMs of cross sections of bleached hair fibers showing penetration of hydrolyzed oat protein: a) before treatment; b) after a single application; c) after 10 applications

is an amber solution in water and butylene glycol.

Data from clinical studies indicates that oat extract serves as a powerful anti-irritant against UV and chemically induced irritation. In vitro studies show that oat extract can significantly reduce the damage associated with UVA/UVB irradiation.

Reduce chemically induced irritation: Lactic acid (15%) was applied to selected sites on the volar surface of the forearms of six subjects and left in place under an occluded patch for 4 h. Then erythema was graded visually on a five step scale (no visible erythema, minimal erythema, defined erythema, moderate erythema and severe erythema). When the chemically insulted sites achieved a score indicating minimal, defined or moderate erythema, metered readings[f] of skin redness (a° in the L°a°b° color notation system) were taken.

Randomly selected sites were treated with oat extract products, leaving one site untreated as the designated control. Via the meter and visual observation, the sites were evaluated for changes in erythema at 4 h and 24 h after treatment. Table 1 shows a significant decrease in the erythema caused by chemical irritation the oat extract treated sites as compared to the untreated control site.

Reduce UV induced irritation: To determine oat extract's ability to alleviate UV induced irritation, six subjects were selected and the minimal erythema dose (MED) for each was determined according to the Federal monograph.[4]

---

[f] Chroma Meter 200b, Minolta, Osaka, Japan

Table 1. Average erythema (a*) changes (%) due to oat extract treatment following chemically induced irritation

Product	After 4 h	After 24 h
Oat extract in 1:1 w/w propylene glycol and water	−29.3	−33.1
Oat extract in 1:1 w/w butylene glycol and water	−20.9	−28.1
Untreated control	+5.6	−8.5

− = decrease in eythema
+ = increase in eythema

Table 2. Average erythema (a*) changes (%) due to oat extract treatment following UV induced irritation

Product	After 4 h	After 24 h
Oat extract in 1:1 w/w propylene glycol and water	−13.3	−27.7
Oat extract in 1:1 w/w butylene glycol and water	−13.3	−32.1
Untreated control	+2.4	−1.0

− = decrease in erythema
+ = increase in erythema

Then each subject was exposed to twice that subject's MED on five sites defined between the scapulae and belt line, lateral to the midline of the back. Twenty-four hours following the irradiation, the test sites were evaluated for erythema on a scale of 0 to 3+ using the meter[e] and visual observation. The irradiated test sites were treated with oat extract products, leaving one site untreated as the control. Via the meter and visual observation, the test sites were evaluated for changes in erythema at 4 h and 24 h after treatment. Table 2 shows a significant decrease in the erythema caused by UV irritation of the oat extract treated sites as compared to the untreated control site.

Another in vitro study used the skin model[b] to assess oat extract's ability to alleviate UV induced irritation. The results of the MTT colorimetric assay showed that oat extract powder (1%) produced a statistically significant reduction in the damage associated with UVA/UVB irradiation, compared to the irradiated control.

Summary

The benefits of using oats, specifically oatmeal, in personal care products has been known for some time. On October 3, 1989, the FDA proposed a tentative final monograph for the use of colloidal oatmeal as a safe and effective skin protectant when labeled with the following claim: "Provides temporary skin protection and relieves minor irritation and itching due to poison ivy, poison oak, poison sumac and insect bites."

Today, for a specific application, the cosmetic chemist obtains better benefits by using oat fractions than by using the entire oatmeal. These materials are safe and non irritating. They have excellent cosmetic stability. Derived from a natural renewable resource, they protect and repair the skin and hair from damaging environmental effects such as UVA/UVB irradiation, pollution, smoke, bacteria and free radicals. They also help repair damage from other chemicals such as AHAs, surfactants and bleaches.

The science and research behind oats and its fractions are opening new doors to their myriad of novel uses in efficacious personal care formulas, functional foods and health care products.

---

[e] Spurr's Low Viscosity Resin
[b] MatTek Epiderm skin model, MatTek Corp., Ashland, Massachusetts, USA

—Janice Hart and Christopher Polla, *Canamino Inc., Northport, New York, USA*
—Judith C. Hull, *Judith C. Hull Consultants Inc., Wyomissing, Pennsylvania, USA*

References

1. GO Aspinall and RC Carpenter, Structural investigations on the non-starchy polysaccharides of oat bran, *Carbohyd Polym* 4 271-282 (1984)
2. AE Clarke and BA Stone, Enzymic hydrolysis of barley and other β-glucans by a β-(1→4)-glucan hydrolase, *Biochem J* 99 582-588 (1966)
3. *Fed Reg* 62(15) 3584-3601 and amendment in *Fed Reg* 62(61) 15343-15344 (1977)
4. Food and Drug Administration of the US Department of Health, Education and Welfare,

Sunscreen products for over-the-counter human drugs, proposed safety, effective and labelling conditions, *Fed Reg* 43(166) 38206-38269 (Aug 25, 1978)
5. D Kim and T Lo, β-Glucan collagen matrix dressing: A fresh concept against stagnant wounds, abstract presented at the Advances in Wound Healing, Burn Care and Infection Control meeting sponsored by the International Burn Foundation in cooperation with the International Society for Burn Injuries, the Australia-New Zealand Burns Association and the Japanese Society for Burn Injuries, Wailea, Maui, February 17-21, 1997
6. JJ Leyden, Clinical features of aging skin, *British J Derm* 122 Sup 35 1-3 (1990)
7. FW Parrish et al, Selective enzymolysis of poly-β-D glucans, and the structure of polymers, *Can J Chem* 38 2094-2104 (1960)
8. DM Peterson and AC Brinegar, Oat storage proteins, in *Oats: Chemistry and Technology*, FH Webster, ed, St Paul: American Association of Cereal Chemists (1986) 153
9. RL Rizer, ML Sigler and DL Miller, Evaluating performance benefits of conditioning formulation on human skin, in *Conditioning Agents for Hair and Skin*, R Schueller and P Romanowski, eds, New York: Marcel Dekker (In press)
10. MB Taubman et al, *Science* 186 1115 (1974)
11. PJ Wood, Oat β-glucan: Structure, location, and properties, in *Oats Chemistry and Technology,* FH Webster, ed, St. Paul, Minnesota: American Association of Cereal Chemists (1986) 121-127

New Active Ingredient for Aging Prevention

Keywords: Quebracho extract, DNA damage, DNA repair, free radicals

Fermented Quebracho is tested for its anti-free radical properties and its protecting effect on DNA

Cutaneous aging is linked to a transformation of connective tissue and a decrease of cell regeneration ability. During aging, genetic material is modified through enzymes, proteins and DNA alterations, and cell proliferation decreases. Consequently, tissue loses its elasticity and capacity to regulate water exchange, and the tissue replication becomes less efficient.[9,12]

Chemical and enzymatic oxidations involve free radical formation that accelerate this aging phenomenon. In fact, free radicals are considered to be the most active compounds in aging. They damage DNA and assist in dehydrogenation, hydroxylation and protein glycation. The last reaction involves a loss of biological functions of proteins, such as collagen and proteoglycans, which results in alterations of membrane structure and increase of skin stiffness.

The source of these free radicals can be endogenous, associated with metabolic reactions (oxidation reaction in mitochondria with disruption of electron transport, excessive phagocytosis, activation of arachidonic acid metabolism) and exogenous because of UV radiation, pesticides, air pollution, anti-tumoral drugs and unhealthy lifestyles (Figure 1).

At cutaneous level, oxidative stress leads to photoaging and a decrease of endogenous defense systems.

The body naturally protects against free radicals by chemical or enzymatic detoxification systems. Nevertheless, the protective capacity of these systems decreases during aging. Exogenous compounds like enzymes, antioxidants (including some vitamins and metals) and phenolic compounds (including some flavonoids and tannins from plants) reinforce the natural protection by limiting oxidative reactions. Plants are an important source of natural antioxidants and particularly of tannins.

Two tannin groups that are usually identified in higher plants are hydrolysable tannins, which give gallic acid and ellagic acid, and condensed tannins, which give catechins (also called proanthocyanidols). They consist of flavan units, such as catechol, epicatechol, gallocatechol and epigallocatechol, referred to collectively as proanthocyanidin oligomers (PCOs).

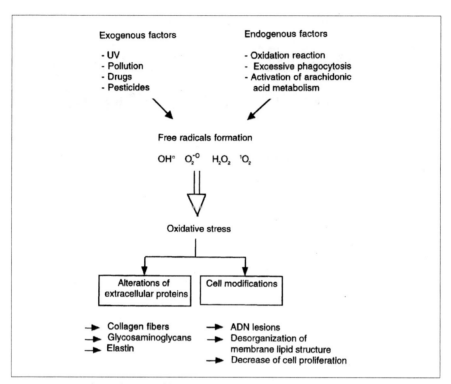

Figure 1. Recapitulative diagram of free radical formations and effects

Most biological properties of PCOs are linked to their ability to form complexes with various free radicals,[4] stabilizing them and making them harmless. They are essential free radical scavengers by inhibiting the formation of anion superoxide. They stabilize ascorbic acid,[15] inhibit lipid peroxidation,[8] metabolize arachidonic acid[5] and inhibit cytotoxicity in primary cultured hepatocytes. It has been shown that they inhibit the enzymatic proteolysis of collagen induced by superoxyde radicals, which helps preserve collagen stability.

A new type of free radical scavenger that contains PCO, fermented Quebracho,[a] has been tested for its anti-free radical properties as well as its protecting effect on DNA. Its stimulating activity on DNA repair system has also been determined.

Properties of Fermented Quebracho

Quebracho blanco from Anacardiaceae family, a tropical plant extract obtained from barks of the tropical tree Quebracho blanco. It is bioconverted into PCO by a fermentation process with specific microorganisms that concentrate PCO and increase the efficiency of the ingredient.

---

[a]Nucleolys, Greentech, Saint Beauzire, France. The complete INCI name is Aspergillus/Aspidosperma Quebracho ferment and water, glycerin, alcohol, lecithin, acrylates/C10-30 alkyl acrylate crosspolymer, triethanolamine and phenoxyethanol.

Free Radical Formation

Free radicals are constituted of one atom (or an association of atoms) that has an unpaired electron on its outer orbital. This status implies that free radicals are energetically and kinetically unstable. They return to a stable state by giving one electron to another molecule or taking an electron from another molecule.

These free radicals can be reducing (give one electron) or oxidizing (take one electron) agents. The main source of free radicals is the oxygen molecule, which is also beneficial for organism survival.

Oxygen Reduction

Oxygen is incorporated into organic molecules through different mechanisms. During the successive univalent reduction of dioxygen, anion superoxide (O_2^{-0}), hydrogen peroxide (H_2O_2), hydroxyl radical ($OH°$) and finally water are formed (see Equation 1).

Through the catalytic action of enzymes, the two electron reduction (H_2O_2) and the four electron reduction (giving water) are carried out without detectable free radical intermediates. As a result of the the electron transfer involving the reduction of oxygen, free radicals are produced.

$$O_2 \xrightarrow{+e-} O_2^{-0} \xrightarrow{+e- \ +2H+} H_2O_2$$

$$H_2O_2 \xrightarrow{e- + H+} OH° (+H_2O) \xrightarrow{+e- \ + H+} H_2O$$

Equation 1

To prepare fermented Quebracho, barks of the tree are pulverized to an aqueous extract. The dried extract is a red powder of condensed tannins of various molecular weights with soluble and unsoluble parts.

The biofermentation process implies a polymerization of monomers and dimers for getting water soluble polymers and partial depolymerization of unsoluble polymers.

HPLC analysis shows that, after fermentation, the extract obtained has been concentrated in proanthocyanidin oligomers. This powder contains 99% proanthocyanidin oligomers and 1% gallic acid. To preserve the stability of PCO, it is microencapsulated in liposomes; the final product is an emulsion containing around 2% of PCO.

Anti-Free Radical Activity

The fermented Quebracho's effect has been compared to superoxide dismutase (SOD) activity by using the xanthine/xanthine oxydase enzymatic method (Figure 3).

Reactive Oxygen Species

Superoxide anion ($O_2^{-\circ}$): Anionized form of oxygen, superoxide anion is produced through the univalent reduction of O_2 or the univalent oxidation of H_2O_2. It is also produced by enzymatic reaction of some flavoprotein deshydrogenase through the autooxidation of substrates such as ferredoxins, hydroquinones, thiols and reduced hemoproteins.

Superoxide is involved with damage such as lipid peroxidation, cellular toxicity and single strand breaks of DNA. Many toxic effects attributed to $O_2^{-\circ}$ could be due to its metal catalyzed interaction with H_2O_2 to produce $OH°$ (Figure 2).

Hydrogen peroxide (H_2O_2): H_2O_2 can be generated by divalent reduction of oxygen or by the enzymatic dismutation of superoxide radicals by superoxide dismutase (SOD). This compound has a low reactivity, and its damaging effects are not associated to a direct attack on cellular components. It decomposes spontaneously or enzymatically (by catalase or peroxidase) into water.

Hydroperoxide ROOH: Oxygen insertion can occur in molecular structure with carbon hydrogen bond. This kind of reaction produces hydroperoxide ROOH, which can be captured by molecular oxygen in a chain reaction generating free radicals.

Singlet oxygen: Singlet oxygen is obtained under the action of light in the presence of a photosensitiser. The photosensitiser is excited by a photon. It transfers its energy to oxygen which is photo excited. A new reactive species is formed: singlet oxygen, which has two unpaired electrons. It can react with chromophors and attack the photosensitizer itself, which leads to a photodynamic effect. Singlet oxygen reacts with several compounds containing a carbon-carbon bond such as carotene or fatty acids.

Hydroxyl radical ($OH°$): The highest reactive radical $OH°$ is the result of the univalent reduction of H_2O_2. It can react with molecules which are able to give an electron such as enzymes, sugars, aminoacids, nucleic acids or membrane phospholipids.

$OH°$ has a large destructive and mutagenic potential in biological systems. It mainly reacts with fatty acids (particularly with polyinsaturated acids) of membranes which leads to membrane disorganization. As the membrane plays a major role in cellular function, the effects can be cellular destruction or a wrong transmission of messages inside the cell.

Proteins are also sensitive to $OH°$, as it can modify their structure, making disulfide bonds and agregation. $OH°$ denatures the proteins of the connective tissue such as collagen, elastin and glycosaminoglycans. These alterations are irreversible.

DNA is very sensitive to $OH°$. This hydroxyl radical causes strand breakage which can damage several DNA bases and sugars. These damages result in mutations and death. [3]

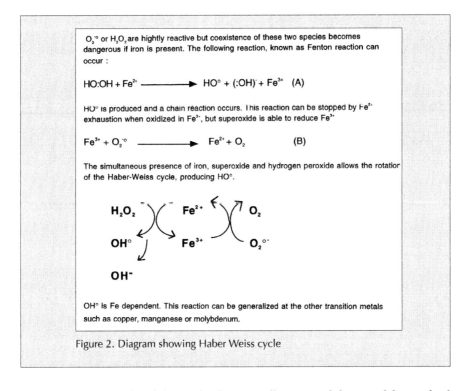

Figure 2. Diagram showing Haber Weiss cycle

In microencapsulated form, the ferment allows an inhibition of free radical formation, which is superior to SOD's.

A concentration of 2.5% is enough to inhibit free radical formations by 80%. The inhibition of free radicals formation is 100% at a final concentration of fermented Quebracho of 5%.

DNA Protection

DNA lesions can be due to a direct reaction with oxygen radicals or to a reaction between DNA and malondialdehyde (MDA) and 5-hydroxynomenal (5 HN), which

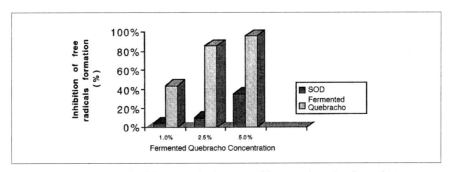

Figure 3. Comparative study of the anti-radical activity of fermented Quebracho and SOD

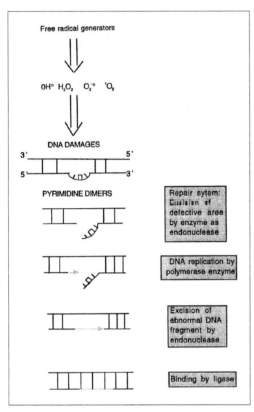

Figure 4. Diagram of DNA repair system

are secondary products of lipid peroxidation. These reactions lead to modifications of nucleobases, DNA deterioration, such as cross linking of purines, pyrimidines and strand breaks.[12,13] These lesions can lead to a decrease in ATP levels and to cell death. If the cell survives, mutations are produced and lead to the formation of pyrimidine dimers. Some of them are associated with malignant transformations in naturally occuring human cancers of the colon, pancreas and lung.[1]

Cells have a protective system for repairing their DNA damages. More particularly, pyrimidine dimers formed after external agressions can be recognized by enzymes which remove the strand damage and use the healthy DNA strand as a template to rebuild a matching piece (Figure 4).

Unfortunately, such mending is not always complete, and additional lesions can occur involving risk factors for health.

It is possible to protect DNA strands by using certain antioxidant substances to modulate the DNA replication and repair system. A new method of molecular biology has been perfected to prove this claim.[11]

Detecting DNA Disorders

The DNA lesions caused by free radicals or UVB are recognized by repair enzymes of a purified cellular extract. The lesion repair process involves an excision phase which is followed by a new synthesis of DNA. The method microplates coated with DNA plasmid (Figure 5).

DNA absorption and DNA damage detection are done as previously described.[11] Protein extract was prepared from frozen pellet cells purchased from Computer Cell Culture Center[6] and mixed with the DNA coating, including labeled nucleotide DIG 11 dUTP. The recognition was done by anti-DIG antibody coupled to an alkaline phosphatase. A luminescent signal is sent out.

The signal intensity is measured with a luminometric apparatus. This signal is proportional to the number of repaired lesions. A dose dependent relationship is observed up to 15 lesions for 16 kilobases.

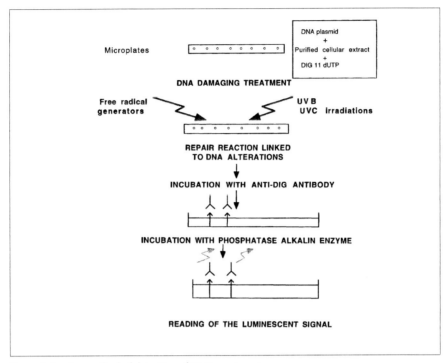

Figure 5. In vitro method of detection of DNA repair system

The relative light unit (RLU) calculated determines the percentage of inhibition of residual repair which corresponds to an inhibition of lesion formations on DNA.

Fermented Quebracho and DNA

Protecting DNA from ROS: It has been shown that fermented Quebracho has no genotoxic effect. There has not been observed any inhibition of the repair system that could be associated to interactions between the ferment and DNA.

To determine the protecting effect of the extract on DNA, we used 10 mg/ml of methylene blue solution diluted with 25 mg/ml with water.[b] This stock solution was mixed with an equal volume of fermented Quebracho used at several concentrations. Because it generates reactive oxygen species, methylene blue causes modifications in or loss of DNA bases, which activates the repair system.

The ferment inhibits the formation of lesions which are measured as follows (RLU: Relative Light Unit, MB: Methylene blue) (Equation 3).

It has been shown that fermented Quebracho is active on DNA protection at low concentration (0.2%) (Figure 6).

---

[b]MilliQ filter, Millipore Corp., Bedford, Massachusetts, USA.

Protection Against Free Radical Toxicity

Skin is particularly exposed to oxidative free radical agressions, but cells possess an array of protective enzymes and compounds.

Protective enzymes: The enzymes involved in $O_2^{\cdot -}$ dismutation (superoxide dismutase), destruction of H_2O_2 (catalase) and reduction of lipoperoxides (glutathione peroxidase) are extremely efficient. Superoxide dismutases are a group of metalloenzymes that catalyze the disproportionation of two superoxide molecules according to Equation 2. Catalase and peroxidase remove H_2O_2 in the biological systems.

Anti-catalytical agents: The immobilization of transition metal ions at a reduced state is an important way to avoid oxygen reactions towards organic molecules. Chelating agents as well as some proteins (transferrin, ferritin, ceruloplasmin) are able to play this role. Inhibitors of oxidase such as allopurinol, oxypurinol deferroxamine are also efficient.

Nonenzymatic systems include lipophilic and hydrophilic antioxidants. Endogenous small molecules also play an important role in the removal of toxic oxygen species. Substances which react quickly with free radicals are commonly named radical scavengers; they act stoechiometrically with radicals, and a molecule of these protectors can only activate one or two free radicals. As a consequence, important concentrations are needed for being effective.

Glutathione (GSH) is an important water-soluble antioxidant and reducing agent. Through its ability to serve as a co-substrate for peroxidases or as a substrate for oxyradicals, it represents a strong defense against toxic oxygen species. It reacts with many oxidants such as H_2O_2 to form the oxidized form (GSSG). This reaction is catalysed by glutathione peroxidase. Glutathione can also react without an enzyme with $O_2^{\cdot -}$. The GSSG formed can be reduced back into GSH by the NADPH dependent enzyme glutathione disulfide reductase.

Among vitamins, ascorbic acid can act as a protector of lipids from oxidative damages by effective scavenging lipid peroxidation initiating radicals.[2] Vitamin E is a very efficient lipophilic antioxidant. It reacts with $O_2^{\cdot -}$ to give phenoxy radical. Ascorbate seems to be implied in a mechanism which could accelerate tocopherol regeneration. Carotenoids are also considered as efficient protectors against $^1O_2$ and photochemical aggressions.

$$O_2^{\cdot -} + O_2^{\cdot -} + 2H^+ \longrightarrow H_2O_2 + O_2$$

$$O_2 \xrightarrow{+e^-} O_2^{\cdot -} \xrightarrow{+e^- \; +2H^+} H_2O_2$$

$$H_2O_2 \xrightarrow{e^- + H^+} OH^\circ \; (+H_2O) \xrightarrow{+e^- \; + H^+} H_2O$$

Equation 2

$$\text{Equation 3:} \quad \frac{(\text{RLU MB (without Quebracho)})_{\text{fermented}} - (\text{RLU MB (+ Quebracho)})_{\text{fermented}}}{\text{RLU MB (without fermented Quebracho)}} \times 100$$

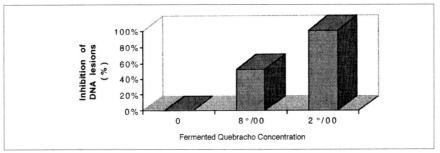

Figure 6. Percentage of inhibition of DNA lesions by fermented Quebracho after free-radical aggressions

At this concentration, 100% of DNA integrity is maintained in comparison with the control, and protection is increased up to 50% with the addition of 80 ppm of fermented Quebracho.

Protecting DNA from UVB: The protecting effect of the ferment on DNA has also been proved after UV irradiation. UVB has a direct toxic effect on superficial layer of the skin. It penetrates only epidermis, is highly involved in the photocarcinogenesis process and has immunosuppressive potential. UVB exposure involves a decrease of the immune defence of skin.

For studying fermented Quebracho's UVB protecting effect, DNA is irradiated in vitro by a lamp which emits at $\lambda=312$ nm (0.4 J/cm^2).

The ferment has been tested at several concentrations. It has been observed that 10% of the ferment completely protects DNA. In this condition, the repair signal is comparable to a nonirradiated system (Figure 7).

Restoring the DNA repair system: The restoration of the DNA repair system by fermented Quebracho has been determined with a saturated repair system obtained after UV hyperirradiation.

The plasmid was previously irradiated by UVC, which creates some lesions. Another irradiation by UVB implies some additional lesions, which have an inhibitory effect on the DNA repair system. In this condition, the repair system is saturated. When applying the ferment, the ability of the repair system to protect DNA after hyper-aggressions is restored.

The restoration effect of fermented Quebracho against UVB irradiation is dose dependent (Figure 8).

This effect is observed at a concentration of 1%. After using 10% of the ferment, the ability of the DNA repair system is restored 40%.

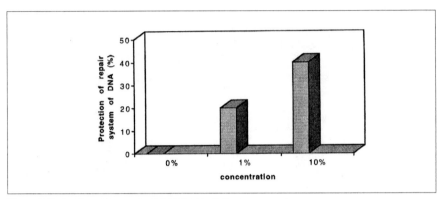

Figure 7. Protection of fermented Quebracho on the DNA repair system

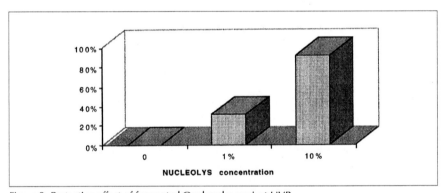

Figure 8. Protecting effect of fermented Quebracho against UVB

Cell culture tests: Fermented Quebracho's protection of DNA has been tested on HELA cells. HELA cells are obtained from human epithelial carcinom. Cells are trypsined and put into wells in the presence of medium culture.

For determining the ferment's protection effect of DNA, the product has been tested by incubation with the cells (1 h at 37°C). After rinsing, H_2O_2 is added at the final concentration of 200 mM. After 30 min of incubation, the medium is removed, cells are broken and the DNA repair signal is measured. The repair signal depends of the intensity of lesions done on DNA by H_2O_2.

When fermented Quebracho is used, the oxidative damage decreases. The product contains less DNA lesions than its control that corresponded to an inhibition of the repair signal.

At 2% concentration of the ferment, the inhibition of damage is 100% (Figure 9).

With this experiment, we show at first a penetration of fermented Quebracho in the cell and then that the ferment keeps its biological properties after its penetration into the cells up to 1 h after incubation.

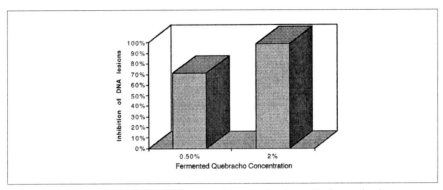

Figure 9. Inhibition of DNA lesions by fermented Quebracho determined on cell culture

Conclusion

When the skin's endogenous system is overworked, topical application of exogenous substances can make up for its deficiencies. Fermented Quebracho appears to be an innovative product for preventing cutaneous disorders and the skin's DNA damage. It would be interesting to compare the fermented Quebracho's efficiency with other known antioxidant products such as tocopherol. This work will be the object of future investigations.

> —Véronique Jay and Jean Yves Berthon, *Greentech S.A., Saint-Beauzire, France*
> —Daniel Hagege, *Laboratories of Biology of Orleans University, France*
> —Marie-Paule Pouget and Brigitte Lejeune and Henri Pourrat, *Laboratories of Galenic and Pharmacotechnic, Pharmacy University Of Clermont, Ferrand, France*

References

1. GC Cochrane, Cellular injury by oxidants, *Amer J Med* 91 23-30 (1991)
2. B Frei, L England and B Ames, Ascorbate is an outstanding antioxidant in human blood plasma, *Proc Natl Acad Sci USA*, 86 6377-6381 (1989)
3. B Halliwelland JMC Gutteridge, Free radicals in biology and medicine, B Halliwell and JMC Gutteridge, eds, 346 *Oxford University Press*, New York (1987)
4. E Haslam, Vegetable Tannins Revisited, Cambridge University Press, Cambridge (1989)
5. Y Kimura, H Okuda, T Okuda, S Arichi, *Planta Med*, 337 (1986)
6. JL Manley, A Fire, M Samuels and PA Sharp, *Methods Enzymol* 101 568-582 (1983)
7. T Okuda, T Yoshida, T Hatano, In Economic and Medicinal Plant Research, vol. 5, H Wagner and NR Farnsworth, eds, Academic Press, London, (1983) 129
8. T Okuda, K Mori, H Hayatsu, *Chem Pharm Bull* 32 3755, (1984)
9. R Sakata, Biochemical studies on collagen metabolism, *Kumatomo Med J* 13, 27-40 (1960)
10. B Salles and P Calsou, *Anal Biochem* 215 304-306 (1993)
11. B Salles, C Provot, P Calsou, I Hennebelle, I Gosset, GJ Fournié, A chemiluminescent microplate assay to detect DNA damage induced by genotoxic treatments, *Analytical Biochemistry* 232, 37-42 (1995)
12. ML Tanzer, Cross linking of collagen, *Science* 180, 561-6 (1973)
13. OW Teelor, RJ Boorstein, J Cadet, The repairability of oxidative free radical mediated damage

to DNA: A review, *Int J Radiat Biol* 54 13111-150 (1988)
14. R Teoule, Radiation induced DNA damage and its repair, *Int J Radiat Biol* 51 573 (1987)
15. T Yoshida, S Koyama, T Okuda *Yakugaku zasshi* 101 695 (1981)

Galactomannan and Xyloglucan: Bio-Active Polysaccharides

Keywords: β-glucan, galactomannan, xyloglucan

Discussion of the potential of Cassia angustifolia *galactomannan and* Tamarindus indica *xyloglucan for skin and hair care*

Carbohydrates are, with proteins, lipids and nucleic acids, one of the four major classes of constituents of living matter. Carbohydrates are the most prevailing constituents among plants and animals. Glucose is the most important sugar in cells because of its role in cellular energy metabolism and as a precursor of other sugars: ribose and deoxyribose.

Glucose molecules may be combined by different types of bonds to form long chains, becoming polysaccharides or gums. Among the most familiar biopolymers are starch, reserve polysaccharide and cellulose, a constituent of the rigid cellular walls surrounding plant cells. Pectins, guar and chitins are other biopolymers. Other well-known molecules, such as xanthan, scleroglucan, dextran, gellan and curdlan, are usually prepared by biotechnology.

For a long time, the notion of glycosides has been linked to the notion of reserve substances, energy components and support structures (glycosaminoglycans of the connective tissue). Since the beginning of the 1970s, researchers have acknowledged the major biological role played by glyco-conjugated elements, either as glycoproteins or glycolipids. Glyco-conjugated elements act as recognition signals of cells by other cells or by molecules. They serve multiple biological functions: induction and maintenance of peptide chains in a biologically active conformation, maturation of some proteins by proteolysis, participating in the activation of pro-proteins, contribution to the immunogenicity of proteins, control of cellular plasma membrane permeability, and participation in cellular recognition and association.[16]

The immunomodulating activity of some polysaccharides, especially β-(1-3)-glucan structures, is well known. Their effect on the immune system is unspecific. The activating effect on macrophages (for example, by the β-glucan from the cellular wall of *Saccharomyces cerevisiae*), which play an essential role in cellular immunity, has potential applications in wound healing.[15] Other polysaccharides of botanical origin, with α, 1-6 and 1-4 bonds, such as mannan-polymannose of *Aloe vera*[4] and lentinan, a polysaccharide from *Lentinus edodes*[3], may also have immunostimulating activity.

Scientists creating treatment cosmetics are always searching for new active ingredients, including molecules structurally identical to those already known but having a different origin, or new effects from previously identified molecules with known structure. Our work has identified two natural biomolecules: galactomannan from *Cassia angustifolia* and xyloglucan of *Tamarindus indica*. We describe their chemical structures and the manufacturing processes together with objective tests of their cosmetic benefits and efficacy. From these, we suggest some applications.

Cassia angustifolia β-Glycan

Origins and Manufacture: *Cassia* β-(1-4)-glycan, or galactomannan, is obtained from the seeds of *Cassia angustifolia* Vahl, an exotic leguminous plant (Caesalpiniaceae) native to India. The common names for these are Indian senna Tinnevelly senna or Meca senna. *Cassia angustifolia* is an annual herbaceous plant measuring about 0.75 m high. Its fruits are thin, flat pods, 4-7 cm long and 2 cm wide, containing five to seven dark brown seeds (Figure 1). Galactomannans, which are mainly present in the endosperm of the leguminous plant seeds, have a double physiological function: They help prevent, in times of extreme heat and dryness, a total dehydration of seeds, thus avoiding the denaturation of enzyme proteins necessary for germination, while also serving as a nutritious reserve used during germination.[11]

Figure 1. Fruit and seeds of *Cassia angustifolia* Vahl (L.S. photographs)

For preparing of the *Cassia* β-glycan, the seeds of *Cassia angustifolia* were ground and their colored shells were discarded. The ground seeds were then subjected to classical techniques of extraction and isolation of polysaccharides, such as extraction in a warm or cold state by aqueous, hydroalcoholic or alcoholic solutions, in an acid or basic medium, followed by precipitation using an organic solvent.

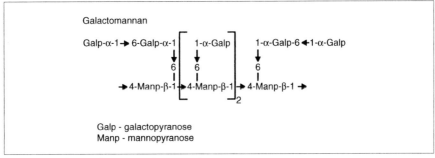

Figure 2. Biochemical structure of *Cassia* galactomannan

C. angustifolia galactomannan (INCI name: *Cassia angustifolia* seed polysaccharide)[a] is a native purified polysaccharide presented as a pure powder or an aqueous solute ready for use. The latter feels like a hydrogel with a thick consistency.

Structure and Biochemical Characteristics

C. angustifolia galactomannan, a tridimensional polysaccharide, is hydrosoluble and consists of D-galactose and D-mannose in a 1.5:3.5 molecular ratio. The main chain of the galactomannan was found to consist of β-(1-4) linked mannopyranosyl units while α-(1-6) linked α-glycosidically bonded galactopyranosyl units form the branching point.[1] The biochemical structure of this component is displayed in Figure 2.

The β-glycan content has been purified to 90%. The constituent sugars of this extract, determined after hydrolysis and gas chromatographic analysis, are distributed as follows: mannose 64%, galactose 27%, glucose 2%, xylose 3%, arabinose 1%. The molecular mass determined by universal standardization is 170,000 daltons.

Cosmetic Properties

Galactomannan, because of its structure and its molecular weight, exhibits the same properties as hyaluronic acid:

- Leaves the skin feeling smooth and soft; gives the effect of emollience without being oily and occlusive
- Exhibits sustained moisturizing effect: improves the capacity of stratum corneum to hold water
- Corrects and repairs rough, dry skin
- Exhibits bio-substantivity to skin and hair
- Possesses film-forming properties.

In vitro test of long-term skin moisturizing effect: The moisturizing activity of *Cassia* galactomannan was evaluated compared to the effect of hyaluronic acid. Both active ingredients were incorporated at equivalent concentrations into the same placebo formulation.

[a]Indinyl CA, Laboratoires Sérobiologiques, Pulnoy, France

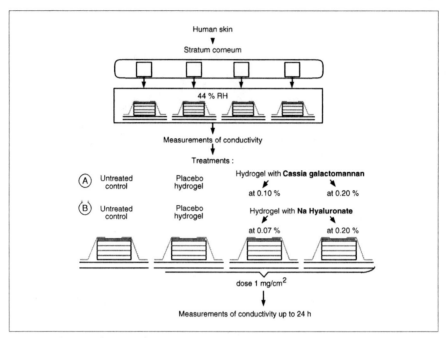

Figure 3. Schematic drawing of the experimental protocol used to test moisturizing effect by increase in conductivity on excised skin

Obata and Tagami described the simulated stratum corneum model we used.[17,18] The isolated horny layer was placed on buffer-saturated filter paper in an occlusive manner. The protocol used is illustrated in Figure 3.

We tested each active ingredient (either *Cassia* galactomannan or sodium hyaluronate) at two different concentrations and compared them to placebo treatment and control conditions. The dose topically applied was 1 mg/cm^2. The stratum corneum hydration, evaluated by conductance measurement under the controlled conditions of relative humidity, has been followed up to 24 h after treatment. The product application caused an increase of dielectric conductivity, hence an increase in moisturization. The efficacy of hydrogels containing the active ingredients has been expressed as a percent of conductivity increase referring to placebo treatment (Figure 4).

The increase of conductivity compared to the placebo 24 h after application was +20% and +35% for *Cassia* galactomannan at 0.10% and 0.20%, respectively. The moisturizing effect of sodium hyaluronate at 0.07% was slight and not significantly different from placebo treatment. At 0.20%, sodium hyaluronate increased the moisturization by +30% compared to the placebo 6 h after application. No significant effect compared to the placebo was demonstrated by sodium hyaluronate 24 h after application.

Strengthening the skin barrier function in vitro: The skin barrier strengthening effect of *Cassia* galactomannan was tested compared to a placebo by evaluation of transepidermal water loss (TEWL) through the horny layer (control or damaged by surfactant treatment). Excised horny layer was placed on stainless-steel cells

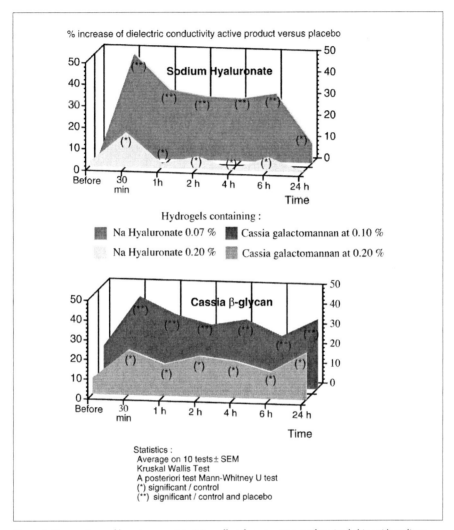

Figure 4. Comparison of long-term moisturizing effect from treatment of excised skin with sodium hyaluronate and *Cassia* galactomannan. The galactomannan shows superior moisturizing after 24 h.

containing water. The use of steel cells has been advised by the work of Blanc,[2] Lévêque et al.[13] and Ribaud et al.[20] This ensures that the water may be lost only through the horny layer (Figure 5).

After treatment (no treatment, placebo cream or cream containing 0.20% of *Cassia* galactomannan) and equilibration for 24 h in a humidity-controlled environment, we determined the quantity of water loss through the horny layer gravimetrically.

Better skin hydration can be achieved by the reduction of transepidermal water loss. The treatment by galactomannan creams reduced the water loss whether the horny layer was damaged or not. The film-forming activity of the galactomannan on skin was thus demonstrated by the reduction of transepidermal water loss measure-

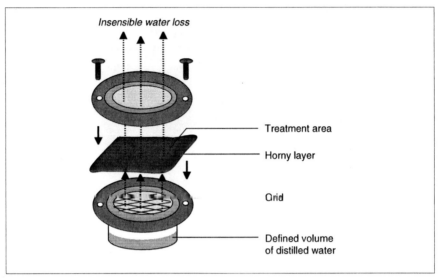

Figure 5. In vitro model of the evaporative water loss through the stratum corneum

ment in vitro. The effect of *Cassia* galactomannan treatment on transepidermal water loss was greater than the placebo effect. The skin barrier function has been significantly strengthened after *Cassia* galactomannan treatment (-30.2% on the undamaged horny layer and -27.2% on the damaged horny layer). The reduction of water loss was less important and not significant after placebo treatment (-22.1% on the undamaged horny layer and -21.8% on the damaged horny layer).

The beneficial effect on TEWL and the skin barrier function has been confirmed by an in vivo test. The colloidal film has protected the skin against water loss and consequently the skin became more hydrated, less squamous and its microrelief was improved.

Increasing skin softness, suppleness and moisturization in vivo: The beneficial effect on the skin properties shown by a cream containing 0.20% of *Cassia* galactomannan versus a cream placebo was evaluated in a study carried out on 15 female volunteers (average age 55 years) having very dry, scaly skin. The volunteers evaluated their skin for a complete range of properties before, immediately after and 45 min after a single application of either the test or placebo cream. Three properties (softness, suppleness and moisturization) have been evaluated for both products according to the following unstructured scale, which has only three gradations:

Cassia angustifolia seed polysaccharide or *Cassia* galactomannan displayed a long-lasting water-retaining effect in vitro – its moisturizing activity was detectable

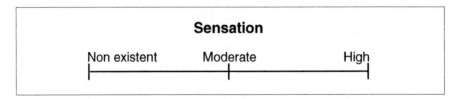

up to 24 h after application. The sensory study on human volunteers confirmed this moisturizing activity. The three principle parameters evaluated by the volunteers–skin softness, suppleness and moisturization–all improved. The beneficial effect on skin softness, suppleness and moisturization was demonstrated by the volunteers' self-evaluation. The areas treated with cream containing 0.20% of *Cassia* galactomannan were considered softer (+21%), more supple (+27%) and more hydrated (+23%) by comparison with placebo-treated areas.

Repairing cutaneous microrelief in vivo: The skin replicas obtained before and after four weeks of treatment twice a day with a cream containing 0.20% of *Cassia* galactomannan were examined using a scanning electron microscope. Four weeks of treatment with *Cassia* galactomannan at 0.20% considerably improved the cutaneous microrelief and reduced skin desquamation. The restructuring of microrelief and the reappearance of its multidirectionality have been observed (Figure 6).

Substantive and conditioning effect on hair: The galactomannan's film-forming capacity on hair was investigated. The method we used, initially described by Garcia and Diaz,[6] involves the continuous recording, by a tensile device, of the forces which oppose the motion of a hair swatch through a comb. Dry-combing curves have been recorded using the swatches previously treated either with a placebo shampoo or with a shampoo containing 0.06% of *Cassia* galactomannan. Combability has been evaluated by means of two parameters that can be directly obtained from the combing curves: peak combing force and combing work. The combing work was reduced by 22.7% (S) and the peak of the combing force by 33.9% (S) with *Cassia* galactomannan treatment compared to placebo treatment. Moreover, the hair swatches treated with *Cassia* galactomannan have a softer, smoother feel.

The hair-surface evaluation by scanning electron microscopy was done on human hair damaged by acetone/ether treatment of control hair. The results of electron microscopy observation confirm the beneficial effect of *Cassia* galactomannan on hair cuticle (Figure 7). The control hair is smooth, the scales are regularly imbricated and their free edges are of smooth contour. The morphology of damaged hair is modified. The cuticle is less regular, the scales edges are lifted up.

Hair morphology was somewhat improved after treatment with a placebo shampoo. The improvement was more significant after treatment with shampoo containing 0.06% of *Cassia* galactomannan. After using the latter, the scale edges are not lifted up, the hair is very smooth and it gives impression of being re-covered with its protecting film.

Tamarindus indica β-glycan

Origins and Manufacturing Processes: *Tamarindus* β-(1-4) glycan is a xyloglucan obtained from the seeds of *Tamarindus indica* Linn., a high tree with a persistent foliage, belonging to the family of leguminous plants (Caesalpiniaceae). The fruits are large, dehiscent pods with a pulpy mesocarp, each containing four to 12 seeds (Figure 8).

Xyloglucans get bound between each other and link cellulose microfibrillae quite specifically to make a composite network which seems to play a major part in regulating plant-cell growth.[14]

The pulp of pods and seeds already have a variety of uses. The fruit pulp, rich in organic acids (free acids or their salts) and having a high content of simple sugars, is used

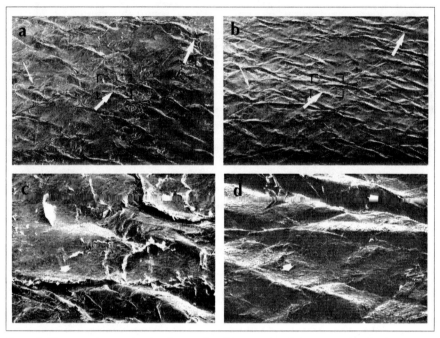

Figure 6. Electron micrographs showing results of four weeks of treatment with *Cassia* galactomannan on skin replicas from one volunteer subject.
A. Before treatment: unidirectionality of skin microrelief, presence of numerous scaly structures indicated by arrows.
B. After four weeks: restructuring evident as secondary furrows emerge; arrows indicate disappearance of previous scaly structures
C. Before treatment: Bold arrows indicate exfoliating scales
D. After four weeks: Bold arrows indicate disappearance of obvious scaling, reappearance of healthier microrelief patterns

in the food industry (sauces, drinks). The pulp or flesh of the seeds contains primarily polysaccharides, but also proteins, lipids and mineral materials. A powder obtained from the seeds is already used in the paper, cardboard and textile industries.

Xyloglucan-rich extracts of tamarin seeds have been obtained by classical techniques of polysaccharide extraction, such as extraction from powders (delipidated or not) in warm aqueous, hydroalcoholic or alcoholic solutions, in an acid or neutral medium.

Structure and Biochemical Characteristics

T. indica xyloglucan, a tridimensional polysaccharide, has a main chain of the cellulose β-D(1-4) glucose type. On the C6, some short lateral chains of α-D xylopyranose are attached, sometimes carrying non-reducing ends of β-D galactopyranose (1–2). The biochemical structure of this component appears in Figure 9.[7,12] *Tamarindus* xyloglucan is thus a heteropolysaccharide or biocopolymer that, by hydrolysis, gives several sugars, mainly glucose, xylose and galactose.

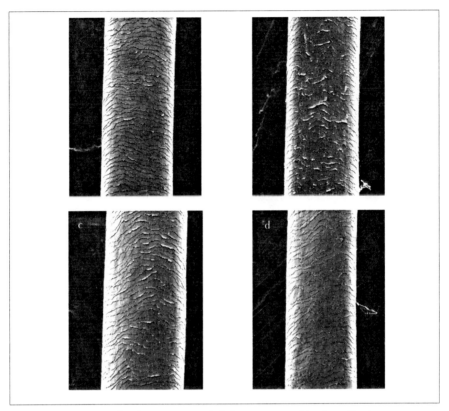

Figure 7. SEM micrographs of improvements in hair structure resulting from *Cassia* galactomannan treatment
A. Control hair
B. Hair damaged by acetone/ether treatment: lifting of scale edges visible
C. Hair damaged by acetone/ether treatment: some improvement after placebo shampoo
D. Hair damaged by acetone/ether treatment: distinct improvement after *Cassia* galactomannan shampoo; hair is as smooth as control hair

The native polysaccharide of *T. indica* xyloglucan has the form, in the presence of water, of a highly opalescent microparticulate dispersion of polymer. The molecular mass of the polymer is larger than 400,000 daltons.

To make this xyloglucan bio-functional, we submitted it to microbial fermentation, which gives us a xyloglucan polysaccharide with a much lower molecular weight (MW ~ 7,000). The constituent sugars of this lower MW xyloglucan, determined after hydrolysis and chromatographic analysis on the 90% pure polysaccharide, are glucose 49%, xylose 31%, galactose 18%. This functionalized xyloglycan, the purified, partly depolymerized polysaccharide of *T. indica* (INCI name: *Tamarindus indica* seed polysaccharide)[b] is available either as a concentrated powder form or a slightly opalescent, ready-to-use aqueous solution.

Immunostimulating effect in vitro: β-glycans, particularly β-(1-3)D-glycans of fungal origin, are known to have an immunostimulating activity by a direct macrophage activation.[15] Some activity has also been found in the series of β-(1-4) glycans of botanical origin. There exists yet no clear relationship between structure and efficacy, but polysaccharides of vegetable origin appear to have a larger efficacy profile as compared to the polysaccharides derived from mushrooms or fungi. They act on granulocytes, macrophages and the complement system.[5]

Among the plant β-glucans that we have tested, only the xyloglucan of *Tamarindus indica* has exhibited a significant immunostimulating effect. Its tolerance has been excellent.

Figure 8. Ripe fruit of *Tamarindus indica* Linn.

Immunostimulating effect on human monocytes in vitro: The immunostimulating effect of *Tamarindus* xyloglucan has been demonstrated on human monocytes incubated either simultaneously or preventively with the yeasts. Phagocytosis, a fundamental function of mononuclear phagocytes, is a two-step process, in which large particles, such as yeasts, become bound to a cell before being internalized. Autoclaved (killed) yeasts are stained pink by May-Grünwal Giesma (MGG). If treated with tannic acid solution just before staining, they display a deep blue color. The staining by MGG can be used to discriminate between extracellular and intracellular yeasts.[8,9]

Monocytes prepared from human blood have been set in culture for 17 h at 37°C in a defined medium and treated as summarized in Figure 10: Simultaneous treatment of monocytes by yeast and *Tamarindus* xyloglucan was accomplished by incubating monocytes for 1 h at 37°C with killed yeasts and *Tamarindus* xyloglucan at different concentrations. For deferred treatment, the monocytes were preincubated for 24 h at 37°C with *Tamarindus* xyloglucan. After washing the cells to remove the xyloglucan from the medium, the killed yeasts were added. After incubation with the killed yeast (with or without the xyloglucan present), the cells were washed with culture medium to remove unbound yeasts. Tannic acid solution was then added in both cases. Only the autoclaved yeasts still unabsorbed by the monocytes are accessible to tannic acid solution and thus appear dark blue after MGG staining; ingested yeasts remain pink.

β-glucan receptors on human monocytes are implicated in receptor-mediated phagocytosis of *Saccharomyces cerevisiae* yeasts.[9] The simultaneous incubation of

[b]Imindinyl, Laboratoires Sérobiologiques, Pulnoy, France

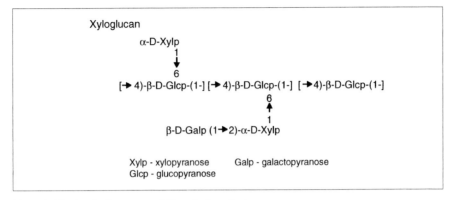

Figure 9. Biochemical structure of *Tamarindus* xyloglucan

monocytes and *Tamarindus* xyloglucan inhibited the ingestion of *Saccharomyces cerevisiae* by the monocytes. The number of ingested yeast has decreased by 85%.

The aim of the deferred treatment was to evaluate the stimulating capacity of *Tamarindus* xyloglucan on phagocytosis of yeasts by monocytes. At the concentration of 160 mg/l, *Tamarindus* xyloglucan has strongly stimulated the capacities of the phagocytosis (+58%). The image of phagocytosis is illustrated in Figure 11.

Immunostimulating effect on human PMNs in vitro: The immunostimulating capacities of *Tamarindus* xyloglucan compared to *Saccharomyces cerevisiae* and *Klebsiella pneumoniae* extracts have been demonstrated in an in vitro model using human polymorphonuclear neutrophiles (PMNs). An important activity of mononuclear and polymorphonuclear phagocytes is their ability to respond to appropriate stimuli by activation of the "respiratory burst," which comprises increased oxygen uptake, and production of both the superoxide anion and hydrogen peroxide. The oxygen reactive species participate in the destruction of microorganisms. We measured respiratory burst activity based on the finding that stimulated phagocytes generate chemiluminescence. By measuring chemiluminescence, we can quantitate the oxygen intermediates released into the extracellular medium.[21]

PMNs prepared from human blood by decantation and hemolysis have been incubated during 30 min with compounds to be tested. Thereafter, freshly prepared luminol and the Zymosan particles (activating agent) were added. The mixture was kept at 37°C and the chemiluminescence was measured up to 30 min. Chemiluminescence was the parameter for evaluation of immunostimulating activity.

Tamarindus xyloglucan has strongly and significatively increased the luminescence peak of Zymosan-activated PMN. The observed efficacy was superior to that observed using *Saccharomyces cerevisiae* oligopeptide (Figure 12).

Anti-free-radical activity: The anti-free radical activity of *Tamarindus* xyloglucan has been evaluated compared to an oligopeptide isolated from *Saccharomyces cerevisiae*, by a series of tests covering the reactive oxygen species (ROS): the hydroxy radical (OH°) and the superoxide anion ($O_2^°$) (Figure 13).

Anti-hydroxy-radical activity has been evaluated using Fenton's reaction, generating hydroxyl radicals by reaction of an iron-EDTA complex with H_2O_2 in the presence of

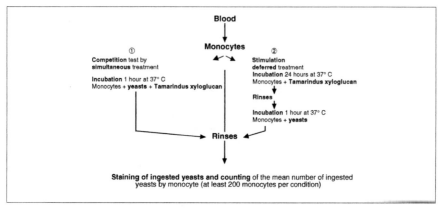

Figure 10. Schematic representation of the experimental protocol for testing the immunostimulating effect of *Tamarindus* xyloglucan on monocytes

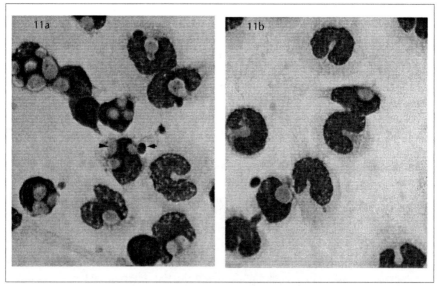

Figure 11. Photograph showing the results of simultaneous competition for β-glucan receptors on monocytes. The arrowhead on the left points to a pink (ingested) yeast; the arrow on the right a blue (attached but not yet ingested) yeast cell. 11a is the control, 11b the culture with *Tamarindus* xyloglucan added along with the killed yeast.

ascorbic acid. When the radicals thus produced attack deoxyribose, they form products that, upon heating with thiobarbituric acid, yield a pink chromogen. Added hydroxyl radical scavengers compete with deoxyribose for the hydroxyl radicals produced and thus diminish the chromogen formation to the extent that they successfully scavenge ROS.[10] The reaction with salicylic acid is based on the same principle.[23]

Superoxide anion is produced during the oxidative stress by xanthine oxidase induction. The superoxide anion activity was evaluated by biochemical tests conducted

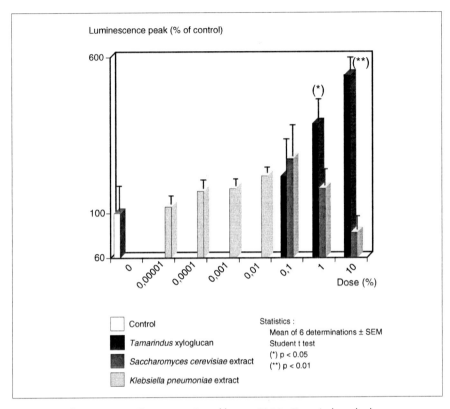

Figure 12. Induction test on Zymosan-activated human PMNs, *Tamarindus* xyloglucan compared to *Saccharomyces cerevisiae* and *Klebsiella pneumoniae* extracts

with a hypoxanthine/xanthine oxidase mixture in the presence of $O_2°$ and luminol (anti-radical activity measured by change in luminescence of the luminol).[19]

The results of anti-free-radical activity testing are expressed as the percentage of inhibition of the measured (radical-indicating) reaction. The effect of *Tamarindus* xyloglucan on both $OH°$ and $O_2°$ elimination is dose-dependent and comparable to the effect using a *Saccharomyces cerevisiae* extract (Figure 14).

Other activities of *Tamarindus* xyloglucan: The regenerating and growth-factor-like activities have been evaluated in vitro on human fibroblasts (MRC5). The tests compared the *Tamarindus* xyloglucan to an oligopeptide isolated from *Saccharomyces cerevisiae*. The test products were added to the fibroblast cultures 24 h after initiating the cultures. Cell counting and ATP analysis were done after the cell cultures had six days of contact with the test products. Cell counting was carried out after trypsinisation[24] and ATP determination was made by luciferase reaction, which produces a luminescence proportional to the ATP concentration.[22] The dose-dependent, growth-factor-like activity of *Tamarindus* xyloglucan has been demonstrated on human fibroblasts, while the *Saccharomyces* oligopeptide has been inactive at all doses tested (Figure 15).

Tamarindus xyloglucan has been perfectly tolerated by human fibroblasts in vitro at concentrations up to 10%. On the other hand, *Saccharomyces* extract became toxic at 1%.

Cosmetic potential for xyloglucan: The immunostimulating activity of *Tamarindus indica* seed polysaccharide (*Tamarindus* xyloglucan) has been demonstrated by two tests: the human monocyte test and human polymorphonuclear neutrophiles (PMN) activation. Both tests have been conducted in comparison with the macrophage activator β-glucan isolated from *Saccharomyces cerevisiae*, a non-specific stimulator of the immune system.

Tamarindus xyloglucan has inhibited the number of ingested yeast by 85% when added simultaneously with the killed yeasts (competitive inhibition) and increased the number of phagocytosed yeasts by 58% (deferred stimulation).

The reactive oxygen species, secreted by the polymorphonuclear neutrophiles, have been considered as a measure of PMN activation and reflection of immunostimulating effect. The immunostimulating activity of *Tamarindus* xyloglucan was comparable to the activity of *Saccharomyces cerevisiae* and *Klebsiella pneumoniae* extracts.

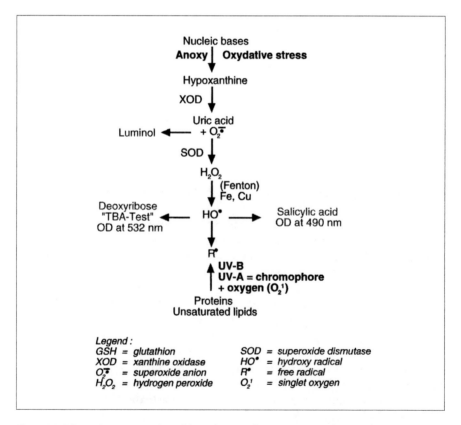

Figure 13. Schematic representation of the induction of reactive oxygen species (ROS) in vivo and the experimental protocol for testing the effect of *Tamarindus* xyloglucan on monocytes

Incorporation of β-Glucans in Products

The powdered form of *Cassia angustifolia* seed polysaccharide contains the *Cassia* galactomannan at 90% purity and should be used in formulations at levels of 0.06-0.20%. The aqueous solution form is ready for use at 3-10% concentrations in products. *Cassia* galactomannan is soluble in water but insoluble in ethanol.

The powdered form of *Tamarindus indica* seed polysaccharide contains the *Tamarindus* xyloglucan at 90% purity and should be used in formulations at levels of 0.10-0.50%. The liquid form of the xyloglucan is also ready for use at 3-10% concentrations in products. *Tamarindus* xyloglucan is soluble in both water and ethanol.

When using the powdered actives, either *Cassia* galactomannan or *Tamarindus* xyloglucan, prepare a concentrated aqueous mother solution immediately before incorporating it in the emulsion during the cooling phase. The fluid active, being

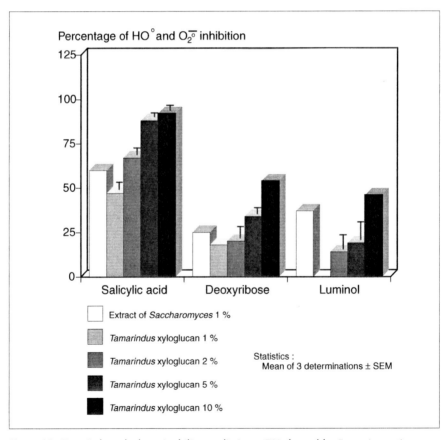

Figure 14. *Tamarindus* xyloglucan's ability to eliminate OH° formed by Fenton's reaction, measured by both salicylic acid and deoxyribose tests, and elimination of $O_2°$ formed by xanthine oxidase in the presence of hypoxanthine, evaluated by luminol fluorescence. Results reported as % inhibition of ROS

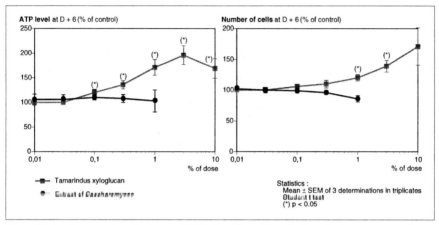

Figure 15. In vitro growth activation on MRC5 fibroblasts, *Tamarindus* xyloglucan compared to *Saccharomyces cerevisiae* extract. (Left) ATP levels measured. (Right) Number of cells six days after inoculation at various dose levels

Formula 1. Repairing, structuring antiage face emulsion

A. Glyceryl stearate (and) PEG-100 stearate	1.50
Cetearyl alcohol (and) ceteth-20	1.50
Cetyl alcohol	1.00
Caprylic/capric triglyceride	8.00
Cetearyl isononanoate	4.00
Octyldodecanol	3.00
Dimethicone	0.50
Preservative	qs
B. Water (*aqua*)	68.55
Glycerin	5.00
Xanthan gum	0.50
Preservative	qs
C. Polyacrylamide (and) isoparaffin (and) laureth-7	0.75
D. *Tamarindus indica* seed polysaccharide (Imindinyl, Laboratoires Sérobiologiques)	5.00
	100.00

Procedure: Prepare A and B separately at 80°C. Prepare D by stirring at room temperature. Holding both A and B at 80°C, add A to B using turbine stirring. Cool to 60°C and add C, still with turbine stirring. Cool to 45°C. Add D under turbine stirring; continue stirring with planetary blade to room temperature.

Formula 2. Body shampoo

A. Water (*aqua*)	52.45
Acrylates/C_{10-30} alkyl acrylate crosspolymer	1.20
B. Sodium laureth sulfate	35.00
Methylchloroisothiazolinone (and) methylisohiazolinone	
(Kathon CG, Rohm and Haas)	0.10
Propylene glycol	5.00
Preservative	qs
C. *Cassia angustifolia* seed polysaccharide	
(Indinyl CA, Laboratoires Sérobiologiques)	3.00
D. Triethanolamine	<u>0.75</u>
	100.00

Procedure: Prepare A at 40°C with stirring; cool to room temperature. Prepare B, stirring at room temperature. Add B to A using turbine stirring. Switch to planetary stirring to add C and D.

Formula 3. Face serum

A. Water (*aqua*)	49.750
Carrageenan (*Chondrus crispus*)	0.100
Preservative	qs
B. Carbomer	0.300
Water (*aqua*)	31.955
C. Propylene glycol	2.000
Dimethicone copolyol	3.000
D. Triethanolamine, 20% aq. soln.	2.145
E. Preservative	qs
F. *Cassia angustifolia* seed polysaccharide	
(Indinyl CA, Laboratoires Sérobiologiques)	<u>10.000</u>
	100.000

Procedure: Prepare A and B separately at 75°C under turbine stirring; cool each to room temperature. Add B, C, D, E and F to A at room temperature using turbine stirring. Switch to planetary stirring, continue mixing until homogenous.

Formula 4. Body emulsion

A. Stearic acid	1.20
Glyceryl stearate SE	2.00
Ceteareth-12	0.80
Ceteareth-20	0.40
Cetyl alcohol	1.20
Isopropyl myristate	2.00
Octyl methoxycinnamate	0.40
Hexyl laurate	3.00
Dimethicone	0.80
B. Water (*aqua*)	57.00
Glycerin	10.00
Preservative	qs
C. Triethanolamine, 20% aq. soln.	1.00
D. Carbomer	0.20
Water (aqua)	9.80
E. Hydrolyzed wheat protein (and) barley (*Hordeum vulgare*) extract (and) *Arnica montana* extract (Firmogen, Laboratoires Sérobiologiques)	2.50
F. *Cassia angustifolia* seed polysaccharide (Indinyl CA, Laboratoires Sérobiologiques)	5.00
G. SD alcohol 39C	2.00
H. Fragrance (*parfum*)	<u>0.30</u>
	100.00

Procedure: Prepare A and B at 80°C under stirring. Add A into B with strong turbine stirring. Cool to 60°C using turbine stirring; add C. Cool to 55°C; add D. Cool to 50°C; add E, F and H. Cool to room temperature; add G.

ready for use, can be added without further preparation by stirring into the cosmetic product during the final stages of manufacture.

Some examples of formulations using the *Cassia angustifolia* or *Tamarindus indica* seed polysaccharides are given in Formulas 1 through 5.

Summary

In summary, *Cassia* galactomannan can be recommended for various applications: face and body care products (repair of rough, dry, uncomfortable skin, antiage and moisturizing products), hair care products (shampoos, after-shampoos, lotions, leave-on products, rinse-off products).

Formula 5. Face emulsion

A. Sorbitan palmitate	3.50
Glyceryl stearate	1.50
Cetyl alcohol	2.50
Cetearyl isononanoate	7.00
Mineral oil	3.00
Octyldodecanol	5.50
Dimethicone	2.00
Preservative	qs
B. Water (*aqua*)	64.10
Glycerin	4.00
Sodium cetearyl sulfate	1.20
Preservative	qs
C. *Cassia angustifolia* seed polysaccharide	
(Indinyl CA, Laboratoires Sérobiologiques)	5.00
	100.00

Procedure: Prepare A and B separately at 80°C. Holding both at 80°C, add A to B under turbine stirring. Cool AB to 50°C and add C under turbine stirring. Cool to room temperature under planetary stirring.

The growth-factor-like activity of *Tamarindus* xyloglucan has been demonstrated. The regeneration and the growth of fibroblasts have been influenced in a dose-dependent way. Moreover, an anti-free-radical activity of xyloglucan has been reported. Concerning its cosmetic application, *Tamarindus* xyloglucan is recommended for the stimulation of skin repair, for the protection of skin being damaged by environmental factors of various origins, and for prematurely aged skin which has weakened mechanisms of immunological defense and repair.

— **M. Pauly, O. Freis and G. Pauly,** *Laboratoires Sérobiologiques S.A., Pulnoy, France*

References

1. N Alam and C Gupta, Structure of a water-soluble polysaccharide from the seeds of *Cassia angustifolia, Planta Medica* 4 308-310 (1986)
2. IM Blanc, Factors which influence the water content of the stratum corneum, *J Invest Dermatol* 18 433-440 (1952)
3. G Chihara, Immunopharmacology of Lentinan, a polysaccharide isolated from *Lentinus edodes, Int J Oriental Med* 17 57-77 (1992)
4. RH Davis, WL Parker, RT Samson and DP Murdock, Isolation of a stimulatory system in Aloe extract, *JAPMA* 81 473-478 (1991)

5. H Eggensperger and M Wilker, Multiaktiv wirksame Polysaccharide. Teil II: Pflanzliche Polysaccharide, *SÖFW Journal* 123 838-842 (1997)
6. ML Garcia and J Diaz, Combability measurements on human hair, *J Soc Cosmet Chem* 27 379-398 (1976)
7. MJ Gildey, PJ Lillford, DW Rowlands et al, Structure and solution properties of Tamarind Seed polysaccharide, *Carbohydrate Research* 214 299-314 (1991)
8. J Giaimis, Y Lombard, M Makaya-Kumba, P Fonteneau and P Poindron, A new simple method for studying the binding and ingestion steps in the phagocytosis of yeasts, *J Immunol Methods* 154 185-193 (1992)
9. J Giaimis, Y Lombard, P Fonteneau, CD Muller, R Levy, M Makaya-Kumba, J Lazdins and P Poindron, Both mannose and β-glucan receptors are involved in phagocytosis of unopsonized, heat-killed *Saccharomyces cerevisiae* by murine macrophages, *J Leucocyte Biology* 54 564-571 (1993)
10. B Halliwell, JMC Gutteridge and OI Aruoma, The deoxyribose method: A simple "Test-tube" assay for determination of rate constants for reactions of hydroxyl radicals, *Analyt Biochem* 165 215-219 (1987)
11. ME Henderson, L Hough and TJ Painter, Mannose containing polysaccharides. Part V. The isolation of oligosaccharides from Lucerne and Fenugreek galactomannans, *J Chem Soc* 3519-3522 (1958)
12. P Lang, F Masci, M Dentini, V Crescenci, D Cook, MJ Gidley, C Fanutti and JSF Reid, Tamarind seed polysaccharide: preparation, characterisation and solution properties of carboxylated, sulphated and alkylaminated derivates, *Carbohydrate Polymers* 17 185-198 (1992)
13. JL Lévêque, L De Rigal, D Saint-Léger and D Billy, How does Sodium Lauryl Sulfate alter the skin barrier function in man? A multiparametric approach, *Skin Pharmacol* 6 111-115 (1993)
14. S Levy and LA Staehelin, Synthesis, assembly and function of plant cell wall macromolecules, *Current Biology* 4 856-862 (1992)
15. PWA Mansell, Polysaccharides in skin care, *Cosm Toil* 109 (9) 67-72 (1994)
16. M Monsigny, Les glycoconjugués: signaux de reconnaissance, *Biofutur* 125 20-24 (1993)
17. M Obata and H Tagami, A rapid in vitro test to assess skin moisturizers, *J Soc Cosmet Chem* 41 235-242 (1990)
18. M Obata and H Tagami, Electrical determination of water content and concentration profile in simulation model of in vivo stratum corneum, *J Invest Dermatol* 92 854-859 (1989)
19. GM Oyamburo, CE Prego, E Prodanov and H Soto, Study of the enzyme-catalyzed oxidation of hypoxanthine through the chemi-luminescence of luminol, *Biophys Biochim Acta* 205 190-195 (1970)
20. Ch Ribaud, JC Garson, J Doucet and JL Lévêque, Organisation of stratum corneum lipids in relation to permeability. Influence of Sodium Lauryl Sulfate and Preheating, *Pharm Res* 11 1414-1418 (1994)
21. RE Schopf, J Mattar, W Meyenburg, O Scheiner, KP Hammann and EM Lemmel, Measurement of the respiratory burst in human monocytes and polymorphonuclear leukocytes by nitro blue tetrazolium reduction and chemiluminescence, *J Immunol Methods* 67 109-117 (1984)
22. H Spielmann, U Jacob-Müller and P Schulz, Simple assay of 0.1-1.0 pmol of ATP, ADP and AMP in single somatic cells using purified luciferin luciferase, *Anal Biochem* 113 172-178 (1981)
23. M Tien and SD Aust, Comparative aspects of several models of lipid peroxydation systems, in *Lipid Peroxides In Biology And Medicine*, K Yagi, ed, (1982) 23-39
24. P Vasseur and C Aerts, Appréciation de la cytotoxicité par la mesure de l'ATP, *J Franç Hydrologie* 9 149-156 (1978)

Take Tea and See

Keywords: green tea, antioxidant, antiaging

Green tea may provide antioxidant antiaging properties

Yet another use has been found for green tea, the popular antioxidant, antibacterial and beverage from the tea shrub (*Camellia sinensis* L.). Green tea has been grown in China, Japan and Indonesia from time immemorial. Now its scent alone is being used in cosmetics and toiletries.

Green tea's medicinal properties are well known. In recent years, several studies in animals and humans have suggested that green tea may help prevent certain human illnesses, such as heart disease and cancer. "Statistics suggest that it may at least postpone these diseases for five years or more," said Lester A. Mitscher, Ph.D., university distinguished professor of medicinal chemistry at the University of Kansas in Lawrence, Kansas.

Antioxidant: Mitscher's research on green tea in 1997 is believed to be the first to quantify the effectiveness of green tea's disease-fighting capabilities and measure it against other popular antioxidants. (Antioxidants appear to help protect human cells from attack by free radicals, which are unstable molecules generated by the body.) His research showed that an antioxidant found in green tea is at least 100 times more effective than vitamin C and 25 times better than vitamin E at protecting cells and their DNA from damage believed linked to cancer, heart disease and other potentially life-threatening illnesses. It is also twice as effective as resveratrol, the antioxidant found in red wine.

The green tea antioxidant Mitscher identified is epigalloca-techin gallate (EGCG). "Our research shows that the EGCG in green tea actually gets into cells to exert its protective effect," Mitscher said.

EGCG is one of several catechins found in green tea, which also includes alkaloids (such as caffeine) and phenolic acids. Could we isolate the EGCG from the catechins and make a more powerful antioxidant in cosmetics? Probably not on a practical basis. The separation would be done by chromatography, which is tedious and expensive on an industrial scale. "For cosmetics purposes, I think most people are using the entire catechin fraction. It's conveniently available by simple chemical steps from the tea plant itself or from the dried tea leaves. Because the mixture is active, and the main component in the mixture is EGCG, most people stop at that stage because they can make it economically," Mitscher said. However, EGCG is being isolated for pharmaceutical applications by San Francisco-based Pharmanex, Inc.

Mitscher noted that black tea, the tea most commonly consumed by Americans, and oolong tea, another popular variety, do not have the same antioxidant potency as green tea. "That's because green tea is steamed immediately after it is picked, which prevents the leaves from oxidizing, thus preserving the EGCG. In comparison, black and oolong tea have less than half the levels of EGCG," Mitscher said.

Antiaging: Green tea's antioxidant activity supports antiaging claims made for products containing green tea. It is well known that under stress from sun or the environment, the human body generates free radicals that damage the skin and hair. This damage accumulates over time, leading to conditions interpreted as aging. Antioxidants appear to help protect human cells by trapping the free radicals so they do less damage. Less damage means less appearance of aging.

"I think the baby boomers, as they approach middle age, don't want to get any older," said Gene Berube, vice president of the cosmetics division of Bio-Botanica, a Hauppage, New York-based supplier of natural ingredients. "They're looking for things that have antiaging effects. Certainly green tea with its antioxidant capability has some antiaging effects. So they're using it. I know in the last year our sales have picked way up on green tea."

Berube reports that in addition to its use as an antioxidant, green tea is added to AHA preparations because its caffeine content helps it reduce inflammation. It is also used in sunscreen preparations because it extends the SPF. Many large companies, including Elizabeth Arden, Esteé Lauder and Revlon, are already using it, according to Berube.

"It's also a good antibacterial. It has antiplaque properties when used orally because it interferes with the ability of bacteria to stick to the teeth. Tom's of Maine once had a green tea toothpaste, for example," Berube said.

Aroma: The latest use of green tea relies less on its health benefits than on its aroma. On July 5, 1999 Elizabeth Arden launched its first bath and body franchise—a fragrance collection called *Green Tea*. The initial launch includes five products: an overall scent spray, an exfoliating sea salt body rub, a bath and shower gel, a body lotion and a tub tea.

Elizabeth Arden plans to introduce additional *Green Tea* products in the first half of 2000. This is reported to be the first time the company has created an extensive product line around a single scent.

Actual green tea leaves are used in the tub tea. Otherwise, the only green tea appearing in the first five products is in the scent, which all products will share. The scent was developed by Quest in its first partnering with Arden. Company documents describe the scent as having a "spirited heart" made up of green tea, jasmine, celery seed and carnation.

The *Green Tea* franchise "is Elizabeth Arden's take on the modern-day movement toward natural herbal remedies to enhance the body and soul," according to a company statement.

In 1998 the French toiletries manufacturer Roger & Gallet launched a bath soap named *Green Tea*, described as an elegant relaxing fragrance. And in the summer of 1999, green tea extract made its appearance in facial care products added to the aromatherapy hair and body care line offered by San Francisco-based Essential Elements. These cleansers, toners, masks and moisturizers use green tea extract along with vitamins C and E for their antioxidant properties.

Are there other benefits from this well-regarded antioxidant, anti-inflammatory, antibacterial, and anti-cancer agent? The research goes on, mounted by those who believe the only way to find out is to take tea and see.

—**Bud Brewster,** *Allured Publishing Corp., Carol Stream, Illinois USA*

Sweet White Lupine Extract as a Skin Restructuring Agent

Keywords: skin barrier function, skin protection/repair/restructuring/ hydration, keratinocyte differentiation, lipid/protein synthesis

Studies on the skin's barrier function with sweet white lupine extract

White lupine (*Lupinus albus*) is a plant belonging to the legume family. Known and grown from antiquity, white lupine seeds have been found in Egyptian pyramids and in sites belonging to the Mayan culture. Its cultivation in the Mediterranean countries and in South America only ceased around the Middle Ages. In the early 19th century, Europe again started to grow this crop. Because white lupine is capable of fixing nitrogen from the air as soybean does, it is useful as a "green fertilizer." Its culture thus fertilized soils of low productivity while producing a cheap forage source for cattle breeding at the same time.

The grain has a high protein content: 47% of the dried matter. Its essential amino acid content, along with a low content of free sugars together with a high percentage of dietary fiber, makes white lupine a well balanced legume for human nutrition as well. It has already been used by nutritionists and dietitians and now, we introduce its uses in the cosmetic industry.

Epidermal Differentiation

The barrier function of the skin is localized in the stratum corneum (SC) and consists of protein-enriched corneocytes and lipid-enriched intercellular domains.[1] The barrier forms during terminal keratinocyte differentiation, which results in corneocytes and crosslinked cytoskeletal proteins.[2,3]

Epidermal differentiation is a continuous process that results from following cellular events.

- Keratinization, involving the synthesis of most fibrous keratinocyte proteins, such as keratin.
- Keratin synthesis, a phase depending on a key molecule, filaggrin, a shortening of "filament aggregating protein" - this cationic protein is able to combine with keratin filaments to form keratohyalin grains.[4]
- Formation of the cornified envelope, a complex step involving an enzyme,

transglutaminase, and a protein, involucrin (40% glutamic acid), both of which play major roles in membrane formation. The residues are crosslinked by the transglutaminase 1 and form a network of linked proteins that constitutes the molecular arrangement of the envelope.[5]
- Synthesis of epidermal lipids, primarily ceramides, cholesterol and free fatty acids–the lipids found in the intercellular spaces of SC. They are formed from polar lipid precursors (phospholipids and glycosphingolipids) supplied by the stratum granulosum via the exocytosis of lamellar bodies. These precursors undergo a series of transformations mediated by lytic enzymes released into the intercellular spaces along with the lamellar bodies. Modifications in the polarity and structure of these structures will lead to the creation of lipid bilayers. The arrangement of lipids in bilayers is as important as their overall composition in supplying the primary means of controlling the SC's barrier function, as well as controlling desquamation.

When the skin is subjected to severe chemical or physical aggressions, these natural regulation systems are saturated or disturbed. The SC is no longer renewed and the skin becomes dry and rough, the "barrier function" is no longer ensured. In these circumstances, SC renewal, keratinocyte differentiation and the skin barrier must be reinforced.

Our studies on the skin's barrier function led us to focus on a new botanical active ingredient obtained from sweet white lupine (*Lupinus albus*). It encourages keratinocyte differentiation and reinforces the epidermal structure of the skin by stimulating the synthesis of lipids and proteins of the epidermis.

Keratinocyte Differentiation

We investigated the action of our lupine extract on keratinocyte differentiation by measuring transglutaminase 1 activity, transglutaminase 1 being the key enzyme in the formation of cornified envelopes. This activity measurement was investigated in cultures of human keratinocyte by the expression of transglutaminase 1 messenger. Total ribonucleic acid (RNA) isolated from cells was subjected to reverse transcriptase, polymerase chain reaction (RT-PCR) amplification with the synthetic oligonucleotide primer complementary to a sequence coding for the genes of the transglutaminase 1 protein studied.

The results are expressed as percent of expression of the transglutaminase 1 mRNA over the expression of mRNA in the control, keratinocytes not treated with our active.

Figure 1. White lupine flower

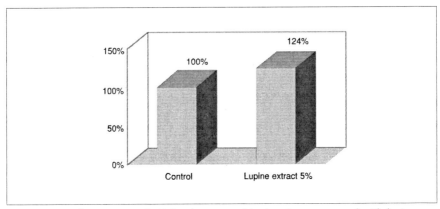

Figure 2. Transglutaminase 1mRNA levels expressed by keratinocytes treated with lupine extract

When the lupine extract is used at 5% (Figure 2), the expression of RNA messengers (mRNA) coding for the transglutaminase 1 increases by 124%. This experiment allows us to suggest that sweet lupine extract plays an important role in the keratinocyte differentiation.

To corroborate this result, we studied the influence of this extract on the synthesis of two major constitutive entities of the SC, structural proteins and lipids.

Synthesis of Structural Protein Activity

The formation of the cornified envelope is a complex process involving enzymes and a large number of precursor proteins, including involucrin, filaggrin and many others.

To validate this activity, we chose to measure the filaggrin synthesis rate, a key protein in the keratinization process, responsible for the aggregation of keratin filaments in the corneocytes.

We measured filaggrin synthesis on human keratinocytes by assessing the expression of filaggrin messenger using the protocol described for measuring transglutaminase 1 synthesis. In this case, total RNA isolated from keratinocytes was subjected to RT-PCR amplification with the synthetic oligonucleotide primer complementary to a sequence coding for the genes of filaggrin. The results are expressed as a percent of expression of the filaggrin mRNA over the mRNA expression of the control, keratinocytes not treated with the lupine extract.

As shown in Figure 3, the extract increased the expression of filaggrin mRNA by a factor of 3.9. Its effect is compared to that of a reference molecule, calcium chloride, that is known to activate keratinocyte differentiation. Thus, this lupine extract stimulates the synthesis of epidermal structural proteins and favors the formation of corneocytes.

Epidermal Lipid Synthesis

The epidermal lipids account for 10-12% of the total dry weight of the epidermis. They help regulate water flux and the SC's barrier function.[6] We conducted studies

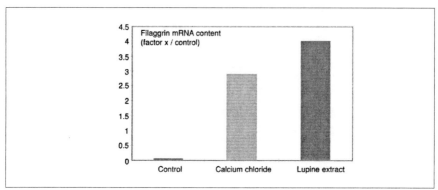

Figure 3. Effect of Lupine extract on the expression of filaggrin mRNA

on human skin explants, using analysis by thin layer chromatography to separate the main classes of neosynthesized, radiolabeled epidermal lipids: polar lipids, ceramides, cholesterol, di- and triglycerides, cerebrosides and cholesterol sulfate.

The analysis showed that there was a significant increase in ceramides (+70%), cholesterol (+30%) and di- and triglycerides (+20%) after incubation with 2% sweet lupine extract (Figure 4).

Increased SC Thickness

We measured SC thickness on a 14-day-old sample of reconstructed epidermis.[a] The treatment thus occurred at the end of cell proliferation. The extract was applied to reconstructed epidermis preparations at concentrations of 3%, 5% and 7% over an interval of 48 h.

Following treatment, we placed the preparations in a fixating solution (Bouin's liquid), dehydrated in toluene, and embedded them in resin. We cut 4-6 μm sections with a microtome and stained them with hematien/erythrosin/saffron trichrome.

We measured SC thickness in the sections using a light microscope and a 20x objective. The thickness parameter is expressed in microns (μ) from the stratum granulosum to the most superficial fiber in the most compact zones.

With this experiment, we conclude that our sweet lupine extract causes the epidermal layers and the SC to thicken by 37% after 48 h of treatment with 7% extract (Figure 5).

Improving Skin Barrier Function

In order to correlate the biological activity we found for the sweet white lupine extract with the reinforcement of its barrier effect, we directly tested the in vivo impact of our restructuring agent on epidermal hydration, as well as its protecting and repairing effects on the SC in vivo, using six human volunteers.

Hydration power: The skin barrier plays a regulatory role in the water balance of the skin. Damage to the barrier interferes with the control of water exchange. Water, therefore, can migrate more easily to the exterior and transepidermal water loss

[a]Episkin, a registered trademark of Saduc SA, Lyon, France

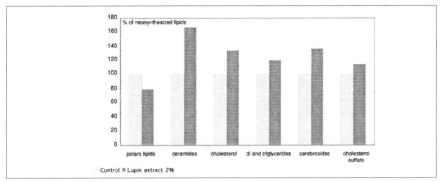

Figure 4. Effect of lupine extract on the synthesis of different classes of epidermal lipids

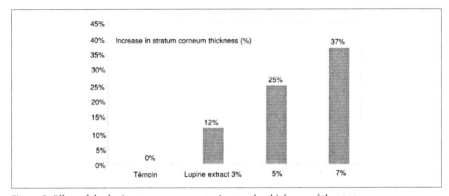

Figure 5. Effect of the lupine extract concentration on the thickness of the stratum corneum

(TEWL) increases. If the status of the skin barrier improves, water loss will decrease as exchanges will be correctly regulated. Water losses by evaporation are calculated by determining the pressure gradient in the water vapor layer surrounding the skin.

We determined the effect of our active ingredient on TEWL by applying it in a cream to in vivo human skin that had been washed with 10% (v/v) sodium lauryl sulfate. The volunteers applied the cream twice daily for two weeks. The results were compared to an untreated control zone that was only washed with 10% sodium lauryl sulfate. We did not have time to complete study using a placebo cream control. However, from our in vitro experiments, we have reason to believe that the improvement is due to the presence of the lupine extract, not the cream base.

Sodium lauryl sulfate increases TEWL, thus amplifying the effects of treatments causing restructuring and assisting restoration of the SC's barrier function. In a healthy epidermis, water losses are so low that it is difficult to measure and significantly quantify an additional decrease in these water losses. After 14 days of twice-daily application at a dose of 7%, the extract reduced TEWL by 19.5% and thus reinforced cutaneous hydration (Figure 6).

SC protection and repair: We determined the effect of our active ingredient at 5% on reconstituted epidermis preparations[b] with or without lactic acid treatment.

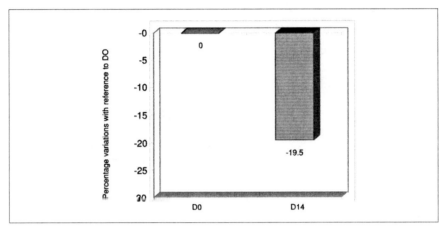

Figure 6. Effect of the lupine extract on TEWL after 14 days of application

We applied a product containing the lupine extract on reconstituted epidermis 48 h before an 18 h lactic acid treatment (as a preventive treatment) or after treatment with lactic acid (as a repairing treatment). The extract's effect was determined by comparing the thickness of the SC on the reconstituted epidermis that had received our active ingredient with samples that had not.

Following treatment, the preparations were placed in fixating solution (Bouin's liquid), dehydrated in toluene and embedded in resin. As with our previous measurements of SC thickness, we then cut 4-6 µm sections with a microtome and stained with hematien/erythrosin/saffron trichrome.

We measured SC thickness in the sections using a light microscope and a 20x objective. The parameter is expressed in microns (μ) from the stratum granulosum to the most superficial fiber in the most compact zones.

We found that, for either the preventive treatment or the repairing treatment used in combination with a lactic acid treatment, the extract at a dose of 5% restores or maintains the thickness of the SC (Figure 7).

Lactic acid applied alone leads to a peeling effect: the thickness of the SC decreases. With the lupine extract applied before lactic acid, the SC maintains its original thickness (80 µm). We interpret this to mean that the lupine extract protects the SC and increases its resistance to lactic acid aggression.

Applying the lupine extract after lactic acid restores and even increases the total thickness of the SC (90 µm). We interpret this to mean that the lupine extract helps repair the SC.

Discussion

As a result of its richness in oligosaccharides and low molecular weight peptides with a high glutamic acid content, the lupine extract appears to penetrate to the basal layers of the epidermis. Our experiments have indicated that it stimulates cellular activity, revitalizing keratinocyte differentiation and thus assuring SC renewal. The

[b]Skinetic, a registered trademark of Skinethic, Nice, France

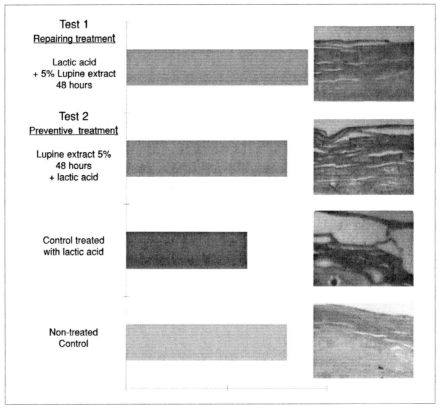

Figure 7. Chemical aggression of the stratum corneum; Repair and preventive treatment of the lupine extract

lupine extract specifically stimulates (+390%) the synthesis of the filaggrin type proteins that are responsible for the formation of corneocyte membranes.

The lupine extract also stimulates the synthesis of epidermal lipids, in particular ceramides (+70%), cholesterol (+30%) and the di- and triglycerides (+20%), the major lipids in the intercellular spaces of the SC.

By stimulating the successive synthesis of two constitutive entities of the SC–proteins and lipids–the lupine extract is capable of improving or restoring the barrier function of the SC.

Following chemical aggression (treatment with 2% lactic acid), a considerable decrease in the thickness of the SC is observed. A 48 h treatment with the lupine extract applied before or after such aggression enables the SC to be preserved or restored and thus maintains or re-equilibrates the skin barrier function.

According to our experiments, the lupine extract reinforces the natural restructuring systems of the epidermis, limits water loss and regenerates the barrier function of the skin. The lupine extract is thus suitable for all restructuring, repair and hydrating skin care products.

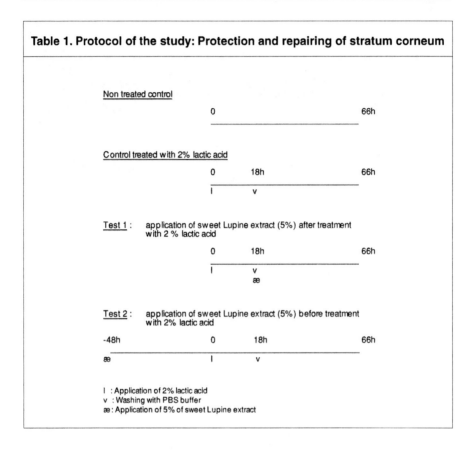

—Brigitte Closs and Jean Paufique, *Silab, Saint-Viance, France*

References

1. PM Elias, Epidermal lipids barrier function and desquamation, *J Invest Dermatol* 80 44-49 (1983)
2. DT Downing, Lipid and protein structures in the permeability barrier of mammalian epidermis, *J Lipid Res* 33 301-313 (1992)
3. D Roop, Defects in the barrier, *Science* 267 474-476 (1995)
4. B Weidenthaler, I Hauber and I Anton-Lamprecht, Is filaggrin really a filament-aggregating protein in vivo, *Arch Dermatol Res* 285 111-120 (1993)
5. RL Eckert, MB Yaffe, F Crish, S Murthy, EA Rorke and JF Welter, Involucrin: structure and role in envelope assembly, *J Invest Dermatol* 100(5) 613-617 (1993)
6. M Danda, J Hori, J Koyama, S Yoshida, R Nanba, M Takahashi, I Horii and A Yamamoto, Stratum corneum sphingolipids and free amino acids in experimentally induced scaly skin, *Arch Dermatol Res* 284 363-367 (1992)

Additional background

WN Holleran, KR Feingold, M Maoquang, WN Gao, JM Lee and PM Elias, Regulation of epidermal sphingolipid synthesis by permeability barrier function, *J Lipid Res* 32 1151-1158 (1991)

Centella asiatica and Skin Care

Keywords: Centella asiatica, skin firmness, connective tissue

New derivatives of Centella asiatica are developed for restoring skin firmness and damaged connective tissue

The beneficial effects of the *Centella asiatica* plant (an herb) have been known for a long time and its stimulation of collagens is used even in pharmaceutical applications. Recent studies reveal the activity mechanism of the plant's active ingredients. Various extracts are now available to provide the cosmetic scientist with the best opportunities to fully utilize the richness of *Centella asiatica*.

Background and History

Centella asiatica (Hydrocotyle asiatica, Gotu Kola) is a plant found in the subtropical regions of the world such as Madagascar and regions of East Africa.

The plant's properties in promoting wound healing have long been recognized in India. It was even mentioned in the Ayurveda, the vedic medicine of primitive Brahminism. Nowadays, it continues to be widely used in traditional medicine where it is known as the "Herb of the Tiger." It received this name when it was observed that Bengal tigers cared for their wounds by rubbing up against the plant and eating its leaves.

Centella asiatica has long been used empirically, however it was not until the mid-19th century that it began attracting scientific attention. The first extract of the *Centella asiatica* was prepared in 1941. Since then, major studies have been undertaken revealing astonishing curative properties for the treatment of wound healing, burns, leprous ulcers, hypertrophic eschars, keloids and venous-lymphatic disorders.

Botanical Data

Centella asiatica is a creeping plant belonging to the Umbelliferae family, genus *Centella*. It is now differentiated from the genus *Hydrocotyle* by its truncated calyx with non-dentale petals, five overlapping petals in the buds, a cancellated mericarp with seven to nine ridges and an involucre with protuberant bracts.

Centella asiatica is a polymorphous herb which sometimes develops an extensive tap root. The cylindrical stem bears one to five leaves per node. The 0.8 to 30 cm long petioles have a glabrous or pubescent sheathing base with laminae that are reniform, oval or orbicular. They range in diameter from 1 to 7 cm, with numerous veins and a cordate base (Figure 1).

Figure 1. *Centella asiatica* grows in subtropical regions.

Its polymorphism indicates how adaptable the plant is to diverse climates and soils, resulting in variable biological activity. Three botanical varieties of *Centella asiatica* are differentiated on the basis of their location: *abyssinica* from East Africa, *typica* from southern Asia and the western islands of the Indian Ocean and *floridana*, an American variety characterized by leaves longer than they are wide, faint pubescence and a well-marked sinus.

The source of the plants must be carefully monitored to ensure their efficacy. *Centella asiatica* leaves from Madagascar and East Africa are the most reliable — regions where the plant grows wild. When cultivated, *Centella* loses the pharmacological properties described above, with a low content of pentacyclic triterpenes which are the active ingredients.

Chemical composition

The major active ingredients are asiatic acid,[1] madecassic acid[2] and corresponding heterosides formed with the acid and three sugar moieties (asiaticoside,[3,4] isolated by Bontems, and madecassoside[5]) (Figure 2).

Several extracts have been developed and manufactured[a] to offer either pure, active molecules or titrated extracts with various ratios of the different components. All the operations (collection of the leaves, drying, extraction and quality control) are carried out under control of the manufacturer.

Each year 400 to 800 tons of fresh leaves are collected and dried yielding up to 100 tons of dried leaves. *Centella asiatica* extract is produced in amounts ranging from 2.5 to 3.5 tons in a manufacturing plant dedicated to the extract (Figure 3).

Medicinal Extracts

One of the extracts (T.E.C.A.) is used in proprietary medicinal products and marketed in many countries around the world. The most well-known brand is probably

Asiatic Acid
2 α, 3 β, 23-trihydroxyurs-12-en-28-oic acid

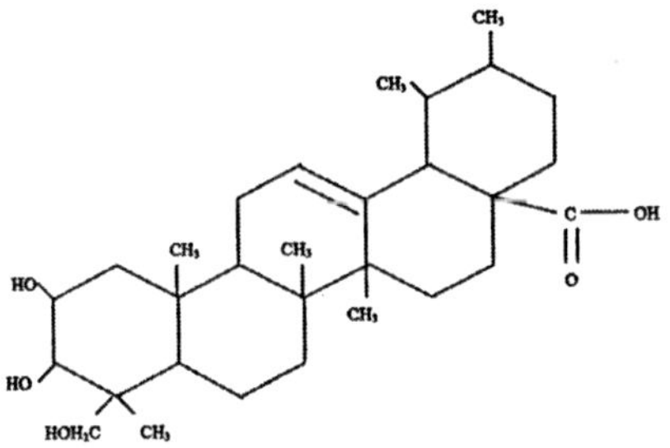

Madecassic Acid
2 α, 3 β, 6 β, 23 α-tetrahydroxyurs-12-en-28-oic acid

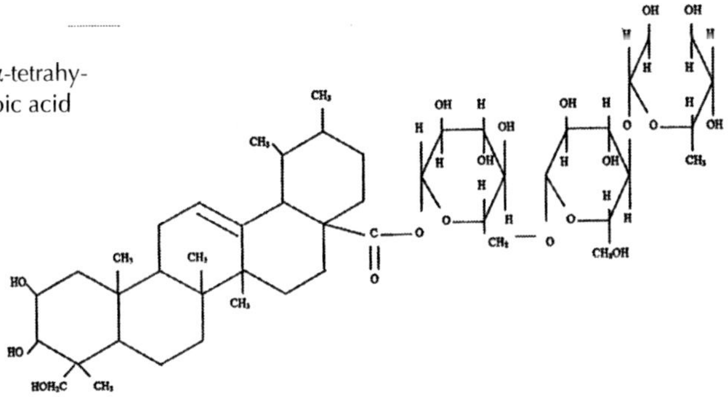

Asiaticoside
2 α, 3 β, 23 trihydroxyurs-12-en-28-oic acid O-6-deoxy-α-L-rhamnopyranosyl-(1->4)-O-β-D-glucopyranosyl-(1->6)-O-β-D-glucopyranosyl ester

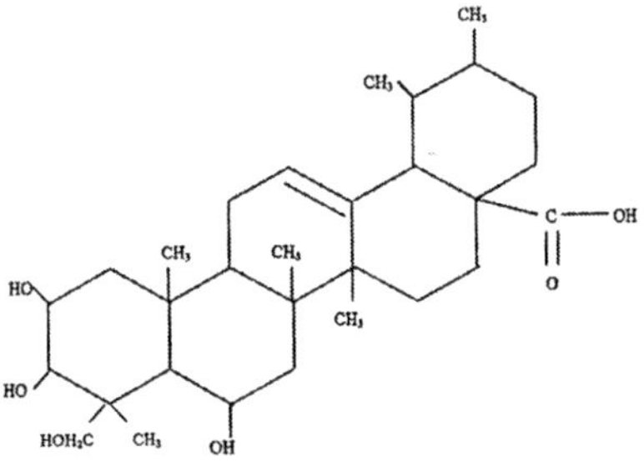

Madecassoside
2 α, 3 β, 6 β, 23 α-tetrahydroxyurs-12-en-28-oic acid O-6-deoxy-alpha-L-rhamnopyranosyl-(1->4)-O-β-D-glucopyranosyl-(1->6)-O-β-D-glucopyranosyl ester

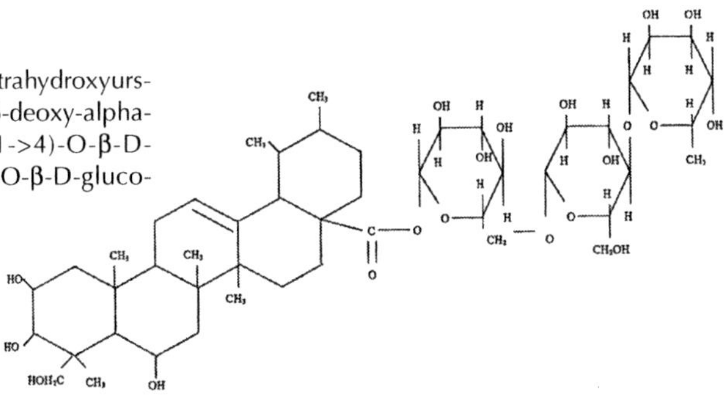

Figure 2. Pentacyclic triterpenes of *Centella asiatica*

Figure 3. Roche-Nicholas plant in Pau, France, manufactures *C. asiatica* extracts.

Madecassol.[b] The dosage forms are creams (dose: 1%), powders (dose: 2%), tablets (dose: 10 mg), capsules (dose: 30 mg) and dressings. They are prescribed for indications such as: treatment of cutaneous ulcers, hypertrophic scars, keloids and wound healing disorders; and treatment of venous-lymphatic disorders (heavy legs, pain).

The available extracts or molecules are described in Table 1.

The composition affects the solubility in the solvents commonly used in cosmetic formulations and offers the cosmetic scientist a wide range of possibilities for the formulation of aqueous, alcoholic or glycerol-based preparations (see Table 2). As an example, it can be seen that heterosides are water soluble and therefore particularly suitable for formulation of gels or serums.

Mechanism of action and applications

French scientists from the National Centre for Scientific Research showed that the synthesis of physiological collagen (types I and III) by fibroblasts in the dermis is stimulated by the three terpenes composing the titrated extract of *Centella asiatica* (T.E.C.A.): asiatic acid, asiaticoside and madecassic acid.[6]

An earlier study already described the mechanism of action of the triterpenoid fraction of *Centella asiatica* and effect on collagenesis.[7] These results were confirmed in other publications.[8,9] There is not only an increase of the collagen synthesis, but also changes in disrupted tissue where cutaneous material was lost (wounds, grafts). Some studies also show a regulatory action on the excessive proliferation of connective tissue (keloids, hypertrophic scars).[10] In poor connective tissue conditions, the triterpenes of *Centella asiatica* are able to renew the collagen, in quantity as well as in quality, restoring tissue firmness and skin elasticity, improving skin appearance and comfort.[11,12]

---

[a]Roche-Nicholas (a subsidiary of Hoffmann-LaRoche)
[b]Hoffman–LaRoche

Table I. Available extracts of *Centella asiatica*			
Denomination of the extract	**Composition**		**CTFA name**
Asiatic acid	Asiatic acid	>95%	Asiatic acid
Asiaticoside	Asiaticoside	>95%	Hydrocotyl (*Centella asiatica*) extract
T.E.C.A. (Titrated Extract of *Centella asiatica*)	Asiaticoside	40%	
	Asiaticoside and madecassic acids	60%	Hydrocotyl (*Centella asiatica*) extract
Heterosides	Madecassoside	>55%	
	Asiaticoside	>1%	Hydrocotyl (*Centella asiatica*) extract
E.M.A.C.A.	Asiaticoside	≥55%	
	Asiatic acid	≥5%	
	Madecassic acid	≥10%	Hydrocotyl (*Centella asiatica*) extract
P.E.C.A. (Purified extract of *Centella asiatica*)	Asiaticoside	≥15%	
	Madecassoside	≥15%	
	Asiatic	≥15%	
	Madecassic acids	≥15%	Hydrocotyl (*Centella asiatica*) extract
R.E.C.A. (Refined Extract of *Centella asiatica*)	Asiaticoside	≥ 20 %	
	Madecassoside	≥20%	
	Asiatic acid	≥4%	
	Madecassic acid	≥ 4 %	Hydrocotyl (*Centella asiatica*) extract
Amel Genins	Asiatic acid	≥20%	
	Madecassic acid	≥55%	Hydrocotyl (*Centella asiatica*) extract

These studies are beneficial to cosmetic applications. On their basis, it clearly appears that components of *Centella asiatica* can be used in the following applications:

- anti-wrinkles, antiaging
- firming and tensing
- anti-cellulitis, slimming, draining
- dark undereye circles
- sunburn healing

Table 2. Solubility of *Centella asiatica* extracts in cosmetic solvents

Extract	Solubility				Use level (Suggested)
	Ethanol	Water	Glycerol	Butylene glycol	
Asiatic acid	Sparingly soluble	Very slightly soluble	Slightly soluble	Sparingly soluble	0.03-0.05%
Asiaticoside	Sparingly soluble	Very slightly soluble	Practically insoluble	Sparingly soluble	0.05-0.1%
T.E.C.A.	Sparingly soluble	Very slightly soluble	Slightly soluble	Sparingly soluble	0.1-0.2%
Heterosides	Soluble	Freely soluble	Sparingly soluble to soluble	Sparingly soluble to soluble	0.05-0.1%
E.M.A.C.A.	Sparingly soluble	Very slightly soluble	Sparingly soluble	Sparingly soluble	0.2-0.5%
P.E.C.A.	Slightly soluble	Sparingly soluble	Very slightly soluble	Sparingly soluble	0.2-0.5%
R.E.C.A.	Sparingly soluble	Sparingly soluble	Slightly soluble	Sparingly soluble to soluble	0.2-0.5%
Amel Genins	Soluble	Very slightly soluble	Slightly soluble	Sparingly soluble	0.05-0.1%

Freely soluble: 1 to 10 mL of solvent to dissolve 1 g material
Soluble: 10 to 30 mL of solvent to dissolve 1 g material
Sparingly soluble: 30 to 100 mL of solvent to dissolve 1 g material
Slightly soluble: 100 to 1,000 mL of solvent to dissolve 1 g material
Very slightly soluble: 1000 to 10,000 mL of solvent to dissolve 1 g material
Practically insoluble: more than 10,000 mL of solvent to dissolve 1 g material

Formula 1. Skin care cream

PEG-100 stearate (and) glyceryl stearate	5.0%
Polysorbate 80	2.0
Cetyl alcohol	3.0
Sweet almond (*Prunus Amygdalus Dulcis*) oil	3.0
Caprylic/capric triglycerides	5.0
Mineral (*Paraffinum liquidum*) oil	10.0
Preservatives	qs
Carbomer	0.1
Water (*aqua*)	qs
Preservatives	qs
Triethanolamine	qs pH
Algae extract	5.0
Asiatic acid	0.05
Butylene glycol	10.0

Formula 2. Body cream

Polysorbate 60	0.8%
Sorbitan stearate	1.2
Isopropyl Isostearate	8.0
Silicone oil	0.5
Cetyl alcohol	0.8
Stearyl alcohol	0.5
Hazel (*Corylus americana*) nut oil	2.0
Shea butter (*Butyrospermum parkii*)	1.0
Preservatives	qs
Water (*aqua*)	qs
Carbomer	0.15
Water (*aqua*)	50.0
Butylene glycol	5.0
Asiaticoside (Hydrocotyle (*Centella asiatica*) Extract)	0.1
Preservatives	qs

Formula 3. Gel

Glycerin	10%
Water (*aqua*)	qs
Heterosides (Hydrocotyle (*Centella asiatica*) extract)	0.1
Carbomer	0.6
Butylene glycol	10
Triethanolamine	qs pH
Preservatives	qs

- strengthening of areas of very fine skin
- healing wounds and stretch marks

Restoring tissue firmness and elasticity is also beneficial to venous walls because *Centella asiatica* extract is used as a venous tonic in the treatment of venous-lymphatic disorders.

Other effects are currently under study, such as treatment of hair loss, and even its antibacterial activity attributed to asiatic acid or asiaticoside.

Conclusion

Extracts of *Centella asiatica* were used for many years in the pharmaceutical industry and have demonstrated their great efficacy and safety. New derivatives from *Centella asiatica* have been developed to meet the cosmetic requirements and for advanced formulations targeted to restore skin firmness and restructure damaged connective tissue.

—**Alain Loiseau,** *Laboratoires Roche Nicholas SA, Division Serdex, France*
—**Michel Mercier,** *MMP International Development and Manufacturing, USA*

References

1. J Polonsky, Sur la constitution chimique de l'acide asiatique, aglycone de l'asiaticoside, *Bull Soc Chim* France 2 173-180 (1953)
2. H Pinhas, D Billet, S Heitz and M Chaigneau, Structure de l'acide madécassique, nouveau triterpène de Centella asiatica de Madagascar, *Bull Soc Chim* France 6 1890-1895 (1967)
3. P Boiteau, A Buzas, E Lederer and J Polonsky, Sur la constitution chimique de l'asiaticoside, hétéroside naturel utilisé contre la lèpre, *Bull Soc Chim* Biol 31 46-51 (1949)
4. J Polonsky, E Sach and E Lederer, Sur la constitution chimique de la partie glucidique de l'asiaticoside, *Bull Soc Chim* France 6 880-887 (1959)
5. H Pinhas and JC Bondiou, Sur la constitution chimique de la partie glucidique du madécassoside, *Bull Soc Chim* France 6 1888-1890 (1967)
6. FX Maquart, G Bellon, P Gillery, Y Wegrowski and JP Borel, Stimulation of collagen synthesis in fibroblast cultures by a triterpene extracted from Centella asiatica, *Conn Tiss Res* 24 107-120 (1990)
7. R Tenni, G Zanaboni, MP De Agostini, A Rossi, C Bendotti and G Cetta, Effect of the triterpenoid fraction of Centella asiatica on macromolecules of the connective matrix in human skin fibroblast cultures, *Ital J Biochem* 37 69-77 (1987)
8. F Bonte, M Dumas, C Chaudagne and A Meybeck, Influence of asiatic acid, madecassic acid and asiaticoside on human collagen I synthesis, *Planta Med* 60 133-135 (1994)
9. FX Maquart, F Chastang, A Simeon, P Birembaut, P Gillery and Y Wegrowski, Triterpenes from Centella asiatica stimulate extracellular matrix accumulation in rat experimental wounds, *Eur J Dermatol* 9 289-296 (1999)
10. HG Vogel, NJ De Souza and A D'sa, Effect of terpenoids isolated from Centella asiatica on granuloma tissue, *Acta Therapeutica* 16 285-298 (1990)
11. A Hachem and JY Bourgoin, Histological and clinical study of the effects of titrated extract of Centella asiatica in localised lipodystrophy, *Med Prat suppl* 2 7-13 (1979)
12. J Mallol, MA Belda, D Costa, A Noval and M Sola, Prophylaxis of striae gravidarum with a topical formulation. A double blind trial, *Int J Cosm Sci* 13 51-57 (1991)

Tinged Autumnal Leaves of Maple and Cherry Trees as Potent Antioxidant Sources

Keywords: maple tree leaves, cherry tree leaves, antioxidant, wrinkles

Tinged autumnal maple and cherry leaves are evaluated as good new antioxidative materials for o/w skin care products

Oxidative damage to cell constituents is assumed to be one of several causal factors for many medical conditions, including cancer and aging. Lipid peroxidation especially is a problem in the human body. Thus, the role of polyunsaturated fatty acids in biological systems has been extensively studied to understand the damage to cells caused by various oxidation systems. Lipid peroxidation initiated by free radical reactions is believed to damage cells by the loss of polyunsaturated fatty acids of cell membranes.[1-3] Moreover, formed lipid peroxides and their secondary products such as reactive carbonyl compounds may damage cellular constituents, including various enzymes.[1,4-6]

Importance of Antioxidants

Basically, free radicals and other reactive oxygen species (hydroxyl radical, superoxide and other singlet oxygens) are formed in the human body continually because biological systems require them for metabolism. They serve useful physiological functions, but can be toxic when generated in excess. This toxicity is often aggravated by the presence of ions of such transition metals as iron or copper. Excess generation of reactive oxygen species within tissues can damage DNA, lipids, proteins and carbohydrates. Whichever is the most important target of damage depends upon the cell type subjected to the oxidative stress and how it is imposed.[7]

In addition, many researchers have reported free radicals derived from the environment are important extrinsic factors accelerating many diseases and aging. In particular, the excess exposure to UV rays and various atmospheric singlet oxygen species is believed to cause the formation of lipid peroxides. Therefore, various antioxidants have been investigated and used to protect the human body from these extrinsic factors.

Intrinsic oxidation-protecting enzyme systems, including superoxide dismutase (SOD), catalase and glutathione peroxidase, and food-derived substances such as tocopherols, flavonoids, ascorbic acid and carotenes are known to diminish the undesired effects caused by oxidation processes in organisms.[1] Recently, some antioxidants occurring naturally in plants have begun to receive much attention because people and animals regularly consume them. In particular, flavonoids and related compounds, which are widely distributed in the plant kingdom, show remarkable promise for a wide range of pharmacological use for allergies, inflammation, antiviral, antitumor and diabetes.[8-10]

In this study, we investigated the possibility of tinged autumnal leaves as an antioxidant source. The growing process of plants is varied with seasonal changes. Temperature and sunshine duration can affect the color change of leaves, which is due to the variation of pigments during the change of season. In fall, green leaves gradually turn red or yellow, because red and yellow colors originate from xanthophyll, carotenoids and flavonoids which appear with the destruction of chlorophyll for photosynthesis. They have a high concentration of secondary metabolites compared to green leaves, which contain primary metabolites such as water, amino acids, carbohydrates and proteins. Autumnal leaves were expected to show not only the increase of desired activity, but also a decrease of unexpected toxicities.

We screened various tinged autumnal leaves to select target materials with potent antioxidant activities, identified the major active compound contained in them and describe their applications to cosmetics.

Methods and Materials

Melting points were determined with Mel-temp II[a] but not corrected. IR spectra were obtained with Jasco FT-IR 5300[b] and UV-VIS spectra were obtained with Varian carry IE spectrophotometer[c]. ^{1}H-(300 MHz) and ^{13}C-(75MHz) NMR spectra were obtained with Bruker 300 MHz NMR spectrometer[d]

Preparation of extracts: Through a series of screening works, the tinged autumnal leaves of maple tree (*Acer Palmatum THUNBERG.* (Aceraceae))[11] and cherry tree (Prunus Donarium Sieb. Var. spontanea Makino (Rosaceae))[12] were selected as our final antioxidant materials.

Two types of leaves were collected between October and November in the middle part of South Korea. The 100 g of each leaf was extracted with 900g of ethanol for 10 days at room temperature. After filtration, each ethanol extract was concentrated in vacuo 40% aqueous 1,3-butylene glycol[e.] Extracts were prepared by the same method for the application to cosmetics.

---

[a]Laboratory Devices, Holliston, Massachusetts, USA
[b]Jasco Co., Tokyo, Japan
[c]Varian, Inc., Palo Alto, California, USA
[d]Bruker Instruments, Inc., Billerica, Massachusetts, USA
[e]1,3-BG; Hoechst Marion Roussel Pharmaceutical Co., Frankfurt, Germany

Antioxidative activities of tinged autumnal leaf extracts: The ethanol and 40% 1,3-BG aqueous extracts were diluted to 3% with ethanol to prepare samples. Three percent ethanol solutions of *dl*-alpha-tocopherol[f] 3% aqueous solution of green tea extract[g] and 3% aqueous solution of *l*-ascorbic acid[h] were prepared as comparative standards.

Thereafter, the antioxidative activity of each sample was evaluated by nitroblue tetrazolium chloride (NBT) mono hydrate test and 1,1-diphenyl-2-picrylhydrazyl (DPPH) test with suitable dilution as follows. First, in NBT test, 0.1mL sample, 2.4 mL Na_2CO_3 buffer (0.05M, pH 10.2), 0.1 mL xanthine, 0.1 mL ethylene diamine tetraacetic acid, 0.1 mL NBT solution(0.75 mM) and 0.1 mL bovine serum albumin solution were added in turn to each test tube and reacted at 25°C for 20 min. To each test tube, 0.1 mL xanthine oxidase was added and reacted at 25°C for 20 min. The reaction was quenched with 0.1 mL $CuCl_2$. Finally inhibitory effect (%) was calculated as compared with blank control after measuring absorbance at 560 nm. In DPPH test, 0.1 mL of the above sample and 3.9 mL of DPPH alcoholic solution (0.06 mM) was mixed in a tube for 30 sec and incubated for 30 min at room temperature. The absorbance was measured at 515 nm. Then, its antioxidative activity (%) was calculated with respect to blank control.

Cytotoxicities of tinged autumnal leaf extracts: This experiment was carried out to determine a suitable dosage for formulations. Human fibroblast (ATCC, Hs68), which was cultured in Dulbecco's Modified Eagle's Medium[i] supplemented with 10% fetal bovine serum[j] under atmosphere of 5% CO_2 and humidity of 100% at 37°C, was used to assay the cytotoxicities of tinged autumnal leaf extracts. 1 x 10^6 cells/mL were seeded into a 96-well microplate and incubated at 37°C for 24 h. Samples sterilized with 0.2 μm filter were added at various concentrations to determine LD_{50} values and incubated at 37°C for 24 h. 3-(4,5-dimethylthiazol-2-yl)-2,5-diphenyl tetrazolium bromide (MTT) reagent was added and incubated at 37°C for 4 h. After removing culture medium, 1N NaOH iso-propanol solution was added and stirred for 20 min and then absorbance was measured at 565 nm. Thereafter, the cytotoxicities of the extracts were determined as compared with non-sample treated controls by observing viability of them. The criterion of evaluation was the LD_{50} value of each sample.

Cell proliferation effects of tinged autumnal leaf extracts: To evaluate the cell proliferation effects of the extracts, we first screened the maximum concentration to avoid cell death. The same cells as above were used to assay the cell proliferation effects. 2 x 10^4 cells/mL were seeded into a 96-well microplate and incubated at 37°C for 24 h. After replacing the medium with 0.5% FBS DMEM, samples at a concentration to avoid cell death were applied and incubated at 37°C for 48 h. They were stirred for 20 min after bicinchoninic acid[k]

---

[f] Sigma-Aldrich Co., St. Louis, Missouri, USA
[g] Bioland Co., Chungbuk, Korea
[h] Sigma-Aldrich Co., St. Louis, Missouri, USA
[i] DMEM; GibcoBRL, Grand Island, New York, USA
[j] FBS; TerraCell International S.A., Ontario, Canada
[k] BCA; Pierce, Rockford, Illinois, USA

solution was added and incubated at 37°C for 30 min. Finally, the cell proliferation effects (%) of tinged autumnal leaf extracts were determined by comparing negative and positive controls after measuring absorbance at 565 nm.

Preparation of emulsion containing tinged autumnal leaf extracts: To investigate the emulsion types with compatibility with these extracts, several types of emulsions such as o/w, w/o, w/s and multi-phase emulsions were prepared. To examine the thermal stability for six weeks, each emulsion was incubated in 4°, 25° and 42°C incubators and variable circulator which are able to vary temperature from −10°C to 42°C every three days. Both UV and fluorescent radiation were applied to each emulsion for six weeks to observe the behavior of each emulsion on light. Finally, we selected an optimized type of emulsion from all the data acquired.

Human use study of anti-wrinkle effect of tinged autumnal leaf extracts: To determine, in vivo, on human subjects, the effect of the selected emulsion sample on cutaneous relief, we requested a human use study to Laboratoire DERMSCAN[l]. The experimental procedure is as follows: We first considered an open and intra-individual study. Volunteers consisted of seven healthy Caucasian females, aged between 43 and 51(47±1), selected strictly by study criteria. The volunteers also had wrinkles around the eye area (crow's-feet). The sample was applied to the crow's feet zone twice daily at home for 56 days. The experimental conditions during measurements were maintained constantly under the ambient temperature of 24±2°C and relative humidity between 40-60%.

The treated zone was compared to the non-treated zone after twice-daily use for two months. Quantitative measurements of the effect on cutaneous relief were done using the Skin Image Analyzer[m]. Oblique lighting (35°) brings shadows from the replica to the fore, which are then observed with CCD camera linked to a computer. An area of 1 cm^2 is studied. It produces a digitized image enabling a roughness index to be obtained by analyzing the shades of gray. The roughness index characterizes skin surface relief.[13,14] Studied parameters include the total wrinkled surface, the number and average depth of cutaneous microrelief furrows, median wrinkles and deep wrinkles using Quantrides[n] software.

Microrelief furrows have a depth inferior to 55 μm. Median wrinkle furrows have a depth between 55 and 110 μm and deep wrinkle furrows have a depth larger than 110 μm. All furrows with a minimum surface of 0.03 mm^2 are detected. Finally, the data gathered was used to judge the efficacy of sample by study criteria.

General separation procedure of the major active substances in tinged autumnal leaves: The active substances in tinged autumnal leaves were roughly separated using solvent fractionation method from the concentrates of the ethanol extracts. This method separates a group of substances having similar polarity from an extract according to polarity by using several organic solvents which are not mixed with the aqueous phase.

---

[l]Lyon/Villeurbanne, France
[m,n]Monaderme Co., Monaco, Monaco

Tinged Autumnal Leaves of Maple and Cherry Trees as Potent Antioxidant Sources

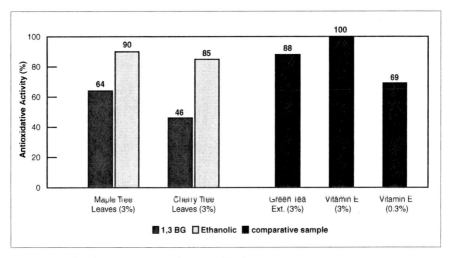

Figure 1. Results of NBT test on tinged autumnal leaf extracts

First, non-polar mixtures in the concentrates were extracted with hexane and chloroform. The polar residues were separated with ethyl acetate and n-butanol in turn. Finally, the aqueous residue was concentrated in vacuo. Antioxidative activity of each solvent layer was examined by NBT test to select the layer containing the most active substances. Selected solvent layers were investigated using thin layer chromatography[o] to confirm the composition of mixtures in the solvent layers. Silica gel was added into the solvent layers to absorb the solutes in solvents on the surface of silica gel and then were evaporated in vacuo. Each powdery mixture was applied on the silica gel column which was prepared with chloroform, and eluted with eluents or a solvent (or gas) of the mobile phase passing through a chromatography column.

Identification of major active substance in maple tree leaves: A yellow powdery precipitate obtained from ethyl acetate fraction was filtered and recrystallized from methanol. We named the compound antioxidant A (110 mg, 0.11%). It was identified using FT-IR, ^{1}H, ^{13}C-NMR (solvent: dimethyl sulfoxide (DMSO)-d_6) and UV-VIS spectra.

Identification of major active substance in cherry tree leaves: n-Butanol fraction which possessed the potent activity was briefly examined to confirm the constitution, chemical and structural properties of constituents using a series of visualizing reagents such as $FeCl_3$, Chloramin T and N-bromo-succinimide. The fraction was adsorbed with silica gel 60 (230-400 mesh) and separated into various subfractions on silica gel column chromatography using chloroform-ethyl acetate-methanol (1:1:1) as an eluent. After the antioxidativity of each subfraction was examined by NBT test, the subfraction that had the potent activity was concentrated and further purified by flash column chromatography. The purity was checked with

[o](TLC, Silica gel 60F-254, Merck, Rahway, New Jersey, USA
Sigma-Aldrich Co.

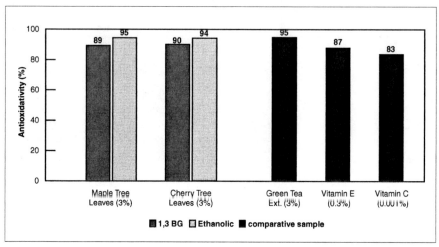

Figure 2. Results of DPPH test on tinged autumnal leaf extracts

TLC and ^{1}H-NMR. The compound was named antioxidant B (70 mg, 0.07%). Its chemical structure was identified using FT-IR, ^{1}H, ^{13}C-NMR (solvent: DMSO-d_6) and UV-VIS spectra.

Antioxidative activities of active substances: The antioxidative activities of antioxidant A and B were compared with other well-known flavonoids and vitamin E to evaluate the efficacy by NBT test.

Results and Discussion

Antioxidative activities of tinged autumnal leaf extracts: Figure 1 represents the results of NBT test on the extracts compared with other well-known antioxidants. In preliminary tests, maple and cherry tree leaves showed the most potent antioxidative activities among the screened autumnal leaves. The activities of extracts were comparatively as good as or higher than green tea extracts but not as high as vitamin E. Antioxidant activity of vitamin C was too high to be evaluated by this method because vitamin C reacts directly with nitroblue tetrazolium, not superoxide produced from reaction between xanthine and xanthine oxidase.

Figure 2 represents the results of DPPH test on the extracts to validate the correlation with NBT test. As a result, all the samples showed very high activities. However, those samples didn't have the antioxidative activities as high as vitamin E and C. The ethanolic extracts showed typically more potent effects than 40% aqueous 1,3-BG extracts as illustrated in Figures 1 and 2.

These results are due to ethanol being the most suitable solvent to extract active constituents from plants. But ethanol can't be used directly in cosmetics because of the destruction of emulsion and mild toxicity. The 1,3-BG is generally used as an excellent extracting solvent as well as a humectant in cosmetics. Based on these results, the maple and cherry tree leaves were selected to use as antioxidative materials for cosmetics and to investigate the active substances.

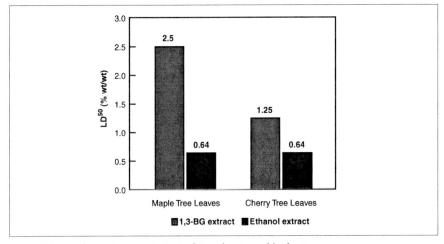

Figure 3. LD$_{50}$ values (% concentration) of tinged autumnal leaf extracts

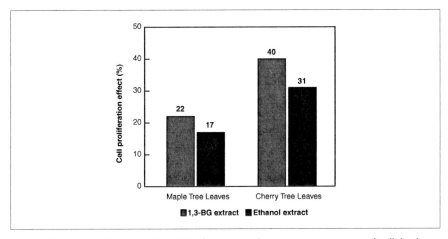

Figure 4. The cell proliferation effects (%) of extracts at the concentrations to avoid cell death

Cytotoxicities of tinged autumnal leaf extracts: As shown on Figure 3, a series of 1,3-BG extracts showed generally far lower cytotoxicities than ethanol extracts. The LD50 values of 1,3-BG maple and cherry leaf extract were 2.4% and 1.6% respectively, while those of ethanol extracts were 0.64%. From these results, two types of 1,3-BG extracts were evaluated to be capable of being applied safely to skin care products within 5-10%(wt/wt) without any problem.

Cell proliferation effects of tinged autumnal leaf extracts: As shown on Figure 4, 1,3-BG cherry leaf extract showed considerably higher proliferation effect than the other samples. Ethanol extracts showed low effects on the cell because of those low LD$_{50}$. On the other hand, 1,3-BG maple leaf extract didn't show any remarkable effects.

Formula 1. o/w Emulsion containing tinged autumnal leaf extracts

40% 1,3-butylene glycol extract of tinged autumnal leaf of maple tree	5.00%
40% 1,3-butylene glycol extract of tinged autumnal leaf of cherry tree	5.00
Cetanol (Kokyu Alcohol Kogyo Co., Chiba, Japan)	2.50
Mineral oil (Penreco Co., Pennsylvania, USA)	10.00
Soybean oil (ICD Horizon Co., New York, USA)	5.00
Dimethicone (Dow Corning Toray Silicone Co., Chiba-Ken, Japan)	2.00
1,3-Butylene glycol (Hoechst Marion Roussel Pharmaceutical Co., Frankfurt, Germany)	5.00
Glyceryl stearate (Nihon Emulsion Co., Tokyo, Japan)	2.00
PEG20 Methylglucose sesquistearate (Amerchol Co., New Jersey, USA)	1.00
Sorbitan monooleate (Uniqema Co., Delaware, USA)	0.30
Carbomer(2% *aqua*) (BFGoodrich, Ohio, USA)	5.00
Preservatives	qs
Water(*aqua*)	qs 100.00

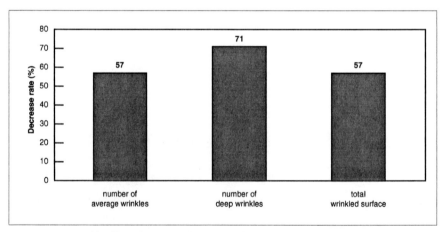

Figure 5. Anti-wrinkle effect of tinged leaf extracts in human use study

Preparation of emulsion containing tinged autumnal leaf extracts: These extracts showed broad compatibility with all types of emulsions such as o/w, w/o and w/s emulsions without the disruption or degeneration under harsh conditions. But in cases of a multi-phase emulsion, severe instabilities on thermal and other physical shocks were observed. W/o type was thought to have a problem for daily use because of oiliness and thickness. W/s type was not suitable for using long term because of giving a dry feeling to skin immediately after application. As a result, we selected o/w type emulsion (Formula 1), giving a fresh, moisturizing effect. Users could make use of the sample without any inconvenience.

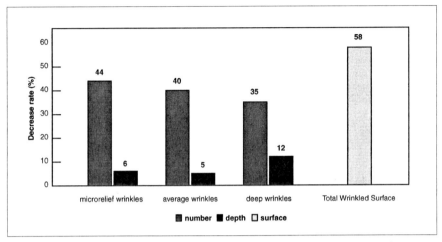

Figure 6. Quantitative results on cutaneous wrinkle relief effect of tinged autumnal leaf extracts

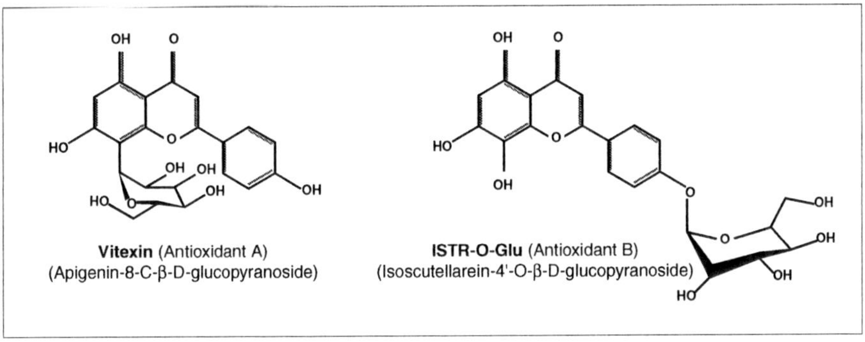

Figure 7. The structures of vitexin and ISTR-O-Glu (isoscutellarein-4'-O-beta-D-glucopyranoside)

Human use study on anti-wrinkle effect of tinged autumnal leaf extracts: In this study, no intolerance reaction was observed related to the treatment such as allergy or irritation. Figure 5 represents the results of volunteers showing effects. Figure 6 shows the quantitative results on cutaneous wrinkle relief effect of tinged autumnal leaf extracts in each measured item. The results, however, are only the first tendencies due to the few number of volunteers (n=7).

After 56 days, the sample showed a decrease in the number of deep wrinkles for five volunteers out of the seven (71%). Moreover the number of average wrinkles and the total wrinkled surface decreased for four volunteers (57% and 57% respectively). Quantitative results with the confidence interval at 95% showed the decrease of the number and depths of wrinkles as well as total wrinkled surface. From these results, tinged autumnal leaf extracts are thought to be considerably effective on wrinkle relief.

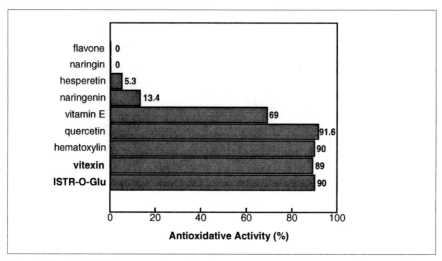

Figure 8. Results of NBT test on various substances

Major active substance in maple tree leaves: Antioxidant A obtained from ethyl acetate fraction of maple leaves showed a dark brown color in $FeCl_3$ test and yellow color in chloramin T test.

Its R_f value on TLC was 0.74 ($CHCl_3$-ethyl acetate-EtOH=1:1:1). From these results, antioxidant A was estimated as a flavonoid. The detailed chemical structure of antioxidant A was identified as vitexin (apigenin-8-C-β-D-glucopyranoside) on data obtained from FT-IR, ^{1}H, ^{13}C-NMR and UV-VIS, and was compared with reported data[13-17] (Figure 7). The spectrum data and physicochemical properties of antioxidant A are as follows.

Mp 250~252°C, IR (KBr): 3383, 1655 (α-, β-unsaturated carbonyl), 1614, 1508, 1429 (aromatic double bond) cm^{-1}, ^{1}H-NMR (DMSO-d_6): 8.13 (d, J=8.32Hz, H-2', 6'), 7.02 (d, J=8.32Hz, H-3', 5'), 6.73 (s, H-3), 6.39 (s, H-6), 4.91 (d, J=9.7Hz, β-anomer proton), 3.4-4.0 (m, D-glucose's protons) ppm. ^{13}C-NMR (DMSO-d_6): 183.1 (α-, β-unsaturated carbonyl), 165.0 (C-2), 163.7 (C-7), 162.0 (C-4'), 161.5 (C-5), 157.1 (C-9), 129.5 (C-2', C-6'), 122.7 (C-1'), 116.6 (C-3', C-5'), 105.4 (C-8), 105.0 (C-10), 103.1 (C-3), 99.0 (C-6), 82.6 (C-3"), 79.9 (C-3"), 74.4 (C-1"), 71.9 (C-2"), 71.6 (C-4"), 62.4 (C-6"), 74.4 ppm. UV-VIS (methanol): lambda$_{max}$ 270, 336 nm.

Major active substance in cherry tree leaves: Antioxidant B obtained from n-butanol fraction of cherry tree leaves showed a strong brown color in $FeCl_3$ test and brown color in chloramin T test. And its R_f value on TLC was 0.66 ($CHCl_3$-ethyl acetate-EtOH=1:1:1). From these results, antioxidant B was estimated as a flavonoid. The detailed chemical structure of antioxidant B was estimated as isoscutellarein-4'-O-beta-glucopyranoside (ISTR-O-Glu) on data obtained from FT-IR, ^{1}H, ^{13}C-NMR and UV-VIS, and was compared with reported data[15-17] (Figure 7). The spectrum data and physicochemical properties of antioxidant B are as follows.

Mp 163~166°C, IR (KBr): 3383, 1657 (α-, β-unsaturated carbonyl), 1610, 1493, 1444 (aromatic double bond) cm^{-1}, ^{1}H-NMR (DMSO-d_6): 7.93 (d, J=8.85Hz, H-2',

6'), 6.99 (d, J=8.85Hz, H-3', 5'), 6.66 (s, H-3), 6.55 (s, H-6), 4.91 (d, J=9.78Hz, β-anomer proton), 3.4-3.9 (m, D-glucose's protons) ppm. ^{13}C-NMR (DMSO-d$_6$): 183.1 (α-, β-unsaturated carbonyl), 165.1 (C-2), 164.4 (C-7), 162.5 (C-4'), 161.3 (C-5), 158.4 (C-8), 158.3 (C-9) 129.4 (C-2', C-6'), 122.5 (C-1'), 116.9 (C-3', C-5'), 109.4 (C-6), 105.0 (C-10), 103.3 (C-3), 95.6 (C-1"), 81.7 (C-3"), 80.0 (C-5"), 75.0 (C-2"), 71.1 (C-4"), 61.7 (C-6") ppm. UV-VIS (methanol): lambda$_{max}$ 272, 337 nm.

Antioxidative activities of active substances: The antioxidative activities of 0.3% ethanol samples of two flavonoids separated from tinged autumnal leaf extracts were compared with the 0.3% ethanol solutions of reagent-grade flavonoids and vitamin E by NBT test. As shown on Figure 8, ISTR-O-Glu and vitexin showed the antioxidative activities of about 90%. These results indicate that these two types of flavonoids have antioxidative activities as good as vitamin E and other flavonoids known to have potent antioxidative activities.

Conclusion

Various green and tinged autumnal leaves were investigated as new antioxidative materials for cosmetics. As a result of testing, the crude tinged leaf extracts showed somewhat higher antioxidativities than green leaf extracts in general, especially the extracts of *Acer Palmatum THUNBERG.* (Aceraceae: maple tree) and *Prunus Donarium Sieb.* var. *spontanea Makino* (Rosaceae: cherry tree).

In this study we also isolated the major antioxidants contained in those leaves, and then identified the chemical structure of them on the basis of their physico-chemical properties and spectroscopic evidences. From the data, we identified vitexin, a C-glycosyl flavonoid, from maple tree leaves, and isoscutellarein-4'-O-β-glucopyranoside (ISTR-O-Glu) from cherry tree leaves.

We confirmed that those compounds were contained in about 0.1%(wt/wt) in the above leaves and have as high antioxidative activities as other flavonoids and vitamin E. In addition, we found that other active compounds such as various alkaloids and other glycosyl flavonoids exist in tinged autumnal leaf extracts in great quantities. We hypothesize that the potent antioxidative activities of the extracts result from the synergistic effect caused by those various compounds contained in the extracts.

Finally, we optimized the method of extraction by using 40% 1,3-butylene glycol aqueous solution (widely used as a humectant in cosmetics) as a solvent system in order to apply to cosmetics without the loss of activities, the disruption of emulsions and cytotoxicities. The emulsions containing these extracts were prepared and studied for physical properties. The emulsion was proven to be effective on cutaneous relief through human use study.

—Jeong Jae Lee, Chung Woo Lee, Young Ho Cho, Sung Min Park, Bum Chun Lee and Hyeong Bae Pyo, *Hanbul Cosmetics Co., R & D Center, Chungbuk, Korea*

References

1. T Shibamoto, Y Hagiwara, H Hagiwara and T Osawa, *American Chem Soc* 154-165 (1994)
2. P Hochstein et al., *Biochem Biophys Res Comm* 100 1537 (1981)
3. AY Potapenko et al., *Studia Biophysica* 124 239 (1988)
4. HW Gardner, *J Agric Food Chem* 27 220 (1979)
5. HK Nielsen et al., *Brit J Nutr* 53 61 (1985)
6. U Reiss et al., *Biochem Biophys Res Comm* 48 921 (1972)
7. IO Aruoma, H Kaur and B Halliwell, *J Roy Soc Health* 172-177 (1991)
8. HK Kim et al., *Arch Pharm Res* 16 18-24 (1993)
9. B Harsteen, *Biochem Pharmacol* 32 1141-1148 (1983)
10. MJ Laughton et al., *Biochem Pharmacol* 42 1673-1681 (1991)
11. TJ Kim, *Korean Resources Plants Vol 3,* Seoul: Seoul National University Publishing (1996) pp 48-59 (In Korean)
12. TJ Kim, *Korean Resources Plants Vol 2,* Seoul: Seoul National University Publishing (1996) pp 158-174 (In Korean)
13. S Makki, JC Barbenel and P Agache, Acta Dermato Venereologica 59 285-291 (1979)
14. P Corcuff, F Chatenay, JL Leveque, *International Journal of Cosmetic Science* 6 167-176 (1984)
15. SH Shin, SS Kang and KS Kwon, *Yakhak Hoeji* 34 282-285 (1990) (In Korean)
16. IS Oh, WK Whang and IH Kim, *Arch pharm Res* 17 314-317 (1994)
17. KR Markham et al., Carbon-13 NMR Spectroscopy of Flavonoids, Chapter 2 in The *Flavonoids,* JJ Marbry et al, eds, London: Champmann & Hall (1982) 19-134
18. BG Osterdahl, *Acta Chem Scand* 32 93-97 (1978)
19. JB Harborne and H Baxter, Chapter 37 and 38 in *Phytochemical Dictionary,* JB Harborne et al, eds, London: Taylor & Francis (1993) 388-434

β-(1,6)-Branched β-(1,3)-Glucan in Skin Care

Keywords: β-glucan, Schizophyllum commune, UV, anti-irritant, anti-inflammatory

Liquid cultivated glucan from Schizophyllum commune shows effectiveness in antiaging and anti-irritant formulations

The skin is the first line of the body's nonspecific defense system. Unlike the immune system, which defends by forming antibodies against specific invaders, the skin provides a nonspecific barrier of cells and an environment that most bacteria, viruses and other pollutants cannot penetrate.[1] Skin is no different than any other organ in the body with a whole host of immune cells. Perhaps the most important of these defenses is the activity of the Langerhans cells, the skin-resident macrophage that engulfs and digests most foreign objects.

Activating the macrophage sets up a cascade reaction in the skin, resulting in fibroblast activation and the production of cytokines (such as IL-1, IL-6, IL-15, GM-CSF), epidermal cell growth factor (ECGF) and angiogenesis factor (AF) which aid the healing of wounds. In aged or wrinkled skin, increased production of ECGF increases the production of collagen and elastin, thus improving the skin's appearance and causing fine lines and wrinkles to disappear.[2]

β-Glucans have been known for a long time to be nonspecific stimulators of the immune system, generally as a result of macrophage activation. In the last two decades the role of β-glucans in stimulating the immune system of the human body has been confirmed.[3,4] In 1991, Czop and Kay identified a macrophage cell surface receptor that is specific for a small oligosaccharide of the β-(1,3)-D-glucan series.[5]

Glucans are natural polysaccharides found in oat, barley, wheat, yeast and fungi. The term "glucan" refers generically to a variety of naturally occurring homopolysaccharides or polyglucoses, including polymers such as cellulose, amylose, glycogen, laminarians and starch. Glucan encompasses branched and unbranched chains of glucose units linked by 1-3, 1-4, and 1-6 glucosidic bonds that may be of either the α or β type. It is generally known that the glucan with β-(1,3) linkages has more potent macrophage stimulatory activity than any other linkage type.[6]

β-(1,3)-Glucan from *Schizophyllum commune*

There are many different types of β-(1,3)-glucans. Depending on their source, they vary in physical and chemical characteristics such as their uniformity, the degree of branching, the molecular weight and the tertiary structure (Table 1).

β-Glucan of *Schizophyllum commune* Fries has a β-1,6 branch on every third glucose residue of the β-1,3 main chain (Figure 1).[7] In contrast with β-glucans of other mushrooms, the β-glucan secreted from the cells of *S. commune* characteristically is a stable neutral polysaccharide containing a branched, uniform and unique structure. These β-glucans consist of only glucose. The molecular weight of several million is very great in comparison to the molecular weight of the other mushrooms, which are in the range of hundreds of thousands to a million. Therefore, β-glucan of *S. commune* has quite different properties and different functions entirely.[4,8]

β-Glucan now in use as a component of various cosmetics is mainly produced from *Saccharomyces cerevisiae* as a water-insoluble particulate or its chemically modified soluble forms such as carboxymethyl or phospholyated glucan.[9] In contrast with this, β-glucan from *S. commune* is both water-soluble and natural, which is more suitable for cosmetic application.

Liquid Cultivated β-Glucan from *S. commune*

The β-glucan from *S. commune* is generally known as schizophyllan. It has been used as an immunotherapeutic agent for cancer treatment in Japan since 1986. It is used in conjunction with chemotherapy or radiotherapy. Clinical studies have demonstrated that administration of schizophyllan, along with antineoplastic drugs, prolongs the lives of patients with lung or gastric cancers. Schizophyllan has no antigenicity or mitogenic effect on the T-cell. Its antitumor activity is exerted only through the activation of macrophages, which subsequently augment the T-cell cascade.[10,11]

Table 1. Comparison of physicochemical characteristics of β-(1,3)-glucan depending on origin

Origin	Characteristics
Ganoderma lucidium	A complex form linked with glucose, mannose, galactose and others
Coriolus versicolor	One type with a β-(1,3)-bond; another type with a β-(1,4)-bond
Letinus edodes	β-(1,3)-glucan which has two β-(1,5)-branches in every fifth residue of the β-(1,3) main chains
Pleurotus ostreatus	Linked type of glucose, mannose, galactose and others
Phellinus linteus	Contains glucose at 70-90%, and also includes mannose, galactose and others
Saccharomyces cerevisiae	β-(1,3) backbone on which occasional (1,6) branching occurs; water insoluble
Cereal grains	Linked by β-(1,4) and β-(1,3) bonds

One of the hallmarks of the aging process is a general decline in the body's immune defense mechanisms.[12] This applies as much to the skin as to any other aspect of the immune defenses. In this regard, because it was known to be a very potent stimulator of the immune system, schizophyllan would be a good candidate for blocking the skin aging process. However, there have been difficulties in using schizophyllan in cosmetics because of its high production cost.

In our laboratory, we developed a process for liquid cultivating of *S. commune* Fr. for efficient isolation of β-glucan. We add activated charcoal and glucose into the liquid medium gradually, and purify it with a proprietary process. The resulting material[a], which we will call LC-Glusc (liquid cultivated glucan from *S. commune*) in this article, had the ^{13}C-NMR spectrum shown in Figure 2 and the IR spectrum shown in Figure 3. Its molecular weight was calculated by aqueous size exclusion chromatography with multi-angle laser light scattering photometry. We view the molecular weight and basic structure of LC-Glusc as equivalent to those of schizophyllan.

Thermostability and pH stability of LC-Glusc were tested. There was no change of viscosity at 121°C for at least 8 h, and also it was stable within the pH range of 2 to 12 (data not shown).

Activity of LC-Glusc

β-Glucan stimulates macrophage and activates the immune system, producing cytokines for use in wound healing, collagen synthesis and cell protection. β-(1,6)-Branched β-(1,3)-glucan, such as LC-Glusc, has greater immune stimulating and anti-cancer effects than β-(1,3)-glucan because of its branched residue and tertiary structure.

Proliferation of skin cells: LC-Glusc's ability to affect proliferation of skin cells was compared to that of yeast glucan. Human fibroblasts were cultured on Dulbecco's Modified Eagle's Medium (DMEM) containing 2.5% of bovine

[a] SC-Glucan is a registered trademark of Pacific R&D Center, Yongin, Korea.

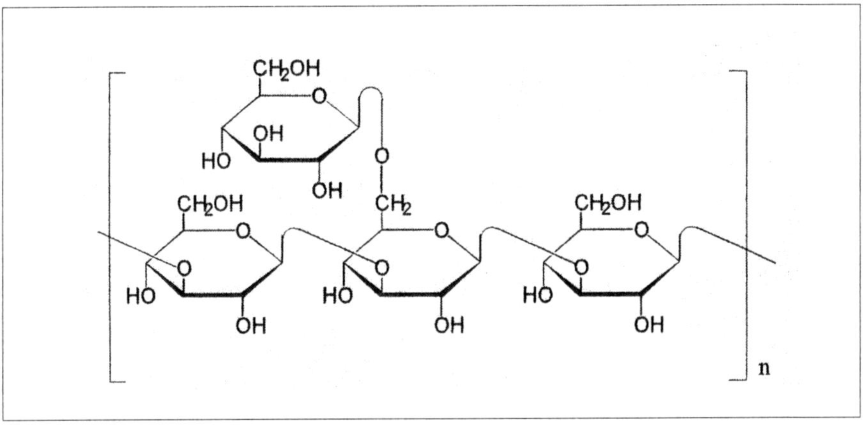

Figure 1. Structure of schizophyllan

fetal serum and then 5,000 cells/well were added on a 96-well microtiter plate. After addition of either LC-Glusc or the yeast glucan into each well, the 96-well plate was maintained at 37°C for 4 days. After the cultivation was completed, 0.2% of 3-(4,5-dimethylthiazol-2-yl)-2,5-diphenyltetrazolium bromide (MTT) was added, 50 µL/well. Again, the plate was maintained at 37°C for four days to give formazan. The quantity of formazan produced can be regarded as an indicator of cell density or viability. After dissolving the formazan in dimethyl

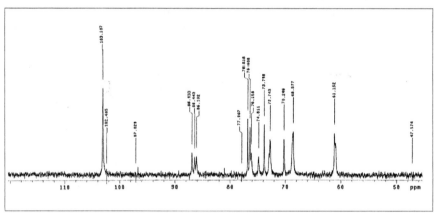

Figure 2. ^{13}C-NMR spectrum. LC-Glusc, measured by Varian Gemini-2000 300 MHz FT-NMR; 50 mg of LC-Glusc (300 min, sonic irradiation) was dissolved into 0.6 mL of dimethyl sulfoxide and then it was measured at 100°C. Spectral width was in the range of 18.67 kHz (scan=10,000 accumulation).

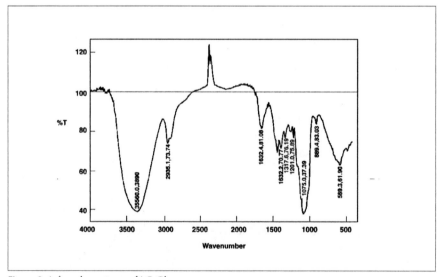

Figure 3. Infrared spectrum of LC-Glusc
10 mg of LC-Glusc was mixed into KBr. The mixing material was ground by mortar to make fine powder and then was measured by FT-IR for obtaining IR spectrum.

sulfoxide, the absorbance at 570 nm was measured with microplate reader. The proliferation of fibroblasts was evaluated by comparing the absorbance with that of the untreated control.

As shown in Figure 4, LC-Glusc at a concentration of 0.04% increased the proliferation of fibroblasts by approximately 40%, while treatment with yeast glucan produced only a 7% increase.

Collagen biosynthesis: We used the fibroblasts created above to test LC-Glusc for its ability to increase synthesis of collagen, and we compared that ability to that of yeast glucan.

Fibroblasts were incubated with LC-Glusc or yeast glucan for 24 hours on a 24-well microtiter plate containing DMEM. After the incubation was completed, 5 mL of DMEM and 10 µCi of L[2,3,4,5-^{3}H]-proline were added into each well. Twenty-four hours later, the medium and the cells contained in each well were recovered and washed with 5% trichloroacetic acid, and then divided between two test tubes. To one of the two test tubes, we added 1 unit/mL of type I collagenase and then maintained the test tube at 37°C for 90 min. The other test tube was incubated at 4°C. Then 0.05 mL of 50% trichloroacetic acid was added to each of two test tubes, and they were maintained at 4°C for 20 min. The resulting solution was centrifuged at 12,000 rpm for 10 min, after which the decay per min (dpm) of the supernatant and of the precipitate was measured with a liquid scintillation counter.

We calculated the Relative Collagen Biosynthesis (RCB) with the following general equation:

$$RCB = \frac{\text{collagen dpm}}{(\text{total collagen} - \text{collagen dpm}) \times 5.4 + \text{collagen dpm}} \times 100$$

where (total collagen – collagen dpm) x 5.4" represents non-collageneous protein of the sample.

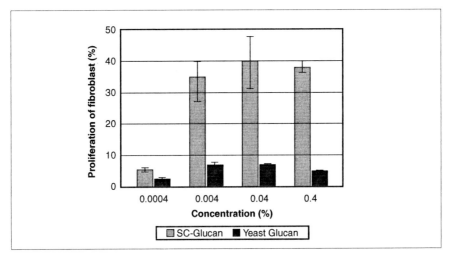

Figure 4. Influence of LC-Glusc on proliferation of human skin cell fibroblasts. Figures shown are expressed as percentages relative to an untreated control.

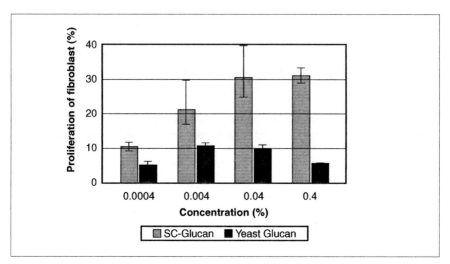

Figure 5. Increased collagen synthesis after treatment with LC-Glusc or yeast glucan. Figures shown are expressed as percentages relative to an untreated control.

It is generally regarded that noncollageneous protein is 5.4 times more abundant than collagen protein.

As shown in Figure 5, the treatment with yeast glucan at 0.04% concentration produced at most a 10% increase in the biosynthesis of collagen, whereas the treatment with LC-Glusc at 0.04% produced a 32% increase.

Sunburn recovery: Using a solar simulator, we measured the minimal erythema dose on the upper arms of 10 volunteers, and then irradiated the upper arms with UV to induce erythema. At 8, 16, 24 and 48 h after the irradiation, we applied 1% LC-Glusc to one arm of each volunteer and distilled water to the other arm. Before the irradiation and after each treatment, we measured[b] the skin color. The relative value, when the skin color before the UV irradiation was converted to 100, was calculated as an erythema index. Its statistical significance was determined by Anova test ($p < 0.05$). The number of volunteers was small, so we increased the number of irradiation sites on each arm for more accuracy.

We used this erythema index to evaluate LC-Glusc's ability to reduce erythema, compared to distilled water. As shown in Table 2, the areas treated with LC-Glusc showed a marked tendency to reduce erythema.

Moisturizing efficacy: The moisturizing activity of SC-Glucan was evaluated compared to the effect of hyaluronic acid. Ten male and ten female volunteers participated in this experiment.

We applied 60 µL of LC-Glusc solution (0.4%) or hyaluronic acid solution (0.5%) to the forearm skin and covered it with tape[c]. The skin hydration was evaluated by conductance measurements immediately before treatment, immediately after treatment and two hours after treatment under controlled conditions of temperature

[b] Chromameter CM2002, Minolta, Japan
[c] Tegaderm tape, 3M, Canada

Table 2. Sunburn recovery effect of LC-Glusc compared to that of a control (distilled water) (n=10)

Erythema index

Test group	8 h	16 h	24 h	48 h
Distilled water	150.67±19.0	197.85±23.2	193.58±25.1	174.98±26.7
LC-Glusc	131.98±19.8	163.60±18.1	155.01±23.7	135.87±32.5

Table 3. Anti-inflammatory activities of LC-Glusc (n=10)

	Carageenan foot edema Inhibition of edema(%)	Mouse ear edema Inhibition of edema(%)	Inhibition of neutrophil influx(%)
Aspirin	34	73	67
LC-Glusc	68	55	40

and relative humidity. To calculate the change in skin moisturizing, we defined a Skin Moisturizing Rate (SMR) as follows:

$$SMR\% = \frac{\Delta CT - \Delta CU}{\Delta CU - \text{intial CT}} \times 100$$

where ΔCT is the change of conductance at a treated site and
ΔCU is the change of conductance at an untreated site

This equation can be interpreted as follows:

$$SMR\% = \frac{(CT_t - CT_i) - (CU_t - CU_i)}{(CU_t - CU_i) - CT_i} \times 100$$

where CT and CU are conductance at the treated site and the untreated site, respectively, and t = conductance value at some time after treatment, which in our case was either immediately after treatment (t=0) or 2 hours after treatment (t=2). For each t, we took three measurements and used the average value in the equation. i = conductance value initially before treatment

Figure 6 shows that 0.4% LC-Glusc is a much better moisturizer than 0.5% hyaluronic acid.

Anti-inflammation: We evaluated the anti-inflammatory efficacy of LC-Glusc and aspirin by two methods: carrageenan foot edema and mouse ear edema.

For carrageenan foot edema, we injected the test compounds intra-peritoneally (dose: 30 mg/kg) 1 h before inducing inflammation by injecting 0.1% carrageenan solution into Sprague Dawley (SD) rat foot. We evaluated the anti-inflammatory

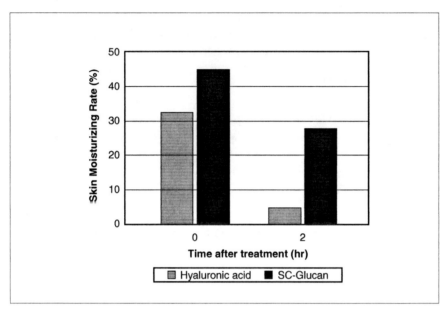

Figure 6. Moisturizing efficacy of LC-Glusc and hyaluronic acid

activity of the compounds by comparing the foot volume before and after injection of carrageenan solution.

For mouse ear edema, we administered the test compounds 15 min after applying the edema-inducing agent tetradecanolphorbol acetate (2 µg/ear in 20 µL acetone). Edema measurements were taken 6 h later. Anti-inflammation activity was evaluated by observing the degree of inhibition of edema and neutrophil influx.

Table 3 shows that LC-Glusc has an anti-inflammatory activity similar to that of aspirin, which is known to be an efficient anti-inflammatory ingredient.[13]

Anti-irritation: For anti-irritation evaluation of LC-Glusc, the yeast glucan and an untreated control, we used three groups of ten New Zealand white rabbits. Retinol skin-irritant compositions were formulated in an appropriate vehicle and then applied to the shaved skin of the rabbits' backs. Then, the test formulations were applied to a defined portion of each rabbit's irritated skin daily for nine days.

Erythema and redness were graded visually on a five-step scale (1 = No visible irritation, 2 = Slight irritation, 3 = Definite irritation, 4 = Moderate irritation and 5 = Severe irritation). Assessment scores were recorded for each treated site on the rabbit's skin. Each day's scores were summed for each group, and the irritation index was expressed as the value of the group's total score for the day divided by the number of sites treated that day.

Figure 7 shows that LC-Glusc significantly decreased the irritation caused by retinol, compared to the irritation reduction produced by yeast glucan or no treatment at all.

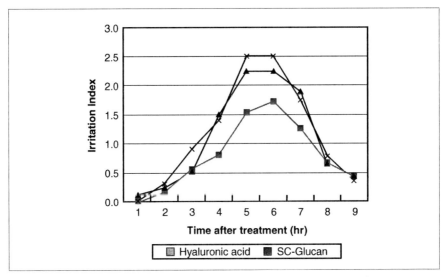

Figure 7. Change of irritation index due to treatment with LC-Glusc or yeast glucan or no treatment, following chemically induced irritation

Conclusion

Skin protects several organs in the body against harm from changes of temperature and humidity, ultraviolet rays and pollutants. But excessive external stresses, such as physical and chemical stimuli and under-nutrition, cause a decrease of the normal functioning of skin and stimulate the skin aging phenomenon.

Researchers have attempted to maintain the intrinsic function of skin and to inhibit skin aging by activation of skin cells. Various physiologically active materials have been obtained from animals, plants and microorganisms to use in cosmetics for this purpose.

Our work shows that β-1,6-branched β-(1,3)-glucan from *Schizophyllum commune* is an active ingredient that can help to increase skin cell proliferation, collagen biosynthesis and sunburn recovery. It can effectively reduce skin irritation. In addition, this β-glucan is water soluble and stable under conditions of changing pH and temperature.

Therefore, we suggest that β-1,6-branched β-(1,3)-glucan from *S. commune* is suitable for cosmetic and dermatological applications and will be an effective ingredient for use in antiaging and anti-irritant formulations.

— Moo-Sung Kim, Kyung-Mok Park, Ih-Seop Chang, Hak-Hee Kang and Young-Chul Sim, *Pacific R&D Center, Yongin, Korea*

References

1. A V Benedeto, The environment and skin aging, *Clinics in Dermatology* 16 129-139 (1998)
2. PWA Mansell, Polysaccharides in skin care, *Cosmet Toil* 109(9) 67-72 (1994)
3. RC Goldman, Biological response modification by β-glucans, *Annual Reports in Medicinal Chemistry* 30 129-136 (1995)
4. JA Bohn and JN BeMiller, (1-3)-β-glucans as biological response modifier: a review of structure-functional activity relationships, *Carbohydrate Polymers* 28 3-14 (1995)
5. JK Czop and J Kay, Isolation and characterization of β-glucan receptors on human mononuclear phagocytes, *J Exp Med* 173 1511-1520 (1991)
6. S Yoshio, H Katsuhiko and M Kazumasa, Augmenting effect of sizofiran on the immunofunction of regional lymph nodes in cervical cancer, *Cancer* 69(5) 1188-1194 (1992)
7. T Yanaki, W Ito and T Kojima, Correlation between the antitumor activity of a polysaccharide schizophyllan and its triple-helical conformation in dilute aqueous solution, *Biophys Chem* 17 337-342 (1983)
8. K Tabata, W Ito and T Kojima, Ultrasonic degradation of schizophyllan, an antitumor polysaccharide produced by *Schizophyllum commune* Fries, *Carbo Res* 89 121-135 (1981)
9. F Zülli, F Suter, H Biltz, HP Nissen and M Birman, Carboxymethylated β-(1-3)-glucan, *Cosmet Toil* 111(12) 91-98 (1996)
10. Y Kimura, H Tojima, S Fukase and K Takeda, Clinical evaluation of sizofilan as assistant immunotherapy, *Otolaryngol* (Stockh), Suppl 511 192-195 (1994)
11. I Sugawara, KC Lee and M Wong, Schizophyllan (SPG)-treated macrophages and anti-tumor activities against syngeneic and allogeneic tumor cells: I. Characteristics of SPG-treated macrophages, *Cancer Immunol Immunother* 16 137-144 (1984)
12. M Jabbar, Immuno-cosmetics, *Happi* 33(8) 41-46 (1996)
13. G DiPasquale, C Rassaert, R Richiter, P Welaj and L Tripp, Influence of prostaglandin (Pg) E2 and F2 on the inflammatory process, *Prostaglandins* 3 741-757 (1973)

Artemia Extract: Toward More Extensive Sun Protection

Keywords: UV, DNA, Artemia extract

UV protective properties of Artemia extract and the potential of its use in sun product

In recent years, people have become increasingly aware that UV exposure can cause skin damage, such as photoaging and carcinogenesis. Therefore, people are exhibiting a growing interest in protecting their skin from the harmful effects of UV radiation. It is well known that UVB induces a variety of cellular damage, including DNA photodamage with formation of cyclobutane pyrimidine dimers,[1] and consequent mutation.[2] Moreover, there is increasing evidence that UVA contributes to skin carcinogenesis and premature aging.

Sunscreens are now widely used in attempts to block and/or decrease UV radiation's undesirable effects, such as photoaging and skin cancer. Also widely used is SPF (sun protection factor) to designate the protective properties of sunscreens. In fact, SPF indicates only the sunscreen's ability to protect skin from UV induced erythema. Studies have suggested that because erythema and sunburn are absent in sunscreen protected skin, sunscreens may lead to a more intensive exposure to sun irradiation, which in turn increases the risk of DNA damage, immunosuppression and skin cancer.[3,4]

Recent studies have found the rate of tumor formation in populations that use sunscreens to be higher than tumor formation in populations that have no sun exposure, thus raising questions about the failure of sunscreens.[4] Concerned with this problem, we were interested in finding a new, natural way of helping protect human skin from UV damage. In this article, we describe the UV protective properties of Artemia extract and, consequently, the potential of Artemia extract use in sun protective products.

Artemia Extract

Artemia extract is prepared by marine biotechnology from a specific plankton, the *Artemia salina*, that lives in hyper-mineralized lakes. Artemia is characterized by its great capability of withstanding and surviving various sudden environmental aggressions. Artemia extract is rich with phosphorylated nucleotides and, in particular, with diguanosine tetraphosphate (GP4G), which our recent studies have demonstrated to have a great protective effect on DNA.[5]

Materials and Methods

Viability of cultured cells: Human fibroblasts were treated with Artemia extract 3% for 24 h, then exposed to UVB irradiation. The MTT (3-(4,5-cimethyl-thiazol-2-2,5-diphenyl tetrazolium bromide) colorimetric assay[6] was carried out 24 h later.

Skin organ culture and staining: We used the established skin organ culture method.[7,8] Test substances (a cream containing Artemia extract and a placebo cream) were applied topically on the skin samples. Designated skin samples were irradiated and all skin samples were submitted to standard hematoxylin and eosin (H&E) or to immunostaining 24 h later.

Immunostaining studies: Cultured skin samples were studied by routine direct immunofluorescence (DIF) microscopy following standard methods.[9,10] For heat shock protein studies, rabbit polyclonal antibody anti-Hsp70 was used. Mouse anti-CD1a antibody was used for Langerhans cell staining.

Immunoblotting studies: Immunoblotting studies[11] were performed using NuPAGE 12% gel and polyclonal rabbit anti-human Hsp70 antibody, monoclonal mouse anti-human p21, or anti-mutant and wild-type human p53 antibody.

Comet assay: Comet assay[12,13] was conducted on cultured fibroblasts. In brief, after irradiation, cells were harvested and cell lysate was subjected to electrophoresis, then stained with propedium iodide, after which the individual cell comet aspect "head and tail" was examined.

Tunel assay: Cultured human keratinocytes received 1% of Artemia extract 24 h before and after their irradiation with 100 mJ/cm² of UVB. Twenty-four hours later, the Tunel assay[14] was performed.

Enzyme-linked immunosorbent assay: Standard ELISA method[15,16] was carried out on human HaCat cells. Cells received Artemia extract 3% for 24 h, followed by UVB irradiation. Total cytokine level was assessed for IL-1 alpha, IL-8 and TNF-alpha. The level of the cytokines was expressed in pg/mg protein.

Results and Discussion

Artemia extract and UV protection of human cells: MTT viability test demonstrated that cell preparation with Artemia extract prior to stress protected the cells from UVB (100 mJ/cm²) aggression and increased their viability by 20% (Figure 1). Likewise, observation of the morphology and aspect of the cells showed that Artemia extract also preserved their morphology from UV induced stress signs.

Artemia extract and protection of DNA: The principal constituent of Artemia extract is a molecule called GP4G. Previous studies had demonstrated that the GP4G molecule has a remarkable effect on cellular DNA.[5] This strongly suggests that the Artemia effect we see on DNA is due mainly to the GP4G molecule in the extract.

With the Tunel method we studied the effect of GP4G on DNA protection from UVB induced DNA damage and consequent programmed cell death or apoptosis. Our studies demonstrated that adding Artemia extract into the cells after UVB exposure (100 mJ/cm^2) decreased UVB induced apoptosis by 35%. Interestingly, GP4G protection reached 45% when the extract was added before and after UVB irradiation (Figure 2), which suggests its role in both DNA protection and repair.

Comet assay: It is known that there is a linear dose response curve for thymine dimers' formation in UV irradiated cells,[17] and that depending on the UV dose and DNA damage, the cell uses its repair mechanism or goes into apoptosis. Therefore, we performed comet assay using UVB doses of 30, 60, and 100 mJ/cm^2 on human fibroblasts previously treated or not treated with Artemia extract. Our results showed that, at the low UVB dose (30 mJ/cm^2), there was very little UVB induced DNA degradation in the cells treated with the extract, compared to the untreated cells (Figure 3). Similarly, with higher UVB doses of 60 and 100 mJ/cm^2, Artemia extract decreased DNA degradation, as shown in Figures 4 and 5.

p53 and p21 expression: Other researchers have reported that after UV irradiation, tumor suppressor gene p53 wild-type is involved in two mechanisms: a) both G_1 cell cycle arrest and DNA repair, and b) apoptosis.[18-20] The mechanism in play depends on the degree of UV damage. Moreover, it

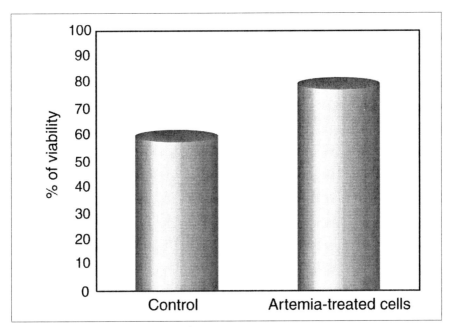

Figure 1. Cell viability after irradiation with 100 mJ/cm^2 UVB

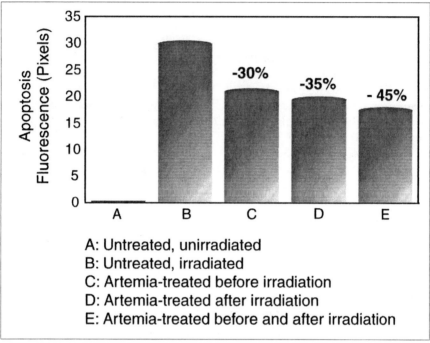

Figure 2. Tunel assay. GP4G-Artemia extract effect on UVB induced apoptosis.

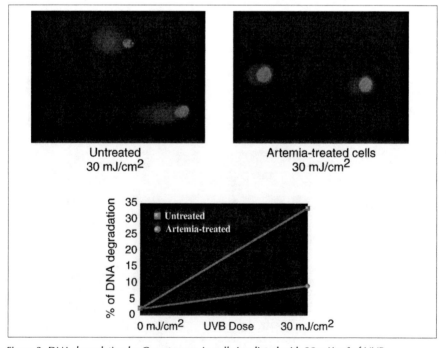

Figure 3. DNA degradation by Comet assay, in cells irradiated with 30 mJ/cm² of UVB

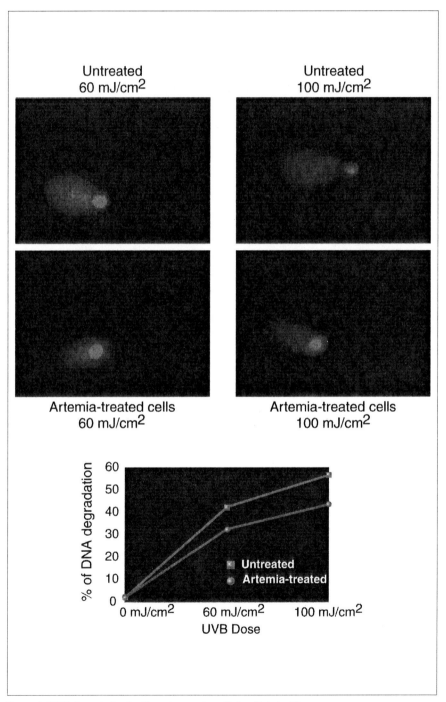

Figure 4. DNA degradation by Comet assay, in cells irradiated with 60 and 100 mJ/cm² of UVB

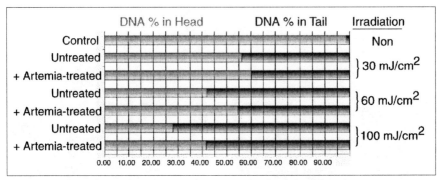

Figure 5. Analysis of DNA degradation percentage in "head & tail" of the comet

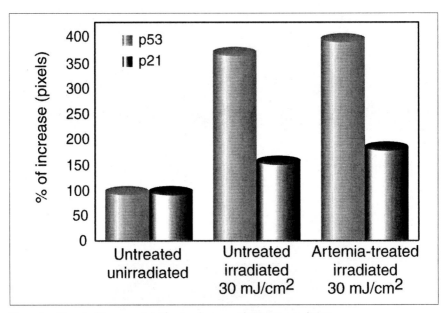

Figure 6. p53 and p21 expression after irradiation with 30 mJ/cm² of UVB

has also been demonstrated that p21 helps protect the cell from UV induced apoptosis.[21-23]

Our results showed that at low UVB doses, both p53 and p21 are upregulated in Artemia treated cells more than in control (Artemia untreated) cells, which align the activation of DNA repair pathway by transient cell cycle arrest through p53 and p21. Interestingly, with higher UVB doses, our results were similar and showed that in irradiated cells treated with Artemia extract, both p53 and p21 were higher than in control cells that failed to repair their damaged DNA (Figure 6). These results, together with the data above, suggest a consistent activation of the repair pathway rather than the apoptosis pathway, in cells treated with Artemia extract.

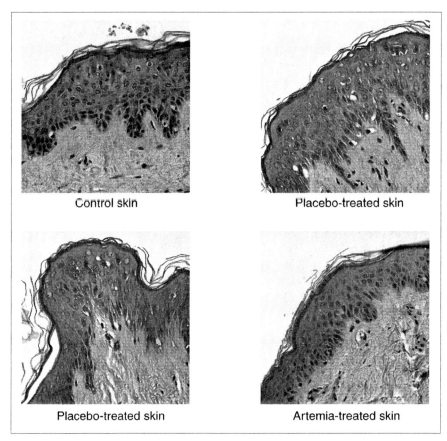

Figure 7. H&E staining of human skin irradiated with 200 mJ/cm² of UVB

Skin protection from UVB by Artemia extract: In order to confirm the above results on human skin, we treated skin samples either with Artemia extract 3% in a cream or with placebo, prior to exposure to UVB at selected doses in the range 100-200 mJ/cm². Skin structure and morphology, evaluated after H&E staining, revealed that skin samples treated with Artemia extract exhibited great skin structure preservation with rare occurrence of sunburn cells,[24,25] in contrast to the control skin which exhibited extended signs of damage with much sunburn cell formation, as shown in Figure 7.

Moreover, CD1a immunostaining of Langerhans cells revealed that Artemia treated skin samples were similar to unirradiated control skin, and exhibited no decrease in the number or distribution of Langerhans cells in the skin. On the other hand, UV irradiation of placebo treated skin induced Langerhans cell depletion from the epidermis (Figure 8), as is usually seen after UV exposure.[25,26]

Hsp70 induction by Artemia extract in cultured human cells and skin: It is well known that heat shock proteins (Hsp) or molecular chaperons, and in

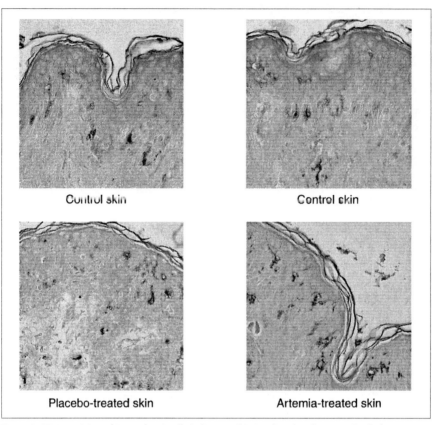

Figure 8. CD1a staining of Langerhans cells in human skin irradiated with 200 mJ/cm² of UVB

particular Hsp70, play an important role in protecting the cell from different types of stress.[27-29] Because Artemia is able to withstand various environmental stresses, we investigated whether the induction of Hsp70 was one of Artemia extract's essential mechanisms of action.

Different studies, conducted on cultured human fibroblasts, showed that, within 3-6 hours, Artemia extract 3% significantly induced Hsp70 synthesis in these cells. These results were confirmed by Hsp70 mRNA studies (data not shown).

Moreover, our recent studies[30] have shown that Artemia extract induction of Hsp70 is stress-free. Similar studies on human skin showed that the application of Artemia extract 3% in a cream formula significantly induced and increased Hsp70 protein expression in the skin, compared to the placebo-treated control skin where Hsp70 was found at the basic level (Figure 9).

Furthermore, we completed the studies above using 1 and 2 J/cm² of UVA alone or followed by UVB irradiation (50-150 mJ/cm²) on cultured keratinocytes and fibroblasts in order to study the influence of UVA. These results confirmed

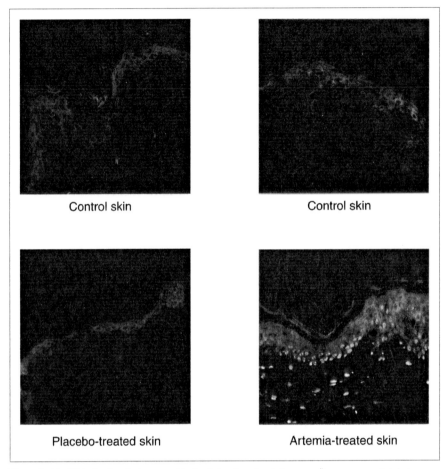

Figure 9. Hsp70 induction in human skin by Artemia extract. Immunofluorescence staining

the above findings concerning the protection Artemia extract offers to the cells. Our results also showed the induction of Hsp70 in the UV stressed cells, and that this induction was higher when the cells were treated with Artemia extract, prior to their irradiation.

Artemia extract's effect on inflammatory cytokines: Prior data have shown that the induction of Hsp70 inhibits the biosynthesis of IL-1 and other cytokines in human cells and, thereby, serves a protective role by suppressing the proinflammatory response.[31] Because our studies have demonstrated that Artemia extract induces Hsp70 in human cells, we looked at Artemia induced Hsp70 and irradiated HaCat cells, and used ELISA to investigate the anti-inflammatory effect of Artemia extract on the following cytokines:

- *IL-1 alpha and IL-8 in dose course studies*. Total IL-1 alpha and IL-8 levels were determined by ELISA after exposure of the cells to UVB doses of 0, 10, 20,

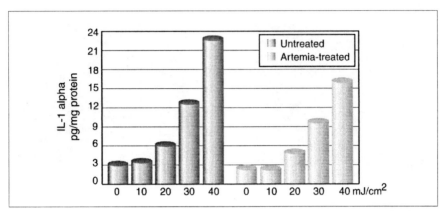

Figure 10. Artemia extract downregulation of total IL-1 alpha synthesis in cells irradiated with different UVB doses

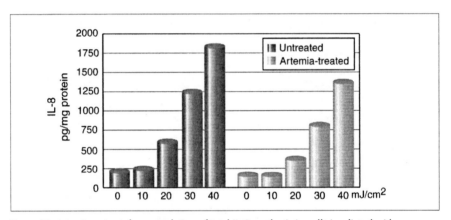

Figure 11. Artemia extract downregulation of total IL-8 synthesis in cells irradiated with different UVB doses

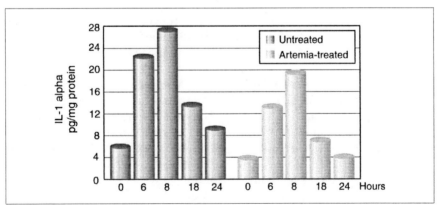

Figure 12. Time course studies of Artemia extract effect on total IL-1 alpha synthesis in cells irradiated with 30 mJ/cm² of UVB

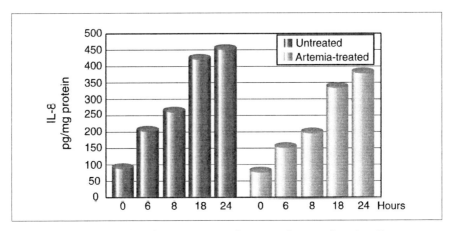

Figure 13. Time course studies of Artemia extract effect on total IL-8 synthesis in cells irradiated with 30 mJ/cm² of UVB

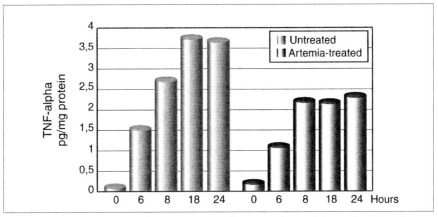

Figure 14. Time course studies of Artemia extract effect on total TNF-alpha synthesis in cells irradiated with 30 mJ/cm² of UVB

30, 40 mJ/cm². Our studies demonstrated that IL-1 alpha and IL-8 level of synthesis decreased considerably in the Artemia-Hsp70 induced cells by 20-30% for IL-1 alpha and 25-40% for IL-8, compared to the control cells (Figures 10 and 11).
- *IL-1 alpha and IL-8 in time course studies.* Total levels of IL-1 alpha and IL-8 were evaluated at the following time points: 0, 6, 8, 18 and 24 h. Interestingly, a significant decrease in UV induced IL-1 alpha level was found in the Artemia-Hsp70 induced cells at different time points, compared to the control cells (Figure 12). A maximum decrease in IL-1 alpha was observed at 18-24 h after UV irradiation. Similarly, a decrease in IL-8 level in irradiated

Artemia-Hsp70 induced cells was found over all time points. Interestingly, the maximum decrease of UV induced IL-8 was observed within 6-8 h after UV irradiation (Figure 13), which is of interest for 'early' control of the inflammatory process.

- *TNF-alpha studies.* TNF-alpha is another potent multifunctional mediator of inflammation and immune responses. As shown in Figure 14, our results demonstrate that TNF-α level is decreased in Artemia-Hsp70 induced cells at all the time points, and this decrease is statistically significant at the 24 h time point, after irradiation. This result is very interesting and corresponds to the decrease in the maximum inflammatory effect of TNF-α.

Conclusion

This data demonstrates that Artemia extract displays two unusual characteristics. It has a high content of GP4G, which protects DNA and decreases DNA damage. It induces Hsp70 in a stress free setting, while also having an anti-inflammatory effect on inflammatory cytokines.[32]

These studies reveal that Artemia extract offers an interesting contribution in the search to meet the present desire for new "agents and natural mechanisms" that offer more comprehensive skin protection from sun damage, and that reinforce the protection of sunscreens. Therefore, Artemia extract has potential applications in sun care products.

—N. Domloge, E. Bauza, K. Cucumel, D. Peyronel and C. Dal Farra, *Vincience Research Center, Sophia Antipolis, France*

References

1. ML Kripke, PA Cox, LG Alas and DB Yarosh, Pyrimidine dimers in DNA initiate systemic immunosuppression in UV-irradiated mice, *Proc Natl Acad Sci USA* 89(16) 7516-7520 (1992)
2. A Ziegler et al, Sunburn and p53 in the onset of skin cancer, *Nature* 372(6508) 773-776 (1994)
3. J Westerdahl, C Ingvar, A Masback and H Olsson, Sunscreen use and malignant melanoma, *Int J Cancer* 87(1) 145-150 (2000)
4. MF Naylor, HW Lim, JK Robinson, MA Weinstock, DS Rigel and MR Verschoore, Photoprotection, the *AAD 2001*, Proceedings of the 59th Annual Meeting of the American Academy of Dermatology, Washington DC, March 2-7, 2001
5. C Dal Farra, K Cucumel, D Peyronel and N Domloge, DNA protective effect of diguanosine tetraphosphate, *J Invest Dermatol* 114(4) 808 (2000)
6. E Borenfreund, H Habich and N Martin-Alguacil, Comparison of two *in vitro* cytotoxicity assays – the neutral red and tetrazolium MTT tests, *Toxic in vitro* 2(1) 1-6 (1988)
7. J van de Sandt, J van Shoonhoven, W Maas and A Ruten, Skin organ culture as an alternative to *in vivo* dermatotoxicity testing, *ATLA* 21 443-449 (1993)
8. S Boisnic, MC Branchet-Gumila, L Benslama, Y Le Charpentier and J Arnaud-Battandier, *Eur J Dermatol* 7 271-273 (1997)
9. N Domloge-Hultsch, P Bisalbutra, R Gammon and KB Yancey, Direct immunofluorescence microscopy of 1 mol/L sodium chloride-treated patient skin, *J Am Acad Dermatol* 24(6) 946-951 (1991)

10. N Domloge-Hultsch, P Benson, R Gammon and KB Yancey, A bullous skin disease patient with autoantibodies against separate epitopes in 1 mol/L sodium chloride split skin, *Arch Dermatol* 128 1096-1101 (1992)
11. S Muller, V Klaus-Kovtum and JR Stanley, A 230-kd basic protein is the major bullous pemphigoid antigen, *J Invest Dermatol* 92 33-39 (1989)
12. EJ Morris, JC Dreixler, KY Cheng, PM Wilson, RM Gin and HM Geller, Optimizing of single-cell gel electrophoresis (SCGE) for quantitative analysis of neuronal DNA damage, *BioTechniques* 26(2) 282-289 (1999)
13. O Ostling and KJ Johanson, Micro-electrophoresis study of radiation-induced DNA damage in individual mammalian cells, *Biochem-Biophys Res Commun* 123 291-298
14. F Washio, M Ueda, A Ito and M Ichihashi, Higher susceptibility to apoptosis following ultraviolet B irradiation of xeroderma pigmentosum fibroblasts is accompanied by upregulation of p53 and downregulation of bcl-2, *Br J Dermatol* 140(6) 1031-1037 (1999)
15. WG Phillips, M Feldmann, SM Breathnach and FM Brennan, Modulation of the IL-1 cytokine network in keratinocytes by intracellular IL-1 alpha and IL-1 antagonist, *Clin Exp Immunol* 101 177-182 (1995)
16. A Guéniche, J Viac, G Lizard, M Charveron and D Schmitt, Effect of nickel on the activation state of normal human keratinocytes through interleukin 1 and intercellular adhesion molecule 1 expression, *Br J Dermatol* 131 250-256 (1994)
17. AR Young, CS Potten, O Nikaido, AG Parsons, A Boenders, JM Ramsden and CA Chadwick, Human melanocytes and keratinocytes exposed to UVB or UVA in vivo show comparable levels of thymine dimers, *J Invest Dermatol* 111 936-940 (1998)
18. G Li and VC Ho, p53-dependent DNA repair and apoptosis respond differently to high- and low-ultraviolet radiation, *Br J Dermatol* 139 3-10 (1998)
19. J Cotton and DF Spandau, Ultraviolet B-radiation dose influences the induction of apoptosis and p53 in human keratinocytes, *Radiat Res* 147 148-155 (1997)
20. ML Smith, IT Chen, Q Zhan, PM O'Connor and AJ Fornace, Involvement of the p53 tumor suppressor in repair of UV-type DNA damage, *Oncogene* 10 1053-1059 (1995)
21. N Bissonnette and DJ Hunting, p21-induced cycle arrest in G_1 protects cells from apoptosis induced by UV-irradiation or RNA polymerase II blockage, *Oncogene* 16 3461-3469 (1998)
22. S Inohara, K Kitagawa and Y Kitano, Coexpression of p21$^{waf1/cip1}$ and p53 in sun-exposed normal epidermis, but not in neoplastic epidermis, *Br J Dermatol* 135 717-721 (1996)
23. M Loignon, R Fetni, AJE Gordon and EA Drobetsky, A p53-independent pathway for induction of p21$^{waf1/cip1}$ and concomitant G_1 arrest in UV-irradiated human skin fibroblasts, *Cancer Res* 53 3390-3394 (1997)
24. F Daniel, L Brophy and W Lobitz, Histochemical response of human skin following ultraviolet irradiation, *J Invest Dermatol* 37 351-357 (1961)
25. BA Gilchrest, NA Soter, JS Stoff and MC Mihm, The human sunburn reaction: Histologic and biochemical studies, *J Am Acad Dermatol* 5 411-422 (1981)
26. V Rae, T Yoshikawa, W Bruins-Slot, JW Streilein and JR Taylor, An ultraviolet B radiation protocol for complete depletion of human epidermal Langerhans cells, *J Dermatol Surg Oncol* 15 1199-1202 (1989)
27. M Jäättelä, Heat shock proteins as cellular lifeguards, *Ann Med* 31 261-271 (1999)
28. TKC Leung, MY Rajendran, C Monferies, C Hall and L Lim, The human heat-shock protein family, *Biochem J* 267 125-132 (1990)

29. MG Santoro, Heat shock factors and the control of the stress response, *Biochem Pharmacol* **59** 55-63 (2000)
30. TJ Hall, Role of Hsp70 in cytokine production, *Experientia* **50** 1048-1053 (1994)
31. K Cucumel, JM Botto, E Bauza, C Dal Farra, R Roetto and N Domloge, Artemia extract induces Hsp70 in human cells and enhances cell protection from stress, the *SID 2001*, Proceedings of the 62nd Annual Meeting of the Society for Investigative Dermatology, Washington DC, May 9-12, 2001
32. French Pat 0007606, Préparations cosmétiques ou dermo-pharmaceutiques renfermant un extrait de zooplancton qui contient et induit des HSP, N Domloge, D Peyronel and C Dal Farra (Jun 15, 2000)

Topically Applied Soy Isoflavones Increase Skin Thickness

Keywords: isoflvone (soy), genistein, postmenopausa/skin

Genistein (soy isoflavone) shows improved effect on skin thickness without swelling in aged skin.

Epidemiological studies indicating an association between diet and disease led to the investigation of a series of bioactive plant compounds. In Japan, for example, the incidences of cardiovascular disease, hormone-dependent cancers (breast, uterus and prostate) and menopausal symptoms (osteoporosis and hot flashes) are all substantially lower than in western countries. The traditional Japanese diet includes a significant amount of soy-based foods such as tofu and tempeh.

The physiologically-relevant compounds in soy are the isoflavones, a subgroup of polyphenolic plant compounds. Isoflavones, or phytoestrogens, adopt a chemical structure very similar to that of the human hormone estrogen after hydrolysis of the sugar moiety (Figure 1). The isoflavones contained in soybeans appear predominantly in the form of polar, water-soluble glycosides, such as genistin.

Many dietary supplements with soy isoflavones already exist. In most cases, supplements contain only isoflavone glycosides, the molecular form that is not biologically active because there is no cellular uptake of glycosides. However, after ingestion, intestinal glucosidases and intestinal bacterial metabolism transform the glycosides into the physiologically active form.

There is an interest in isoflavones in the cosmetic industry because these compounds may have potential in the treatment of skin aging. We assume that two different activities of isoflavones could affect the skin by binding estrogen receptors and inhibiting the protein tyrosine kinase-signaling pathway.

Since the skin does not have the hydrolytic activity of the intestine, the isoflavones must be in the active aglycone form for skin care applications. In this rather apolar form, isoflavones penetrate easily into deeper skin layers and into the skin cells. Unfortunately, these aglycones have poor solubility in water and in oil. We have therefore developed a process to produce a water-soluble isoflavone aglycone preparation based on liposomes that is suitable for cosmetic applications, which

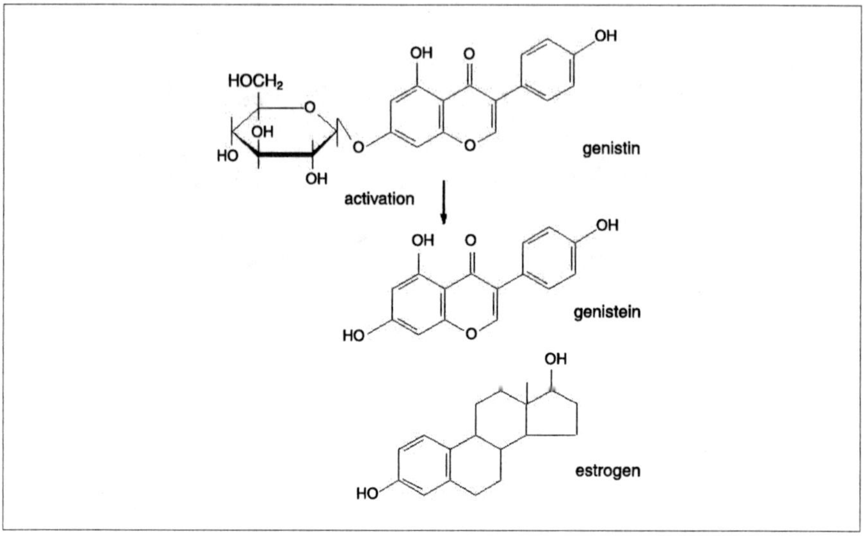

Figure 1. Transformation of the isoflavone glycoside genistin to the bioactive isoflavone aglycone, genistein. Comparison of the molecular structure of the hormone estrogen to genistein

we have reported previously.[1] This article summarizes the first study conducted on the efficacy of isoflavones in the prevention of skin aging.

Skin Aging

The skin is composed of two cell layers, the epidermis and the dermis, with the corresponding major cell types being keratinocytes and fibroblasts, respectively. The function of the outer epidermis is to provide a barrier against water loss and external chemical injury. The underlying dermis is rich in an extracellular matrix (ECM) that is essential for providing strength and elasticity to the skin. The ECM is composed of fibrillar collagen bundles and elastic fibers in a complex array of proteoglycans and other matrix components. During aging, the skin becomes thinner and less elastic (Figure 2). As collagen accounts for the biggest component of the matrix, regulation of collagen biosynthesis and degradation is central to the skin aging process.

Several factors ultimately lead to cutaneous aging. The Hayflick Phenomenon predicts that fibroblasts undergo cellular senescence with advanced age.[2] This results in a switch from a matrix-producing to a matrix-degrading phenotype. Another mechanism in skin aging is based on reactive oxygen species (ROS) as by-products of the oxidative cell metabolism (Free Radical Theory). A general decline in hormone production is a further important aging mechanism; the effect of declining estrogen production on skin has been well documented. At menopause, when the production of estrogen drops drastically, skin aging is more pronounced than before. The principal external factor that causes premature skin aging is UV light.

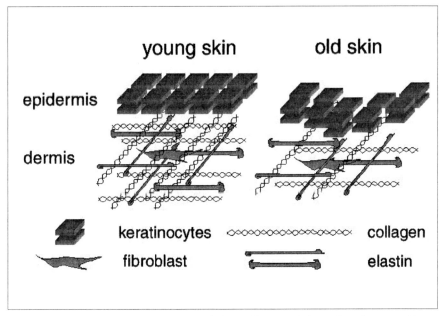

Figure 2. Schematic drawing of young and old skin. Collagen and elastin degradation in old skin leads to a decrease in skin thickness.

Cellular Regulation of Collagen Breakdown

A basic knowledge of collagen formation and breakdown is necessary to fully understand the effect of genistein in the treatment of skin aging. The fibroblasts synthesize and secrete individual polypeptide chains of types I and III collagens as procollagen precursors. Regulation of procollagen expression is controlled at the transcriptional level, with two transcription factors involved in down-regulation of collagen: the activator protein-1 (AP-1) and nuclear factor kappa B (NF-kB) (Figure 3). Both factors are induced by pro-inflammatory cytokines (IL-1alpha and TNF-alpha) or UV light. AP-1 can directly suppress collagen expression. But both AP-1 and NF-kB reduce collagen by up-regulation of the expression of matrix metalloproteinase (MMP) genes. MMPs comprise a large family of zinc-dependent endopeptidases that are specific for collagen degradation.

The epidermal growth factor (EGF) receptor on the surface of fibroblast cells is the cellular gate in the signaling pathway. The intrinsic part of the EGF receptor is a tyrosine kinase activity, which is activated in response to receptor binding. UV light and the inflammatory cytokine IL-1 have been found to induce tyrosine phosphorylation in a manner similar to EGF.[3] Following activation, the kinase phosphorylates other kinases in the signal transduction cascade, like those of the mitogen-activated protein (MAP) kinase pathway. The signaling pathway ends in the cellular nucleus with the activation of AP-1 and NF-kB. It has been suggested that H_2O_2 as well as other ROS are produced upon UVA radiation. ROS, produced

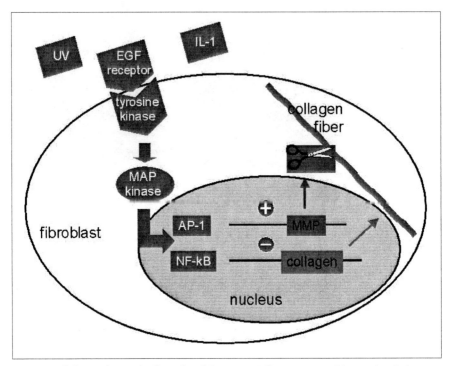

Figure 3. Cellular regulation of collagen breakdown. Pro-inflammatory cytokines such as IL-1 α and UV light activate the epidermal growth factor (EGF) receptor on the surface of fibroblasts. The tyrosine kinase on the intracellular side phosphorylates the mitogen-activated protein (MAP) kinase, which finally activates the transcription factors, activator protein-1 (AP-1) and nuclear factor kappa B (NF-kB). Both factors increase the expression of matrix metalloproteinases (MMP), which degrade collagen. AP-1 additionally down-regulates the expression of collagen.

either by UVA or as by-products of the oxidative cell metabolism, may account for the expression of matrix metalloproteinases through induction of inflammatory cytokines.

Biochemical Properties of Isoflavones in Skin

Isoflavones as selective estrogen receptor modulators: Postmenopausal skin aging has been found to be the result of lower collagen production.[4] Studies show that topical estrogen replacement therapies can increase the skin collagen content.[5] Although estrogen receptors have been identified in skin cells, it is not known how estrogen receptor binding ultimately results in increased skin collagen.

Classical hormone replacement therapies to fight postmenopausal symptoms, including hormone-related skin aging, are highly disputed because estrogen appears to increase the risk of breast and uterine cancers. Thus pharmaceutical companies have recently developed selective estrogen receptor modulators (SERMs), synthetic estrogen-like compounds, that are safer because they exert an estrogenic effect only in selected tissues, such as in bones to prevent osteoporosis (Table 1).

Although isoflavones are not steroids, they fit estrogen receptors very well. In humans, two different estrogen receptors exist, ERα and ERβ. Compared with estrogen, the soy isoflavone genistein (the active form of genistin) has a lower affinity for ERα but about the same affinity for ERβ. Unlike estrogen, genistein has been shown to even reduce the risk of breast and uterine cancers. In this sense, genistein can be regarded as a natural SERM with a high potential in prevention of hormone deficiency-related disorders such as skin aging.

Isoflavones as protein tyrosine kinase inhibitors: Genistein is a well-known inhibitor of protein tyrosine kinases as it competes at the ATP side. Several reports describe a regulatory effect of genistein on collagen metabolism by the inhibition of protein tyrosin kinases.

This is the case for certain cancerous cells, where genistein has been found to down-regulate the expression of MMPs and up-regulate the tissue inhibitor of metalloproteinase (TIMP).[6,7] Inhibition of MMPs (collagenases) is considered an important step in prevention of growth and spread of a metastatic tumor.

But also in normal cells genistein stimulates collagen production by interacting with kinases. This has been shown in bone cells by Yoon et al.[8] and in skin fibroblasts by Ravanti et al.[9] Wang et al.[10] found that genistein blocks the signaling pathway from UVB to activation of AP-1 by specific inhibition of the kinase at the growth factor receptor site.

All these publications indicate a high potential for genistein in up-regulation of collagen production by inhibition of the protein tyrosine kinase.

Soy Isoflavones in Antiaging Cosmetics

The scientific literature on the physiological effects of genistein shows interesting possibilities for its application in cosmetics, particularly in antiaging cosmetics. The decrease in skin collagen caused by chronological aging is even

Table 1. Selective tissue effects of the synthetic estrogen-like compound raloxifene and soy isoflavones compared to estrogen[12]

Target tissue	Estrogen	Raloxifene	Isoflavones
Brain	++	−	+
Uterus	++	−	−
Breast	++	−	−
Bone	++	++	+
Cardiovascular	++	++	+

++ = strong
+ = medium
− = no effect

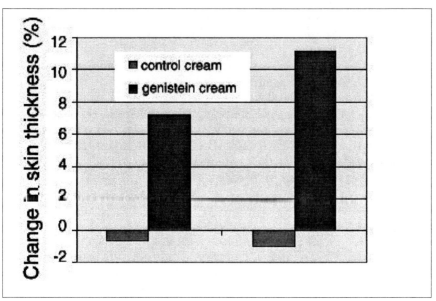

Figure 4. Comparison of skin thickness when treated with cream formulated with and without genistein

more pronounced after menopause as collagen decrease accelerates, due to hypoestrogenism.

A consequence of this skin aging is reduced skin thickness. In our study, we therefore used skin thickness as a major parameter to measure the theoretical skin benefits of topical application of genistein. Only two substances, estrogen and retinoic acid, have previously been shown to significantly increase skin thickness upon topical application. However, neither agent is allowed for use in cosmetics in most countries for safety reasons.

Clinical Study of Genistein

The safety of genistein as a cosmetic ingredient was first confirmed by a repeated patch test in human subjects for the determination of the irritation and photo-irritation potential.

A cream containing 90 mg/kg genistein in a suitable carrier, as described by Schmid et al.,[1] was tested in a study with 20 women between age 55 and 64. The product was applied twice daily on the inside of the forearm. The same cream without genistein served as a control. Several skin parameters were measured after two and three months of product application, including skin thickness by ultrasonic measurement, skin firmness by cutometry, skin roughness by a digital micro-mirror device and skin hydration by corneometry. All skin parameters were determined eight to 12 h after the last product application.

After three months, skin thickness in subjects using the genistein cream had increased by 11%. In comparison, skin thickness in subjects using the control cream did not change significantly (Figure 4). This improvement was statistically significant by the Wilcoxon matched pairs signed rank test ($p < 0.05$). The application of genistein cream also improved other skin parameters slightly, such as elasticity, roughness and hydration.

Genistein therefore appears to have a specific effect on skin thickness that is not based on a swelling of the stratum corneum. No side effects were reported. Furthermore, the effect of genistein cream on skin thickness as determined in our study was better than the 7.7% increase after one year found in a study[11] with oral estrogens.

As skin thickness is principally determined by the concentration of collagen in the dermis, this study clearly suggests a positive effect of genistein on collagen content. To determine whether this result was based on genistein interacting with the estrogen receptors in the skin or with the receptor protein tyrosine kinase on skin cells or with both needs further study.

Conclusion

We conclude that isoflavone aglycones from soy, such as genistein, are very interesting molecules for cosmetic products. We therefore are studying new applications of our genistein preparation such as the treatment of UV-stressed skin and adipose tissue in cellulite.

—**Daniel Schmid and Fred Zülli,** Mibelle AG Cosmetics, Buchs, Switzerland

References

1. D Schmid, R Muggli and F Zülli, Dermatological application of soy isoflavones to prevent skin aging in postmenopausal women, *Cosmetics and Toiletries Manufacture Worldwide* (Aston Publishing Group) 1 146-151 (2001)
2. L Hayflick, The limited in vitro lifetime of human diploid cell strains, *Exp Cell Res* 37 614-636 (1965)
3. YS Wan, ZQ Wang, J Voorhees and G Fisher, EGF receptor crosstalks with cytokine receptors leading to the activation of c-Jun kinase in response to UV irradiation in human keratinocytes, *Cell Signal* 13 139-144 (2001)
4. P Affinito, S Palomba, C Sorrentino, C Di Carlo, G Bifulco, MP Arienzo and C Nappi, Effects of postmenopausal hypoestrogenism on skin collagen, *Maturitas* 33 239-247 (1999)
5. JB Schmidt, M Binder, G Demschik, C Bieglmayer and A Reiner, Treatment of skin aging with topical estrogens, *Int J Dermatol* 35 669-674 (1996)
6. Z-M Shao, J Wu, Z-Z Shen and SH Barsky, Genistein exerts multiple suppressive effects on human breast carcinoma cells, *Cancer Res* 58 4851-4857 (1998)
7. MH Kim, P Albertsson, Y Xue, U Nannmark, RP Kitson and RH Goldfarb, Expression of neutrophil collagenase (MMP-8) in Jurkat T leukemia cells and its role in invasion, *Anticancer Res* 21 45-50 (2001)
8. HK Yoon, K Chen, DJ Baylink and KHW Lau, Differential effects of two protein tyrosine kinase inhibitors, tyrphostin and genistein, on human bone cell proliferation as compared with differentiation, *Calcif Tissue Int* 63 243-249 (1998)
9. L Ravanti, J Heino, C Lopez-Otin and V-M Kähäri, Induction of collagenase-3 (MMP-13) expression in human skin fibroblasts by three-dimensional collagen is mediated by p38 mitogen-activated protein kinase, *J Biol Chem* 274 2446-2455 (1999)
10. Y Wang, Y E X Zhang, M Lebwohl, V DeLeo and H Wei, Inhibition of ultraviolet B (UVB)-induced

c-fos and *c-jun* expression *in vivo* by a tyrosine kinase inhibitor genistein, *Carcinogenesis* 19 649-654 (1998)
11. R Maheux, F Naud, M Rioux, R Grenier, A Lemay, J Guy and M Langevin, A randomized, double-blind, placebo-controlled study on the effect of conjugated estrogens on skin thickness, *Am J Obstet Gynecol* 170 642-649 (1994)
12. G B Maroulis, Alternatives to estrogen replacement therapy, *Ann N Y Acad Sci* 900 413-415 (2000)

Whitening Complex with *Waltheria indica* Extract and Ferulic Acid

Keywords: melanin, Waltheria indica, ferulic acid, skin lightening

Complex with Waltheria indica *extract, ferulic acid and certian other ingredients show skin lightening efficacy*

Skin color and hair color are determined by melanins, which are black polymeric pigments. An abnormal increase in the amount of melanin in the epidermis causes hyperpigmentation such as cloasma and freckles. Melanin is synthesized by specialized cells, the melanocytes, which are located in the basal layer of the epidermis (Figure 1).

Stored in melanosomes (granules in the melanocytes), the melanins are distributed to keratinocytes surrounding the melanocytes. The enzymes involved in the melanin synthesis are tyrosinase, TRP 1 (tyrosinase related protein 1) and TRP 2 (tyrosinase related protein 2).

Tyrosinase is found mainly in the melanosomes and is regarded as the key enzyme in melanin synthesis. The enzyme belongs to the mono-oxygenases and catalyzes the oxidation of tyrosine into DOPA (L-3,4-dihydro-phenylalanin) and further into DOPA-quinone. DOPA-quinone is either transformed to reddish pheomelanin or brown black eumelanin. Cu^{++} is one of the catalysts in this reaction.

During exaggerated sun exposure or aging, skin pigmentation is disturbed and skin spots may appear on the backs of the hands and other uncovered areas (Figure 2). The more or less dark coloring of the skin and excessive pigmentation or spots are sometimes considered to be aesthetic deficiencies. Therefore, there is a high interest in depigmenting agents.

Tyrosinase inhibition plays a key role in the inhibition of melanin synthesis and thus tyrosinase inhibitors are widely used in whitening cosmetic products. Arbutin (hydroquinone-beta-D-glucopyranoside), kojic acid, vitamin C and vitamin C derivatives are widely known as tyrosinase inhibitors. Due to their low toxicity to melanocytes, these agents are used in cosmetics. Hydroquinone has a strong effect but due to its side effects, such as skin irritation, it is prohibited in the U.S. and Europe.

A Whitening Complex

To maximize tyrosinase inhibition and reduce hyperpigmentation, we developed,

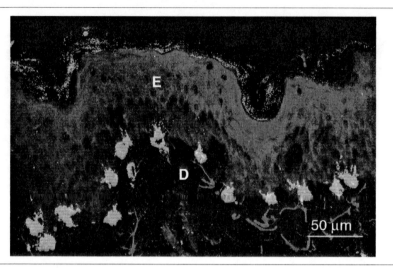

Figure 1. Microphotograph of epidermis (E) and dermis (D) showing general features of melanins (brown)

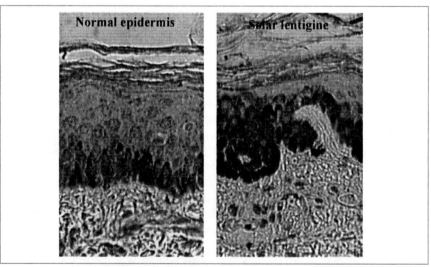

Figure 2. Normal epidermis and solar lentigine

in the year 2000, a complex of compounds that act on the tyrosinase activity synergistically in three ways:

- By direct inhibition of tyrosinase
- By indirect inhibition of tyrosinase via Cu^{++}-chelating compounds
- By mild exfoliation to enhance cell renewal and to enhance the elimination of melanin residues

Figure 3. Action mode of the synergistic complex

The INCI name for this complex[a] is mannitol (and) sodium gluconate (and) citric acid (and) sodium citrate (and) *Waltheria indica* leaf extract (and) dextrin (and) ferulic acid.

The direct inhibition of tyrosinase is due to the action of *Waltheria indica* extract and ferulic acid. *Waltheria indica* is a flowering plant (Figure 3) that belongs to the family Sterculiaceae. Among the main components identified in the *Waltheria indica* extract by thin layer chromatography are flavonoids and catechic tanins, along with phenolic acid and its derivatives.

Gluconic acid supports the inhibition by chelating Cu^{++}. Citric acid promotes desquamation of the superficial horny layers and the melanin residues contained therein (Figure 4).

Tests and Results

In tubo and in vitro studies were conducted to determine the inhibition of tyrosinase. A clinical study was also performed.

In tubo tyrosinase inhibition: In tubo studies are in vitro studies using chemical compounds in a chemical reaction that is measurable. The term is meant to distinguish them from in vitro studies that use cell lines.

Dose-dependent in tubo tyrosinase inhibition by the synergistic complex containing *Waltheria indica* extract and ferulic acid was proven in the presence of different concentrations of the complex. The complex has an IC_{50} (the concentration at which the activity of tyrosinase is decreased by 50%) of 2.7% (w/v) (Figure 5).

In vitro inhibition of melanin synthesis in B-16 melanocytes: A dose-dependent inhibition of melanogenesis was observed when B-16 melanocytes in culture were incubated in the presence of 0.2%, 0.5% and 1% of the complex containing *Waltheria indica* extract and ferulic acid. The control was cultivation of B-16 melanocytes in the absence of the synergistic complex. Homogenization was accomplished with NaOH. The melanin was measured after three days by photometry at 405 nm.

---

[a] Dermawhite NF is a trade name owned by Laboratoires Serobiologique, a division of Cognis France, in Nancy, France.

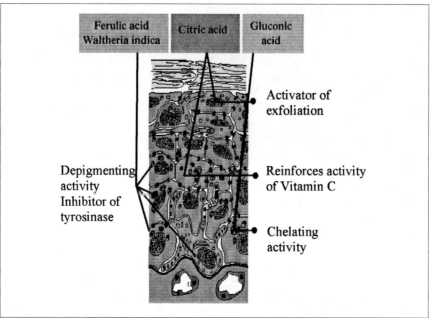

Figure 4. *Waltheria indica* is indigenous to Hawaii and can also be found in some regions of India and Africa. The whole plant including the roots is used for treating cough, fever, infertility, ophthalmia and diarrhea depending on the region of origin.

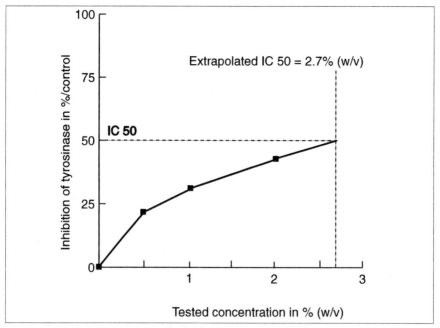

Figure 5. Inhibition of tyrosinase in tubo by the synergistic complex

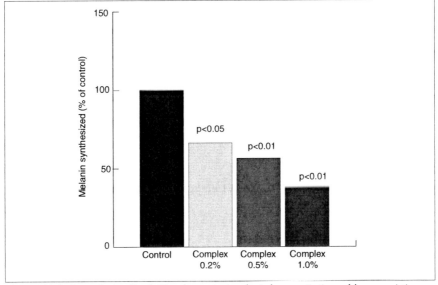

Figure 6. Melanin synthesized by B-16 melanocytes in selected concentrations of the synergistic complex at three days, shown as a percentage of the control. Statistics: Average ± SEM; three tests in triplicate; p values from Student's t test

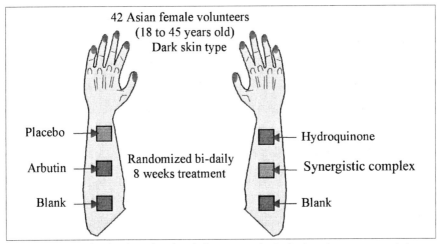

Figure 7. Protocol for the clinical trial with the active (arbutin, hydroquinone or synergistic complex) at 2% in each case

In the presence of 0.2% of the synergistic complex, the melanin synthesis decreased to 71% of the control. In the presence of 0.5% of the synergistic complex, the melanin synthesis was reduced to 53% of the control (B-16 melanocytes being cultivated only with medium). One percent of the synergistic complex decreased the melanin synthesis to 36% melanin compared to the control (Figure 6).

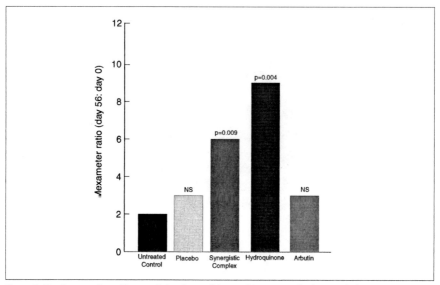

Figure 8. Depigmentation efficacy of placebo and three whiteners applied twice daily at 2% concentration on forearms of 42 humans for 56 days, shown as a ratio of day 56 Mexameter values to day 0 Mexameter values. Statistics: TSEM; p values from Student's t test, NS = not significant

Clinical trial: To evaluate the depigmenting efficacy of the complex, an in vivo, double blind randomized clinical trial (Figure 7) was conducted at the Philippine Dermatological Research and Testing Foundation. Fifty female volunteers were treated on the forearms twice daily with a cream containing 2% of the complex. Another area on the forearms was treated with a placebo. A third area was treated with a cream containing 2% arbutin or 2% hydroquinone as a reference. A fourth area was marked, but left untreated (control). After 56 days, the melanin index was measured[b] on the 42 volunteers remaining in the study.

Figure 8 shows the depigmentation results as a ratio of the melanin index on day 56 to the melanin index on day zero.

The synergistic complex containing *Waltheria indica* extract and ferulic acid was proven to be less efficient than hydroquinone (3.8 and 4.3 times as efficient as the untreated control, respectively) but more efficient than arbutin (1.65 times as efficient as the untreated control).

Conclusion

Two in vitro tests (the inhibition of tyrosinase in tubo and inhibition of melanin synthesis in B-16 melanocytes) showed that the synergistic complex was effective. Furthermore the results of the clinical test showed that the synergistic complex has good skin lightening efficacy.

[b] Mexameter, Courage + Khazaka electronic GmbH, Cologne, Germany

—**Liye Maeyama,** *Cognis Japan, Care Chemicals Asia Pacific, Tokyo, Japan*

A Botanical Anti-Sagging / Firming Blend

Keywords: botanical blend, skin firmness, skin sagging, collagen growth, collagen degradation

A proprietary blend of herbal extracts is designed to assist in the alleviation of the visible signs of aging.

This chapter reports research to develop a synergistic blend of natural plant extracts that can assist in restoring the "imbalanced" skin of an aged individual to a state of equilibrium.

A proprietary blend of enriched extracts of *Centella asiatica* (gotu kola), *Rosmarinus officinalis* (rosemary) and *Echinacea angustafolia* (echinacea) has been clinically evaluated and shown to exhibit unique cosmetic properties that result in a decrease in sagging of skin, reduction of fine lines and wrinkles and an increase in firmness of skin. In balancing both the rates of growth and degradation in aged skin, it stands to reason that the visible signs of aging could be reduced and equilibrium restored.

Balancing Growth and Degradation

The appearance of human skin is potentially influenced by the balance or equilibrium between two important actions – the rate of growth versus the rate of degradation. At birth, the rate of synthesis exceeds the rate of degradation, resulting in growth and the youthful appearance of skin. In aged individuals, the reverse is true – the rate of degradation exceeds the rate of synthesis, resulting in visible signs of aging. During the early stages of maturity, the rate of growth is essentially balanced by the rate of degradation. At this point, the body approaches a state of equilibrium and the visible signs of aging are difficult to detect.

The purpose of a cosmetic treatment can be twofold. In the early stages of maturity, a cosmetic treatment can be preventive in that it assists in maintaining the balance between growth and degradation, resulting in an extended period of equilibrium and healthy, youthful skin. In aged individuals, a cosmetic treatment can be restorative in that it assists in returning the skin to a state of equilibrium where the balance between growth and degradation is restored.

Firmness: Collagen is the structural component of the skin responsible for tensile strength, which is the ability to resist pulling forces. Studies have shown that, as we age, the rate of collagen degradation exceeds the rate of collagen production.[1]

Consequently, the total collagen content of the skin decreases, leading to a lack of tensile strength in the skin and a potential decrease in firmness.

As collagen undergoes damage over time–a process known as bundling–protection of the collagen content of the skin may result in further problems. It stands to reason that stimulation of collagen production is necessary in aged individuals in order to maintain a balance between the rates of collagen synthesis and degradation.

Published articles have demonstrated that enriched extracts of gotu kola have the ability to stimulate fibroblasts and increase collagen synthesis (in vitro).[2] Although extracts of gotu kola have been available for some time, few have been demonstrated as extracts having significant cosmetic effects.

Sagging: Elastin is the structural component of the skin responsible for resiliency, the ability of an object to return to its original shape after stretching forces have been applied.

Studies have shown that the quality of the elastin content of the skin deteriorates as we age. This is due in part to the long-term effect of inflammatory responses and the subsequent release of degradative enzymes including Human Leukocyte Elastase (HLE).[3] This enzyme is moderately specific for the proteolytic destruction of elastin. Thus, it can be responsible for a lack of resiliency and elasticity of the skin, potentially leading to "skin sagging." Therefore, it stands to reason that the protection of elastin from the degradative effects of HLE is necessary to protect the structural resiliency of skin.

Studies have documented that enriched extracts of rosemary have the ability to effectively inhibit HLE.[4] Although extracts of rosemary have been available for some time, few have demonstrated significant cosmetic effects.

Fine lines and wrinkles: Glycosaminoglycans/hyaluronic acid are the structural components of the skin responsible for hydration. These large molecular weight polysaccharides exhibit an immense capacity for water binding, and their increased presence in the skin leads to increased internal hydration.

As we age, we experience a reduction of the content of these structures in the skin – a reduction of the water-binding capacity of the skin. This may lead to the appearance of fine lines and wrinkles that are otherwise concealed by healthy, hydrated, "plump" skin.

Published observations have documented the fact that enriched extracts of echinacea have the ability to inhibit hyaluronidase, an enzyme implicated in the degradation of those polysaccharide structures and the potential loss of hydration.[5] Although extracts of echinacea have been available for some time, few have demonstrated significant cosmetic effects.

Clinical Studies

The reported antiaging properties of these three botanical extracts suggested their use in a synergistic blend aimed at restoring the "imbalanced" skin of an aged individual to a state of equilibrium. In our laboratories, we developed a proprietary blend[a] of enriched extracts of gotu kola, rosemary and echinacea. In this article we will use the term GRE to refer to that proprietary blend.

---

[a]Actifirm Ultra is a registered trademark of Active Organics, Inc., Lewisville, Texas, USA

In clinical studies, the ability of the GRE blend was demonstrated to significantly modify the cosmetic properties of the skin. In vitro results previously suggested that such an enriched blend of extracts could in fact influence the metabolism of the skin in a synergistic and beneficial manner. Thus, in vivo tests were conducted to determine if a formula containing 5% addition of the GRE blend could assist in the reduction of the visible signs of aging. Such visible signs tested include firmness, fine lines and wrinkles and skin sagging.

Volunteers were chosen–25 female subjects (age 35+) with dry skin and exhibiting some of the visible signs of aging (wrinkles, lack of firmness and sagging)–to participate and complete an eight-week product efficacy in-use study. Subsequent to a "drying-out" stage, subjects were to apply each of the two treatment products, a control (containing no botanical blend) and a test formula (containing 5% GRE blend) to one side of their face after cleansing, twice a day (morning and evening). They were then asked to visit the test center at weeks zero, four and eight for "skin firmness" and "skin sagging" evaluations by trained technicians.

In a separate study, subjects were to apply the formulas as stated above in order to determine the effects on the appearance of fine lines and wrinkles.

Measuring firmness: Ballistrometry was used to measure skin firmness – mainly epidermal (superficial) firmness – according to the methods developed by Hargens.[6] Although a number of different analyses have been used to determine skin firmness, we chose to analyze the value of the height of first rebound peak (H1) to the second rebound peak (H2). Previous results have determined this value changes as one examines subjects of greater age. In fact, decreases of up to 40% in the H1/H2 values have been observed with a corresponding increase in the age of the subject tested (20-30 vs. 50-60 years old).

Measuring skin sagging: As skin ages, the loss of integrity of collagen, elastin and other connective tissue results in a change in the biomechanical properties of the skin. A variety of techniques can be used to assess such changes including indentometry, cutometry or ballistometry. Each of these methods assesses properties such as firmness and elasticity. While these are useful methods and can document product effect, the one limitation of such techniques is that they are not consumer-relatable and cannot be directly correlated with a visual effect on the skin.

In this study, we have utilized a method to assess changes in skin properties related to firmness by measuring the sagging or deformation of facial skin due to different forces, but mainly the force of gravity which is responsible for sagging observed under real life conditions. In our method, subjects are first placed in a reclining position, with their face directed towards the ceiling. One or more marks are placed on the cheek and lateral chin area, and the distance is precisely measured, using a modified Vernier caliper, from a fixed feature on the face such as the center of the eye when focused directly forward. Subjects are then either inclined into a normal upright position, or further declinated so their body axis is –45° in orientation with respect to a horizontal body axis. The change in distance from the fixed reference point is remeasured.

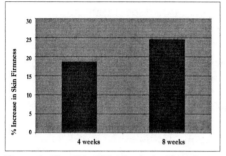

Figure 1. The 5% GRE blend significantly increased skin firmness.

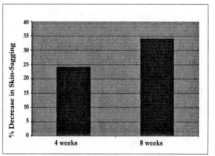

Figure 2. The 5% GRE blend significantly decreased skin sagging.

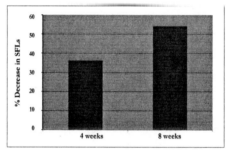

Figure 3. The 5% GRE blend significantly decreased SFLs (superficial facial lines).

From these measurements, a sagging quotient (SQ) for each point marked on the face is determined as follows:

$$SQ = (\Delta d1)^2 + (\Delta d2)^2$$

where d1 is the distance of the marked point measured at rest and then remeasured when subject is upright, and d2 is the distance from the marked point measured at rest and then remeasured in the declinated position.

To help visualize, the marked point is usually a point defined in the center of the cheek. Distance is measured to the center of the eye in the reclined position, a position in which gravity is not pulling the mark away or towards the eye. When the subject is re-positioned upright, the marked point moves away from the eye. The more the skin sags, the greater the point moves. The increase in distance the point moves is defined as the change in d1 (distance at rest minus distance inclined, i.e. $\Delta d1$).

Next the subject is declinated, which means the head is placed lower than the rest of the body. In this case the marked point will move towards the eye. The poorer the skin condition, the more the skin sags and the closer the point moves towards the eye. The distance to the eye is defined as d2. The change in d2 is the distance at rest minus distance inclined. Greater changes in d1 and d2 upon change in subject inclination result in greater changes in SQ.

The SQ values obtained before treatment are compared with those obtained after treatment to determine the significance of product effects. Using this analysis, data on facial skin from various age groups of subjects show dramatic increases in the SQ values with corresponding increases in age.

Measuring superficial facial lines and wrinkles: In a separate study, researchers clinically assessed the visible appearance of lines and wrinkles via the method of Packman and Gans.[7]

Results on firmness: Figure 1 presents the results of superficial skin firmness measurements on 25 subjects treated with the test product described above. With 5% GRE blend, significant increases in firmness were observed. The GRE-blend-containing formula used in the study exhibited levels of improvement as follows: at four weeks, 18.5% ($p< 0.0001$); at eight weeks, 24.6% ($p< 0.0001$).

Results on sagging: Figure 2 presents the results of sagging quotient (SQ) measurements on 25 subjects treated with the test product described above. With 5% GRE blend, significant decreases in skin sagging were observed. The formula containing the botanical blend used in the study exhibited levels of improvement as follows: at four weeks, 24% ($p< 0.0001$); at eight weeks, 34% ($p<0.0001$).

Results on superficial facial lines and wrinkles: Figure 3 presents the results of fine lines and wrinkle evaluation on 10 subjects treated with the test product described above. With 5% GRE blend, significant decreases in SFLs (superficial facial lines) were observed at four weeks (36%) and eight weeks (> 50%) of use. Results are normalized against controls.

Conclusions

In aged skin, the concentration and quality of those components that provide structural integrity to the skin is in decline. Restoring the balance between growth and decline (synthesis and degradation) reduces the appearance of the visible signs of aging, once again making the proverbial glass appear "half-full."

The purpose of the cosmetic treatment of aged skin is to assist in the restoration of that balance between growth and decline. The aim should be to reach a constant state of equilibrium, which would maintain a healthy and attractive appearance to aged skin.

The clinical results presented herein show that the consistent use of synergistic blends of enriched herbal extracts can assist in the decrease of the appearance of the visual signs of aging. In essence, cosmetic treatments such as the described GRE blend may assist in the restoration of the balance/equilibrium of skin.

—Glen Gillis, Scott J. Norton and Michael Bishop,
Active Organics Inc., Lewisville, Texas

References

1. CL Phillips, SB Combs and SR Pinnell, *J Invest Dermatol* 103(2) 228-232 (1994)
2. FX Maquart, G Bellon, P Gillery, Y Wegrowskiand and JP Borel, *Connect Tissue Res* 24(2) 107120 (1990)

3. HM Morrison, HG Welgus, CA Owen, RA Stockley and EJ Campbell, *Biochim Biophys Acta* 1430(2) 179-190 (Mar 19, 1999)
4. QL Ying, AR Rinehart, SR Simon and JC Cheronis, *Biochem* J 277 (Pt 2) 521-526 (Jul 15) 1991
5. RM Facino, M Carini, G Aldini, C Marinello, Arlandini, L Franzoi, M Colombo, P Pietta and P Mauri, *Farmaco* 48(10) 1447-1461 (1993)
6. CW Hargens, in *Non-Invasive Methods and the Skin*, J Serup and GBE Jemee, eds, Boca Raton, Florida: CRC Press (1995) 359-364
7. E Packman and EH Gans, Topical moisturizers: Quantification of the effect on superficial facial lines, *Soc Cos Chem* 29 79-90 (1992)

Anti-Wrinkle Activity of Hydrolyzed Ginseng Saponins

Keywords: ginseng, wrinkles, collagen

Hydrolyzed ginseng saponinsare evaluated for skin penetration and in vivo activity

Ginseng (the roots of *Panax ginseng C. A. Meyer*) is a well-known oriental crude drug and has been used as a tonic for enhancing body strength, recovering physical balance and stimulating metabolic function. A number of saponins called ginsenosides have been isolated,[1] and are regarded as the main biological active principles of ginseng. Ginseng saponins are classified into three groups according to their chemical structures: protopanaxadiol, protopanaxatriol and oleanolic acid. The majority of saponins existing in the vegetable kingdom are of the oleanane family, while ginseng saponin is known to be of the triterpenoid saponin, a family of dammarane, which is rarely seen in other herbs.

Recently, many studies on absorption, distribution and excretion of ginsenosides (the biologically active ingredients of ginseng) after oral administration have been reported.[2-4] Studies conducted especially on the metabolites of ginseng saponins show various pharmacological activities in vivo as well as in vitro—the inhibition of tumor growth and invasion, the reversal of multidrug resistance in tumors and bacteria, and the prevention of tumor metastasis[5-8]—which were undertaken by some researchers using acids, enzymes and intestinal bacteria.

Han et al. examined the decomposition products of ginsenosides Rb1, Re, and Rg1 using 0.1 N HCl,[9] while Kanaoka et al. investigated the metabolism of ginsenosides Rb1 and Rg1 by human intestinal flora.[3] According to these studies, actual physiological effects in the human body arise from the metabolites of ginsenosides as well as compound K (20-O-β-D-glucopyranosyl-20(S)-protopanaxadiol)[3,10] produced by gastric acid and intestinal bacteria etc.

However, thus far in cosmetic formulations, ginsenosides have only been used as an extract from ginseng and contain other undefined components, and the biological activity of each ginsenoside in the skin was not completely understood.

Therefore, in order to utilize ginsenosides on the skin as cosmetic active ingredients, we have to consider, firstly, the measurement of penetration rate through the skin surface and, secondly, evaluation of in vivo activities. To retain a certain degree of the actual effect in the skin, the active ingredients have to penetrate through the stratum corneum, the outermost layer of the skin structure. But human skin has several permeability-disturbing factors.

The major barrier function of the skin is generally believed to be the result of the organized distribution of intercellular lamellar lipids within the stratum corneum.[11] These lipids mainly are comprised of ceramides, cholesterol and cholesterol derivatives. The stratum corneum, therefore, presents a considerable lipophilic barrier to the permeation of chemicals. As a result, if the active ingredients have lipophilic characteristics, penetration into the skin will be greatly enhanced.

Another factor that determines the penetration of a compound through the stratum corneum is its molecular size.[12] The smaller the molecule, the easier the penetration.

Ginsenosides have hydrophilic properties and larger molecule sizes due to their glycosidic linkage composed of several glucose, arabinose, rhamnose units, etc. to their corresponding aglycone base units. So the passage of ginsenoside molecules through the stratum corneum layer in the skin is not feasible.

Therefore, in the present study, we prepared hydrolyzed ginseng saponins (HGS) containing high concentrations of compound K and ginsenoside F1 and the metabolites of ginsenosides, which overcome the hurdle of skin penetration and retain the major physiological effects of ginseng. We examined the contribution to the total collagen synthesis, MMP-1 inhibition in vitro, and collagen synthesis in vivo in order to estimate its anti-wrinkle activity. Moreover, we prepared a nano-emulsion containing HGS and measured its skin penetration and collagen synthesis in vivo.

Preparation of HGS

We purchased Rb1, Rd, Rc of protopanaxadiol ginsenosides and Rg1, Re of protopanaxatriol ginsenosides. After these ginsenosides were reacted with several crude enzymes, we identified components in the reaction mixture. As a result, compound K and ginsenoside F1 were produced by crude naringinase, crude pectinase, crude celluase, etc.[13] To obtain pure compound K and ginsenoside F1,[10] the reaction products were separated through silicagel column chromatography (chroloform : methanol = 9:1 → chroloform : methanol = 4:1) and their identification was performed by a comparison of their FAB-MS, H-, and C13-NMR spectra with those of standard compounds. Their purity was assessed by high performance liquid chromatography (HPLC). The structures of these compounds are shown in Figure 1 and the thin layer chromatography (TLC) chromatogram is shown in Figure 2.

Primarily, ginseng extract containing about 30% saponin was prepared from red ginseng, white ginseng, tiny ginseng roots and ginseng leaves grown in Korea. HGS, containing more than 30% compound K and more than 7% ginsenoside F1, was prepared by the enzymatic hydrolysis of ginseng extracts with selected crude naringinase and crude pectinase etc. The HPLC chromatogram of HGS is shown in Figure 3.

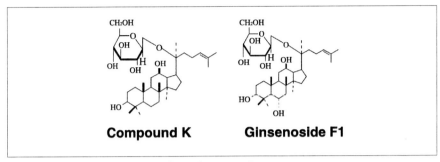

Figure 1. The structures of compound K and ginsenoside F1

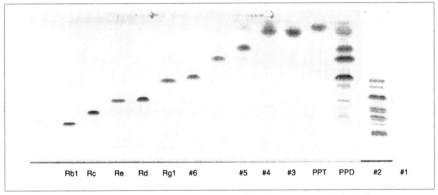

Figure 2. TLC chromatogram of ginsenosides produced by the enzymatic hydrolysis
PPD: Protopanaxadiol, PPT: Protopanaxatriol, #1: total ginseng saponins before enzymatic hydrolysis, #2: HGS, #4: Compound K, #5: Ginsenoside F1

Anti-Wrinkle Activity on Skin

Total collagen synthesis in vitro: Incorporation of radioactive precursor into extracellular matrix proteins was measured after labeling confluent cells in serum-free medium for 24 h with 5[^{3}H]proline (5μCi, 28Ci/mmol), as described previously.[14] Incorporation into collagen was determined by digesting proteins with purified clostridium histolyticum collagenase according to the method of Peterkofsky et al.[15] Results are shown as combined values for cell plus medium fractions.

The total collagen synthesis in normal human skin fibroblast treated with 0.1 ppm and 1 ppm concentrations of HGS was measured. In both cases, HGS increased the total collagen synthesis in comparison with the control (untreated normal human skin fibroblast) and the result is shown in Figure 4.

Expression of MMP-1: Expression of matrix metalloproteinase-1 (MMP-1) was measured using an enzyme-linked immunosorbent assay kit[a].[16] We treated 0.1 ppm, 1 ppm HGS and 1 mM retinoic acid (RA) in normal human skin fibroblast and measured the expression of MMP-1.

---

[a] ELISA (enzyme-linked immunosorbent assay) kit purchased from Amersham Bioscience, Seoul, Korea.

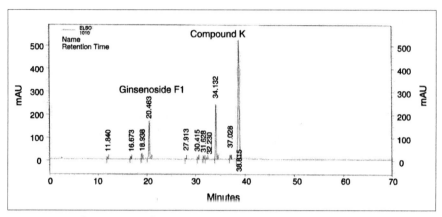

Figure 3. HPLC chromatogram of HGS

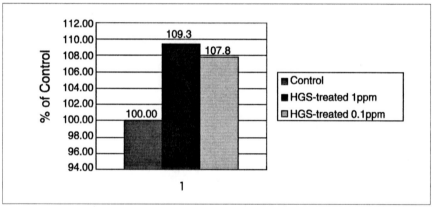

Figure 4. Collagen synthesis

As shown in Figure 5, 0.1 ppm HGS decreased the expression of MMP-1 more than 1 µM retinoic acid.

In vivo efficacy: Hairless mice, 50 weeks old, photo-aged by UV radiation for 30 weeks, were prepared. We performed patch tests for seven days (patch for three days, rest for 24 h, twice repeat) on the hairless mice with HGS 0.1% and HGS 1%. After the patch was finished, collagen synthesis was estimated by immunostaining type-1 procollagen. The result is shown in Figure 6.

HGS-treated mice showed a significant increase in comparison with control. Moreover, the higher concentration of HGS increased more collagen.

Activity of Nano-Emulsion Containing HGS

To investigate the variation of the skin penetration rate according to the particle size and contact area, HGS was included within nanometer-sized fine emulsion particles.

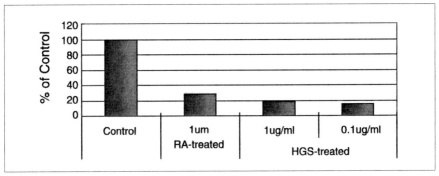

Figure 5. Expression of MMP-1

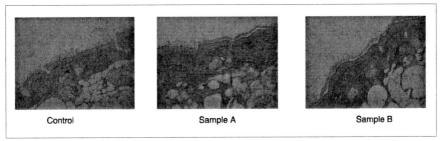

Figure 6. Collagen synthesis in a hairless mouse after patch testing with HGS 0.1% (A) and 1% (B) solutions

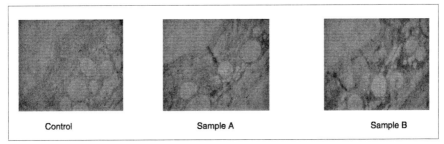

Figure 7. Collagen synthesis in a hairless mouse after patch testing with HGS 1% in propylene glycol and ethanol mixture (A) and HGS 1% in a nano-emulsion (B).

We used lecithin having excellent affinity on the skin as an emulsifier and high-pressure homogenizer to prepare the nano-emulsion. The particle size of this emulsion was from 300 nm to 50 nm, and the lecithin content was 2% ~ 3% (wt/wt). HGS was dissolved in 1% concentration in the propylene glycol and ethanol (7:3) mixture (Sample A), and the nano-emulsion was prepared to contain 1% of HGS (Sample B). The patch test was performed, and collagen synthesis was examined by immunostaining the type-1 procollagen produced from the mice.

As shown in Figure 7, collagen synthesis was increased more by sample B than by sample A. This result suggests that the skin penetration rate of HGS was highly increased by the nano-emulsion, and as the result, the collagen synthesis was also significantly increased.

Conclusion

HGS composed of compound K and ginsenoside F1, the principal metabolites of ginsenoside in the human body, was prepared by enzymatic hydrolysis to maximize the biological activity of ginsenosides - the main active ingredients of ginseng - in the skin. HGS highly increased collagen synthesis and remarkably decreased expression of MMP-1. Additionally, the formation of a nano-emulsion containing HGS significantly increased the collagen synthesis in the dermis.

Therefore, we suggest that HGS will be an effective ingredient for use in anti-aging formulations.

—M.H. Yeom, D.S. Sung, K.S. Woo, B.Y. Kang, D.H. Kim, I.S. Chang, H.H. Kang and O.S. Lee, *Pacific R&D Center, Yongin-si, Kyunggi-do, Korea*

References

1. S Sanada and J Shoji, *Chem Pharm Bull* 26 1694-1697 (1978)
2. Y Takino, *Yakugaku Zasshi* 114 550-564 (1994)
3. M Kanaoka, T Akao and K Kobayashi, *J Traditional Med* 11 241-245 (1994)
4. H Hasegawa, JH Sung and Y Benno, *Planta Medica* 63 436-440 (1997)
5. H Hasegawa, S Matsumiya, M Uchiyama, T Kurokawa, Y Inouye, R Kasai, S Ishibashi and K Yamasaki, *Planta Medica* 60 240-243 (1994)
6. H Hasegawa, S Matsumiya, M Uchiyama, JH Sung, Y Inouye, R Kasai, S Ishibashi and K Yamasaki, *Planta Medica* 61 409-413 (1995)
7. H Hasegawa, S Matsumiya, M Uchiyama, Y Inouye, R Kasai, S Ishibashi and K Yamasaki, *Phytother Res* 9 260-263 (1995)
8. C Wakabayashi, H Hasegawa, J Murata and I Saiki, *Oncology Res* 9 411-417 (1997)
9. BH Han, M Park, YN Han, IK Woo, U Sankawa, S Yahara and O Tanaka, *Planta Medica* 44 146-149 (1982)
10. M Karikura, T Miyase, H Tanizawa, T Taniyama and Y Takino, *Chem Pharm Bull* 39 2357-2361 (1991)
11. KR Brain, KA Walters, VJ James, WE Dressler, D Howes, CK Kelling, SJ Moloney and SD Gettings, *Fd Chem Toxic* 33 315-322 (1995)
12. JJ Hostynek and PS Magee, *Toxicology In Vitro* 11 377-384 (1997)
13. H Kohda and O Tanakak, *Yakugaku Zasshi* 95 246-249 (1975)
14. I Oyamada, EM Schalk, K Takeda, J Palka and B Peterkofsky, *Arch Biochem Biophys* 276 85-93 (1990)
15. B Peterkofsky, M Chojkier and J Bateman, Immunochemistry of the Extracellular Martix, Boca Raton, FL: CRC Press, (1982) 19-47
16. Y Hojo, U Ikeda, T Katsuki, O Mizuno, H Fujikawa, and K Shimada, *Atherosclerosis* 161 185-192 (2002)

Phytoestrogens: Applications of Soy Isoflavones in Skin Care

Keywords: estrogren, collagen, isoflavone (soy), hair growth

Soybean extract containing isoflavones shown to generate collagen, suppress the activity of sebaceous glands and inhibit hair growth in rodents

Estrogen, the female sex hormone that modulates feminization, begins to play its role in a woman's life during puberty (see sidebar). After puberty, estrogen is secreted from the ovary and is kept at a high level. Nevertheless when a woman reaches menopause, the secretion of estrogen decreases with the decline of ovary function. Insufficient estrogen affects a woman's physiological functions. A drastic change of skin condition with aging is common.

It is well known that estrogen stimulates the fibroblast to make collagen and hyaluronic acid,[1,2] and estrogen is thought to prevent the skin from physiological aging. In 1996, it was reported that male and female mice have different UV-sensitivity, and this difference depends on their estrogen levels.[3] This suggested that estrogen may have an ability to prevent UV damage.

Since 1999, several researchers have reported studies on rodents whose ovaries have been removed. These reports have demonstrated an interrelation between photoaging and the decline of estrogen. Estrogen reduction accelerates photoaging. It is associated with the formation of wrinkles, as the skin's elasticity is reduced and the 3D structure of elastin fibers in the dermis is disordered.[4] It is also associated with the decline of collagen[5] by activation of gelatinase.[6] Finally, estrogen affects secretion by the sebaceous glands[7] and prolongs the life span of some cells by the restoration of telomere.[8]

This knowledge of the relation between estrogen and skin aging suggests that insufficient estrogen at menopause not only induces physiological aging by reducing the biosynthesis of collagen and hyaluronic acid, but also accelerates photoaging by increasing the effects of UV damage. Therefore, estrogen is important for preventing skin aging and is useful in the cosmetic field.

Although estrogen is useful in cosmetics, it is a type of steroid hormone and should be handled carefully to avoid side effects. This is why estrogen is restricted for use in cosmetics in many countries, and why safe estrogen, suitable for cosmetics, has been in demand.

Estrogen: The Female Hormone

In general, female hormones are classified under two types: estrogen and progesterone. These hormones are essential for maintaining female functions. Progesterone mainly controls pregnancy while estrogen relates to feminization.

Higher animals including humans are considered to begin life with female characteristics. During the fetal period, androgen is necessary for the fetus to change to a male, however, estrogen is not necessary for the fetus to change to a female. Not until puberty does a female need estrogen.

Isoflavone: The Typical Phytoestrogen

Since initial reports in 1996,[9] blackberry lily (leopard flower) extract has attracted a great deal of attention as the first plant extract with estrogenic efficacy for the cosmetic field. Blackberry lily is a member of the Iridaceae family, and the active elements of the extract are isoflavones. The structure of isoflavone is similar to that of estrogen (Figure 1), therefore isoflavones can bind to the estrogen receptor and be effective as a mild estrogen.[10] A phytoestrogen is a plant constituent that has estrogenic efficacy. Today, isoflavones are considered a typical phytoestrogen.

Isoflavones are members of the flavonoid family, distributed mainly in Leguminosae and Iridaceae. It is mainly from soybeans that isoflavones can be obtained efficiently. The traditional cuisine of Japan contains significant amounts of soybean products, therefore, Japanese women experience menopausal effects that are milder than those of Western women.

The whitening properties of isoflavones were described as early as 1985.[11] Not until recently, however, have the isoflavones of soybeans and arrowroot[12] been described as antiaging actives. Here we report our experiments to evaluate the efficacy of soy isoflavones in the generation of collagen, the suppression of sebaceous gland activity and the suppression of hair growth.

Material and Methods

Preparing soybean extract: Whole soybeans (*Glycine max*) grown in Japan were dried and then extracted in boiling water. After the soybeans were removed, the water of the extract was evaporated with ethanol to obtain a dried material. Table 1 shows the variety of isoflavones in the soybean extract. The isoflavone content of the extract was approximately 50% (w/w).

Generating collagen: The efficacy of soybean extract on the generation of collagen was evaluated by a published method.[13] Collagen is generated by fibroblasts. Estrogen induces the generation of collagen. The newborn collagen called tropocollagen has few cross-links and is highly soluble in isotonic sodium chloride water. During aging, however, tropocollagen changes to labile collagen and then to stabile polymer collagen.

The aged collagen has a high quantity of cross-links from the Maillard reaction, making it either hard to solubilize or insoluble. According to this principle,

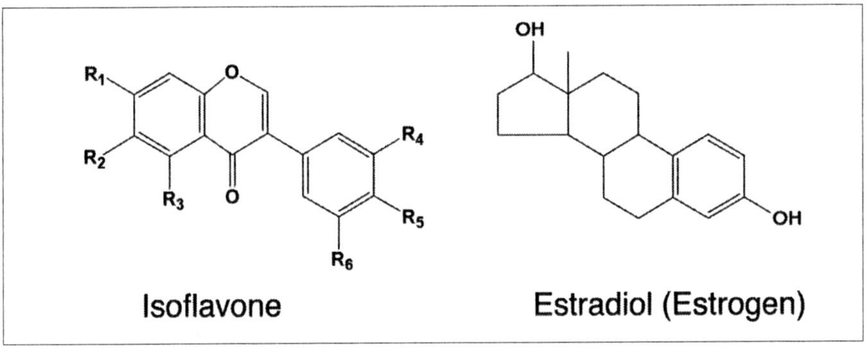

Figure 1. Structure of estrogen and various forms of isoflavone (For radicals, see Table 1.)

Table 1. Isoflavones of soybean extract, with radicals for Figure 1

Isoflavone	R1	R2	R3	R4	R5	R6
Daidzein	H	H	H	H	OH	H
Daidzin	O-Glu	H	H	H	OH	H
Genistin	O-Glu	H	OH	H	OH	H
Glycitein	H	CH_3O	H	H	OH	H
Glycitin	O-Glu	CH_3O	H	H	OH	H

the percentage of tropocollagen in total collagen was measured and used as an index of collagen generation.

A base complex was prepared containing 10% (w/w) hydrophilic petrolatum, 45% propylene glycol and 45% water. For the test sample, soybean extract was added and emulsified to be 1% (w/w) of the base complex. For the positive control, 0.5% (w/w) of an estrogen emulsion was added to the base complex. The estrogen contained 50% (w/w) of β-estradiol and 50% (w/w) of diethylstilbestrol. Estradiol is real hormone that has a steroid structure. Diethylstilbestrol does not have a steroid structure, but it has strong efficacy like that of isoflavones.

Before the experiment, a hairless rat (HWY/Slc rat, female, eight weeks, n=5) was operated on to remove the ovaries and provide a nearly estrogen-free animal as a menopausal disorder model. Fifteen rats prepared in this way were separated into three groups of five. One group was treated with the base complex. A second group received the soybean extract. A third group was treated with the estrogen positive control.

The test sample was administered on the back of the rat, 1 g per day, once a day for three weeks. Forty-eight hours after the last application, the rat's back skin was excised and the subcutaneous tissue was fully removed. The skin was cut closely, and 3 g of the skin was homogenized and sonicated with a small quantity of isotonic sodium chloride water in a water bath with ice. The homogenate was adjusted to be 5 mL with isotonic sodium chloride water, and extracted at 4°C for 24 h. Then the homogenate was centrifuged at 3000 rpm and 4°C for 20 min, removing the supernatant.

At this time, the supernatant contained tropocollagen. After an equal volume of Ringer's solution was added to the precipitate, the complex was sonicated and extracted at 65°C for 45 min. The supernatant prepared the same way contained labile collagen. After an equal volume of purified water was added to the precipitate, the complex was sonicated and extracted at 125°C for 60 min. The supernatant prepared in similar fashion contained stabile polymer collagen. After a little water was added to bring the total weight of the three extracts (tropocollagen, labile collagen and stabile polymer collagen) to 5 g, the content of hydroxyproline was measured with the amino acid analyzer. The hydroxyproline content was used as an indicator of the total collagen content.

Suppressing sebaceous gland activity: Sebaceous secretion depends on sex hormones. Androgen (the male sex hormone) enhances the secretion of sebum, and it is generally known that men secrete more sebum than women. Too much androgen leads to an excess of sebum, which causes acne. On the other hand, the effects of estrogen are opposite from those of androgen.[14]

We devised a procedure based on a previously published finding that measuring the size of the sebaceous gland is a simple method for evaluating the

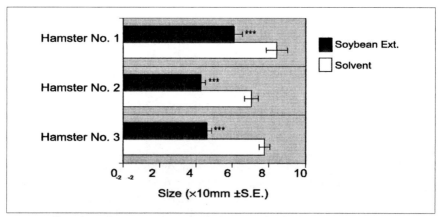

Figure 2. Effect of a soybean extract on size of sebaceous glands in hamsters (***p< 0.01)

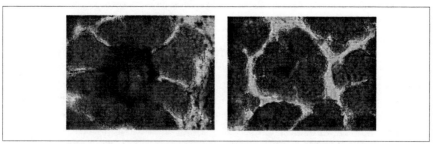

Figure 3. Photomicrograph of sebaceous glands of Syrian golden hamsters treated with a solvent without (left) and with (right) soybean extract

efficacy of the sebaceous gland.[14] Soybean extract was dissolved in 50% (w/w) ethanol to a level of 0.05% (w/w). Testosterone propionate as an androgen was dissolved in dimethyl sulfoxide (DMSO). In order to enlarge the sebaceous gland, 2 mg of testosterone propionate were injected hypodermically into a hamster (Syrian golden female, nine weeks old, n=3) on days one, three, eight and 10. The hamster's left ear was treated once a day, five times/week with 0.1 mL of soybean extract for two weeks. For comparison, the right ear received similar treatments with the solvent.

After the final administration, the ear was cut off at the paunch (6 mm diameter) and dipped in isotonic sodium chloride water. The ear was stored at 4°C for 18 h and then the cartilage was removed. The ear was dipped in 2N sodium bromide at 37°C for 2 h. The epidermis was then removed. The dermis, including the sebaceous gland, was stained with the chemical reagent called oil red O, and 10 pieces of the sebaceous gland were photo-micrographed. The area of each sebaceous gland was measured with a picture analyzer[a].

Suppressing growth of terminal hair: The operations of sex hormones on hair are complicated.[15] Furthermore, the effects of the sex hormones on hair depend on the location of the hair. Androgen, for example, promotes the growth of beards, chest hair, underarm hair, and the hair on the limbs. Nevertheless, it causes male-pattern baldness (*Androgenic alopecia*). On the other hand, estrogen suppresses the growth of hair. Moreover, it is believed to have an antagonistic effect on the causes of male-pattern baldness. In experiments on mice, estrogen suppressed the growth of back hair.

Soybean extract was dissolved in 50% (w/w) ethanol to a level of 0.5% (w/w). Before the experiment, the back hair of 10 mice (C3H mouse, male, eight weeks old, n=10) was shaved (about 8 mm^2) with an electric shaver. Then the mouse's back was treated with 0.2 mL of soybean extract in ethanol solution once per day for 25 days. Ten other mice were given similar treatments with the ethanol solvent containing no soybean extract. Hair regrowth measurements were taken starting at day 12. With a picture analyzer, we determined the ratio of the area of regenerated hair to the area of shaved hair.

Results

Generating collagen: In rats variously treated for three weeks with the base complex with or without either soybean extract or estrogen, we found that skin treated with soybean extract contained an average of 25% more tropocollagen than skin treated with the base complex. Skin treated with estrogen contained 46% more tropocollagen than skin treated with the base complex.

Reducing the size of the sebaceous gland: In three androgen-stimulated hamsters whose ears were treated for two weeks with a solvent either with or without soybean extract, the area of the sebaceous gland was 6.09, 4.25 and 4.59 x 10^{-2} mm^2 for the soybean-treated ears versus 8.44, 7.05 and 7.77 x 10^{-2} mm^2, respectively,

---

[a]Win Roof, Mitani Corporation, Japan

for the solvent-treated ears. Thus in every case the area of the sebaceous gland treated with soybean extract was significantly smaller than the area of the solvent-treated sebaceous gland. Therefore, soybean extract suppressed the expansion of the sebaceous gland induced by androgen. Figures 2 and 3 show the sizes of the sebaceous glands following treatment.

Suppressing the growth of terminal hair: Figure 4 shows the average percentage of shaved area that contains regenerated hair on the backs of mice treated with a solvent either with or without soybean extract. The regenerated area treated with soybean extract was significantly smaller than that of the solvent. Therefore, soybean extract strongly suppressed the regeneration of the hair and prolonged the telogen phase of hair growth.

Conclusion

Soybean extract used in these experiments contained a relatively high level of isoflavones. The efficacy of this extract is based upon its isoflavone content. Because of estrogenic efficacy, isoflavone is suitable for application in cosmetics for older women. Furthermore, isoflavone is useful for preventing skin aging and for improving aged skin.

Soybean extract increased the percentage of tropocollagen. The increase of tropocollagen indicates the activation of collagen synthesis. Moreover, the

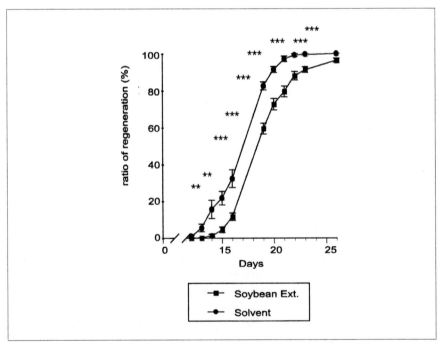

Figure 4. Effect of a solevant with and without soybean extract in suppressing the growth of hair in mice (***$p < 0.01$, **$p < 0.02$)

activation of collagen synthesis can supply the collagen lost by aging and keep the skin looking young. Collagen in the skin is generated by fibroblasts in the dermis. Estrogen stimulates those fibroblasts to produce collagen. It is believed that isoflavones have a similar effect on the fibroblasts. Therefore, we conclude that soybean extract can function like estrogen in stimulating collagen synthesis by fibroblasts.

Soybean extract also reduced the size of the sebaceous glands enlarged by androgen supplementation (Figure 2). This phenomenon apparently depends on the isoflavone suppressing androgen. If that is true, soybean extract may have potential for the treatment of acne. It is not obvious why isoflavone can inhibit the effect of androgen on the sebaceous gland. One can speculate that the mechanism is either the inhibition of 5-α reductase (an enzyme that aids in the conversion of androgen into a more powerful form of androgen) or the stimulation of estrogen receptors.

Soybean extract prolonged the telogen phase of terminal hair in mice, and inhibited the regeneration of hair (Figure 3). This efficacy is useful for after-shave cosmetics to inhibit the regeneration of hair in the beard, under the arms and on the limbs. In humans, the growth and distribution of hair is controlled by sex hormones, and the effect of these hormones varies from location to location on the body. In mice, the effect of sex hormones on hair is not obvious. Thus, it is difficult to compare the hair of humans with the hair of mice. Nevertheless, estrogen inhibits the growth of hair in mice. Indeed, human hair (the beard, underarm hair and the hair of limbs) behaves similarly to mouse hair. Therefore, we conclude that isoflavones, like estrogen, are able to control hair growth.

In summary, isoflavones have an estrogenic activity, which is demonstrated by the generation of collagen, the suppression of the sebaceous gland activity, and the suppression of terminal hair growth. These effects are significantly useful in the cosmetic field.

—**Norihisa Kawai,** *Ichimaru Pharcos Co. Ltd., Gifu, Japan*

References

1. M Uzuka et al, *Biochimica et Biophysica Acta* 673 387-393 (1981)
2. LM Wahl et al, *Endocrinology* 100 571-579 (1977)
3. K Hiramoto et al, *Environ Dermatol* 3 321-325 (1996)
4. K Tsukahara et al, *Phomed Phobiol* in press
5. H Tanaka et al, *J Traditional Medicine* 18 122 (supplement 2001)
6. N Ochiai et al, *Jpn J Dermatol* 111(3) 532 (2001)
7. The Chemical Daily (Sep 25, 2000)
8. Z Wang et al, *Cancer Research* 60 5376-5381 (2001)
9. Shiseido technical letter, http://www.shiseido.co.jp/e/e9608let/html/let00017.htm
10. H Kaneko, *J Jpn Oil Chemists' Society* 48(10) 89-95 (1999)
11. JP 1985-19885 B2, Ichimaru Pharcos (1982)
12. H Mori et al, Stimulating effect of Kakkon (Pueraria lobata) extracts on dermal matrix synthesis of human dermal fibroblasts, in *Summary of 46th Congress, Society Cosmetic Chemists Japan*, Shiseidou (Yokohama): Society Cosmetic Chemists Japan (2000) 25
13. M Yoshida et al, *J Soc Cosmet Chem Jpn* 30(3) 293-299 (1996)

14. Motoyosi et al, *J Soc Cosmet Chem Jpn* 22(3) 185-189 (1988)
15. S Itami, *J Soc Cosmet Chem Jpn* 33(3) 220-228 (1999)

Multifunctional Ingredients: The Novel Face of Natural

Keywords: UV, anti-inflammatory, curcuminoids

A natural extract from turmeric root offers functionality with wrinkling, hyperpigmentation and inflammation.

Multifunctional natural extracts present an attractive option to formulators seeking to develop innovative products for the antiaging personal care market. An all-in-one natural ingredient that has the potential to renew, recondition, soothe and revitalize epitomizes the new face of natural ingredients in the 21st century.

Formulators working under stringent constraints to strike a balance between natural ingredient compatibility, economics and functionality would find such an option immensely beneficial. The educated consumer no longer equates "more" with "better" and would rather like to see the scientific rationale behind ingredients in a formulation. The ingredients described here offer enhanced possibilities.

Antioxidants and Skin Aging

Aging of the skin is a cumulative effect influenced by factors such as environmental pollution, chemicals and atmospheric temperature fluctuations. Twenty-five percent of the lipids in the skin's surface are unsaturated and therefore more prone to attack by free radicals.[1] Ultraviolet radiation from sunlight penetrates the skin and accelerates damage by free radicals. With prolonged exposure, collagen and elastin fibers that maintain the elasticity and integrity of the skin are broken down by inherent enzymes, thus deteriorating the skin texture. Apparent deteriorative oxidative changes triggered by free radicals include wrinkling, hyperpigmentation (excessive tanning) and inflammation (sunburn).

Inflammation is the prime mover in the progression of skin aging. An inflamed site is, in reality, a micro-scar that matures into a wrinkle or blemish under the influence of environmental factors including ultraviolet rays and pollutants. Inflammation also affects skin pigmentation. Various types of inflammatory mediators (such as leukotrienes and prostaglandins, cytokines and growth factors) may influence melanin synthesis by affecting the proliferation and functioning of melanocytes. Protein kinase C, the enzyme that phosphorylates proteins, may also influence the growth and differentiation of melanocytes. Cytokines such as endothelins (also known as vasoconstrictive peptides) are also reported to accelerate melanogenesis.[2]

Tetrahydrocurcuminoids: Versatile "Bioprotectants"

Botanical extracts that support the health, texture and integrity of skin and hair are widely used in cosmetic formulations. Plant materials from which these extracts are prepared have a long history of traditional "cosmeceutical" use, although the term itself is of recent origin. In most cases, these cosmetic applications are adequately supported by efficacy data from scientific literature, as well as documented safety. Among the more popular functional natural ingredients, several antioxidants used in cosmetics are scientifically proven to offer additional benefits in supporting skin texture, appearance and tone.

One example is the turmeric extract, which is rich in curcuminoids. It is well known for its antioxidant properties, antimicrobial effects and beneficial effects on inflammation.[3] Turmeric has traditionally been used by South Asian women in skin care since ancient times. However, its yellow color is unattractive to contemporary formulators because the general consumer trend favors white or light colored products and at higher levels, yellow curcuminoids would stain the skin.

An innovative, colorless (white to very light tan) derivative addresses this drawback and offers effective protection against sun damage. The derivative form is the tetrahydrocurcuminoids. Their antioxidant action is of a comprehensive, so-called "bioprotectant[a]" nature,[4] efficiently preventing the formation of free radicals as well as quenching those that have already formed, thereby protecting the skin cells from damage by UV radiation and the resulting inflammation and injury. This in turn has far-reaching benefits on overall skin health and well being, rendering a healthy glow to the skin.

To better understand the importance and role of skin protection we need to realize that the skin is the largest organ in the body, covering an area of approximately 20 square feet in an adult. The skin offers efficient protection against the harsh environmental elements while simultaneously nourishing the body and maintaining overall health. This nourishment occurs in the skin during the conversion of 7-dehydrocholesterol to provitamin D3, a precursor of vitamin D in the body.

The skin actively participates in the body's homeostasis through abundant microcirculatory and sensory functions. Its appendages participate in the excretion of sweat and metabolic by-products, and may participate in the regulation of melatonin, a hormone that regulates circadian rhythms in the body. Healthy skin therefore reflects overall health and warrants optimal well-being. Thus, care and maintenance of the largest somatic organ literally translates to protecting the "protector" against the excesses of environmental exposure such as UVB rays, temperature and humidity fluctuations, as well as the ravages of time. Such factors trigger cumulative damage to the skin, resulting in aged appearance and operational decline in carrying out critical "good for health" physiological functions.

The obvious fact that healthy skin is needed in order to achieve overall health underscores the significance of the umbrella-like bioprotective action of tetrahydrocurcuminoids against damage to the skin and skin aging. One proprietary

---

[a] US Pat 5,861,415

standardized turmeric extract[b] is a colorless hydrogenated derivative of the natural yellow curcuminoids (curcumin, demethoxycurcumin, bisdemethoxycurcumin) from *Curcuma longa* (turmeric). In this article, we will refer to that extract as THC. Its principal component is tetrahydrocurcumin (INCI: Tetrahydrodiferuloylmethane). Additional components are tetrahydrodemethoxycurcumin (INCI: Tetrahydrodemethoxydiferuloylmethane), and tetrahydro-bisdemethoxycurcumin (INCI: Tetrahydrobisdemethoxydiferuloylmethane). These three tetrahydrocurcuminoids are reported to be the major metabolites of curcuminoids in vivo in experimental studies.[5,10]

Curcumin and its direct derivative tetrahydrocurcumin are the most potent individually. However, our studies show that it is the specific THC combination (the proprietary standardized percentages of each of the three tetrahydrocurcuminoids) that surpasses the biological potential of any given individual component.[4] This THC combination is valued as a topical antioxidant and anti-inflammatory agent, with superior free radical scavenging and lipid peroxidation inhibition efficacy as compared to vitamin E (Figure 1).[6]

In laboratory studies, THC was shown to effectively suppress inflammation and to protect epidermal cells from damage by ultraviolet B (UVB) radiation (Figure 2).[7]

In addition, THC was shown to be an effective inhibitor of tyrosinase, the enzyme that catalyzes melanogenesis. It was shown to be more effective than the well known natural skin lightening agent, kojic acid (Figure 3).[7]

Mechanism of Action

Like their parent curcuminoids, tetrahydrocurcuminoids are well recognized for their anti-inflammatory action. They inhibit a range of pro-inflammatory events, such as the production of free radicals and inflammatory mediators of the arachidonic acid

---

[b] C³ Complex (INCI: Curcumin (and) demethoxycurcumin (and) bisdemethoxycurcumin) is a registered trademark of Sabinsa Corporation, Piscataway, New Jersey.

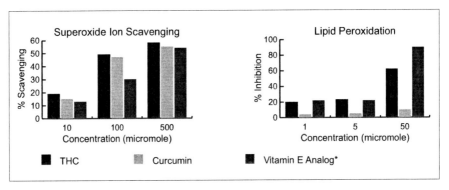

Figure 1. Inhibition of free radicals and lipid peroxidation by THC, compared to vitamin E analog *Trolox is a product of Hoffmann La Roche, Parsippany, New Jersey USA

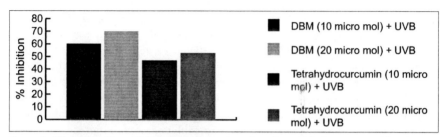

Figure 2. Effect of THC on UVB-induced formation of sunburn cells in mouse epidermis, compared to effect of dibenzoylmethane (DBM)

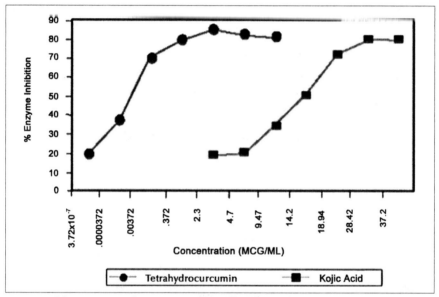

Figure 3. Inhibitory activity of tyrosinase, compared to kojic acid

cascade,[8] such as prostaglandins. These events are critical to the initiation, promotion and sustaining of inflammation, a process that eventually leads to destruction of the delicate fabric of any tissue and organ in the body.

These very same mechanisms at a lower intensity operate in the aging process as well and are responsible for gradual changes in the skin and subcutaneous tissue leading to visible signs of aging such as wrinkles, skin sagging and "liver spots." At the molecular level, the anti-inflammatory action of curcuminoids and tetrahydrocurcuminoids is due to inhibition of the enzyme cyclooxygenase type II (COX-2), which is responsible for the overproduction of mediators of inflammation.[10]

Overview of Documented Efficacy and Safety

The bioprotective effect of curcuminoids and tetrahydrocurcuminoids is further expressed in well-documented experiments related to anti-tumor activity. The

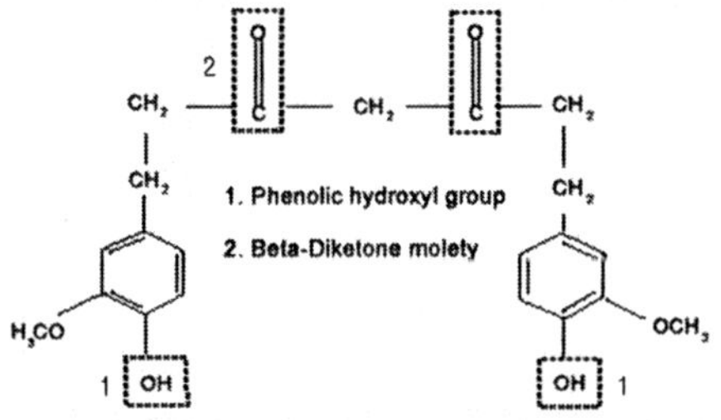

Figure 4. Relationship between molecular structure and biological activity[3] of tetrahydrocurcumins

addition of 0.5-10 µM of curcumin to an in vitro skin cell culture inhibited the activity of the enzyme ornithine decarboxylase, which has been linked to the promotion of tumor cell growth. These compounds have, in experimental conditions, been found to prevent a genetic change (mutation) in the cells that often initiates the aging process and contributes to cancer development.[9]

Traditionally, turmeric from which curcuminoids and tetrahydrocurcuminoids are derived has been used in the ancient system of Indian medicine known as Ayurveda, or the science of life, in a variety of topical products. Documented traditional uses include application to the ear in chronic otorrhea, as a cooling eyewash to relieve purulent ophthalmia, as a topical treatment for pemphigus (an allergic and inflammatory skin condition), in viral conditions such as shingles and chickenpox to facilitate the process of healing, and as a dressing applied to bruises, sprains, cuts and wounds. Recent research at the Sloan Kettering Cancer Research Center found that curcuminoids can be effectively applied in the treatment of precancerous lesions called leukoplakia (usually affecting the oral mucosa).[10]

In view of their long-standing tradition of use in Ayurvedic medicine, and the relatively recent research findings on their safety and efficacy, the multifunctional curcuminoids, and more specifically the tetrahydrocurcuminoids, deserve a special place among natural cosmeceuticals. The broad mechanisms of action observed by these compounds are attributed to the presence of the phenolic hydroxy and beta-diketone moieties (Figure 4).[11]

The UV-protectant, tyrosinase-inhibitory and antioxidant properties of tetrahydrocurcuminoids work together in a formulation to provide multifaceted benefits.

Another proprietary extract is tetrahydropiperine, which is a naturally derived extract[c] from black pepper fruits. A small quantity of this extract in the formulation enhances the transdermal permeation of tetrahydrocurcuminoids, as evidenced by laboratory studies (Figure 5).[12] The mechanism of enhancement is probably related to an increase in diffusion rates and enhanced cell membrane fluidity that occurs when the tetrahydropiperine-active combination is in contact with the stratum corneum.[13]

The inherent antioxidant action of tetrahydrocurcuminoids offers additional benefits to the formulator who needs to protect fat-based compositions from oxidation.

[c] Cosmoperine is a registered trademark of Sabinsa Corporation, Piscataway, New Jersey

Formula 1. Bioprotectant skin cream with tetrahydrocurcuminoids and tetrahydropiperine

A. Cetyl alcohol	10.0% wt
Cetostearyl alcohol	3.0
Glyceryl monostearate SE	4.0
Isopropyl myristate	3.0
Caprylic/capric triglycerides	2.0
Cetyl esters	3.0
Stearyl stearate	0.5
Cetyl palmitate	0.5
Myristyl pyristate	0.5
Mineral (*Paraffinum liquidum*) oil	2.0
B. Carbomer (Carbopol 940, BFGoodich)	0.1
Dimethiconol	1.5
Sodium methylparaben	0.25
Sodium propylparaben	0.1
Imidazolidinyl urea	0.15
Sodium benzoate	0.1
Tetrasodium EDTA	0.02
Glycerin	2.5
Water (*aqua*)	qs 100.0
C. Tetrahydrocurcuminoids (C³ Complex, Sabinsa)	0.5
Tetrahydropiperine (Cosmoperine, Sabinsa)	0.1

Procedure: In separate containers, heat A and B to 65-75°C. At this temperature, add C to A, mix thoroughly, and add to B under homogenization. Cool to fill temperature.

Tetrahydrocurcuminoids quench free radicals more efficiently than the commonly used synthetic antioxidant, butylated hydroxytoluene (BHT).[3]

From a safety point of view, the bioprotective role of THC is further enhanced by its very low toxicity (oral LD_{50} is 5000 mg/kg) with a 0.00 irritation score in a skin patch test. Turmeric root, the source of tetrahydrocurcuminoids, is listed by the U.S. Food and Drug Administration (FDA) as an herb generally recognized as safe (GRAS) for its intended use as a spice, seasoning and flavoring agent.[10,14]

Formulation Guidelines

This combination of safety and efficacy makes THC a very versatile compound for use in cosmetic formulations. Stability and compatibility studies have shown that THC can be used within a broad range of cosmetic vehicles and with other active ingredients without being affected by or affecting the stability of the composition.

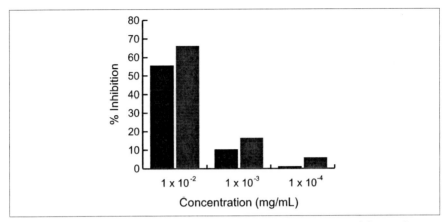

Figure 5. Antioxidant activity of THC is enhanced by tetrahydropiperine (Percent inhibition refers to inhibition of free radicals in a standard DPPH (1,1-diphenyl 2-picrylhydrazyl) radical scavenging in vitro assay.)

Suggested levels of usage in formulations are 0.02-0.5% in skin creams and lotions, and 0.5-1% in oral care formulations to soothe inflammation. Formula 1 presents a typical formulation.

Summary

THC, a colorless compound derived from the yellow curcuminoids of turmeric root, presents a refreshing, versatile approach to formulations for the antiaging cosmetics market. With "bioprotectant" properties that combine the prevention of free radical formation with scavenging pre-formed free radicals, potential skin lightening capabilities, UV protection benefits and anti-inflammatory effects, this ingredient is well-equipped to furnish a multifunctional advantage. Additionally, the off-white colored THC was shown to preserve all the biological properties of the parent yellow curcuminoids. The safety and efficacy of this composition for cosmetic use are well documented.

—Lakshmi Prakash, Ph.D., K.S. Satyan, Ph.D. and Shaheen Majeed,
Sabinsa Corporation, Piscataway New Jersey USA

References

1. H Black, Potential involvement of free radical reactions in ultraviolet light-mediated cutaneous damage, *Photochem Photobiol* 46 213–22 (1987)
2. M Masuda et al, Skin lighteners, *Cosmet Toil* 111(10) 65-77 (1996)
3. M Majeed, V Badmaev, U Shivakumar and R Rajendran, *Curcuminoids Antioxidant Phytonutrients*, Piscataway, New Jersey: NutriScience Publishers, Inc (1995)
4. US Pat 5,861,415, Bioprotectant composition, method of use and extraction process of curcuminoids, M Majeed, V Badmaev and R Rajendran (Jan 19, 1999)
5. JK Li and SY Lin-Shia, Mechanisms of cancer chemoprevention by curcumin, *Proc Natl Sci Counc Repub China B* 25(2) 59-66 (2001)

6. Comparative Antioxidant Properties of Tetrahydrocurcuminoids and Vitamin E, www.tetrahydrocurcuminoids.com
7. Research Report, Sabinsa Corporation (2000-2001)
8. K Fogh and K Kragballe, Eicosanoids in inflammatory skin diseases, *Prostaglandins Other Lipid Mediat* 63(1-2) 43-54 (2000)
9. Y Nakamura, Y Ohto, A Murakami, T Osawa and H Ohigashi, Inhibitory effects of curcumin and tetrahydrocurcuminoids on the tumor promoter-induced reactive oxygen species generation in leukocytes in vitro and in vivo, *Jpn J Cancer Res* 89(4) 361-70 (1998)
10. V Badmaev and M Majeed, Tetrahydrocurcuminoids (THC) as a skin bioprotectant, *Agro-Food-Industry Hi-Tech* 25-27 (Jan-Feb 2000)
11. Y Sugiyama, S Kawakishi and T Osawa, Involvement of the beta-diketone moiety in the antioxidative mechanism of tetrahydrocurcumin, *Biochem Pharmacol* 52(4) 519-525 (1996)
12. Research Report, Sabinsa Corporation (2003)
10. V Badmaev and M Majeed, Skin as a delivery system for nutrients, nutraceuticals and drugs THP a natural compound with the potential to enhance the bioavailability of nutrients and drugs through the skin, *Agro-Industry Hi-Tech* 6-10 (Jan-Feb 2001)
14. Code of Federal Regulations:21 CFR 582.10, 582.20, 182.10, 182.20 (April 2003)

Innovative Natural Active Ingredient with Anti-Inflammatory Properties

Keywords: Centella asiatica, madecassoside, terminoloside, anti-inflammatory

Madecassoside and terminoloside can reduce inflammation, normalize skin homeostasis and protect cells

Homeostasis belongs to the main principles of biological process as it deals with maintaining our organism functions and constants (temperature, hydrous balance). In the skin, homeostasis occurs in the epidermis and corresponds to the keratinocytes' proliferation and differentiation equilibrium: the continuous renewal of the basal keratinocytes layers is offset by the loss of corneocytes in the stratum corneum and the intermediate differentiation effects on cohesion of cells in epidermis. As a result, homeostasis helps to preserve skin barrier function, either for external aggressions such as penetration of antigens or for protection of the inner equilibrium by regulating the transepidermal water loss (TEWL).

Many molecules (growth factors, hormones and trace elements such as zinc or copper) are involved in regulating homeostasis. Disorders can happen biologically (aging, physiological dysfunction such as psoriasis) or can be induced by contact with irritant products. These conditions lead to a modification of the multiplication speed of the basal keratinocytes and can influence the differentiation process to corneocytes and affect the stratum corneum thickness. These effects come from the inflammation induction and the associate immunological disorders. This state is very difficult to treat, because it is an auto-induced phenomenon that follows an amplifying cycle. For cosmetic applications, inflammation is a very important concern because it can be initialized by many causes, including can provoke its initialization (aging, UV or pollution).

A new extract of *Centella asiatica* shows very interesting anti-inflammatory properties: keratinocytes physiological growth is maintained, but their abnormal development in an inflammatory state is regulated. Tests were performed to evaluate the efficiency of this active ingredient in the reduction of inflammation, the normalization of epidermal homeostasis and the restructuring of the extra-cellular matrix.

Centella asiatica

Centella asiatica is a small herbaceous plant of the Umbelliferae family, which grows wild in tropical moist and shady regions, at an ideal altitude of 800 to 1,200 meters. The plant has long been used in American, Asian, African and Madagascan traditional medicine, mainly in the treatment of skin and mucosal diseases.

Therapeutic properties are mainly due to the presence of molecules that belong to the triterpene series, which participate in the natural defense of the plant. The parasitic attacks allow the plant to develop its natural defenses and the biosynthesis of saponins.

In its natural state, the plant converts saponins in triterpenic acids in order to acidify its cytoplasmic pH and exert the antimicrobial properties (principally gram-positive) of free aglycones to protect itself against yeasts and molds, among other aggressors. In order to ensure that the cell tolerates these substances, sufficient concentrations of substances providing anti-inflammatory cellular protection are required. These substances are heterosides, which include madecassoside.

To date, three of the four known triterpenes (asiaticoside, asiatic acid and madecassic acid) have been widely used in pharmacy and cosmetics, especially for their capacity to stimulate collagen synthesis[1-3] and their dermis and venous wall-restructuring properties.[4-5]

The study presented here concerns our experiments with the fourth known triterpene—madecassoside—which is, in reality, composed of two isomer molecules: madecassoside and terminoloside (Figure 1).

Inflammation State

The skin faces daily urban attacks (from solar radiation, pollution, artificial light, air conditioning and other harmful effects) that, in the long term, progressively induce an insidious irritative condition that is marked by a series of visible imbalances within the skin:

- A generally weak, but constant activation of a pro-inflammatory condition marked by an almost permanent production of pro-inflammatory ligands (such as cytokines and prostaglandins) by the keratinocytes in response to these repeated attacks.
- An acceleration of the keratinocytes' renewal process, whose functionality is affected, thus seriously depleting their natural youth reserves.
- Keratinization deficiencies resulting from cellular hyper-proliferation result in an acceleration of the differentiation of keratinocytes into corneocytes. The latter have less time to organize properly, leading to irregularities in the horny layer, which in turn increase the risk of penetration of external aggressors into the skin.

In general, this activation of the epidermis results in a generalized disorder of the auto-regulation capacities. The skin is then incapable of keeping intact the physiological tissue barrier (homeostasis), or maintaining its natural defense capacities. The skin, thus weakened, loses the capacity to protect itself against internal disorders and external pathogen attacks.

Inflammatory Process

Daily aggression by environmental conditions helps the penetration of antigens into the dermis, which is the origin of the inflammation cycle (Figure 2). After having captured

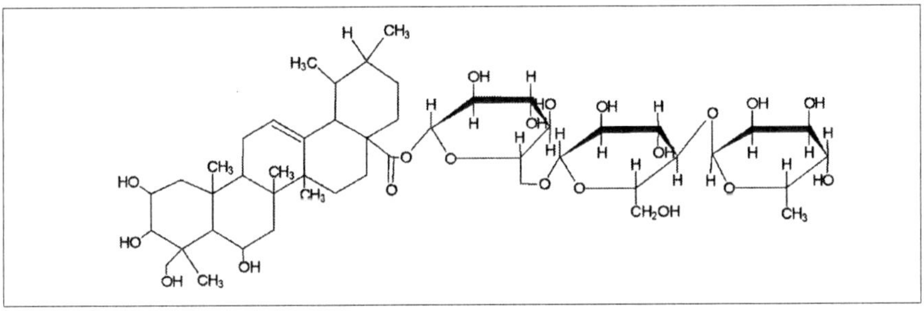

Figure 1. Structure of madecassoside

antigens in the epidermis, the dendritic cells (Langerhans cells in particular) migrate into the dermis and activate the T lymphocytes and defensive chain reactions in the skin. IFN gamma (released by activated T lymphocytes) stimulates the pro-inflammatory neuromediators by keratinocytes. This leads to a hyperproliferation of T lymphocytes and to the invasion of the tissues by polynuclears. These events provoke the increase of the interferon gamma (IFN gamma) and protease levels. The high level of cytokines produced by keratinocytes causes the synthesis of mediators (autocrin effect): interleukins 1 and 8 (IL1, IL8), prostaglandine 2 (PGE2) and Skin AntiLeicoProteases (SKALP).

These molecules are at the origin of the induction but they also participate in maintaining the inflammation cycle. As the multiplication of T lymphocytes increases (IL1 participate in the activation) and the inflammation-involved cells (T lymphocytes, polynuclear cells) migrate by chemotactism into the epidermis (PGE2, IL8), and the concentration of ligands produced by keratinocytes rises continually. These conditions lead to an inflamed, thick and dull skin.

Prevention and Correction of Immune Disorders

Material and methods: All the tests were performed on normal human cultured keratinocytes. The assay is based on the evaluation of soluble pro-inflammatory cytokines production

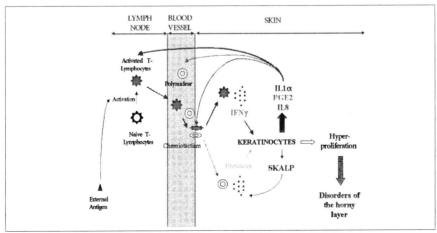

Figure 2. Inflammatory process

and secretion by keratinocytes in response to a stress induced by a non-sensitizing contact irritant (Phorbol-12-myristate 13-acetate (PMA)) or by the stimulation of a neuromediator (IFN gamma) biologically released by T lymphocytes in the irritation cycle.

IL1 alpha production and PGE2 release by keratinocytes after stimulation by PMA are evaluated on control and treated cultures. IL8 production was evaluated after the addition of IFN gamma to the medium on control and treated cultures.

Results: Madecassoside prevents and corrects immune disorders and irritative attacks by attenuating the release of the main pro-inflammatory epidermal ligands (such as cytokines, chemokines, prostaglandins, IFN gamma). This mixture of isomers modulates, in a dose-dependent manner, the expression and release of interleukins (including IL1 alpha and IL8) by human culture keratinocytes, as well as the expression of arachidonic acid derivatives (such as PGE2) during irritative stress induced either by phorbol ester or IFN gamma (Figure 3).

By continuously modulating the strong inflammatory response at various levels of the immune disorder previously described, madecassoside succeeds in normalizing renewal of keratinocytes.

Protection of the Cells from Inflammatory Stress

Material and methods: The assay seeks to compare the SKALP expression by keratinocytes in a normal medium or in a hyperproliferative medium. The normal medium is a keratinocytes growth medium (KGM); the hyperproliferative medium is obtained by addition of 5% of fetal calf serum (FCS) to KGM. The SKALP production is evaluated after treatment by madecassoside at different concentrations.

Results: The impact of a strong proteolytic environment constituted by FCS, whose high concentration normally leads to a defensive secretion of SKALP by the skin, is decreased when madecassoside is added to the medium. Madecassoside protects the cells from inflammatory stress by reducing the keratinocytes' response to stress and also by decreasing the protease sensitivity of keratinocytes.

The protective and/or desensitizing activity by madecassoside on keratinocytes is proven by a 40% decrease in the hyperproliferative medium (Figure 4).

Preservation of Renewal Potential of Keratinocytes

Method: The aim is to make sure that the action of madecassoside doesn't play a role on the cells' normal metabolism. A test is done by stimulating keratinocytes either by KGF or by human leucocytar elastase (HLE) with or without different concentrations of madecassoside. KGF represents a normal proliferation medium whereas HLE corresponds to a pro-inflammatory medium, which induces hyperproliferation. The cellular density is then measured after two and five days.

Results: The keratinocytes' renewal potential is totally preserved, as shown by the normal growth of keratinocytes under the effect of KGF in the presence of madecassoside. On the other hand, the abnormal growth caused by EGF is regulated by this active ingredient. The proteins and ligands synthesis capacity of the cells remains intact. The action of madecassoside does not have any effect on the normal metabolism (growth and multiplication) and immune function of the cell itself (biosynthesis and release of ligands) (Figure 5).

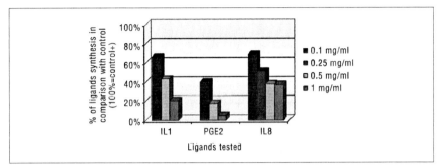

Figure 3. Effect of different concentrations of madecassoside on ligands biosynthesis

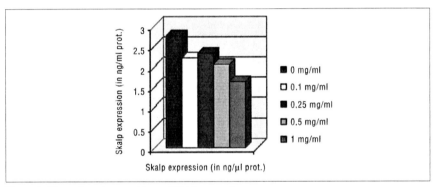

Figure 4. Effect of madecassoside on SKALP expression

Other tests show that madecassoside preserves all the nucleus functionalities. For example, there is no inhibition of the NFkappaB–involved in nuclear transcription–and no significant decrease of TNF α (tumor necrosis factor, which acts in cellular vigilance against anomalies).

Restructuring of the Extra-Cellular Matrix

Method: This test was performed on normal human cultured fibroblasts. The fibroblasts were grown in a medium supplemented with trypsin and EDTA solutions. The collagen synthesis–especially for collagen Types I and III–was measured by the ELISA system.

Results: Madecassoside assists in maintaining the structure and quality of the extra-cellular matrix to a very high degree, and contributes to the improvement of the link between the epidermis and dermis by assuring an increased expression of type I and type III collagens (Figure 6).

Pre-Clinical Study

Efficiency of madecassoside was confirmed in vivo in a pre-clinical study on psoriasis, a severe chronic inflammatory disease. For six weeks, patients used daily an o/w emulsion containing 3% madecassoside. The results show a great improvement of the

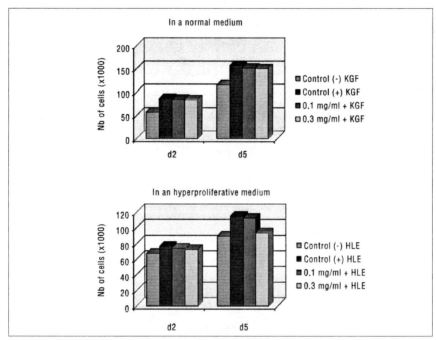

Figure 5. Effect of madecassoside on multiplication of keratinocytes multiplication

skin appearance (Figures 7 and 8), validating the activity of this active ingredient on the inflammation state and supporting its interest for cosmetic applications.

Formulation

As previously mentioned, inflammation and its consequences can appear in many skin problems and can be treated in a preventive or a curative way by madecassoside. Environmental conditions, especially pollution, can rupture immunological equilibrium because these conditions bring antigens that activate T lymphocytes and the inflammation process.

In aging skin or for physiological evolutions, immunological disorders are biological and the break of the self-induced inflammation cycle can avoid homeostasis deregulation and hypersensitivity; moreover the activity on collagens reduces wrinkles and line formation.

Another environmental condition is UV radiation. It causes proteinic degradation and chromosomic alteration in cells, generating inflammation that must be balanced to reduce photo-induced aging. Formula 1 shows an antiaging composition containing madecassoside.

Conclusion

The patented mixture of madecassoside and terminoloside is positioned as a true ally of the skin, which is under environmental attack and, as a result, is becoming

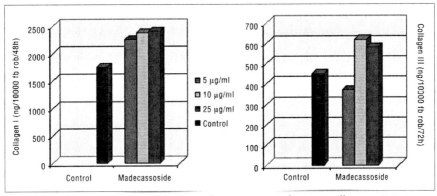

Figure 6. Effect of madecassoside on Type I collagen (+30%) and Type III collagen (+49%)

prematurely aged or simply weakened and incapable of self-defense. It allows regulation at three levels:

- Modulation of skin's chronic inflammatory condition by controlling and regulating the overproduction of inflammation mediators in keratinocytes linked to immuno-stimulations
- Normalization of keratinocytes hyperproliferation by re-balancing their natural renewal with the attenuation of the auto-amplification loop of the inflammatory process. Thus the natural epidermal homeostasis, characterized by a better organization of the horny layer, is reestablished, restructuring the natural protection of the skin.
- Restructuring of the extra-cellular matrix of the dermis through activating collagen expression by fibroblasts, in order to preserve the functional and physical qualities of the skin, during chronological aging (antiaging).

The immuno-regulating activity is demonstrated here. Madecassoside doesn't play a role on the cells' essential physiological functions and allows the use of madecassoside in cosmetic products. The efficiency has also been confirmed by a clinical study carried out on an auto-immunological disease. The improvement of the state of psoriatic patients offers promising perspectives for the treatment of this illness. Other studies will deal with extending the clinical study and with improving our knowledge of the activity mechanism.

Figure 7. Before treatment

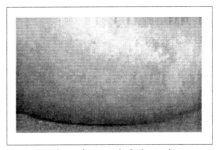

Figure 8. After a five-week daily application

Formula 1: Anti-aging formulation

A. Beheneth-10	1.5%
Beheneth-25	1.5
Hexyl laurate	5.0
Dicaprylyl carbonate	5.0
Isohexadecane	5.0
Cetearyl isononanoate	5.0
Dimethicone	1.0
Hydrogenated vegetal glycerides	2.0
Behenyl alcohol	2.0
Tocopheryl acetate	0.5
Phenoxyethanol and parabens	0.5
B. Madecassoside	0.5
Water (*aqua*)	qs to 100.0
Carbomer	0.2
Xanthan gum	0.1
Butylene glycol	2.0
Glycerin	3.0
C. Triethanolamine	qs pH 5.5

Procedure : Heat A and B separately to 80°C. Add B to A under continuous stirring and cool to 30°C. Add C under stirring.

Note: Madecassoside can be introduced in the aqueous phase before heading. The introduction is easier in water heated at 50°C with gentle agitation.

—Caroline Segond, Eric Théron, Virginie Petit and Alain Loiseau,
Laboratoires Roche Nicholas S.A.S., Division Serdex, Saint Ouen, France

References

1. FX Macquart et al, *Connect Tiss Rev* 24 107-120 (1990)
2. FX Macquart et al, *Eur J Dermatol* 9 289-296 (1999)
3. F Bonté et al, *Ann Pharmaceutiques Françaises* 53 38-42 (1995)
4. C Allegra, *Clin Terap* 99 507 (1981)
5. M Apperti et al, *Quad Chir Prat* 3 115 (1982)

Low Molecular Weight Tannins of *Phyllanthus emblica*: Antiaging Effects*

Keywords: Phyllanthus emblica, UV, tannin

Tannin-based P. emblica extract in skin or sun products applied before UV irradiation may help prevent skin photoaging

Ultraviolet irradiation from the sun has deleterious effects on human skin, including sunburn, immune suppression, cancer and premature aging (photoaging). Sunburn and immune suppression occur acutely in response to excessive exposure to the sun, whereas skin cancer and photoaging result from accumulated damage caused by repeated exposures. Skin cancer, the most prevalent form of cancer in humans, typically occurs in skin that is photoaged.[1] Photoaged skin is characterized by wrinkles, uneven pigmentation, brown spots and a leathery appearance.[2] In contrast, chronologically aged skin that has been protected from the sun is thin and has reduced elasticity but is otherwise smooth and unblemished.[3]

The unifying pathogenic agents for these changes are UV-generated reactive oxygen species (ROS) that deplete and damage non-enzymatic and enzymatic antioxidant defense systems of the skin, and the release of matrix metalloproteases (MMPs) such as MMP-1 and MMP-3 that damage the extracellular matrix proteins.[4] Photoaging of the skin is a complex biological process affecting various layers of the skin with major damage seen in the connective tissue of the dermis. The dermis lies below the epidermis, and in conjunction with the basement membrane at the dermal-epidermal junction, provides mechanical support for the outer protective layers of the epidermis. Any damage to the dermal components is seen predominantly on the sun-exposed body areas, especially on the face.

This chapter focuses on the major causes of the photoaging of skin and the use of low molecular weight tannins from *Phyllanthus emblica* fruits (Syn. *Emblica officinalis*) in reducing some of the causes of premature skin aging.

Major UV-Induced Chemical and Biochemical Changes

Generation of superoxide, singlet oxygen and hydrogen peroxide: There is now ample evidence showing that reactive oxygen species (ROS), generated in vitro and in vivo after UVA and UVB irradiation, cause serious damage to skin.[5] Besides direct absorption of UVB photons by DNA and subsequent structural changes, generation of ROS following irradiation with UVA and UVB requires the absorption of photons by endogenous photosensitive molecules.

Recently, researchers[6] identified the epidermal UVA-absorbing chromophore transurocanic acid that quantitatively accounts for the action spectrum of photoaging. The excited photosensitizer subsequently reacts with oxygen resulting in the generation of ROS including superoxide anion and singlet oxygen.

Superoxide anion and singlet oxygen are also produced by neutrophiles that are present in increased quantities in photo-damaged skin, and contribute to the overall pro-oxidant state. Superoxide dismutase (SOD) converts superoxide anion to hydrogen peroxide. Hydrogen peroxide is able to cross cell membranes easily and, in conjunction with Fe^{2+}, generates a highly toxic hydroxyl radical. Both singlet oxygen and hydroxyl radical can initiate lipid peroxidation.

To counteract the harmful effects of ROS, the skin is equipped with antioxidant defense systems consisting of a variety of low molecular weight antioxidants (such as vitamin C and vitamin E) and antioxidant defense enzymes (such as superoxide dismutase, glutathione peroxidase and catalase) forming an "antioxidant network."

The antioxidant network is responsible for maintaining the equilibrium between pro-oxidants and antioxidants. However, the antioxidant defense can be overwhelmed by increased exposure to exogenous sources of ROS. Such a disturbance of the pro-oxidant/antioxidant balance may result in oxidative damage to lipids, proteins and DNA. A review of the protective effects of topical antioxidants in humans has recently been published.[7]

Release of iron and copper: Iron and copper play ambivalent roles in biology because they are required as cofactors for many biological reactions, even though their toxicity threatens cellular integrity.[8] In mammalian cells, the level of iron-storage protein is tightly controlled by the iron-regulatory protein-1 at the post-transcriptional level. This regulation prevents iron from acting as a catalyst in reactions between ROS and biomolecules.

Recently, it has been shown that both UVB and UVA can generate lipid peroxidation induced by iron.[9,10] The iron content is substantially elevated over basal levels in the skin of mice exposed to UVB irradiation and in the sun-exposed skin of healthy individuals.[11,12] The underlying mechanism appears to be the UVB-induced formation of superoxide radical and its attack on ferritin, resulting in the release and mobilization of free iron.[13]

Brenneisen et al. have identified the iron-dependent Fenton reaction and lipid peroxidation as the central mechanisms underlying signal transduction of the UVB response.[9] Singlet oxygen and hydrogen peroxide are presently considered to be the most important reactive oxygen species generated intracellularly by UVA light promoting biological damage in exposed tissues via iron-catalyzed oxidative stress.[14]

Gutteridge et al. have found copper in sweat samples from the arm or trunk of athletes immediately after exercise.[15] Arm samples also contained much greater concentration of iron after exercise. It is also easy to see from these data how athletes following an intensive training might become anemic by loss of iron.

Unfortunately, skin does not have any defense against oxidative stress induced by free iron and copper. Application of metal chelators having chelating ability to occupy all the coordination sites in iron and copper may be a route to prevent or reduce oxidative damage to skin.[16] The iron-chelating agents have been shown as protectants against UV-radiation-induced free radical production.[12,17]

Release of matrix-degrading MMPs: Matrix metalloproteases (MMPs) are enzymes able to degrade most components of the extracellular matrix (ECM). Among these components are collagens, elastins, fibronectin and proteoglycans.

At this time more than 20 different MMPs have been identified and classified.[18] They show consistent sequence homology and in general share a pre-domain (which is a signal peptide for secretion), a pro-domain (important for maintaining latency) and a catalytic domain (with a highly conserved zinc-binding site). Based on sequence homology and substrate specificity, MMPs can be classified into five groups: collagenase, gelatinases, stromelysins, membrane type and others. This classification is somewhat arbitrary, because the true physiological substrates are a matter of debate.

The ECM proteins not only provide a supportive function for the development and organization of tissues, but also serve as a physical barrier to limit the migration of most normal cells away from their sites of origin. The ECM is not a homogeneous structure. It can include any of several classes of biomolecules, including the following: structural proteins such as collagens and elastin; adhesion proteins, including fibronectins, laminins and entactin; proteoglycans; and glycosaminoglycans. Further, the precise compositions of the ECM vary between tissues, and perhaps even in a cell state-specific manner. This complex mixture does not simply surround cells, hold them together and provide an environment in which interesting events occur; it also directly or indirectly mediates a number of critical biological processes.

Several studies carried out by Scharffetter-Kochanek's group using dermal fibroblast cells show that both UVA and UVB cause a fourfold to fivefold increase in the production of MMP-1 and MMP-3.[9,19,20] In contrast, the synthesis of tissue inhibitory metalloprotease-1 (TIMP-1), the natural inhibitor of matrix metalloprotease, increases only marginally. This imbalance is one of the causes of severe connective tissue damage resulting in photoaging of the skin.

The damage caused by excessive MMP on the ECM proteins does not appear overnight, but results from the accumulation of successive molecular damages, especially in the case of overexposure to UV light. However, the degradation of the ECM proteins has consequences for the skin. These consequences may be revealed in many ways depending on age, genetic predisposition, lifestyle and, of course, on the general health of the individual.[21] Application of MMP inhibitors may be a route to prevent or minimize damage to ECM proteins.[22]

Low Molecular Weight Tannins of *Phyllanthus emblica* Fruits

A tannin-based* P. emblica *extract: *P. emblica* is one of the important Ayurvedic (Science of Life) herbs in India and has been used for thousands of years in a wide variety of human ailments. Its status ranges from insignificant in the western world to highly prized in tropical Asia. The fruits are selected, harvested and processed according to strict criteria to ensure a consistently high quality product.

A particular *P. emblica* extract[a], protected by a U.S. patent[23] and other pending patents, is extracted from premium quality fruits using a water-based process. This extract is distinctly different from other commercially available extracts of *P. emblica* fruits because it is defined to the extent of well over 50% (typically, 60-75%) in terms of its key active components, which are low molecular weight hydrolyzable tannins. None of the other *P. emblica* extracts in the market compares to this tannin-based extract in terms of composition and consistency of composition, aqueous stability and color. In this chapter, we will refer to the product as a tannin-based *P. emblica* extract, or a TBPE extract, because this extract is based to such a large degree (60-75%) on these tannins.

Tannins: Tannins are chemically complex substances widely distributed in the plant kingdom and employed in medicine as astringents. They possess an abundance of polyphenols, have molecular masses in the range of 500-5000 Dalton and display a diversity of structures that form the basis of their classification into two families, of which the hydrolyzable tannins are of interest in this chapter.

As its key ingredients, the patented TBPE extract contains four of these hydrolyzable tannins (Figure 1): Emblicanin A, Emblicanin B, pedunculagin and punigluconin. In nature, Emblicanin A and Emblicanin B have only been found in *P. emblica* plants.[24]

Standardization and stability: The TBPE extract has been standardized[25] by using either high performance thin layer chromatography (HPTLC) or high performance liquid chromatography (HPLC).

It is stable in aqueous as well as in formulated products for well over two years. It is extremely photostable, which was determined by exposing a 1% aqueous solution and a formulated product of this extract to UVA and UVB irradiation separately for a period of 4 h (about 8 minimum erythemal dose, MED). Photostability of the product was determined from the λ_{max} value at 271 nm and normalized with respect to the time zero. A loss of only 3% was observed after irradiating the product for 2 h (4 MED). On the contrary, a vitamin E-containing formulated product (where the vitamin E was natural tocopherol and λ_{max} was 263 nm) lost well over 70% in 2 h (4 MED).

The photostability of an active in a formulation under UV irradiation is tested by ultrathin film transmission of light. An ultrathin film of the formulation is prepared between glass or quartz slides to allow a minimum of 90% light transmission at 600 nm. Formulations are subject to two difficulties: (1) light scattering by the emulsion (mostly multiple scattering) and (2) a nontransparent background over which

---

[a] Emblica Antioxidant (INCI: *Phyllanthus emblica* fruit extract) is a product of Rona, a division of EMD Chemicals, Inc., Hawthorne, New York, USA

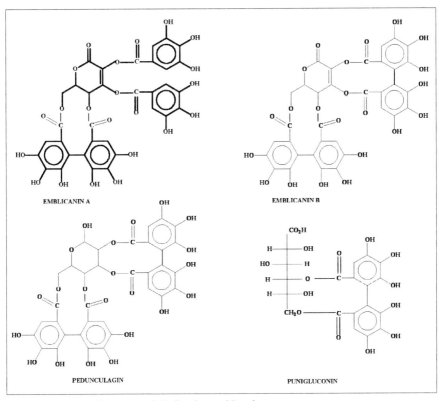

Figure 1. Hydrolyzable tannins of Phyllanthus emblica fruits

to measure the UV absorption of the active. While the laws of light scattering are known, these effects are limited by measuring the properties of an ultrathin film of product. In this case, chromophore absorption becomes clearly visible above the strongly reduced light scattering background.

The UV photodegradation of an active can therefore be measured from its decrease in the characteristic UV absorption band after correction for the light scattering background. This method requires a minimum concentration of active in order to provide a good signal-to-background ratio and is performed for identical formulation bases in order to compare actives.

TBPE extract has been found to have broad-spectrum antioxidant activity,[26] excellent iron- and copper-chelating ability and MMP-1 and MMP-3 inhibitory activity. These multifunctional attributes of TBPE extract are described in the following sections.

Quenching of ROS

There are dozens of testing methods available for determining the ROS quenching ability of a substance. We are describing here results of three tests–superoxide anion, hydroxyl radical quenching and singlet oxygen—carried out for TBPE extract and a few commercially available antioxidants.

Cell protection against superoxide damage: A hypoxanthine-xanthine oxidase test evaluates cell protection against superoxide damage. About 90% protection of fibroblast cells against superoxide damage was observed using 20 µg/mL of TBPE extract.

This study uses a human skin fibroblast cell culture to determine the cell viability under superoxide anion [generated by using hypoxanthine and xanthine oxidase (HX-XO) system[27]] and the ability of TBPE extract to protect cells under these conditions. Two different batches (1 and 2) of TBPE extract were used for this study. Cell survival was determined with a colorimetric method[28] using 2,3-bis(2-methoxy-4-nitro-5-sulfophenyl)-5-[(phenylamino)carbonyl]-2H-tetrazolium-hydroxide (XTT). The test is an indication of living cells, which is determined by measuring the optical density at 570 nm (Figure 2).

Protection of dye against superoxide damage: Another hypoxanthine-xanthine oxidase test evaluates protection of dye against superoxide damage. Superoxide anions are generated in an aqueous solution of hypoxanthine (100 µM) in the presence of EDTA (1 mM) by adding 0.014 units of xanthine oxidase freshly diluted in 100 µL of phosphate buffer (pH 7.4). The production of superoxide anions is determined through its reduction of NBT (nitroblue tetrazolium) into nitroblue diformazan by monitoring the light absorption of NBT at 560 nm.[29] Spectrophotometric measurements are thus used to determine the antioxidant efficiency of various antioxidants. Efficient superoxide anion scavenging results in reduced NBT destruction and, thus, in a stronger light absorption at 560 nm.

Results of this study are given in Figure 3. IC_{50} of the TBPE extract was found to be 12 µg/ml, compared to 26 µg/ml for vitamin C and 360 µg/mL for a vitamin E water-soluble analog[c].

Hydroxyl radical quenching: The deoxyribose test evaluates hydroxyl radical quenching. Results demonstrate that TBPE extract possesses the strongest hydroxyl radical scavenging ability, significantly better than pine antioxidant, grape antioxidant and a vitamin E product[c] with the alkyl chain substituted by a carboxylic group. Vitamin C and green tea extract have been shown to be pro-oxidant by this testing method, following the protocol developed by Halliwell et al.[30]

The hydroxyl radical scavenging efficiency of products was obtained at respective concentrations of 0.3 mM of $FeCl_3$, 1.2 mM of EDTA, 33.6 mM of H_2O_2, 33.6 mM of deoxyribose in pH 7.4 phosphate buffer (20 mM) and 0.2-10 mM of chelator. The amount of hydroxyl radicals was determined from the deoxyribose test by using 1% w/v of thiobarbituric acid (TBA) and 2.8% w/v of trichloroacetic acid (TCA). Results are summarized in Figure 4.

Singlet oxygen quenching: Singlet oxygen quenching can be evaluated by photooxidation of a sensitizer. Results show that TBPE extract (IC_{50} 61 µg/ml) is an excellent singlet oxygen quencher and is superior to the vitamin E water-soluble

---

[c] Trolox C (Chemical name: 6-Hydroxy-2,5,7,8-tetramethylchroman-2-carboxylic acid), Sigma-Aldrich, Milwaukee, Wisconsin

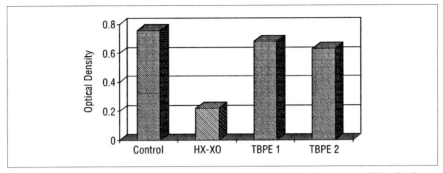

Figure 2. Protective effect of TBPE extract on skin fibroblast cells against superoxide radical damage

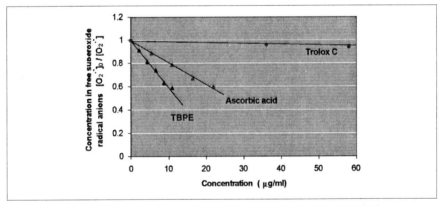

Figure 3. Comparative superoxide radical quenching ability of TBPE extract and other antioxidants

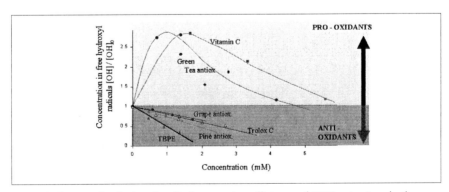

Figure 4. Comparative hydroxyl radical scavenging efficiency of TBPE extract and other antioxidants

analog^c (IC_{50} 84 µg/ml). In this test, vitamin C was found to be a strong enhancer of singlet oxygen (pro-oxidant). Results are summarized in Figure 5.

Singlet oxygen was produced by UVA irradiation of a sensitizer (methylene blue, 10^{-5} M). Histidine (10^{-2} M) was used as a substrate, which reacts with singlet oxygen to form trans-annular peroxide. This reaction product, in turn, bleaches N,N-dimethyl-p-nitrosoaniline (RNO, 5×10^{-5} M in a pH 7.0 phosphate buffer). The bleached form of RNO was measured spectrophotometrically at 440 nm. The singlet oxygen scavenging efficiency of an antioxidant will therefore reduce the amount of free singlet oxygen and thus prevent the bleaching of RNO.

The test was performed under experimental conditions to insure less than 15% of RNO bleaching. This is needed to limit secondary product accumulation, which could contribute to RNO bleaching. The present product concentrations and irradiation time (20 min with a total energy of 12 J/m^2) were chosen to fulfill this condition.[31,32]

Chelation of Iron and Copper

Recognizing the importance of available coordination sites in transition metal catalysis,[33] and the crucial role iron plays in initiating oxidative stress to skin, we envisioned that an antioxidant can be a true photoprotective agent provided it chelates all the coordination sites in iron and copper.[34] This is particularly critical because the formation of hydroxyl radicals from superoxide anion or hydrogen peroxide and iron requires only one empty coordination site or a site occupied by a readily dissociable ligand such as water. Water may be completely displaced by stronger ligands like the azide (N_3^-) anion. We have applied this principle to determine the presence of free coordination site(s) in the Fe^{3+}-antioxidant complexes by the UV spectrophotometric method.[34]

Of all the Fe^{3+}-chelates tested, only the complex of TBPE extract and iron (or copper) showed the absence of any water coordination (that is, the complex is fully

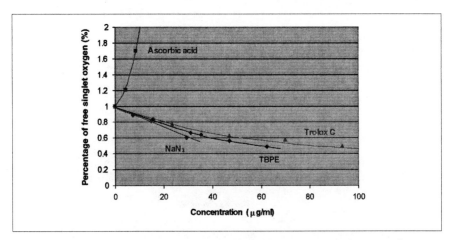

Figure 5. Comparative singlet oxygen quenching ability of TBPE extract and other antioxidants

and firmly saturated and there is no room for any pro-oxidant activity via the oxo-ferryl or oxo-cupryl radical formation). All other chelators showed disparate coordination site(s), thereby making room for opening oxo-ferryl or oxo-cupryl radical formation, manifesting pro-oxidant effect, particularly at low concentrations.

Table 1 shows that all complexes except TBPE extract contain at least one coordinated water molecule as shown by the presence of spectral shift(s) induced by sodium azide; only TBPE extract shows no spectral shift induced by sodium azide. The maximum wavelength, the extinction coefficient of the complex, and the association constant between N_3^- and Fe^{3+}-antioxidant/chelator complex are highly variable and depend on the type of bonding between Fe^{3+} and chelator, the stereochemistry of the complex and the number of coordination positions. A complete interpretation of these spectral data, however, is beyond the scope of this report.

MMP Inhibitory Activity

As described earlier, elevated levels of MMP-1, MMP-2 and MMP-3 in skin connective tissue under low-dose UV-irradiation are responsible for the breakdown of various connective tissue components. Although MMP-1 cleaves collagen type I, MMP-2 is able to degrade elastin as well as basement membrane compounds including collagen type IV and type VII. MMP-3 reveals the broadest substrate range for proteins such as type IV, proteoglycans, fibronectin and laminin.

We have investigated the enzyme inhibitory activity of TBPE extract against MMP-1 and MMP-3. Also, we investigated the inhibitory effect of MMP-1 expression using TBPE extract. Results with protocols are given next.

Collagenase (MMP-1) inhibitory activity: A dose-dependent inhibition of gelatinase/collagenase activity by approximately 55-70% was observed with the TBPE extract at 150-300 µg/ml. Quantification of gelatinase/collagenase inhibitory activity of the extract was determined by using a gelatinase/collagenase kit[d] and measuring the substrate fluorescence emission at 515 nm. 1,10-Phenanthroline was used as a positive control and collagenase without inhibitor was used as a negative control. Results of this study are summarized in Figure 6.

Collagenase expression inhibition: An inhibition of about 40% in collagenase expression was observed using only 50 µg/ml of TBPE extract. Human skin fibroblast cell lines were used for this study. Quantification of collagenase expression was done by using a nonisotopic immunoassay[e] for the in vitro quantification of human matrix metalloprotease (MMP-1, interstitial collagenase).

Experiments were done using human skin fibroblast cells of fifth and eighteenth passages. Data shows a comparable effect of the TBPE extract on MMP-1 (collagenase) expression after 48 h of incubation in two different experiments. Results are presented in the Figure 7.

Stromelysin 1 (MMP-3) inhibitory activity: An inhibition of stromelysin 1 activity by more than 50% was observed with the TBPE extract at 100 µg/ml.

---

[d] E-12055 EnzChek gelatinase/collagenase kit, Molecular Probes, Inc.)EnzChek is a registered trade name of Molecular Probes, Inc.)
[e] MMP-1 ELISA (Catalog # QIA55) from Oncogene Research Products, Cambridge, Massachusetts, USA

Table 1. Ultraviolet spectral data of Fe^{3+}- and Cu^{2+}-chelators (No shift is seen with the TBPE extract)

Absorption Maxima of Complex (λ_{max} in nm)

Chelator	With Fe^{3+}	N$_3$-Induced Shift	With Cu^{2+}	N$_3$-Induced Shift
EDTA	241, 283	241, 283, 410	240, 278	241, 279, <u>354</u>
TBPE extract	241, 294, 353, 377	241, 294, 353, 377	240, 272, 313	240, 272, 313
Pine Antioxidant	241, 294, 353, 384	241, 294, 353, 400, 440	239, 279, 302, 331	239, 280, 307, <u>430</u>
Vitamin C	238, 262	241, 266, <u>295</u>	239, 263	239, 263, <u>284</u>, <u>364</u>
Grape Antioxidant	247, 295, 353, 396	247, 295, 353, 415, <u>430</u>	240, 277, 328	240, 277, 328, <u>359</u>
Green Tea Antioxidant	240, 272, 324, 390	240, <u>277</u>, 325, 390	241, 276, 327, 403	240, 277, <u>336</u>, 404
Trolox C	240, 284	240, <u>273</u>, 284, <u>360</u>	241, 288	241, <u>261</u>, 352, <u>440</u>
Gallic Acid	247, 295, 337	247, 295, <u>353</u>, <u>412</u>	240, 258, 321	240, 258, <u>331</u>, <u>463</u>

*The peak positions are obtained from differential spectroscopic scans of 1.0 mM Fe^{3+} or Cu^{2+} and 5 mM chelator, 1.0M NaN$_3$, 50 mM phosphate buffer, pH 7.4, versus the same solution without sodium azide.

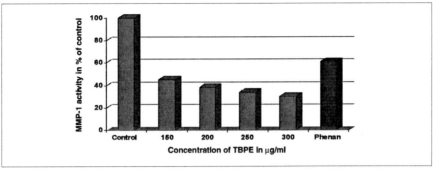

Figure 6. Collagenase (MMP-1) inhibitory activity of TBPE extract. (Phenan = 1,10-Phenanthroline)

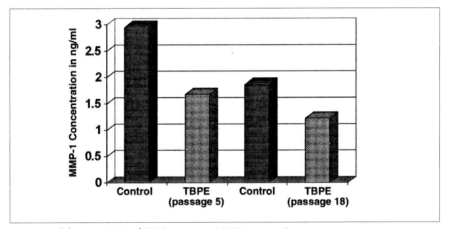

Figure 7. Inhibitory activity of TBPE extract on MMP-1 expression

Quantification of MMP-3 inhibitory activity of TBPE extract was determined by using a stromelysin activity assay kit[f]. The principle of the assay is based upon fluorescence measurement of substrate fragments released upon cleavage of a substrate by MMP-3. Fluorescence intensity of the resulting product is measured and correlated with MMP-3 activity. Results of this study are summarized in the Figure 8.

Clinical Study

Background on UV-induced erythema: Erythema, the most familiar manifestation of UV radiation exposure, occurs in a biphasic manner. UVA and certain visible light mediate the early part of this reaction, known as immediate pigment darkening (IPD), which lasts for about one-half hour. Delayed erythema, a function primarily of UVB dosages, begins 2-8 h after exposure and reaches a maximum after 24-36 h, with erythema, pruritus and pain in the sun-exposed areas.[35]

---

[f] Chemicon MMP-3 / Stromelysin Activity Assay Kit (ECM 481), Chemicon International, Temecula, California, USA

Microscopically, changes are detectable as early as 30 min after UV radiation exposure. Epidermal changes include intracellular edema, vacuolization and swelling of melanocytes, and the development of characteristic sunburn cells. In the dermis, UV radiation initially leads to interstitial edema and endothelial cell swelling. Later, there is perivenular edema, degranulation and loss of mast cells, decrease in number of Langerhans cells, neutrophil infiltration and erythrocyte extravasation.[35]

The protocols: We used reduction in UV-induced erythema as a criterion for photoprotection (Protocol A) and reversal (Protocol B) of photo-damaged skin by the TBPE extract. Test sites were areas 4 x 2.5 cm on the backs of human subjects. Test substances were creams with 0.2% and 0.5% TBPE extract, 0.5% magnesium ascorbyl phosphate (MAP) or 0.5% vitamin E. The creams were applied once daily at a dose of 2 mg/cm².

Results were represented by using the individual typology angle ITA° (COLIPA SPF test method) obtained by chromometric measurement. ΔE ITA° (difference in skin lightening) was calculated by subtracting the ITA° at the treated irradiated site from the ITA° at the untreated irradiated site. ITA°, the ITA degree, is calculated using the following equation

$$ITA° = [Arc\ Tangent\ (L° - 50)/b°)]\ 180 / 3.1416$$

where L° value is the Lightness value, a° is the color value in the red-green axis, and b° is the color value in the blue-yellow axis.

For Protocol A (photoprotection), we applied product for eight days on 11 humans and then on day 9 induced pigmentation by UV light. We compared the untreated irradiated control site versus the product-treated sites on day 10.

For Protocol B (reversal), we induced pigmentation by UV light on 10 humans and then immediately applied the product and continued product application once a day for 10 days. We compared the untreated irradiated control site versus the product-treated sites every day and found a statistically significant difference in ITA° on day five.

Results: The results are shown in Figure 9. For Protocol A, 0.2% TBPE extract and 0.5% vitamin E showed statistically significant ($p < 0.05$) reduction in erythema. For Protocol B, only 0.2% TBPE extract showed statistically significant reduction in erythema.

Formulation Guidelines

The TBPE extract can be used in formulations ranging from 0.1% to 1.0% (w/w) level. Typical use levels are 0.1-0.2% for antioxidant formulas and 0.3-1.0% for age-defying applications. Acceptable formulation aesthetics can be obtained with use levels of up to 0.5%.

Use of light fragrance may be desired with higher levels of the TBPE extract (>0.5%, w/w). Nonionic or anionic emulsifiers can be used for making stable emulsions. The pH of the formulations must be acidic (preferably below 5.5) to maintain

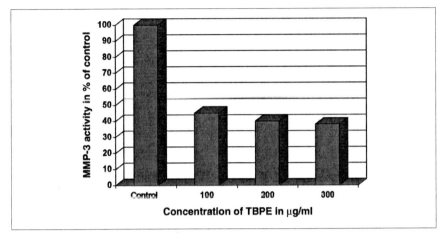

Figure 8. Stromelysin 1 (MMP-3) inhibitory activity of TBPE extract

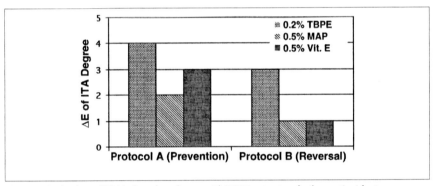

Figure 9. Reduction of UV-induced erythema with TBPE extract and other antioxidants

its antioxidant activity and stability. Addition of a skin penetration enhancer, such as lecithin, may improve the extract's efficacy.

The TBPE extract as a suspension in water can be added to the formulation with moderate agitation at about 40°C. Prolonged heating or exposure to sunlight must be avoided because it causes darkening of formulated products.

Conclusion

Photoaging of skin is a complex biological process affecting various layers of the skin with major changes seen in the connective tissue of the dermis. The natural shift toward a more pro-oxidant state in intrinsically aged skin can be significantly enhanced by UV-irradiation.

A *Phyllanthus emblica* extract composed primarily of low molecular weight tannins has been shown to reduce UV-induced erythema and has excellent free-radical quenching ability, chelating ability to iron and copper (no pro-oxidative activity) and inhibitory activity against MMP-1 and MMP-3. Possibly, this extract modulates UV/ROS-initiated signal transduction pathways of matrix metalloprotease induction, thereby protecting the extracellular matrix proteins from degradation.

A properly constituted *Phyllanthus emblica* extract, such as the one described here, may provide a great value as a stand-alone photoprotective agent or in combination with sunscreens for skin care products of the future. Thus, if skin (or sun) care products containing this extract are topically applied before UV-irradiation, at least partial prevention from skin photoaging can be provided.

—Ratan K. Chaudhuri, Cristina Hwang and Germain Puccetti, *EMD Chemicals, Inc., Hawthorne, New York, USA (An Affiliate of Merck KGaA, Darmstadt, Germany)*
—Gilles Guttierez and Mustafa Serrar, *Texinfine, Lyon, France*

* Adapted from a paper presented by R.K. Chaudhuri at the Active Ingredients Conference – A Perspective on Naturals and their Actives, Paris, 2003

References

1. MS Kaminer, Photodamage: magnitude of the problem, in *Photodamage*, BA Gilchrest, ed, Cambridge, Massachusetts: Blackwell Science (1995) 1-11
2. BA Gilchrest and M Yaar, Aging and photoaging of the skin: observations at the cellular and molecular level, *Br J Dermatol* 127(Suppl 41) 25-30 (1992)
3. RM Lavker, Cutaneous aging: chronologic versus photoaging, in *Photoaging*, BA Gilchrest, ed, Cambridge, Massachusetts: Blackwell Science (1995) 123-135
4. K Scharfetter-Kochanek, P Brenneisen, J Wenk, G Herrmann, W Ma, L Kuhr, C Mewes and M Wlaschek, Photoaging of the skin from phenotype to mechanisms, *Experimental Gerontology* 35 307-316 (2000)
5. BA Jurkiewicz and GR Buettner, Ultraviolet-light-induced free radical formation in skin: an electron paramagnetic resonance study, *Photchem Photobiol* 59 1-4 (1994)
6. KM Hanson and JD Simon, Epidermal trans-uronic acid and the UV-A induced photoaging of the skin, *Proc Natl Acad Sci USA* 95 10576 (1998)
7. F Dreher and H Maibach, Protective effects of topical antioxidants in humans, in *Oxidants and Antioxidants in Cutaneous Biology*, J Thiel and P Elsner, eds, vol 29 in *Current Problems in Dermatology* (2001) 157-164
8. MW Hentze and LC Kuhn, Molecular control of vertebrate iron metabolism: mRNA-based regulatory circuits operated by iron, nitric oxide and oxidative stress, *Proc Natl Acad Sci USA* 93 8175-8182 (1996)
9. P Brenneisen, J Wenk, LO Klotz, M Wlaschek, K Brivia, T Krieg, H Sies and K Scharffetter-Kochanek, Central role of ferrous-ferric iron in the ultraviolet B irradiation-mediated signaling pathway leading to increased interstitial collagenase (matrix-degrading metalloprotease (MMP)-1) and stromelysin-1 (MMP-3) mRNA levels in cultured human dermal fibroblasts, *J Biol Chem* 273 5279-5287 (1998)
10. C Pourzand, RD Watkin, JE Brown and RM Tyrell, Ultraviolet A radiation induces immediate release of iron in human primary skin fibroblasts: The role of ferritin, *Proc Natl Acad Sci USA* 96 6751-6756 (1999)
11. DL Bissett, R Chatterjee and DP Hannon, Chronic ultraviolet radiation-induced increase in skin iron and the photoprotective effect of topically applied iron chelators, *Photochem Photobiol* 54 215-223 (1991)

12. DL Bissett and JF McBride, Iron content of human epidermis from sun-exposed and non-exposed body sites, *J Soc Cosmet Chem* 43 215-217 (1992)
13. PA Cerutti, Oxy-radicals and cancer, *Lancet* 344 862-863 (1994)
14. B Halliwell and JM Gutteridge, Biologically relevant metal ion-dependant hydroxyl radical generation: An update, *FEBS Lett* 307 108-112 (1992)
15. JM Gutteridge, DA Rowley, B Halliwell, DF Cooper and DM Heeley, Copper and iron complexes catalytic for oxygen radical reactions in sweat from human athletes, *Clin Chim Acta* 145 267-273 (1985)
16. RK Chaudhuri and G Puccetti, Transition metal-induced oxidation: Implications for skin care products, *Cosmet Toil* 117(9) 43-56 (2002)
17. BA Jurkiewicz Lange and GR Buettner, Electron paramagnetic resonance detection of free radicals in UV-irradiated human and mouse skin, in *Oxidants and Antioxidants in Cutaneous Biology*, J Thiele and P Elsner, eds, vol 29 in *Current Problems in Cutaneous Biology*, Basel: Karger (2001) 18-25
18. R Hoekstra, FALM Eskens and J Verweij, Matrix metalloprotease inhibitors: Current developments and future perspectives, *The Oncologist* 6 415-427 (2001)
19. P Brenneisen, J Oh, M Wlashek, J Wenk, K Briviba, C Hommel, G Herrmann, H Sies and K Scharffetter-Kochanek, Ultraviolet B wavelength dependence for the regulation of two major matrix-metalloproteinases and their inhibitor TIMP-1 in human dermal fibroblasts, *Photochem Photobiol* 64(5) 877-885 (1996)
20. G Hermann, M Wlaschek, TS Lange, K Prenzel, G Goerz and K Scharffetter-Kochanek, UV-A irradiation stimulates the synthesis of various matrix-metalloproteinases (MMPs) in cultured human fibroblasts, *Exp Dermatol* 2(2) 92-97 (1993)
21. A Oikarinen, The aging of skin: Chronoaging versus photoaging, *Photoderm Photoimmun Photmed* 43 3-4 (1990)
22. A Thibodeau, Metalloprotease inhibitors, *Cosmet Toil* 115(11) 75-82 (2000)
23. US Pat 6,124,268, Natural antioxidant compositions, method for obtaining same and cosmetic, pharmaceutical and nutritional formulations thereof, S Ghosal, assigned to Natreon (Sep 26, 2000)
24. S Ghosal, VK Triphati and S Chauhan, Active constituents of *Emblica officinalis*: Part 1- The chemistry and antioxidative effects of two new hydrolysable tannins, Emblicanin A and B, *Indian J Chem* 35B 941-948 (1996) and references cited therein
25. Emblica, a monograph by Merck KGaA (2001)
26. RK Chaudhuri, Emblica cascading antioxidant: A novel natural skin care ingredient, *Skin Pharmacol Appl Physiol* 15 374-380 (2002)
27. M Richard, JP Guiraud, AM Monjo and A Favier, Development of a simple antioxidant screening system assay using human skin fibroblasts, *Free Rad Res Commun* 16(5) 303-314 (1992)
28. DA Scudiero, RH Shoemaker, KD Paull, A Monks, S Tierney, TH Nofziger, MJ Currens, D Seniff and MR Boyd, *Cancer Research* 48(17) 4827-4833 (1988)
29. M Paya, B Halliwell and JRS Hoult, Interactions of a series of coumarins with reactive oxygen species, *Biochem Pharmacol* 44 205-214 (1992)
30. B Halliwell, JM Gutteridge and OI Aruoma, The deoxyribose method: a simple "test-tube" assay for determination of rate constants for reactions of hydroxyl radicals, *Anal Biochem* 165 215-219 (1987)
31. I Kraljic and S El Moshni, A new method for the detection of singlet oxygen in aqueous solutions, *Photochem Photobiol* 28 577-582 (1978)
32. S Gonzalez and MA Pathak, Inhibition of ultraviolet-induced formation of reactive oxygen species, lipid peroxidation, erythema and skin photosentization by *Polypodium leucotomos*, *Photodermatol Photoimmunol Photomed* 12 45-56 (1996)
33. AE Martel, R Gustafson and S Chaberek, in Advances in Catalysis, IX, A Farakas, ed, New York: Academic Press (1957) 319

34. E Graf, JR Mahoney, RG Byrant and JW Eaton, Iron-catalyzed hydroxyl radical formation, Stringent requirement for free iron coordination site, *J Biol Chem* 259 3620-3624 (1984)
35. NJ Lowe and J Friedlander, Sunscreens: rationale for use to reduce photodamage and phototoxicity, in *Sunscreens, Development, Evaluation and Regulatory Aspects*, 2nd edition, N Lowe, N Shaath and M Pathak, eds, New York: Marcel Dekker (1997)

A New Active from Germinated Seeds Boosts HSP Expression in Skin's Natural Defenses

Keywords: Secale cereale (rye) seed extract, stress proteins, heat shock proteins

Complex containing rye seed extract improves the natural skin defense by boosting HSP expression in response to stress

Stress is implicated in all relations between organisms and their external environment. Endogenous to each cell, there is an active protection system comprising, among other elements, proteins called "Heat Shock Proteins" (HSPs) or stress proteins.

HSPs represent a heterogeneous group of proteins between 10 and 110 kDa, classified according to their molecular weight.[1]

In normal conditions, these HSPs are implicated in transport, translocation and folding of new synthesized proteins in the cell, and they are often referred to as molecular chaperones.[1,2]

After stress, HSP quantity increases and they are involved in the repair of damaged proteins.[3,4] The HSPs are expressed by both epidermal and dermal cells, after any situation, any stress likely to compromise the cellular survival.

Their constant protective presence is needed in the skin, because skin is the first barrier against external aggressions such as heat shock, UV irradiation and pollutants.

Because of their protective and repairing implications, the stress proteins are a topic of great interest for cosmetics.

Controlling HSP Production

Controlling HSP production would enable one to control the cellular stress response and to maintain the cell in an optimal physiological tate. Several facts support the approach of boosting HSP synthesis in order to allow the skin to better defend itself against aggressions. First, in aged subjects the heat shock response is attenuated: the aged cells have less capacity to express HSPs upon stress exposure and lose their self-defense potential.[5-7]

Secondly, exposure to mild stress results in an increased expression of HSPs, which is followed by a transient state of increased resistance of cells to further stress.[7,8] This can be considered as a preconditioning of cells in order to maintain them in an optimal state of reactivity versus stress.

Consequently an "HSP-booster" active ingredient may be used for the protection of the cell in prevention and repair of stress-induced skin aging.

Therefore, we have developed and tested an active ingredient, which in presence of stress, boosts HSP expression improving the skin's natural defense.

Materials & Methods

The test was developed on a primary culture of human keratinocytes. The active ingredient was introduced into the culture medium, then keratinocytes were submitted to stress. Two kinds of stress have been tested. The first one, heat shock, was chosen because it is a reference stress, well described in literature. The second stress, pollution exposure, was chosen because it is close to environmental stress endured by skin all day long. The level of HSP induction in cells was measured after stress achievement.

Since HSP72 is one of the most studied HSPs,[6,9] it was chosen as the marker of HSP induction in this test model.

Active ingredient: Vegetal Germinated Complex, called VGC in this text (INCI name: *Secale cereale* (rye) seed extract (and) dextrin), was obtained under controlled conditions from organically certified germinated rye seeds. We selected the germination phase in order to benefit from the natural biotransformation that endogenous enzymes of the seed perform during this stage, when they hydrolyze storage proteins and saccharides into bio-available oligopeptides and oligosaccharides.

Human primary keratinocyte culture: Human keratinocytes were isolated from normal human surgical skin from healthy adults and seeded on culture chambered slides[a]. Cells were incubated at 37°C with 5% of CO_2 and 95% relative humidity (RH). At confluence, keratinocytes were treated with VGC and immediately submitted to stress.

Stress conditions: Heat shock was achieved by incubating cells at 45°C for different periods.

Pollution exposure was carried out by incubating cells for 20 min, in enclosure, with cigarette smoke at different concentrations.

After stress, cells were incubated in normal conditions at 37°C, before HSP72 evaluation.

Evaluation of HSP72 mRNA by real time RT-PCR: Total ribonucleic acid (RNA) was extracted using an RNA kit[b]. Reverse Transcriptase – Polymerase Chain Reaction (RT-PCR) was performed using HSP72-specific primers and a reaction mix for RT-PCR[c]. Results were expressed in rate of HSP72 mRNA (messenger RNA) present in stressed cells versus control non-stressed cells.

Evaluation of HSP72 protein by immunocytochemistry: Immunocytochemistry (ICC) was performed on methanol-fixed keratinocyte cultures using a HSP72-specific antibody[d], revealed by a fluorescein isothiocyanate-labeled secondary antibody[e]. Quantification of HSP72 was performed by analysis of pictures done with a confocal laser-scanning microscope[f]. Results were expressed in percentage of area occupied by HSP72 or in number of stained nuclei in cell culture.

[a] VWR, France
[b] Nucleospin RNA kit, Macherey Nagel, France

A New Active from Germinated Seeds Boosts HSP Expression in Skin's Natural Defenses

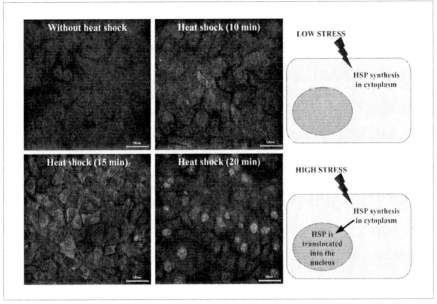

Figure 1. Effect of heat shock on the expression of HSP72 in human keratinocytes. The HSP response depends on the stress intensity.

Results

HSP72 response to stress in cultured keratinocytes: In cultured keratinocytes without stress exposure, no HSP induction was noticed.

Upon stress conditions, we observed an induction of HSP72 in the cultured keratinocytes. By ICC, we visualized that newly synthesized HSP72 proteins appeared first in the cytoplasm (see Figures 1 and 2). For higher intensity of stress, HSP72 were also detectable in nuclei of stressed cells. This observation is applicable for both stresses that we have studied: heat shock (Figure 1) and pollution exposure (Figure 2).

So, cultured human keratinocytes induce a protective response, mediated by HSP synthesis, which is modulated according to the stress dose.

HSP-booster effect of VGC after a heat shock: The effect of our active ingredient VGC on HSP72 protein expression was evaluated in keratinocyte cultures by ICC (Figure 3).

Compared to untreated cells, we observed a higher synthesis of HSP72 after a mild heat shock when cells have been treated by the VGC at 3%. Indeed, after a heat shock of 15 min, the percentage of occupation of HSP72 in the cytoplasm was significantly higher in cells treated by the VGC than in untreated cells. In the same way, after a heat shock of 20 min, the HSP72 expression was significantly increased by the VGC treatment both in the cytoplasm and in nuclei (Figure 4).

[c] Light Cycler Fast Start DNA Master Sybr Green I kit, Roche, France
[d] Tebu, France
[e] Amersham, France
[f] LSM, Germany

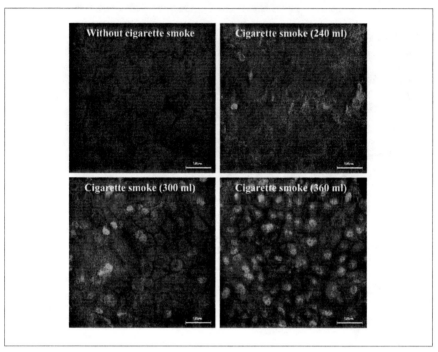

Figure 2. Effect of a pollutant (cigarette smoke at different concentrations) on the expression of HSP72 in human keratinocytes

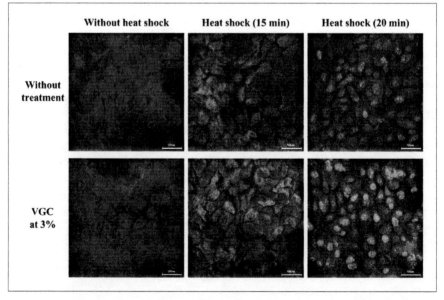

Figure 3. HSP-booster effect of VGC at 3% after a mild heat shock

It is important to underline that in "stress free" conditions, VGC did not induce HSP72 by itself (Figure 4). So VGC can not be considered as a stressing agent. Moreover, the absence of toxicity and the protective effect of VGC were checked by measuring adenosine triphosphoric acid levels (marker of cell viability) and lactate dehydrogenase release (a marker of cell distress) on heat shock cultured cells (data not shown).

HSP-booster effect of VGC after a pollution exposure: The HSP-booster effect of VGC was verified during another kind of stress, pollution exposure, with two concentrations of cigarette smoke (Figure 5).

The same observations as for heat shock were done. We have confirmed that in "stress free" conditions, VGC did not induce HSP72 by itself (Figure 6).

After the lower concentration of cigarette smoke exposure, the HSP72 expression was significantly increased by the VGC treatment in nuclei. Likewise, after the higher concentration of cigarette smoke tested, VGC treatment has significantly increased the HSP72 expression in both cytoplasm and nuclei (Figure 5).

We conclude that VGC has proved its ability to boost protein HSP72 synthesis following stresses as different as heat shock and pollution exposure. The protective response of cells is increased by VGC, according to the intensity of stress.

Characterization of the HSP-booster effect of VGC at mRNA level: Seeing that VGC was able to boost the protein HSP72 in cultured keratinocytes, we also studied this induction at mRNA level (Figure 7).

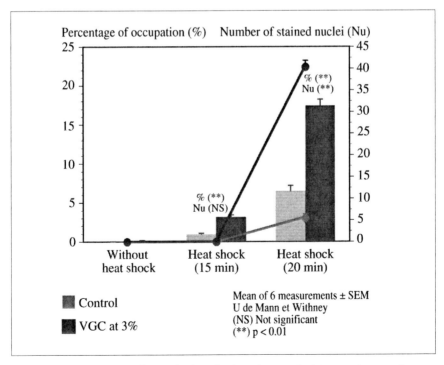

Figure 4. HSP-booster effect of VGC after heat shock. Evaluation of HSP72 protein expression. Columns = percentage of HSP occupation in the cytoplasm; Lines = number of stained nuclei

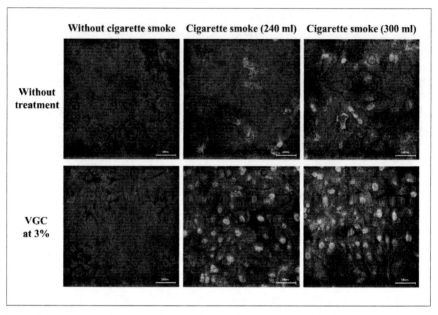

Figure 5. HSP-booster effect of VGC at 3% after a pollution exposure (cigarette smoke)

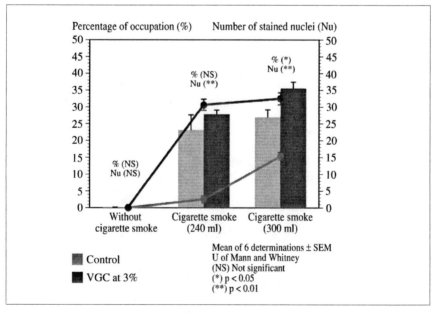

Figure 6. HSP-booster effect of VGC after cigarette smoke exposure. Evaluation of HSP72 protein expression. Columns = percentage of HSP occupation in the cytoplasm; Lines = number of stained nuclei

The HSP72 mRNA level was quantified at different times after stress exposure, by real time RT-PCR technique. It allowed us to follow the kinetics of HSP72 mRNA induction.

Without heat shock, no HSP72 mRNA induction was noticed. After a mild heat shock, the treatment with VGC provoked a quicker synthesis of HSP72 mRNA (Figure 8). Indeed the maximum of the response was obtained 60 min after heat shock in VGC-treated cells and only after 90 min in untreated cells.

We conclude that VGC has proved its ability to boost HSP72 synthesis at both mRNA and protein levels. The speedier HSP72 mRNA induction boosts in turn the HSP72 protein synthesis too.

Conclusion

VGC has been developed and successfully tested as an HSP-booster active ingredient. In stressing conditions, VGC has led to gentle HSP induction improving natural skin defense. VGC maintains the cell in an optimal physiological status, allowing a quick and efficient response to stress via increase of HSP synthesis. VGC strengthens the endogenous defense system of the skin resulting in better self-protection versus daily environmental aggression.

Furthermore, by reinforcing the natural self-defense mechanism of cells, VGC counterbalances age-related deficiency in endogenous protection. This could mean that the level of protection enjoyed by young cells could be restored to aging cells.

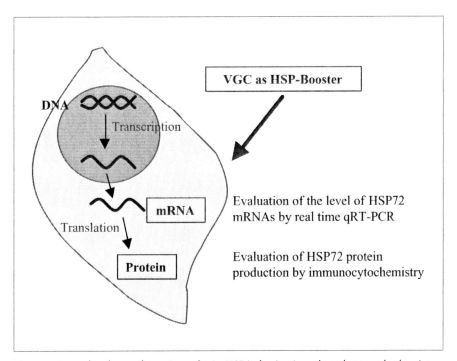

Figure 7. General pathway of protein synthesis. HSP induction is evaluated at two check points.

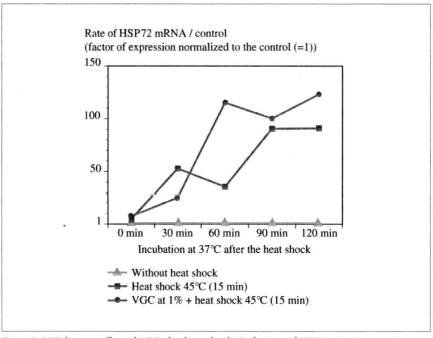

Figure 8. HSP-booster effect of VGC after heat shock. Evaluation of HSP72 mRNA expression

—Christine Jeanmaire, Vincent Bardey, Louis Danoux, Philippe Moussou and Gilles Pauly, *Laboratoires Sérobiologiques, A division of Cognis France, Pulnoy, France*

References

1. BS Polla, Heat (Shock) and the skin, *Dermatologica* 180 113-117 (1990)
2. D Ang, K Liberek, D Skowyra, M Zylicz and C Georgopoulos, Biological role and regulation of the universally conserved heat shock proteins, *J Biol Chem* 266 24233-24236 (1991)
3. MJ Gething and J Sambrook, Protein folding in the cell, *Nature* 355 33-35 (1992)
4. MYS Sherman and AL Goldberg, Involvement of molecular chaperones in intracellular protein breakdown, in *Stress-Inducible Cellular Responses*, U Feige, RI Morimoto, I Yahara and BS Polla, eds, Birkhauser Verlag (1996) 57-78
5. A Gutsmann-Conrad, AR Heydari, S You and A Richarson, The expression of Heat Shock Protein 70 decreases with cellular senescence in vitro and in cells derived from young and old human subjects, *Exp Cell Res* 241 404-413 (1998)
6. T Muramatsu, M Hatoko, H Tada, T Shirai and T Ohnishi, Age-related decrease in the inductability of heat shock protein 72 in normal human skin, *Br J Dermatol* 134 1035-1038 (1996)
7. F Trautinger, Heat shock proteins in the photobiology of human skins, *J Photochem Photobiol* 63 70-77 (2001)
8. F Trautinger, I Kindas-Mügge, RM Knobler and H Honigsmann, Stress proteins in the cellular response to ultraviolet radiation, *J Photochem Photobiol* 35 141-148 (1996)
9. F Trautinger, I Trautinger, I Kindas-Mügge, D Metze and TA Luger, Human keratinocytes in vivo and in vitro constitutively express the 72-kD heat shock protein, *J Invest Dermatol* 101 334-338 (1993)

Hyaluronan: History and Biochemistry

Keywords: hyaluronic acid, hyaluronan

Research from mid-1800s to present shows development of hyaluronan as the skin's natural moisturizing substance

This chapter and the next two review the literature on hyaluronic acid, which is also called hyaluronan. This chapter reviews the history and biochemistry of hyaluronan, a natural moisturizer of skin. The two subsequent articles will discuss hyaluronan's biology, pathology and pharmacology.

Hyaluronan in the Skin

Skin is a large and complex tissue that interfaces with a hostile environment and performs a vast range of functions. Scientists are just beginning to understand the mechanisms that underlie the resilience of skin to the harsh outside world and the extraordinary ability of the skin to protect underlying tissues.

Skin retains a large amount of water but experiences losses of that moisture due to the normal process of aging and to much of the external trauma to which skin is constantly subjected. One key molecule involved in skin moisture is hyaluronic acid (HA) with its associated water-of-hydration. Hyaluronic acid is also called hyaluronan (see sidebar), a term whose meaning will be explained later.

Understanding the metabolism of HA, its reactions within skin and the interactions of HA with other skin components will facilitate the ability to modulate skin moisture in a rational manner. This rational approach differs from the empirical attempts that have been utilized up to now.

Recent progress in the details of the metabolism of HA has also clarified the long-appreciated observations that premature aging of skin has two causes: 1) chronic

"Hyaluronan" or "Hyaluronic Acid"?

The term "hyaluronic acid" has fallen out of favor. The term "hyaluronan" is used in ophthalmology, particularly after cataract surgery, especially with patients who are uncomfortable with an "acid" being injected into the eye.

Another argument for preferring the term "hyaluronan" is that it brings the terminology for this repeating disaccharide glycosaminoglycan in line with the other members of this class of molecule, such as dermatan, keratin and heparan.

inflammation and 2) sun damage caused by ultraviolet light. These processes, as well as normal aging, all use similar mechanisms that cause loss of moisture and changes in HA distribution.

In the past several decades, the constituents of skin have become better characterized. The earliest work on skin was devoted predominantly to the cells that make up the layers of skin: epidermis, dermis and underlying subcutis. Now it is beginning to be appreciated that the materials that lie between cells, the matrix components, have major instructive roles for cellular activities. This extracellular matrix (ECM) endows skin with its hydration properties.

The components of the ECM, though they appear amorphous by light microscopy, form a highly organized structure of glycosaminoglycans (GAGs), proteoglycans, glycoproteins, peptide growth factors and structural proteins such as collagen and, to a lesser extent, elastin. The predominant component of the ECM of skin, however, is HA. It is the primordial and the simplest of the GAGs, and the first ECM component to be elaborated in the developing embryo. It is the water-of-hydration of HA that forms the blastocyst, the first recognizable structure in embryonic development.

Attempts to enhance the moisture content of skin, in the most elemental terms, require increasing the level and the length of time HA is present in skin, preserving optimal chain length of this sugar polymer and inducing expression of the best profile of HA-binding proteins to decorate the molecule.

Historical Perspective

The "ground substance" era: The term "ground substance" was first attributed to the amorphous-appearing material between cells by the German anatomist Henle in 1841.[1] It is a mistranslation of the German "Grundsubstanz" which would be better translated as "basic," "fundamental" or "primordial" substance. By 1855, sufficient information had accumulated for "Grundsubstanz" to be included in a textbook of human histology by Köllicker.[2]

The study of ground substance began in earnest in 1928, with the discovery of a "spreading factor" by Duran-Reynals.[3-7] A testicular extract was shown to stimulate the rapid spreading of materials injected subcutaneously and to function by causing a dissolution of ground substance. Thus, a new field of research was founded. The active principle in the extract was later shown to be a hyaluronidase, one of the class of enzymes that degrade HA.[8,9] The observed dissolution of ground substance stimulated Duran-Reynals to write the following, which is just as applicable today:

"If the importance of a defensive entity is to be judged by the magnitude of the measures taken against it, nature is certainly pointing its finger to the ground substance, as if to invite us to learn more about it."[10]

"Mucopolysaccharide" period: "Ground substance" was subsequently renamed "mucopolysaccharides," a term first proposed by Karl Meyer[11] to designate the hexosamine-containing polysaccharides that occur in animal tissues. It was intended to refer to the sugar polymers alone as well as when bound to proteins. However, the term "ground substance" persisted for many years afterwards, and could be found in textbooks of biochemistry, dermatology and pathology as late as

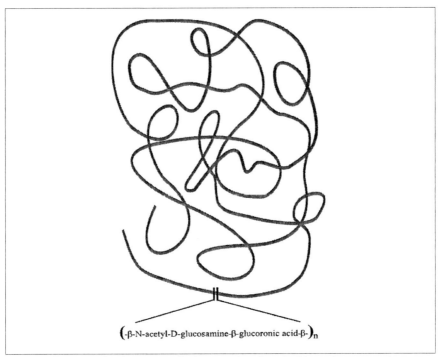

Figure 1. The HA molecule, composed of a repeating disaccharide, and represented as a random coil

the 1970s. It is now established that HA is the predominant mucopolysaccharide of skin and the major component of ground substance.

Discovery of HA: Hyaluronan was identified in 1938 by Meyer,[12] as a hexuronic acid-containing material that also provided the turgor for the vitreous humor of the eye. The name hyaluronic acid was proposed from the Greek hyalos (glassy, vitreous) and uronic acid. Not until nearly 20 years later, however, was the chemical structure of HA established.[13]

The molecule is composed of the repeating disaccharide shown in Figure 1, which represents the molecule as a random coil. Its molecular weight can approach 10 million Daltons. It takes on a huge volume of water-of-hydration that makes it a very expansive molecule. It can expand its solvent domain up to 10,000 times its actual polymer volume and is the mechanism of the edema and swelling associated with inflammation.

Later still it was found that the polymer is present throughout the body, identified in virtually every vertebrate tissue. The highest concentrations occur in the vitreous of the eye, in the synovial fluid in the joint capsule and in the umbilical cord as Wharton's jelly. However, more than 50% of total body HA is present in skin.[14]

The modern era: The modern era of HA biology began with the realization that HA is a critical regulator of cell behavior and has profound effects on cellular metabolism. Thus, it is not merely a passive structural component of the ECM. This understanding was brought into focus by a number of observations (see sidebar).

Cellular Behavior of HA

- HA is prominent in embryogenesis in maintenance of the undifferentiated state. Its removal is required prior to the onset of differentiation as was established by the pioneering work of Toole.[15]
- HA has a dynamic turnover rate. In the circulation HA has a half-life of three to five minutes.[16]
- HA is prominent in the earliest stages of adult wound healing,[17] while elevated levels occur over a long period during scar-free fetal repair.[18-20] The prolonged presence of HA is invoked as the mechanism of such scar-free repair.
- HA is involved in malignant progression,[21] and the aggressiveness of tumors correlates with levels of HA on the cancer cell surface.[22]
- HA is a signaling molecule, and fragmented HA has major influences on angiogenesis[23,24] and inflammation.[25-27]
- HA has receptors on cell surfaces. The predominant HA receptors CD44[28-30] and RHAMM[31,32] have complex variant isoforms. These receptors have the ability to confer motility upon cells with signalling to the cytoskeleton.[33-35] These receptors themselves are regulated and are the substrates for phosphokinases.[33]
- HA is found intracellularly and has intracellular modes of action.[36]

The future: The growth of molecular genetics and progress in the human genome project have facilitated rapid development in the understanding of HA metabolism.

The enzymes that synthesize HA (HA synthases), as well as the HA receptors and the enzymes that catalyze the catabolic reaction (the hyaluronidases), are all multigene families of enzymes with distinct patterns of tissue expression.

The multiple sites for the control of HA synthesis, deposition, cell- and protein-association and degradation reflect the complexity of HA metabolism. Their relationships are becoming clarified through the ability to sequence rapidly using the new techniques of molecular genetics. There promises to be an enormous increase in information and in the understanding of HA biology as the genes for these enzymes and proteins become sorted out.

Biochemistry of Hyaluronan

Overview: Hyaluronan is a high molecular weight, very anionic polysaccharide that promotes cell motility, adhesion and proliferation. These processes require cell movement and tissue organization.[37,38] The tight regulation required of HA expression under such conditions is modulated in part by association of HA with cell surface receptors.

Despite the monotony of its composition, without branch points or apparent variations in sugar composition, HA has an extraordinarily high number of functions. Physicochemical studies indicate that the polymer can take on a vast number of shapes and configurations, dependent on polymer size, pH, salt concentration and associated cations. Hyaluronan also occurs in a number of physiological states: circulating freely, tissue-associated by way of electrostatic interactions but easily dissociated and in equilibrium with the HA in the rest of the body.

Hyaluronan may be bound to proteins termed hyaladherins.[39,40] The HA can be very tightly associated with hyaladherins through electrostatic interactions. The HA in the ECM of cartilage is an organizer of the matrix; there the proteoglycan aggrecan and link proteins decorate the HA in a bottlebrush configuration. The affinity of these molecules for each other is of such magnitude that HA is not easily dissociated and is not in equilibrium with the HA of the surrounding loose connective tissues. HA also occurs covalently bound to proteins such as inter-alpha trypsin inhibitor, a plasma protein that also functions as a stabilizer of HA-rich structures.[41]

Tissues that contain high molecular weight HA are unusually resistant to invasion and penetration.[42] Blood vessels are unable to penetrate joint synovium, cartilage and the vitreous of the eye. It is also unusual for tumor metastases to develop in these structures. It may be the large size of the HA polymer that also protects such structures from invasion by parasites. The mechanism by which such high molecular weight structures resist hyaluronidase degradation and avoid the rapid HA turnover characteristic of the rest of the body is not known. Potent hyaluronidase inhibitors may be involved. Little is known about this class of molecules.

HA and GAGs: Hyaluronan is composed of repeating alternating units of glucuronic acid and N-acetylglucosamine, all connected by β-linkages, GlcAβ(1Æ 3)GlcNAcβ(1Æ 4). The β-linkage is of more than passing interest and not merely a curiosity relevant only to carbohydrate chemists. Glycogen is a polymer of α-linked glucose. Changing to a β-linkage converts the polymer to cellulose. A high molecular weight chain of β-linked N-acetylglucosamine is the structure of chitin. Chitin and cellulose are the most abundant sugar polymers on the surface of the earth. Yet such β-linked sugar polymers are rare in vertebrate tissues and require unusual reactions for their catabolic turnover.

Hyaluronan is the simplest of the GAGs, the only one neither covalently linked to a core protein nor synthesized by way of a Golgi pathway, and it is the only non-sulfated GAG. The current terminology refers to 1) GAGs, the straight-chain hexosamine sugars and 2) proteoglycans, referring to GAG chains together with the core protein to which they are covalently bound. Hyaluronan is thus the only GAG that is not also a component of a proteoglycan.

Existing models suggest that for high molecular mass HA, super molecular organization consists of networks in which molecules run parallel for hundreds of nanometers, giving rise to flat sheets and tubular structures that separate and then join again into similar aggregates.

There is strong evidence that an H_2O bridge between the acetamido and carboxyl groups is involved in the secondary structure. The hydrogen-bonded secondary structure also shows large arrays of contiguous -CH groups, giving a hydrophobic character to parts of the polymer that may be significant in the lateral aggregation or self-association, and for interaction with membranes.[43] This same hydrophobic character is perhaps involved in the extrusion of newly synthesized HA chains from the cytoplasmic surface of the plasma membrane, where the HA synthases are located, through the membrane to the exterior of the cell.[44] The unusually stiff tertiary polymeric structure is also stabilized by such hydrophobic interactions.

Glycosaminoglycans and proteoglycans must be distinguished from "mucins," the branch-chained sugars and their associated proteins. These occur more often on cell surfaces, though they also accumulate in the intercellular ground substance, particularly in association with malignancies. The terms are used carelessly, particularly among pathologists and histologists, and "mucin," "mucinous," "myxomatous," "myxoid" or "acid mucoproteins"—unless they have been defined biochemically—may or may not refer to HA-containing materials. This problem has arisen in part because of the ill-defined or unknown nature of histochemical color reactions. A recent example of this ambiguity is the incorrect assumption that the stain Alcian blue has some specificity for HA at pH 3 and for the sulfated GAGs at pH 1.5.[45]

Structure and function: By electron microscopy, HA is a linear polymer.[46] It is polydisperse but usually has a molecular mass of several millions.

In solution at physiological pH and salt concentrations, HA is an expanded random coil with an average diameter of 500 nm. The molecular domain encompasses a large volume of water and, even at low concentrations, solutions have very high viscosity.

The HA in high concentrations, as found in the ECM of the dermis, regulates water balance, osmotic pressure, functions as an ion exchange resin and regulates ion flow. It functions as a sieve to exclude certain molecules, to enhance the extracellular domain of cell surfaces (particularly the lumenal surface of endothelial cells), to stabilize structures by electrostatic interactions and also to act as a lubricant.

Hyaluronan also acts as an organizer of the ECM. It is the central molecule around which other components of the ECM distribute and orient themselves.[47]

The anomalous ability of HA to be both hydrophobic and hydrophilic, to associate with itself, with cell surface membranes, with proteins or with other GAGs speaks to the versatility of this remarkable molecule.

Summary

As so often happens in biology, the more we know of a molecule, the more we become aware of the complexities and subtleties associated with it. Our own preconceived notions become the greatest hindrance to understanding the realities.

Hyaluronan began as the amorphous and presumably inert ground substance of skin. Now it is recognized as being the skin's natural moisturizing substance and a molecule involved in a myriad of basic biological processes. These will be reviewed in the following two chapters as will other aspects of the role of HA in skin biology.

—Birgit A. Neudecker, Antonei Benjamin Csóka, Susan Stair Nawy, Howard I. Maibach and Robert Stern, *University of California, San Francisco, School of Medicine, San Francisco, California USA*
–Kazuhiro Mio, *Lion Corporation, Kanagawa, Japan*

References

1. F Henle, Vom Knorpelgewebe, *Allgemeine Anatomielehre, Von den Mischungs- und Formbestandteilen des menschlichen Koerpers*, Leipzig: Leopold Voss Verlag (1841) 791
2. A Koelliker, *Von den Geweben, Handbuch der Gewebelehre des Menschen*, Leipzig: Wilhelm Engelmann Verlag, (1852) 51
3. F Duran Reynals, Exaltation de l'activité du virus vaccinal par les extraits de certains organes, *Compt rend Soc Biol* 99 6 (1928)
4. F Duran-Reynals and J Suner Pi, Exaltation de l'activité du Staphylocoque par les extraits testiculaires, *Compt rend Soc Biol* 99 1908 (1929)
5. F Duran-Reynals, The effect of extracts of certain organs from normal and immunized animals on the infecting power of virus vaccine virus, *J Exp Med* 50 327 (1929)
6. F Duran-Reynals and FW Stewart, The action of tumor extracts on the spread of experimental vaccinia of the rabbit, *Amer J Cancer* 15 2790 (1933)
7. F Duran-Reynals, Studies on a certain spreading factor existing in bacteria and its significance for bacterial invasiveness, *J Exp Med* 58 161 (1933)
8. E Chain and ES Duthie, Identity of hyaluronidase and spreading factor, *Brit J Expl Path* 21 324 (1940)
9. GL Hobby, MH Dawson, K Meyer and E Chaffee, The relationship between spreading factor and hyaluronidase, *J Exp Med* 73 109 (1941)
10. J Casals, *Significance and Transcendence of the Scientific Work of Duran-Reynals, Viruses and Cancer*, WM Stanley, J Casals, J Oro and R Segura, eds, Madrid: Span Biochem Soc Press (1971) 416
11. K Meyer, The chemistry and biology of mucopolysaccharides and glycoproteins, *Sympos Quant Biol* 6 91 (1938)
12. K Meyer and JW Palmer, The polysaccharide of the vitreous humor, *J Biol Chem* 107 629 (1934)
13. MM Rapport, B Weissman, A Linker and K Meyer, Isolation of a crystalline disaccharide, hyalobiuronic acid, from hyaluronic acid, *Nature* 168 996 (1951)
14. RK Reed, K Lilja and TC Laurent, Hyaluronan in the rat with special reference to the skin, *Acta Physiol Scand* 134 405 (1988)
15. BP Toole, Proteoglycans and hyaluronan in morphogenesis and differentiation, in *Cell Biology of Extracellular Matrix*, ED Hay, ed, New York: Plenum Press (1991) 14
16. JR Fraser, TC Laurent, H Pertoft and E Baxter, Plasma clearance, tissue distribution and metabolism of hyaluronic acid injected intravenously in the rabbit, *Biochem J* 200 415 (1981)
17. PH Weigel, GM Fuller and RD LeBoeuf, A model for the role of hyaluronic acid and fibrin in the early events during the inflammatory response and wound healing, *J Theor Biol* 119 219 (1986)
18. RL DePalma, TM Krummel, LAD Durham, BA Michna, BL Thomas, JM Nelson and RF Diegelmann, Characterization and quantitation of wound matrix in the fetal rabbit, *Matrix* 9 224 (1989)
19. BA Mast, LC Flood, JH Haynes, RL DePalma, IK Cohen, RF Diegelmann and TM Krummel, Hyaluronic acid is a major component of the matrix of fetal rabbit skin and wounds: implications for healing by regeneration, *Matrix* 11 63 (1991)
20. MT Longaker, ES Chiu, NS Adzick, M Stern, MR Harrison and R Stern, Studies in fetal wound healing. V. A prolonged presence of hyaluronic acid characterizes fetal wound fluid, *Ann Surg* 213 292 (1991)
21. W Knudson, Tumor-associated hyaluronan. Providing an extracellular matrix that facilitates invasion, *Am J Path* 148 1721 (1996)
22. L Zhang, CB Underhill and L Chen, Hyaluronan on the surface of tumor cells is correlated with metastatic behavior, *Cancer Res* 55 428 (1995)
23. DC West and S Kumar, The effect of hyaluronate and its oligosaccharides on endothelial

cell proliferation and monolayer integrity, *Exp Cell Res* 183 179 (1989)
24. P Rooney, S Kumar, J Ponting and M Wang, The role of hyaluronan in tumour neovascularization, *Int J Cancer* 60 632 (1995)
25. MR Horton, CM McKee, C Bao, F Liao, JM Farber, J Hodge-DuFour, E Purae, BL Oliver, TM Wright and PW Noble, Hyaluronan fragments synergize with interferon-gamma to induce the C-X-C chemokines mig and interferon-inducible protein-10 in mouse macrophages, *J Biol Chem* 273 35088 (1998)
26. MR Horton, MD Burdick, RM Strieter, C Bao and PW Noble, Regulation of hyaluronan-induced chemokine gene expression by IL-10 and IFN-gamma in mouse macrophages, *J Immunol* 160 3023 (1998)
27. M Slevin, J Krupinski, S Kumar and J Gaffney, Angiogenic oligosaccharides of hyaluronan induce protein tyrosine kinase activity in endothelial cells and activate a cytoplasmic signal transduction pathway resulting in proliferation, *Lab Invest* 78 987 (1998)
28. C Underhill, CD44: the hyaluronan receptor, *J Cell Sci* 103 293 (1992)
29. J Lesley and R Hyman, CD44: structure and function, *Front Biosci* 3 616 (1998)
30. D Naor, RV Sionov and D Ish-Shalom, CD44: structure, function, and association with the malignant process, *Adv Cancer Res* 71 241 (1997)
31. LM Pilarski, A Masellis-Smith, AR Belch, B Yang, RC Savani and EA Turley, RHAMM, a receptor for hyaluronan-mediated motility, on normal human lymphocytes, thymocytes and malignant B cells: a mediator in B cell malignancy?, *Leuk Lymph* 14 363 (1994)
32. CL Hall and EA Turley, Hyaluronan: RHAMM mediated cell locomotion and signaling in tumorigenesis, *J Neuro-Onc* 26 221 (1995)
33. LY Bourguignon, VB Lokeshwar, X Chen and WG Kerrick, Hyaluronic acid-induced lymphocyte signal transduction and HA receptor, *J Immunol* 151 6634 (1993)
34. J Entwistle, CL Hall and EA Turley, HA receptors: regulators of signalling to the cytoskeleton, *J Cell Biochem* 61 567 (1996)
35. L Collis, C Hall, L Lange, M Ziebell, R Prestwich and EA Turley, Rapid hyaluronan uptake is associated with enhanced motility: implications for an intracellular mode of action, *Febs Letters* 440 444 (1998)
36. B Formby and R Stern, Phosphorylation stabilizes alternatively spliced CD44 mRNA transcripts in breast cancer cells: inhibition by antisense complementary to casein kinase II mRNA, *Molec Cell Biochem* 187 23 (1998)
37. TC Laurent and JR Fraser, Hyaluronan, *Faseb J* 6 2397 (1992)
38. *The Chemistry, Biology and Medical Applications of Hyaluronan and its Derivatives*, TC Laurent, ed, London: Portland Press (1998)
39. BP Toole, Hyaluronan and its binding proteins, the hyaladherins, *Cur Opin Cell Biol* 2 839 (1990)
40. CB Knudson and W Knudson, Hyaluronan-binding proteins in development, tissue homeostasis, and disease, *Faseb J* 7 1233 (1993)
41. M Zhao, M Yoneda, Y Ohashi, S Kurono, H Iwata, Y Ohnuki and K Kimata, Evidence for the covalent binding of SHAP, heavy chains of inter-alpha-trypsin inhibitor, to hyaluronan, *J Biol Chem* 270 26657 (1995)
42. RN Feinberg and DC Beebe, Hyaluronate in vasculogenesis, *Science* 220 1177 (1983)
43. JE Scott, J. E., Secondary structures in hyaluronan solutions: chemical and biological implications, in *The Biology of Hyaluronan*, D Evered and J Whelan, eds, Chichester: John Wiley & Sons (1989) 16
44. P Prehm, Hyaluronate is synthesized at plasma membranes, *Biochem J* 220 597 (1984)
45. W Lin, S Shuster, HI Maibach and R Stern, Patterns of hyaluronan staining are modified by fixation techniques, *J Histochem Cytochem* 45 1157 (1997)
46. JH Fessler and LI Fessler, Electron microscopic visualization of the polysaccharide hyaluronic acid, *Proc Nat Acad Sci USA* 56 141 (1966)
47. TN Wight, DK Heinegard and VC Hascall, Proteoglycans structure and function, in *Cell Biology of the Extracellular Matrix*, ED Hay, ed, New York: Plenum Press (1991) 45

Hyaluronan: Biology, Pathology and Pharmacology

Keywords: hyaluronic acid, hyaluronan, UV, extracellular matrix, wound healing, inflammation, synthetic dermis

Functions and receptors of hyaluronan are discussed as are its effects on aging skin, inflammation and synthetic dermis

Hyaluronan is now recognized as being the skin's natural moisturing substance and a molecule involved in a myriad of basic biological processes. The history and biochemistry of hyaluronon (also known as hyaluronic acid - HA is the abbreviation used for either term) were reviewed in the previous chapter.[1] This chapter continues our review of the literature on this key molecule.

Function of Hyaluronan

The large volume that HA occupies—including its cloud of solvent, the water-of-hydration under physiological conditions—underlies its ability to distend and maintain the extracellular space and preserve tissue hydration. Hyaluronan content increases whenever rapid tissue proliferation, regeneration and repair occur.[2] Its ability to organize the extracellular matrix (ECM) and its voluminous water-of-hydration and its interaction with other macromolecules explain only a portion of the functions with which it is associated.

For example, bursts of HA deposition correlate with mitosis.[3-5] Elevated levels promote cell detachment, in preparation for mitosis, as cells leave tissue organization, and enter the transient autonomy required for the mitotic event to occur. Cells must then degrade that HA after mitosis has occurred to regain adhesiveness, and to reenter the "social contract" with surrounding cells. The prediction is that HA synthesis occurs as cells enter mitosis, and that hyaluronidase activity is activated as cells leave mitosis. To date, such experiments have not been carried out in synchronized cells.

The persistent presence of HA also inhibits cell differentiation,[6,7] creating an environment that instead promotes cell proliferation. The elevated levels of anti-adhesive surface HA that promotes cell detachment, also permits the embryonic cell to migrate[8] or the tumor cell to move and metastasize.[9,10] The water-of-hydration also opens up spaces creating a permissive environment for cell movement.

Hyaluronan is generally produced in the interstitium in the mesenchymal connective tissue of the body, and is largely a product of fibroblasts. It reaches the blood through the lymphatics. Most of the turnover of HA, approximately 85%, occurs in the lymphatic system. This remaining 15% that reaches the blood stream has a rapid turnover, with a $t_{1/2}$ of three to five minutes, being rapidly eliminated by receptors in the liver, and also by unknown mechanisms in the kidney.[11-13] When the hepatic or renal arteries are ligated, there is an immediate rise in the level of circulating HA.[14] Thus, humans synthesize and degrade several grams of HA daily.

During acute stress, such as in shock or with septicemia, there is a rapid rise in circulating HA.[15-18] Such HA may function as a volume expander, as a survival mechanism to prevent circulatory collapse. Some of this rapid rise in HA represents HA recruited from interstitial stores and from lymphatics; thus, the rise is not entirely a reflection of increased synthesis or decreased degradation.[19] However, a higher plasma level of HA does correlate with decreased turnover rates, the $t_{1/2}$ reaching 20 to 45 minutes in situations of acute stress.

The mean serum and plasma level in healthy young people is 20-40 µg/L.[20,21] This value increases with age[22,23] well into the ninth decade, and probably reflects slower clearance and decreased HA degradative capacity, though this has not been carefully investigated. Hyaluronan also increases in the circulation in liver disease, particularly cirrhosis, and in renal failure reflecting aberrant degradation,[24-26] in rheumatoid arthritis[27] and consistently in some malignancies as a result of increased tumor tissue synthesis.[28]

In embryonic development: The developing embryo is rich in HA. The HA creates the spaces permissive for fetal cell migration and proliferation. The HA concentration is high not only in the fetal circulation, but also in amniotic fluid,[29] the fetal tissues, fetal membranes and in the placenta. The HA levels reach a maximum of 20 µg/mL at approximately 20 weeks of gestation, and then drop until, at 30 weeks gestation, they reach the 1 µg/mL adult-like levels. This corresponds approximately to the time when a "switch" from the scar-free fetal wound healing to the adult-like wound healing with scarring occurs.[30] The factors in the fetal circulation that support such high levels of HA synthesis have been explored and partially characterized,[31] but have not yet been isolated or fully identified.

The neural crest cells, as they pinch off from the neuroectoderm, migrate through the embryonic body in a sea of HA.[8] When these cells reach their particular destination, hyaluronidases remove the HA, and cell migration then ceases. In embryology, as parenchymal glands develop, HA can be found in the stroma immediately ahead of the arborizing tips, creating the spaces into which the growing glands can grow.[32,33]

The classic studies of Toole and his laboratory separate embryology into two stages. In the first stage, which is a primary HA-rich phase, undifferentiated stem cells proliferate and migrate. In the second stage, in which the HA is removed, we see the onset of cellular differentiation and morphogenesis.[2] This model can be superimposed on the development of virtually all parenchymal organs and vertebrate structures.

In wound healing: The ECM in the earliest stages of wound healing is rich in HA. There is also an abundance of inflammatory cells, a necessary component for the normal process of wound healing.

In the adult, HA levels rapidly reach a maximum and then drop rapidly,[34] reminiscent of the stages in embryology. Decreasing HA levels are followed by increasing amounts of chondroitin sulfate, the appearance of fibroblasts and then deposition of a collagen-rich ECM.

In the adult, wound healing often results in scar formation. In the fetus, however, wound repair is associated with levels of HA that remain elevated, and the final result is a wound free of scar. Such observations are made in both the experimental fetal rabbit and sheep models, as well as clinically, in term infants following mid-gestational in-utero surgery.

It is on this basis that elevated HA in the wound matrix is invoked as a key to decreased scarring, contractures and adhesions in adult wound repair. Aspects of wound healing appear to be a strategic retreat to an embryonic situation, followed by a rapid recapitulation of ontogeny.

In carcinogenesis: In malignancy, HA also appears to play a critical role.[9,35] Levels of HA on the surface of tumor cells correlate with their aggressiveness.[10] In a study of tumor cell-associated HA, the proportion of tumor HA-positive cells, as well as intensity of HA staining are unfavorable prognostic factors in colorectal cancer.[36] However, over-expression of hyaluronidase also correlates with disease progression, as shown in bladder[37,38] and in breast tumor metastases.[39,40] These apparently diverse scenarios may indicate that HA and hyaluronidase are required at different stages in the multi-step progression of cancer.

In aging: HA levels are high in the fetal circulation and fall shortly after birth. After maintaining a steady level for several decades, circulating levels of HA then begin to increase again in old age.[20,23,41] Elevated levels of circulating HA are also found in the syndromes of premature aging, in progeria[42] and in Werner's Syndrome.[43]

Increased HA levels in the bloodstream decrease immune competence.[44] Various mechanisms have been invoked. An HA coating around circulating lymphocytes may prevent ligand access to lymphocyte surface receptors.[45-48] The increased HA may represent one of the mechanisms for the immunosuppression in the fetus. The reappearance of high levels of HA in old age may be one of the mechanisms of the deterioration of the immune system in the elderly. The increasing levels of HA with aging may be a reflection of the deterioration of hydrolytic reactions, including the hyaluronidases that maintain the steady state of HA. This is a far more likely mechanism than an increase in HA synthase activity.

The increased HA that is often found in malignancy in the bloodstream[49-52] as well as on the surface of tumor cells[10] may be one of the cancer's techniques for compromising host immune function. It is the probable basis of the failure to rosette in the classic sheep red blood cell rosette test, a former laboratory procedure used to diagnose malignancy.[53,54] The rosetting failure may have been due to the HA coating on the cancer patients' lymphocyte surfaces.

Biology of Hyaluronan

Hyaladherins: Hyaluronan exists in a number of states in the vertebrate body. Within the ECM, it can be firmly intercalated within proteoglycans and binding proteins in a bottlebrush-like configuration. It can be bound to cells by means of cell surface receptors. Some of the HA exists in a free form circulating in the lymphatic or cardiovascular system. However, even in this relatively free form, there are a number of binding proteins that decorate HA. These are referred to collectively as hyaladherins, a term coined by Toole.[55]

The hyaladherins associate with HA through electrostatic or covalent bonds.[56] It is likely that some of the unique properties attributed to HA are in fact a function of the hyaladherins that are bound to the HA. Growth factors, collagen[57] and a myriad of other proteins have been identified.

One of the major challenges and opportunities in dermatology is to identify the profile of hyaladherins specific for the HA of epidermis and dermis, to characterize these proteins and to understand their function in relation to age-related changes. In an examination of skin as a function of age, the levels of HA did not decrease, as would be expected. Instead, the binding of HA to tissue proteins became more tenacious, and the HA became increasingly more difficult to extract.[58]

Another challenge is to understand how HA as a substrate for degradation by hyaluronidases is affected by associated hyaladherins. It is also reasonable to assume that the secondary structure of the HA polymer is modulated, in part, by the hyaladherins bound to it.

Hyaluronan in the ECM: The extracellular matrix that surrounds cells and occupies the variable spaces between cells is composed predominantly of structural proteins such as collagen and elastin, as well as proteoglycans and a number of glycoproteins. The basal lamina or basement membrane that separates dermis and epidermis is composed of similar materials, and is therefore also considered an ECM structure.

A number of growth factors are embedded in the ECM, concentrated by ECM components where they are protected from degradation. Such factors are presented to cells as mechanisms for growth control and modulators of cell function. Heparan sulfate-containing proteoglycans bind members of the fibroblast growth factor (FGF) and epidermal growth factor (EGF) family,[59] while HA can bind growth factors such as transforming growth factor beta (TGF-β).[60]

An emerging, complex picture is beginning to suggest that HA and heparan sulfate have opposing functions. An HA-rich environment is required for the maintenance of the undifferentiated, pluripotential state, facilitating motility and proliferation. On the other hand, the heparan-sulfate proteoglycans promote differentiation.

However, the concentration of HA in the ECM can vary widely. Even when the levels are decreased, as in areas of marked fibrosis, HA functions as an organizer of the ECM, as a scaffold about which other macromolecules of the ECM orient themselves. Diameters of collagen fibers can be modulated by levels of HA, the thinner more delicate fibers being favored in regions of high HA concentrations. In fibroblast cultures, the addition of exogenous HA to the medium decreases the diameter of the collagen fibers that accumulate.[61]

The ability of HA to promote cell proliferation is dependent in part on the concentration of the HA molecule,[62] opposite effects being achieved at high and low concentrations. Size is also important. High molecular weight HA is anti-angiogenic,[63] while lower molecular weight HA moieties are highly angiogenic, stimulating growth of endothelial cells,[10] attracting inflammatory cells and also inducing expression of inflammatory cytokines in such cells.[64-66] Partially degraded HA may have the opposite effect, possibly because it is no longer able to retain and release growth factors such as TGF-β.[60]

The intense staining for HA in psoriatic lesions may in part be due to partially degraded HA, and may be the mechanism for the marked capillary proliferation and inflammation that characterizes these lesions.[67-69] Attempts to stimulate HA deposition for purposes of promoting skin hydration must use caution that the HA deposited remain high molecular weight, by preventing free radical-catalyzed chain breaks and by carefully restricting the catabolic reactions of the hyaluronidases.

Intracellular hyaluronan: The most recent development is the realization that HA and associated hyaladherins are intracellular and have major effects on cellular metabolism. Much of the recent advance comes from the ability to remove the ECM of cultured cells using the highly specific *Streptomyces* hyaluronidase. Permeabilizing such cells and using confocal microscopy then makes it possible to use localization techniques for the identification of intracellular HA and its associated proteins.[70]

Such HA complexes also appear to be a component of the nuclear matrix in a wide variety of cells.[71,72] They also have importance in regulating the cell cycle and gene transcription. A vertebrate homologue of the cell cycle control protein CDC37 was recently cloned and found to be an hyaladherin,[73] as was a protein that copurified with the splicing factor SF2.[74] An intracellular form of the HA receptor RHAMM was demonstrated to regulate erk kinase activity.

Changes in function of these intracellular hyaladherins, depending on whether or not they have HA molecules attached, confers another layer of complexity dependent on intracellular hyaluronidase enzymes.

In the HA-rich vertebrate embryo and fetal tissues, there is minimal intercellular ECM. Most of the HA is intracellular, and the role of such intracellular HA in development is unknown. The HA-rich germinal epithelium and pluripotential basal cells of the bone marrow, as well as basal epithelium keratinocytes contain large amounts of HA involved in cell physiology. Such HA should be separated from the HA of the ECM, presumably the more important compartment when dealing with skin moisture.

Hyaluronan Receptors

CD44: HA-binding proteins are varied and broadly distributed in a wide variety of locations. In the ECM, they may be cell surface-associated, intracellular, both cytoplasmic and nuclear. The same molecule may occur in multiple locations. However, it is those that attach HA to the cell surface that constitute receptors.

The most prominent among these is CD44, a transmembrane glycoprotein that occurs in a wide variety of isoforms, products of a single gene with variant exon expression.[75-77] CD44 is coded for by 10 constant exons, plus from zero to 10 variant exons, all inserted into a single extracellular position near the membrane insertion site.[78] Additional variations in CD44 can occur as a result of post-translational glycosylation. This can take the form of addition of various glycosaminoglycans (GAGs), including chondroitin sulfate and heparan sulfate.

CD44 is able to bind a variety of other ligands, some of which have not yet been identified. CD44 has been shown, however, to interact with fibronectin, collagen and heparin-binding growth factors. CD44 is distributed widely, being found on virtually all cells except red blood cells. It plays a role in cell adhesion, migration, lymphocyte activation and homing, and in cancer metastasis.

The appearance of HA in dermis and epidermis parallels the histolocalization of CD44. The nature of the CD44 variant exons in skin at each location has not been described. The ability of CD44 to bind HA can vary as a function of differential exon expression. It would be of intrinsic interest to establish whether modulation occurs in CD44 variant exon expression with changes in the state of skin hydration. Also awaiting description are changes in the profile of CD44 variant exon expression as a result of skin pathologies.

Only one of many possible examples of the importance of CD44-HA interactions in normal skin physiology is given here. The HA in the matrix surrounding keratinocytes serves as an adhesion substrate for the Langerhans cells with their CD44-rich surfaces as they migrate through the epidermis.[79,80] In skin pathophysiology, the effect of local and systemic immune disorders on such interactions between Langerhans cells and keratinocytes awaits explanation.[81]

RHAMM: The other major receptor for HA is RHAMM (Receptor for HA-Mediated Motility),[82,83] which was discovered and cloned by Turley. This receptor is implicated in cell locomotion, focal adhesion turnover and contact inhibition. It also is expressed in a number of variant isoforms. The interactions between HA and RHAMM regulate cell locomotion by a complex network of signal transduction events and interaction with the cytoskeleton of cells. It is also an important regulator of cell growth.[84]

The TGF-β stimulation of fibroblast locomotion uses RHAMM. TGF-β is a potent stimulator of motility in a wide variety of cells. In fibroblasts, TGF-β triggers the transcription, synthesis and membrane expression of not only RHAMM, but also the synthesis and expression of the HA, all of which occurs coincident with the initiation of locomotion.[85]

Both RHAMM and CD44 may be among the most complex biological molecules ever described, with locations in an unusually wide variety of cell compartments, and associated with a spectrum of activities involving signal transduction, motility and cell transformation.

Laboratories differ in their observations regarding the receptors CD44 and RHAMM.[86] This apparent inconsistency reflects the subtle ways HA exerts its broad spectrum of biological effects and the myriad of mechanisms for controlling levels of HA expression and deposition. Particularly in the experimental laboratory situation, numerous factors have major repercussions in expression of HA, its receptors or the profile of hyaladherins that decorate the HA molecule. Among these factors are minor changes in culture conditions, differences in cell passage number and length of time following plating. One could also cite variations in growth factors contained in lots of serum, or differences in stages of cell confluence.

Pathology and Pharmacology

Artifacts of hyaluronan histolocalization in skin: Hyaluronan occurs in virtually all vertebrate tissues and fluids, but skin is the largest reservoir of body HA, containing more than 50% of the total. Earlier studies on the distribution of HA in skin using histolocalization techniques seriously underestimated HA levels. Formalin is an aqueous fixative and much of the soluble tissue HA is eluted by this procedure. The length of time tissue is in the formalin is a variable that may explain the conflicting results that are often encountered. Acidification and addition of alcohol to the fixative causes the HA to become more avidly fixed so that subsequent aqueous steps are unable to elute HA out of the tissue (see sidebar).[87]

Epidermal hyaluronan: Until recently, it was assumed that only cells of mesenchymal origin were capable of synthesizing HA, and HA was therefore restricted to the dermal compartment of skin. However, with the advent of a technique using biotinylated HA-binding peptide[88] for the histolocalization of HA, the evidence for HA in the epidermis became apparent.[58,89-92] In addition, techniques for separating dermis and epidermis from each other permitted accurate measurement of HA in each compartment, verifying that epidermis does contain HA.[93]

Hyaluronan is most prominent in the upper spinous and granular layers of the epidermis, where most of it is extracellular. The basal layer has HA, but it is predominantly intracellular, and is not easily leeched out during aqueous fixation. Presumably, basal keratinocyte HA is involved in cell cycling events, while the secreted HA in the upper outer layers of the epidermis are mechanisms for disassociation and eventual sloughing of cells.

Cultures of isolated keratinocytes have facilitated the study of epithelial HA metabolism. Basal keratinocytes synthesize copious quantities of HA. When Ca^{++} of the culture medium is increased from 0.05 to 1.2 mM these cells begin to differentiate, HA synthesis levels drop,[94] and there is the onset of hyaluronidase activity.[95] This increase in calcium that appears to simulate in culture the natural *in situ* differentiation of basal keratinocytes parallels the increasing calcium gradient observed in the epidermis. There may be intracellular stores of calcium that are released as keratinocytes mature.

Alternatively, the calcium stores may be concentrated by lamellar bodies from the intercellular fluids released during terminal differentiation. The lamellar bodies are thought to be modified lysosomes containing hydrolytic enzymes, and a potential source of the hyaluronidase activity. The lamellar bodies fuse with the plasma membranes of the terminally differentiating keratinocytes, increasing the plasma membrane surface area. Lamellar bodies are also associated with proton pumps that enhance acidity. The lamellar bodies also acidify, and their polar lipids become partially converted to neutral lipids, thereby participating in skin barrier function.

Diffusion of aqueous material through the epidermis is blocked by these lipids synthesized by keratinocytes in the stratum granulosum, the boundary corresponding to the level at which HA-staining ends. This constitutes part of the barrier function of skin. The HA-rich area inferior to this layer may obtain water from the moisture-rich dermis. And the water contained therein cannot penetrate beyond the lipid-rich stratum granulosum. The HA-bound water in both the dermis and in the vital area of the epidermis is critical for skin hydration. And the stratum granulosum is essential for maintenance of that hydration, not only of the skin, but of the body in general. Profound dehydration is a serious clinical problem in burn patients with extensive losses of the stratum granulosum.

Method of Fixation Affects HA Localization in Skin Sections

Figure 1 shows sections of human skin. The sections were stained for HA using a biotinylated HA-binding peptide derived from bovine cartilage aggrecan. Histolocalization of HA is indicated by the blue color developed using an avidin-conjugated alkaline phosphatase. Slides were counterstained with Nuclear Fast Red to visualize cell structures of dermis and epidermis.

In Figures 1a and 1b, the skin section was fixed in acid-formalin/ethanol (1a) and in formalin/phosphate-buffered saline (PBS) (1b).

HA staining in skin is found predominantly in the dermis, rather than in the epidermis, particularly in the papillary dermis. The most intense staining is observed in the sections fixed with the acid-formalin/ethenol (1a), compared to the section fixed with conventional neutral-buffered formalin (1b). Note that in the acid-formalin/ethanol-fixed sample (1a), the color intensity from small scattered foci of staining in the epidermal layer is comparable to the intensity of staining found in the dermis; this relationship does not hold in the conventionally fixed sample (1b).

The staining for HA was blocked by preincubation of the HA-binding peptide with HA (not shown). In addition, preincubation of the HA-binding peptide with other GAGs (such as chondroitin sulfate, dermatan sulfate and keratan sulfate) at the same concentration did not decrease the intensity of subsequent HA staining (not shown). These results demonstrate that the HA-binding peptide staining reaction is highly specific for HA and that the peptide does not react with other tissue GAGs.

In Figures 1c and 1d, sections of human skin were incubated in PBS for 24 h at 37°C prior to undergoing the staining procedure described above. The skin section was fixed in acid-formalin/ethanol (1c) and in formalin/PBS (1d).

As shown, approximately 80-90% of HA is retained in the section fixed in the acid-formalin/ethanol (1c) compared to the section fixed in neutral-buffered formalin (1d). Notably, the staining of the epidermis is almost unchanged (1a and 1c). However, much of the stainable HA in the dermis has been leeched out of the sample fixed with neutral formalin during the overnight incubation in PBS (1d). Therefore, tissue fixed in acid-formalin/ethanol retained HA far better than that fixed in neutral-buffered formalin.

Dermal hyaluronan: The HA content of the dermis is far greater than that of the epidermis, and accounts for most of the 50% of total body HA present in skin.[96] The papillary dermis has the more prominent levels of HA than does reticular dermis.[58] The HA of the dermis is in continuity with both the lymphatic and vascular systems; epidermal HA has no such direct and immediate continuity. Exogenous HA is cleared from the dermis and rapidly degraded.[12]

The dermal fibroblast provides the synthetic machinery for dermal HA, and should be the target for pharmacological attempts to enhance skin hydration. The fibroblasts

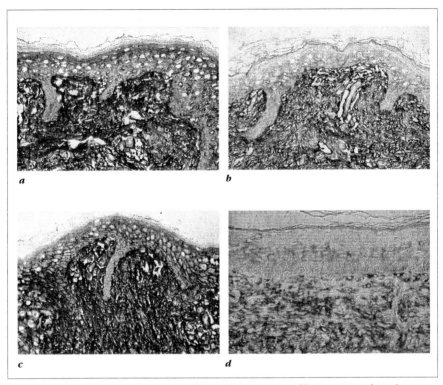

Figure 1. HA localization in skin sections (from Reference 87, used by permission from the *Journal of Histochemistry and Cytochemistry* and the Histochemical Society, Inc.)

of the body, the most banal of cells from a histologic perspective, are probably the most diverse of all vertebrate cells with the broadest repertoire of biochemical reactions and potential pathways for differentiation. Much of this diversity is site specific. What makes the papillary dermal fibroblast different from other fibroblasts is not known. However these cells have an HA synthetic capacity similar to that of the fibroblasts that line the joint synovium responsible for the HA-rich synovial fluid.[97]

Hyaluronan in aging skin: Though dermal HA is responsible for most skin HA, epidermal cells are also able to synthesize HA. The most dramatic histochemical change observed in senescent skin is the marked decrease in epidermal HA.[58] In senile skin, HA is still present in the dermis, while the HA of the epidermis has disappeared entirely.

The proportion of total GAG synthesis devoted to HA is greater in epidermis than in dermis, and the reason for the precipitous fall with aging is unknown. The synthesis of epidermal HA is influenced by the underlying dermis, as well as by topical treatments, such as with retinoic acids, indicating that epidermal HA is under separate controls from dermal HA.

In contrast with previous in vitro[98,99] and in vivo[100,101] observations, recent studies document that the total level of HA remains constant in the dermis with aging.[58]

The major age-related change is the increasing avidity of HA for tissue structures, and the concomitant loss of HA extractability. Such intercollated HA may have diminished ability to take on water of hydration. This decreased volume of water of hydration HA is obviously a loss in skin moisture.

An important study for the future would be to define precisely the hyaladherins that decorate the HA in senile skin, and to compare that profile with the hyaladherins of young skin in both the dermal and epidermal compartments.

Progressive loss in the size of the HA polymer in skin as a function of age has also been reported.[102]

The increased binding of HA with tissue as a function of age parallels the progressive cross-linking of collagen and the steady loss of collagen extractability with age. Each of these phenomena contributes to the apparent dehydration, atrophy and loss of elasticity that characterizes aged skin.

Photoaging of skin: Repeated exposure to UV radiation from the sun causes premature aging of skin.[103,104] Initially, UV damage causes a mild form of wound healing, and is associated first with elevated dermal HA. As little as five minutes of UV exposure in nude mice causes enhanced deposition of HA,[105] indicating that UV-induced skin damage is an extremely rapid event. The initial "glow" after sun exposure may be a mild edematous reaction induced by the enhanced HA deposition. But the transient sense of well being in the long run extracts a high price, particularly with prolonged exposure. Repeated exposures ultimately simulate a typical wound healing response with deposition of scar-like type I collagen, rather than the usual types I and III collagen mixture that gives skin resilience and pliability. The biochemical changes that distinguish photoaging and chronological aging have not been identified.

The abnormal GAGs of photoaging are those also found in scars, in association with the changes found late in the wound healing response, with diminished HA and increased levels of chondroitin sulfate proteoglycans. There is also an abnormal pattern of distribution.[104] The GAGs appear to be deposited on the elastotic material that comprises "elastosis" and diffusely associated with the actinic damaged collagen fibers. These appear as "smudges" on H&E sections[a] of sun-damaged skin, rather than between the collagen and elastin fibers as would be observed in normal skin.

Acute and chronic skin inflammation: Chronic inflammation causes premature aging of the skin, as observed in patients with atopic dermatitis. The constant inflammatory process leads to decreased function of the skin barrier accompanied by loss of skin moisture. Presumably, the skin of such patients contains decreased levels of HA.

Alternatively, the HA may reflect the changes found with chronological aging. With aging, HA becomes more tenaciously associated with tissue structures and becomes progressively more resistant to biochemical extraction.

Demonstration of such changes and the precise histolocalization of this decreased HA deposition would be of intrinsic interest, but that is a study that has not been performed yet.

---

[a] An H&E section refers to a slide on which there is a slice of paraffin-embedded tissue that is stained with hematoxylin and eosin to facilitate observations of cellular details.

The acute inflammatory process is associated initially with increased HA levels, the result of the cytokines released by the polymorphonuclear leukocytes, which are the predominant cells of the acute inflammatory process. The erythema, swelling and warmth of the acute process are followed later by the characteristic dry appearance and the formation of wrinkles. The precise mechanisms are unknown, but may relate to the differences between acute and chronic inflammatory cells and the attendant chemical mediators released by such cells. Alternatively, initiation of a wound healing response, with collagen deposition, may be a mechanism invoked for the premature aged appearance of the skin in chronic inflammation.

Hyaluronan in skin substitutes: There is a requirement for skin substitutes in a great number of clinical situations. In patients with extensive burns, insufficient skin is available for autologous split-thickness skin grafts. Resurfacing of the burned area can occur with autologous cultured epidermal cell autografts. However, this is dependent on a functioning dermal support, a problem that has given rise to a number of reasonable approaches.

Cadaver skin might be the source of a skin substitute in the case of extensive burns, but cadaver skin dermis has the problem of possible contamination and potential infection.

A synthetic dermis has the requirement for an HA content that will support epithelial migration, angiogenesis and differentiation. Various methods have been examined for modifying natural HA to provide materials with properties similar to the native polymer.

Many derivatives of HA have been formulated.[106-108] Such materials could provide flat dressings that can be seeded with fibroblasts. These same artificial dressings could also be seeded with cultured autologous keratinocytes. With laser-drilled microperforations, the keratinocytes can migrate through the membrane onto the wound bed. Such applications are already in use and result in complete healing with a minimum of scarring.

It is anticipated that in the coming years, a number of HA-derivatives will appear for clinical application in dermatology. They will contain cross-linked HA polymers as well as HA-ester derivatives obtained by the conjugation of the carboxylic acid of HA with various drugs in their alcohol forms. The HA polymer, because of its intrinsic biocompatibility, reactivity and degradability, will have many uses in the rapidly expanding field of tissue engineering and in the tissue substitutes of the future.

Summary

Hyaluronan is a deceptively simple molecule. Its complexity and its many modes of action are beginning to be understood. It can exist in many physical states, dependent on associated binding proteins, molecular size and the small molecules in the immediate environment.

Hyaluronan is the molecule chiefly responsible for the hydration of skin. The many different physical states are partly recapitulated as skin ages, from the embryo to late in life. Understanding these physical states, and recognizing the differences between them and the mechanisms involved, can provide important handles for the cosmetic industry. The ability to modulate and manipulate the physical states of HA will facilitate our intervention in some of the inevitable changes that occur in aging skin.

—Birgit A. Neudecker, Antonei Benjamin Csóka, Susan Stair Nawy, Howard I. Maibach and Robert Stern, *University of California, San Francisco, School of Medicine, San Francisco, California, USA*
—Kazuhiro Mio, *Lion Corporation, Kanagawa, Japan*

References

1. BA Neudecker, AB Csóka, SS Nawy, HI Maibach, R Stern and K Mio, Hyaluronan: History and biochemistry, *Antiaging*, Carol Stream, IL: Allured Publishing Corp. (2006)
2. BP Toole, Proteoglycans and hyaluronan in morphogenesis and differentiation, in *Cell Biology of Extracellular Matrix*, ED Hay, ed, New York: Plenum Press (1991) 14
3. M Tomida, H Koyama and T Ono, Hyaluronate acid synthetase in cultured mammalian cells producing hyaluronic acid: oscillatory change during the growth phase and suppression by 5 bromodeoxyuridine, *Biochim Biophys Acta* 338 352 (1974)
4. N Mian, Analysis of cell-growth-phase-related variations in hyaluronate synthase activity of isolated plasma-membrane fractions of cultured human skin fibroblasts, *Biochem J* 237 333 (1986)
5. M Brecht, U Mayer, E Schlosser and P Prehm, Increased hyaluronate synthesis is required for fibroblast detachment and mitosis, *Biochem* 239 445 (1986)
6. MJ Kujawa, DG Pechak, MY Fiszman and Al Caplan, Hyaluronic acid bonded to cell culture surfaces inhibits the program of myogenesis, *Develop Biol* 113 10 (1986)
7. MJ Kujawa and K Tepperman, Culturing chick muscle cells on glycosaminoglycan substrates: attachment and differentiation, *Develop Biol* 99 277 (1983)
8. RM Pratt, MA Larsen and MC Johnston, Migration of cranial neural crest cells in a cell-free hyaluronate-rich matrix, *Develop Biol* 44 298 (1975)
9. W Knudson, Tumor-associated hyaluronan. Providing an extracellular matrix that facilitates invasion, *Am J Path* 148 1721 (1996)
10. L Zhang, CB Underhill and L Chen, Hyaluronan on the surface of tumor cells is correlated with metastatic behavior, *Cancer Res* 55 428 (1995)
11. JR Fraser, TC Laurent, H Pertoft and E Baxter, Plasma clearance, tissue distribution and metabolism of hyaluronic acid injected intravenously in the rabbit, *Biochem J* 200 415 (1981)
12. RK Reed, UB Laurent, JR Fraser and TC Laurent, Removal rate of [^{3}H]hyaluronan injected subcutaneously in rabbits, *Am J Physiol* 259 H532 (1990)
13. UB Laurent, LB Dahl and RK Reed, Catabolism of hyaluronan in rabbit skin takes place locally, in lymph nodes and liver, *Exp Physiol* 76 695 (1991)
14. A Engstroem-Laurent and S Hellstroem, The role of liver and kidneys in the removal of circulating hyaluronan. An experimental study in the rat, *Connect Tissue Res* 24 219 (1990)
15. H Onarheim, RK Reed and TC Laurent, Elevated hyaluronan blood concentrations in severely burned patients, *Scand J Clin Lab Inv* 51 693 (1991)
16. H Onarheim, AE Missavage, RA Gunther, GC Kramer, RK Reed and TC Laurent, Marked increase of plasma hyaluronan after major thermal injury and infusion therapy, *J Surg Res* 50 259 (1991)
17. JJ Ferrara, RK Reed, DL Dyess, MI Townsley, H Onarheim, TC Laurent and AE Taylor, Increased hyaluronan flux from skin following burn injury, *J Surg Res* 50 240 (1991)
18. S Berg, B Brodin, F Hesselvik, TC Laurent and R Maller, Elevated levels of plasma hyaluronan in septicaemia, *Scand J Clin Lab Inv* 48 727 (1988)
19. H Onarheim, RK Reed and TC Laurent, Increased plasma concentrations of hyaluronan after major thermal injury in the rat, *Circ Shock* 37 159 (1992)
20. A Engstroem-Laurent, UB Laurent, K Lilja and TC Laurent, Concentration of sodium hyaluronate in serum, *Scand J Clin Lab Inv* 45 497 (1985)
21. K Chichibu, T Matsuura, S Shichijo and MM Yokoyama, Assay of serum hyaluronic acid in

clinical application, *Clin Chim Act* 181 317 (1989)
22. U Lindqvist and TC Laurent, Serum hyaluronan and aminoterminal propeptide of type III procollagen: variation with age, *Scand J Clin Lab Inv* 52 613 (1992)
23. J Yannariello-Brown, SH Chapman, WF Ward, TC Pappas and PH Weigel, Circulating hyaluronan levels in the rodent: effects of age and diet, *Amer J Physiol* 268 C952 (1995)
24. R Haellgren, A Engstroem-Laurent and U Nisbeth, Circulating hyaluronate. A potential marker of altered metabolism of the connective tissue in uremia, *Nephron* 46 150 (1987)
25. U Lindqvist, A Engstroem-Laurent, U Laurent, A Nyberg, U Bjeorklund, H Eriksson, R Pettersson and A Tengblad, The diurnal variation of serum hyaluronan in health and disease, *Scand J Clin Lab Invest* 48 765 (1988)
26. EH Cooper and BJ Rathbone, Clinical significance of the immunometric measure-ments of hyaluronic acid, *Ann Clin Biochem* 27 444 (1990)
27. G Smedegeard, J Bjeork, S Kleinau and A Tengblad, Serum hyaluronate levels reflect disease activity in experimental arthritis models, *Agents Actions* 27 356 (1989)
28. I Frebourg, G Lerebours, B Delpech, D Benhamou, P Bertrand, C Maingonnat, C Boutin and G Nouvet, Serum hyaluronate in malignant pleural mesothelioma, *Cancer* 59 2104 (1987)
29. L Dahl, JJ Hopwood, UB Laurent, K Lilja and A Tengblad, The concentration of hyaluronate in amniotic fluid, *Biochem Med* 30 280 (1983)
30. MT Longaker, DJ Whitby, NS Adzick, TM Crombleholme, JC Langer, BW Duncan, SM Bradley, R Stern, MW Ferguson and MR Harrison, Studies in fetal wound healing, VI. Second and early third trimester fetal wounds demonstrate rapid collagen deposition without scar formation, *J Ped Surg* 25 63 (1990)
31. M Decker, ES Chiu, C Dollbaum, A Moiin, J Hall, R Spendlove, MT Longaker and R Stern, Hyaluronic acid-stimulating activity in sera from the bovine fetus and from breast cancer patients, *Cancer Res* 49 3499 (1989)
32. MR Bernfield, SD Banerjee and RH Cohn, Dependence of salivary epithelial morphology and branching morphogenesis upon acid mucopolysaccharide-protein, *J Cell Biol* 52 674 (1972)
33. P Gakunga, G Frost, S Shuster, G Cunha, B Formby and R Stern, Hyaluronan is a prerequisite for ductal branching morphogenesis, *Devel* 124 3987 (1997)
34. MT Longaker, ES Chiu, NS Adzick, M Stern, MR Harrison and R Stern, Studies in fetal wound healing. V. A prolonged presence of hyaluronic acid characterizes fetal wound fluid, *Ann Surg* 213 292 (1991)
35. B Delpech, N Girard, P Bertrand, MN Courel, C Chauzy and A Delpech, Hyaluronan: fundamental principles and applications in cancer, *J Intern Med* 242 41 (1997)
36. K Ropponen, M Tammi, J Parkkinen, M Eskelinen, R Tammi, P Lipponen, U Agren, E Alhava and VM Kosma, Tumor cell-associated hyaluronan as an unfavorable prognostic factor in colorectal cancer, *Cancer Res* 58 342 (1998)
37. VB Lokeshwar, C Obek, MS Soloway and NL Block, Tumor-associated hyaluronic acid: a new sensitive and specific urine marker for bladder cancer, *Cancer Res* 57 773 (1997)
38. VB Lokeshwar, MS Soloway and NL Block, Secretion of bladder tumor-derived hyaluronidase activity by invasive bladder tumor cells, *Cancer Let* 131 21 (1998)
39. P Bertrand, N Girard, C Duval, J d'Anjou, C Chauzy, JF Maenard and B Delpech, Increased hyaluronidase levels in breast tumor metastases, *Int J Canc* 73 327 (1997)
40. AK Madan, K Yu, N Dhurandhar, C Cullinane, Y Pang and DJ Beech, Association of hyaluronidase and breast adenocarcinoma invasiveness, *Oncol Rep* 6 607 (1999)
41. A Engstroem-Laurent, Changes in hyaluronan concentration in tissues and body fluids in disease states, in *The Biology of Hyaluronan*, D Evered and J Whelan, eds, Chichester: John Wiley & Sons (1989) p 233
42. WT Brown, Progeria: a human-disease model of accelerated aging, *Am J Clin Nutr* 55 1222S (1992)
43. FJ Kieras, WT Brown, GE Houck Jr and M Zebrower, Elevation of urinary hyaluronic acid in Werner's syndrome and progeria, *Biochem Med Metabol Biol* 36 276 (1986)

44. TC Laurent, UB Laurent and JR Fraser, Serum hyaluronan as a disease marker, *Ann Med* 28 241 (1996)
45. JM Delmage, DR Powars, PK Jaynes and SE Allerton, The selective suppression of immunogenicity by hyaluronic acid, *Ann Clin and Lab Sci* 16 303 (1986)
46. WH McBride and JB Bard, Hyaluronidase-sensitive halos around adherent cells. Their role in blocking lymphocyte-mediated cytolysis, *J Exp Med* 149 507 (1979)
47. JV Forrester and PC Wilkinson, Inhibition of leukocyte locomotion by hyaluronic acid, *J Cell Sci* 48 315 (1981)
48. SJ Dick, B Macchi, S Papazoglou, EH Oldfield, PL Kornblith, BH Smith and MK Gately, Lymphoid cell-glioma cell interaction enhances cell coat production by human gliomas: novel suppressor mechanism, *Science* 220 739 (1983)
49. G Manley and C Warren, Serum hyaluronic acid in patients with disseminated neoplasm, *J Clin Path* 40 626 (1987)
50. CR Wilkinson, LM Bower and C Warren, The relationship between hyaluronidase activity and hyaluronic acid concentration in sera from normal controls and from patients with disseminated neoplasm, *Clin Chim Act* 256 165 (1996)
51. B Delpech, B Chevallier, N Reinhardt, JP Julien, C Duval, C Maingonnat, P Bastit and B Asselain, Serum hyaluronan in breast cancer, *Int J Cancer* 46 388 (1990)
52. H Hasselbalch, D Hovgaard, N Nissen and P Junker, Serum hyaluronan is increased in malignant lymphoma, *Am J Hemat* 50 231 (1995)
53. RL Gross, AG Levin, CM Steel, S Singh, G Brubaker and FG Peers, In vitro immunological studies on East African cancer patients. II. Increased sensitivity of blood lymphocytes from untreated Burkitt lymphoma patients to inhibition of spontaneous rosette formation, *Int J Cancer* 15 132 (1975)
54. RL Gross, A Latty, EA Williams, and PM Newberne, Abnormal spontaneous rosette formation and rosette inhibition in lung carcinoma. *N Engl J Med.* 292 169 (1975)
55. BP Toole, Hyaluronan and its binding proteins, the hyaladherins, *Cur Opin Cell Biol* 2 839 (1990)
56. M Zhao, M Yoneda, Y Ohashi, S Kurono, H Iwata, Y Ohnuki and K Kimata, Evidence for the covalent binding of SHAP, heavy chains of inter-alpha-trypsin inhibitor, to hyaluronan, *J Biol Chem* 270 26657 (1995)
57. DA Burd, JW Siebert, JP Ehrlich and HG Garg, Human skin and post-burn scar hyaluronan: demonstration of the association with collagen and other proteins, *Matrix* 9 322 (1989)
58. LJ Meyer and R Stern, Age-dependent changes of hyaluronan in human skin, *J Invest Derm* 102 385 (1994)
59. M Piepkorn, MR Pittelkow and PW Cook, Autocrine regulation of keratinocytes: the emerging role of heparin-binding, epidermal growth factor-related growth factors, *J Inv Derm* 111 715 (1998)
60. P Locci, L Marinucci, C Lilli, D Martinese and E Becchetti, Transforming growth factor beta 1-hyaluronic acid interaction, *Cell Tissue Res* 281 317 (1995)
61. R Stern, unpublished observations
62. RL Goldberg and BP Toole, Hyaluronate inhibition of cell proliferation, *Arth Rheum* 30 769 (1987)
63. RN Feinberg and DC Beebe, Hyaluronate in vasculogenesis, *Science* 220 1177 (1983)
64. MR Horton, CM McKee, C Bao, F Liao, JM Farber, J Hodge-DuFour, E Purae, BL Oliver, TM Wright and PW Noble, Hyaluronan fragments synergize with interferon-gamma to induce the C-X-C chemokines mig and interferon-inducible protein-10 in mouse macrophages, *J Biol Chem* 273 35088 (1998)
65. MR Horton, MD Burdick, RM Strieter, C Bao and PW Noble, Regulation of hyaluronan-induced chemokine gene expression by IL-10 and IFN-gamma in mouse macrophages, *J Immunol* 160 3023 (1998)
66. M Slevin, J Krupinski, S Kumar and J Gaffney, Angiogenic oligosaccharides of hyaluronan induce protein tyrosine kinase activity in endothelial cells and activate a cytoplasmic signal transduction pathway resulting in proliferation, *Lab Invest* 78 987 (1998)

67. S Kumar and DC West, Psoriasis, angiogenesis and hyaluronic acid, *Lab Invest* 62 664 (1990)
68. R Tammi, K Paukkonen, C Wang, M Horsmanheimo and M Tammi, Hyaluronan and CD44 in psoriatic skin. Intense staining for hyaluronan on dermal capillary loops and reduced expression of CD44 and hyaluronan in keratinocyte-leukocyte interfaces, *Arch Derm Res* 286 21 (1994)
69. S Gustafson, T Wikstreom and L Juhlin, Histochemical studies of hyaluronan and the hyaluronan receptor ICAM-1 in psoriasis, *Int J Tissue React* 17 167 (1995)
70. L Collis, C Hall, L Lange, M Ziebell, R Prestwich and EA Turley, Rapid hyaluronan uptake is associated with enhanced motility: implications for an intracellular mode of action, *Febs Letters* 440 444 (1998)
71. PS Eggli and W Graber, Association of hyaluronan with rat vascular endothelial and smooth muscle cells, *J Histochem Cytochem* 43 689 (1995)
72. SP Evanko and TN Wight, Intracellular localization of hyaluronan in proliferating cells, *J Histochem Cytochem* in press (1999)
73. N Grammatikakis, A Grammatikakis, M Yoneda, Q Yu, SD Banerjee and BP Toole, A novel glycosaminoglycan-binding protein is the vertebrate homologue of the cell cycle control protein, Cdc37, *J Biol Chem* 270 16198 (1995)
74. TB Deb and K Datta, Molecular cloning of human fibroblast hyaluronic acid-binding protein confirms its identity with P-32, a protein co-purified with splicing factor SF2. Hyaluronic acid-binding protein as P-32 protein, co-purified with splicing factor SF2, *J Biol Chem* 271 2206 (1996)
75. C Underhill, CD44: the hyaluronan receptor, *J Cell Sci* 103 293 (1992)
76. J Lesley and R Hyman, CD44 structure and function, *Front Biosci* 3 616 (1998)
77. D Naor, RV Sionov and D Ish-Shalom, CD44: structure, function, and association with the malignant process, *Adv Cancer Res* 71 241 (1997)
78. GR Screaton, MV Bell, DG Jackson, FB Cornelis, U Gerth and JI Bell, Genomic structure of DNA encoding the lymphocyte homing receptor CD44 reveals at least 12 alternatively spliced exons, *Proc Nat Acad Sci USA* 89 12160 (1992)
79. JM Weiss, J Sleeman, AC Renkl, H Dittmar, CC Termeer, S Taxis, N Howells, M Hofmann, G Keohler, E Scheopf, H Ponta, P Herrlich and JC Simon, An essential role for CD44 variant isoforms in epidermal Langerhans cell and blood dendritic cell function, *J Cell Biol* 137 1137 (1997)
80. JM Weiss, AC Renkl, J Sleeman, H Dittmar, CC Termeer, S Taxis, N Howells, E Scheopf, H Ponta, P Herrlich and JC Simon, CD44 variant isoforms are essential for the function of epidermal Langerhans cells and dendritic cells, *Cell Adhes Com* 6 157 (1998)
81. S Seiter, D Schadendorf, W Tilgen and M Zeoller, CD44 variant isoform expression in a variety of skin-associated autoimmune diseases, *Clin Immunol Immunopath* 89 79 (1998)
82. EA Turley, Hyaluronan and cell locomotion, *Canc Metas Rev* 11 21 (1992)
83. E Turley and R Harrison, RHAMM, a member of the hyaladherins, http://www.glycoforum.gr.jp (1999)
84. S Mohapatra, X Yang, JA Wright, EA Turley and AH Greenberg, Soluble hyaluronan receptor RHAMM induces mitotic arrest by suppressing Cdc2 and cyclin B1 expression, *J Exper Med* 183 1663 (1996)
85. SK Samuel, RA Hurta, MA Spearman, JA Wright, EA Turley and AH Greenberg, TGF-beta 1 stimulation of cell locomotion utilizes the hyaluronan receptor RHAMM and hyaluronan, *J Cell Biol* 123 749 (1993)
86. M Hofmann, V Assmann, C Fieber, JP Sleeman, J Moll, H Ponta, IR Hart and P Herrlich, Problems with RHAMM: a new link between surface adhesion and oncogenesis?, *Cell* 95 591 (1998)
87. W Lin, S Shuster, HI Maibach and R Stern, Patterns of hyaluronan staining are modified by fixation techniques, *J Histochem Cytochem* 45 1157 (1997)
88. JA Ripellino, M Bailo, RU Margolis and RK Margolis, Light and electron microscopic studies on the localization of hyaluronic acid in developing rat cerebellum, *J Cell Biol* 106 845 (1988)

89. R Tammi, JA Ripellino, RU Margolis and M Tammi, Localization of epidermal hyaluronic acid using the hyaluronate binding region of cartilage proteoglycan as a specific probe, *J Invest Derm* 90 412 (1988)
90. C Wang, M Tammi and R Tammi, Distribution of hyaluronan and its CD44 receptor in the epithelia of human skin appendages, *Histochem* 98 105 (1992)
91. U Bertheim and S Hellstroem, The distribution of hyaluronan in human skin and mature, hypertrophic and keloid scars, *Brit J Plast Surg* 47 483 (1994)
92. R Tammi and M Tammi, Hyaluronan in the epidermis, http://www.glycoforum.gr.jp (1998)
93. R Tammi, AM Seaeameanen, HI Maibach and M Tammi, Degradation of newly synthesized high molecular mass hyaluronan in the epidermal and dermal compartments of human skin in organ culture, *J Invest Derm* 97 126 (1991)
94. SI Lamberg, SH Yuspa and VC Hascall, Synthesis of hyaluronic acid is decreased and synthesis of proteoglycans is increased when cultured mouse epidermal cells differentiate, *J Invest Derm* 86 659 (1986)
95. OI Frost and R Stern, A microtiter based assay for hyaluronidase activity not requiring specialized reagents, *Analyt Biochem* 251 263 (1997)
96. RK Reed, K Lilja and TC Laurent, Hyaluronan in the rat with special reference to the skin, *Acta Physiol Scand* 134 405 (1988)
97. R Stern, unpublished experiments
98. DO Schachtschabel and J Wever, Age-related decline in the synthesis of glycosaminoglycans by cultured human fibroblasts, *Mech Ageing Develop* 8 257 (1978)
99. G Sluke, DO Schachtschabel and J Wever, Age-related changes in the distribution pattern of glycosaminoglycans synthesized by cultured human diploid fibroblasts, *Mech Ageing Develop* 16 19 (1981)
100. M Breen, HG Weinstein, LJ Blacik and MS Borcherding, Microanalysis and characterization of glycosaminoglycans from human tissue via zone electrophoresis, in *Methods in Carbohydrate Chemistry*, RL Whistler and JN BeMiller, eds, New York: Academic Press (1976) p 101
101. JH Poulsen and MK Cramers, Determination of hyaluronic acid, dermatan sulphate, heparan sulphate and chondroitin 4/6 sulphate in human dermis, and a material of reference, *Scand J Clin Lab Inv* 42 545 (1982)
102. MO Longas, CS Russell and XY He, Evidence for structural changes in dermatan sulfate and hyaluronic acid with aging, *Carbohyd Res* 159 127 (1987)
103. BA Gilchrest, A review of skin ageing and its medical therapy, *Brit J Derm* 135 867 (1996)
104. EF Bernstein, CB Underhill, PJ Hahn, DB Brown and J Uitto, Chronic sun exposure alters both the content and distribution of dermal glycosaminoglycans, *Brit J Derm* 135 255 (1996)
105. JJ Thiele and R Stern, unpublished experiments
106. GD Prestwich, DM Marecak, JF Marecek, KP Vercruysse, and MR Ziebell, Controlled chemical modification of hyaluronic acid: synthesis, applications, and biodegradation of hydrazide derivatives, *J Cont Rel* 53 93 (1998)
107. KP Vercruysse and GD Prestwich, Hyaluronate derivatives in drug delivery, *Crit Rev Therapeut Drug Carrier Syst* 15 513 (1998)
108. F Duranti, G Salti, B Bovani, M Calandra and ML Rosati, Injectable hyaluronic acid gel for soft tissue augmentation. A clinical and histological study, *Dermat Surg* 24 1317 (1998)

Hyaluronan: Metabolism and Modulation of Hyaluronan Levels in Skin

Keywords: hyaluronic acid, hyaluronan, hyaluronidases, UV, free radicals, AHAs

Hyaluronan synthesis and catabolism are discussed as is enhancing skin moisture

Understanding the metabolism of hyaluronan (HA), its reactions within skin and the interactions of HA with other skin components will facilitate the ability to modulate skin moisture. Recent progress in the detail of the metabolism of HA has also clarified the long-appreciated observations that premature aging of skin has two causes: 1) chronic inflammation and 2) sun damage caused by ultraviolet (UV) light. These processes, as well as normal aging, all use similar mechanisms that cause loss of moisture and changes in HA distribution.

The previous two chapters discussed history and biochemistry[1] and biology, pathology and pharmacology[2] of hyaluronan (also known as hyaluronic acid - HA is the abbreviation used for either term). This chapter concludes our review of the literature of this molecule.

Hyaluronan Synthases

Single-protein enzymes are now recognized as being able to synthesize HA, utilizing the two uridine-diphosphate (UDP)-sugar substrates, UDP-glucuronic acid and UDP-N acetylglucosamine. In eukaryotes, the enzyme resides on the cytoplasmic surface of the plasma membrane, and the HA product is extruded by some unknown mechanism through the plasma membrane into the extracellular space, permitting unconstrained polymer growth.[3] Such growth could not occur in the Golgi nor on the endoplasmic reticulum where most sugar polymers are synthesized without destruction of the cell.

Recent work has demonstrated that the HA synthases are a multigene family with at least three members, HAS-1, -2, and -3,[4,5] which are differentially regulated.

In situ expression of the HAS-1 and -2 genes are up-regulated in skin by TGF-β, in both dermis and epidermis, but there are major differences in the kinetics of the TGF-β response between HAS-1 and HAS-2, and between the two compartments, suggesting that the two genes are independently regulated. This also suggests that HA has a different function in dermis and epidermis.

Stimulation of HA synthesis also occurs following phorbol ester (PMA) and platelet-derived growth factor (PDGF) treatment, though a direct effect on HAS has not been demonstrated.[6] Glucocorticoids induce a nearly total inhibition of HAS mRNA in both dermal fibroblasts and osteoblasts. Extracts of dermal fibroblasts indicate that HAS-2 is the predominant HA synthase therein. This may be the molecular basis of the decreased HA in glucocortcoid-treated skin. However, an additional effect on rates of HA degradation has not been examined.

The parallels between chitin, cellulose and HA structures—all being β-chains of hexose polymers—are reflected in the striking similarity in sequence between the HA synthases from vertebrates, cellulose synthases from plants and chitin synthases from fungi. There must have been a shared primordial ancestral gene in the evolution of these enzymes involved in the biosynthesis of all polymers that contain β-glycoside linkages, an ancient β-polysaccharide synthase.

Hyaluronan Catabolism

The hyaluronidases: Hyaluronan is very metabolically active, with a half-life of 3-5 minutes in the circulation, and less than one day in skin. Even in an inert tissue such as cartilage, the HA turns over with a half-life of 1-3 weeks.[7-9] This catabolic activity is primarily the result of hyaluronidases, endoglycolytic enzymes with a specificity in most cases for the β 1-4 glycosidic bond.

The hyaluronidase family of enzymes have, until recently, been neglected,[10-12] in part because of the great difficulty in measuring their activity. They are difficult to purify and characterize. They are present at exceedingly low concentrations. They have very high specific activities, and are unstable in the absence of detergents.

New assay procedures have now facilitated their isolation and characterization.[13,14] The Human Genome Project has also promoted explication at the genetic level, and a virtual explosion of information has ensued.

An entire family of hyaluronidase-like genes has been identified.[15] There are seven hyaluronidases in the human genome. A cluster of three occurs on chromosome 3p, and a similar cluster of three is found on chromosome 7q31. This arrangement suggests that an original ancient sequence arose, followed by two tandem gene duplication events. This was followed by a more recent *en masse* duplication and translocation. From divergence data, it can be estimated that these events occurred over 300 million years ago, before the emergence of modern mammals. A seventh and non-homologous hyaluronidase gene occurs on chromosome 10q.[16] All of the hyaluronidase-like genes have unique tissue-specific tissue patterns.

The biology of hyaluronidases in skin has not been investigated, nor has it been established which of the various hyaluronidases participate in the turnover of HA in dermis and epidermis.

In vertebrate tissues, total HA degradation occurs by the concerted effort of three separate enzymatic activities: hyaluronidase and the two exoglycosidases that remove the terminal sugars; a β-glucuronidase; and a β-N-acetyl glucosaminidase.

Endolytic cleavage by the hyaluronidase generates ever-increasing substrates for the exoglycosidases. The relative contribution of each to HA turnover in either dermis or epidermis has yet to be established. But these classes of enzymes—as well as the hyaluronidases—represent important potential targets for the pharmacological control of HA turnover in skin.

Non-enzymatic degradation: The HA polymer can be degraded non-enzymatically by a free-radical mechanism,[17] particularly in the presence of reducing agents such as thiols, ascorbic acid, ferrous or cuprous ions. This mechanism of depolymerization requires the participation of molecular oxygen. The use of chelating agents in pharmaceutical preparations to retard free-radical-catalyzed breaks of HA chains has validity. However, a carefully monitored effect of such agents on HA chain length in human epidermis has not been attempted. Whether such agents can also affect the integrity of dermal HA in protecting them from free-radical damage, and whether these agents have any substantial effect on the moisturizing properties of skin HA remain important questions to be answered.

Hyaluronidase Inhibitors

Macromolecular inhibitors: The extraordinarily rapid turnover of HA in tissues suggests that tightly controlled modes exist for modulating steady-state levels of HA. The HA of the vertebrate body is of unique importance, and rapid increases are required in situations of extreme stress. Rapid turnover of HA in the normal state indicates constant synthesis and degradation.

Inhibition of degradation would provide a far swifter response to the sudden demand for increased HA levels than increasing the rate of HA synthesis. The ability to provide immediate high HA levels is a survival mechanism for the organism. This might explain the apparent inefficiency of rapid rates of HA turnover that occurs in the vertebrate animal under basal conditions. It can be compared to the need to suddenly drive an automobile much faster in the case of an emergency, not by stepping on the accelerator, but by taking a foot off the break.

If inhibition of HA degradation by hyaluronidase occurs, then the hyaluronidase inhibitors are very important. This class of molecules has not been explored. It can be postulated that with extreme stress, hyaluronidase inhibitors would be found in the circulation as acute phase proteins, the stress response products synthesized by the liver. These would prevent the ever-present rapid destruction and allow levels of HA to quickly increase.

Circulating hyaluronidase inhibitor activity was identified in human serum more than half a century ago.[18,19] Modifications in levels of inhibitor activity have been observed in the serum of patients with cancer,[20,21] liver disease[22] and with certain dermatological disorders.[23] This area of biology is unexplored, and though some early attempts were made,[24-26] and even though a review appeared,[27] these hyaluronidase inhibitors have been neither isolated nor characterized at the molecular level.

Inhibitors of mammalian origin, such as the serum inhibitor or heparin, are far more potent than the relatively weak inhibitors of plant origin. Hyaluronidase inhibitors of animal origin would provide a means for enhancing levels of HA in skin, and represent an important research area in attempting to enhance skin moisture.

Low molecular weight inhibitors: Classes of lower molecular weight inhibitors of hyaluronidase have been identified. Some of these inhibitors come from folk medicines, from the growing field of ethno-pharmacology. Some anti-inflammatories as well as some of the ancient beauty aids and practices for freshening of the skin may have some of these compounds as the basis of their mechanism of action.

Those that have been identified in recent times include flavonoids,[28-30] aurothiomalate,[31] hydrangenol[32] (occurring in the leaves of Hydrangea), tannins[33] and derivatives of tranilast.[34] One could also include curcumin[35] (an extract of the spice turmeric) and glycyrrhizin[36] (found in the roots and rhizomes of licorice and used as an effective anti-inflammatory agent in Chinese medicine).

Clinically, the anticoagulant heparin has potent anti-hyaluronidase activity,[37] as do salicylates[38] and the classic non-steroidal anti-inflammatory agent indomethacin.[39,40]

Oxidative Stress and Skin Hyaluronan

Reactive oxygen species or free radicals are a necessary component of the oxygen combustion that drives the metabolism of living things. Though they are important for generating the life force, they simultaneously are extraordinarily harmful. Organisms thus had to evolve protective mechanisms against oxidative stress. Over the course of evolution, different enzymatic and non-enzymatic anti-oxidative mechanisms were developed. Examples of these mechanisms include various vitamins, ubiquinone, glutathione, and circulating proteins such as hemopexin. In addition, hyaluronan may be one such mechanism, acting also as a free-radical scavenger.[41]

Sunlight (ultraviolet light) is an additional generator of harmful oxygen-derived species such as hydroxyl radicals. These radicals have the ability to oxidize and damage other molecules such as DNA, causing cross-linking and chain breaks. These hydroxyl radicals may also be destructive for proteins and lipid structures, as well as ECM components such as HA. After a very few minutes of UV exposure, disturbance in HA deposition can be detected.[42]

The anomalous situation exists, therefore, that HA can be both protective as a free-radical scavenger and simultaneously a target of free-radical stress. This paradox may be understood by a hypothetical model in which HA protects the organism from the free-radical stress resulting from the oxygen-generated internal combustion, but is itself harmed by the more toxic free radicals generated by the external world, by UV irradiation.

The generation of HA fragments by UV may underlie some of the irritation and inflammation that often accompanies long-term or intense sun exposure.[43-46] As discussed above, HA fragments are themselves highly angiogenic and inflammatory, inducing the production of a cascade of inflammatory cytokines. Further complications have occurred in this assembly of metabolic attack and counter-attack reactions that have been compiled in the selective forces of evolution. Antioxidants are present in the skin at an unusually high level. These antioxidants include vitamins C and E, as well as ubiquinone and glutathione. However, these precious compounds are depleted by exposure to sunlight.[47-49]

To prevent this sun-induced cascade of oxidative injuries, topical preparations containing antioxidants have been developed in the past several decades. Initially, such antioxidants were added as stabilizers to various dermatologic and cosmetic preparations. In particular, lipophilic vitamin E has been a favorite as a stabilizing agent. However, following oxidation, vitamin E is degraded into particularly harmful pro-oxidative metabolites.[50]

In the past several years, increasing concentrations of antioxidants have been used in such skin preparations, in an attempt to create complementary combinations, or to create constant recycling pairs that alternatingly oxidize and reduce each other.[51]

Finally, molecules such as HA should be protected by topical antioxidants to prevent degradation. Topical antioxidants, protecting against free-radical damage as well as maintaining HA integrity, may have major effects against natural aging and photoaging.[52,53]

Enhancing Skin Moisture by Modulating Hyaluronan

α-Hydroxy acids: Fruit compresses have been applied to the face as beauty aids for millennia. The α-hydroxy acids (AHAs) contained in fruit extracts, tartaric acid in grapes, citric acid in citrus fruits, malic acid in apples, mandelic acid in almond blossoms and apricots are thought to be active principles for skin rejuvenation. Such AHAs do stimulate HA production in cultured dermal fibroblasts.[54] The results of such alkaline preparations may depend more on their peeling effects than on the ability of AHAs to stimulate HA deposition.

Lactic acid,[55,56] citric acid,[55,57] and glycolic acid,[55,58-60] in particular, though frequent ingredients in AHA-containing cosmetic preparations, have widely varying HA-stimulating activity in the dermal fibroblast assay. Some of these mildly acidic (pH 3.7-4.0) preparations may owe their effectiveness to their traumatic peeling, astringent properties, with constant wounding of the skin. The cosmetic effects of these preparations of AHAs, including lactic acid, involve increased skin smoothness with the disappearance of lines and fine wrinkles.

Long-term use, however, results in thickening of the skin, in both the epidermal and papillary dermal layers, because of a mild fibrous reaction. This results from a process that could resemble prolonged, classic wound healing. This might explain the increased thickness and firmness of both dermis and epidermis. The increased collagen deposition documented in skin after prolonged use is consistent with a wound healing effect.[61] Preparations of AHAs as would have been found in the fruit compresses of the ancients have yet to find current cosmetic equivalents, though such vehicles are actively being sought.[62]

Upon examining the structure, it is obvious that ascorbic acid is similar in structure to an AHA. This is generally not appreciated. However, ascorbic acid is also present in fruit, and may underlie some of the effects attributed to fruit extracts. It has pronounced HA-stimulating effects in the fibroblast assay. But its antioxidant activity confounds the effects it may induce.

Retinoic acid and its derivatives: Topical applications of retinoic acid derivatives reduce the visible signs of aging and of photodamage[63] though there is little correlation between the histologic changes and the clinical appearance of the skin. Initial improvement in fine wrinkling and skin texture correlates with the deposition of HA in the epidermis.

While vitamin D is considered the "sunshine vitamin," vitamin A has been accepted as an apparent antidote for the adverse effects of sun exposure, and assumed to prevent and repair cutaneous photodamage.[63] Application of vitamin A derivatives does reverse some of the sun damage to skin, the roughness, wrinkling, and irregular pigmentation.[64,65] For the over-40 generation, brought up in an era of "suntan chic," appropriate preparations to restore or to prevent further deterioration of skin are critically important.

Impairment of the retinoid signal transduction pathways occurs as a result of prolonged UV exposure. Down regulation of nuclear receptors for vitamin A occurs,[66] resulting in a functional deficiency of vitamin A. Application of vitamin A derivatives would appear to be an obvious treatment modality. Topical application of vitamin A does increase the HA in the epidermal layer, increasing the thickness of the HA meshwork after prolonged treatment.[67]

Steroids: Topical and systemic treatment with glucocorticoids induces atrophy of skin, bone and a number of other organs, with a concomitant decrease in glycosaminoglycans, in particular HA. In human skin organ cultures, hydrocortisone has a bimodal effect. At low physiological concentrations (10^{-9} M), hydrocortisone maintains active synthesis and turnover of HA in the epidermis, while at high concentrations (10^{-5} M), hydrocortisone reduces epidermal HA content. The effect is achieved through both decreased synthesis as well as decreased rates of degradation.[68] The high concentrations of cortisone also enhance terminal differentiation of keratinocytes and reduces rates of cell proliferation.

Hydrocortisone is a potent inhibitor of HA synthases in fibroblasts. HA synthase 2 is the predominant synthase of dermal fibroblasts of the three HA synthase genes. Glucocorticoids induce a rapid and near total suppression of HA synthase 2 mRNA levels. The inhibition of HA deposition thus appears to occur at the transcriptional level. Progesterone inhibits HA synthesis in fibroblasts cultured from the human uterine cervix.[69] The steroid effect on HA appears to be system-wide.

Edema is one of the four cardinal signs of acute inflammation. The ability of glucocorticoids to suppress inflammation occurs in part by their ability to suppress the deposition of HA, the primary mechanism of edematous swelling that occurs during the inflammatory response.

General Comments from the Dermatology and Cosmetic Perspectives

The natural moisture of skin is attributed to its HA content. The critical property of HA is its ability to retain water, which it can do better than any known synthetic or naturally occurring compound.

Even at very low concentrations, aqueous solutions of HA have very high viscosity.

The advantage of using HA in cosmetic preparations was recognized very soon after its discovery. Difficulties in preparing large enough amounts of HA free of contaminating glycoproteins, lipids and other tissue materials prevented its convenient use in commercial preparations, including its use in cosmetics.

Initially, HA was isolated from rooster combs. This HA was highly purified, and used in ophthalmology as a viscoelastic to replace fluid loss following cataract surgery.

The revolution in biotechnology and molecular genetics made it possible more recently to engineer bacteria with augmented HA production by amplifying the HA synthase genes. This generates a material that has a much lower molecular weight, which is a disadvantage because low molecular weight hyaluronan fragments may induce expression of inflammatory cytokines. The material has the additional disadvantage of frequent contamination by residual bacterial pyrogens. Processed HA from vast fermentation of engineered bacteria has reduced the price of HA drastically, bringing the price into a range that is reasonable for its use in cosmetics. However, this genetically engineered HA of bacterial origin is not of sufficient purity for injectional use.

Many of the cosmetic preparations that contain HA have a concentration of 0.025% to 0.05%, sufficient to give the preparations a very smooth and viscous feel. When such solutions are applied to the skin, they form hydrated films that hold water for considerable periods, and confer the properties of a moisturizer.

Currently, research is underway to modify HA in such a way as to make it more stable and to confer very specific properties. Another direction in this research is to combine HA with other materials, such as chondroitin sulfate and modified sugar polymers, to simulate more closely the associations that HA has in its natural state in vertebrate tissues. Since the low molecular size HA fragments are highly angiogenic, defining the optimal size of the HA polymer for cosmetic purposes should be a major goal of such research.

Future Developments

We are in the very early stages of understanding the biology of HA and its metabolism. The enzymatic steps that constitute extracellular and intracellular HA cycles are beginning to be sorted out. The goals that lie before us are the identification of such reactions, and the development of techniques for modulating these reactions. This will facilitate formulation of strategies that enhance skin appearance and increase the moisture content of photodamaged and aging skin.

—**Birgit A. Neudecker, Antonei Benjamin Csóka, Susan Stair Nawy, Howard I. Maibach and Robert Stern,** *University of California, San Francisco, School of Medicine, San Francisco, California USA*
—**Kazuhiro Mio,** *Lion Corporation, Kanagawa, Japan*

References
1. BA Neudecker et al, Hyaluronan: History and biochemistry, *Antiaging*, Carol Stream, IL: Allured Publishing Corp. (2006)
2. BA Neudecker et al, Hyaluronan: Biology, pathology and pharmacology, *Antiaging*, Carol Stream, IL: Allured Publishing Corp. (2006)

3. P Prehm, Hyaluronate is synthesized at plasma membranes, *Biochem J* 220 597 (1984)
4. N Itano and K Kimata, Molecular cloning of human hyaluronan synthase, *Biochem Biophys Res Commun* 222 816 (1996)
5. PH Weigel, VC Hascall and M Tammi, Hyaluronan synthases, *J Biol Chem* 272 13997 (1997)
6. T Asplund, J Brinck, M Suzuki, MJ Briskin and P Heldin, Characterization of hyaluronan synthase from a human glioma cell line, *Bioch et Biophysic Act* 1380 377 (1998)
7. JR Fraser, TC Laurent, H Pertoft and E Baxter, Plasma clearance, tissue distribution and metabolism of hyaluronic acid injected intravenously in the rabbit, *Biochem J* 200 415 (1981)
8. RK Reed, UB Laurent, JR Fraser and TC Laurent, Removal rate of [3H]hyaluronan injected subcutaneously in rabbits, *Am J Physiol* 259 H532 (1990)
9. UB Laurent, LB Dahl and RK Reed, Catabolism of hyaluronan in rabbit skin takes place locally, in lymph nodes and liver, *Exp Physiol* 76 695 (1991)
10. G Kreil, Hyaluronidases — a group of neglected enzymes, *Prot Science* 4 1666 (1995)
11. GI Frost, T Csoka and R Stern, The Hyaluronidases: A Chemical, Biological and Clinical Overview, *Trends in Glycosci Glycotech* 8 419 (1996)
12. TB Csoka, GI Frost and R Stern, Hyaluronidases in tissue invasion, *Invasion and Metastasis* 17 297 (1997)
13. GI Frost and R Stern, A microtiter-based assay for hyaluronidase activity not requiring specialized reagents, *Analyt Biochem* 251 263 (1997)
14. MW Guntenheoner, MA Pogrel and R Stern, A substrate-gel assay for hyaluronidase activity, *Matrix* 12 388 (1992)
15. TB Csoka, SW Scherer and R Stern, Expression analysis of paralogous human hyaluronidase genes clustered on chromosomes 3p21 and 7q31, *Genomics*, 60 356 (1999)
16. D Heckel, N Comtesse, N Brass, N Blin, KD Zang and E Meese, Novel immunogenic antigen homologous to hyaluronidase in meningioma, *Hum Molec Genet* 7 1859 (1998)
17. L Lapcik Jr, P Chabreacek and A Staasko, Photodegradation of hyaluronic acid: EPR and size exclusion chromatography study, *Biopolym* 31 1429 (1991)
18. E Haas, On the mechanism of invasion. I. Antivasin I, An enzyme in plasma, *J Biol Chem* 163 63 (1946)
19. A Dorfman, ML Ott and R Whitney, The hyaluronidase inhibitor of human blood, *J Biol Chem* 223 621 (1948)
20. B Fiszer-Szafarz, Demonstration of a new hyaluronidase inhibitor in serum of cancer patients, *Proc Soc Exp Biol Med* 129 300 (1968)
21. M Kolarova, Host-tumor relationship XXXIII. Inhibitor of hyaluronidase in blood serum of cancer patients, *Neoplasma* 22 435 (1975)
22. GG Snively and D Glick, Mucolytic enzyme systems. X. Serum hyaluronidase inhibitor in liver disease, *J Clin Inv* 29 1087 (1950)
23. ML Grais and D Glick, Mucolytic enzyme systems. II. Inhibition of hyaluronidase by serum in skin diseases, *J Invest Derm* 257 259 (1948)
24. DH Moore and TN Harris, Occurrence of hyaluronidase inhibitors in fractions of electrophoretically separated serum, *J Biol Chem* 179 377 (1949)
25. JK Newman, GS Berenson, MB Mathews, E Goldwasser and A Dorfman, The isolation of the non-specific hyaluronidase inhibitor of human blood, *J Biol Chem* 217 31 (1955)
26. MB Mathews, FE Moses, W Hart and A Dorfman, Effect of metals on the hyaluronidase inhibitor of human serum, *Arch Biochem Biophys* 35 93 (1952)
27. MB Mathews and A Dorfman, Inhibition of hyaluronidase, *Physiol Rev* 35 381 (1955)
28. UR Kuppusamy, HE Khoo and NP Das, Structure-activity studies of flavonoids as inhibitors of hyaluronidase, *Biochem Pharm* 40 397 (1990)
29. UR Kuppusamy and NP Das, Inhibitory effects of flavonoids on several venom hyaluronidases, *Experientia* 47 1196 (1991)
30. MW Li, AI Yudin, CA VandeVoort, K Sabeur, P Primakoff and JW Overstreet, Inhibition of monkey sperm hyaluronidase activity and heterologous cumulus penetration by flavonoids,

Biol Reprod 56 1383 (1997)
31. S Perreault, LJ Zaneveld and BJ Rogers, Inhibition of fertilization in the hamster by sodium aurothiomalate, a hyaluronidase inhibitor, *J Reprod Fert* 60 461 (1980)
32. H Kakegawa, H Matsumoto and T Satoh, Inhibitory effects of hydrangenol derivatives on the activation of hyaluronidase and their antiallergic activities, *Plant Med* 54 385 (1988)
33. H Kakegawa, N Mitsuo, H Matsumoto, T Satoh, M Akagi and K Tasaka, Hyaluronidase-inhibitory and anti-allergic activities of the photo-irradiated products of tranilast, *Chem Pharm Bull* 33 3738 (1985)
34. H Kakegawa, H Matsumoto, K Endo, T Satoh, G Nonaka and I Nishioka, Inhibitory effects of tannins on hyaluronidase activation and on the degranulation from rat mesentery mast cells, *Chem Pharm Bull* 33 5079 (1985)
35. HH Tonnesen, Studies on curcumin and curcuminoids. XIV. Effect of curcumin on hyaluronic acid degradation in vitro, *Int J Pharmaceut* 50 91 (1989)
36. T Furuya, S Yamagata, Y Shimoyama, M Fujihara, N Morishima and K Ohtsuki, Biochemical characterization of glycyrrhizin as an effective inhibitor for hyaluronidases from bovine testis, *Biol Pharm Bull* 20 973 (1997)
37. RA Wolf, D Glogar, LY Chaung, PE Garrett, G Ertl, J Tumas, E Braunwald, RA Kloner, ML Feldstein and JE Muller, Heparin inhibits bovine testicular hyaluronidase activity in myocardium of dogs with coronary artery occlusion, *Am J Card* 53 941 (1984)
38. G Guerra, Hyaluronidase inhibition by sodium salicylate in rheumatic fever, *Science* 103 686 (1946)
39. A Szary, SH Kowalczyk-Bronisz and J Gieldanowski, Indomethacin as inhibitor of hyaluronidase, *Arch Immun Ther Exp* 23 131 (1975)
40. A Kushwah, MK Amma and KN Sareen, Effect of some anti-inflammatory agents on lysosomal & testicular hyaluronidases, *Indian J Exp Biol* 16 222 (1978)
41. D Foschi, L Castoldi, E Radaelli, P Abelli, G Calderini, A Rastrelli, C Mariscotti, M Marazzi and E Trabucchi, Hyaluronic acid prevents oxygen free-radical damage to granulation tissue: a study in rats, *Int J Tissue React* 12 333 (1990)
42. JJ Thiele and R Stern, unpublished experiments
43. Y Takahashi, O Ishikawa, K Okada, Y Kojima, Y Igarashi and Y Miyachi, Disaccharide analysis of human skin glycosaminoglycans in sun-exposed and sun-protected skin of aged people, *J Dermatol Science* 11 129 (1996)
44. V Uchiyama, Y Dobashi, K Ohkouchi and K Nagasawa, Chemical change involved in the oxidative reductive depolymerisation of hyaluronic acid, *J Biol Chem* 265 7753 (1990)
45. H Saari, Oxygen derived free radicals and synovial fluid hyaluronate, *Ann Rheum Disease* 50 389 (1991)
46. RA Greenwald and WW Moy, Effect of oxygen-derived free radicals on hyaluronic acid, *Arthritis Rheum* 23 455 (1980)
47. JJ Thiele, MG Trabber and L Packer, Depletion of human stratum corneum viamin E: an early and sensitive in vivo marker of UV photooxidation, *J Invest Dermatol* 110 756 (1998)
48. V Kagan, E Witt, R Goldman, G Scita and L Packer, Ultraviolet light-induced generation of vitamin E radicals and their recycling. A possible photosensitizing effect of vitamin E in skin, *Free Radical Res Commun* 16 51 (1992)
49. J Fuchs and R Milbradt, Antioxidant inhibition of skin inflammation induced by reactive oxidants: evaluation of the redox couple dihydrolipoate/lipoate, *Skin Pharmacol* 7 278 (1994)
50. GR Buettner, The pecking order of free radicals and antioxidants: lipid peroxidation, alpha-tocopherol, and ascorbate, *Arch Biochem Biophys* 300 535 (1993)
51. V Kagan, E Serbinova and L Packer, Antioxidant effects of ubiquinones in microsomes and mitochondria are mediated by tocopherol recycling, *Biochem Biophys Res Commun* 169 851 (1990)
52. D Darr, S Dunston, H Faust and S Pinell, Effectiveness of antioxidants (vitamin C and E) with and without sunscreens as topical photoprotectants, *Acta Derm Venerol* 76 264 (1996)
53. J Fuchs, *Oxidative injury in Dermatopathology*, Berlin: Springer-Verlag (1992)

54. R Stern, unpublished experiments
55. CM Ditre, TD Griffin, GF Murphy, H Sueki, B Telegan, WC Johnson, RJ Yu and EJ Van Scott, Effects of alpha-hydroxy acids on photoaged skin: a pilot clinical, histologic, and ultrastructural study, *J Amer Acad Derm* 34 187 (1996)
56. WP Smith, Epidermal and dermal effects of topical lactic acid, *J Amer Acad Derm* 35 388 (1996)
57. EF Bernstein, CB Underhill, J Lakkakorpi, CM Ditre, J Uitto, RJ Yu and EV Scott, Citric acid increases viable epidermal thickness and glycosaminoglycan content of sun-damaged skin, *Derm Surg* 23 689 (1997)
58. N Newman, A Newman, LS Moy, R Babapour, AG Harris and RL Moy, Clinical improvement of photoaged skin with 50% glycolic acid. A double-blind vehicle-controlled study, *Derm Surg* 22 455 (1996)
59. K Ash, J Lord, M Zukowski, and DH McDaniel, Comparison of topical therapy for striae alba, *Derm Surg* 24 849 (1998)
60. W Bergfeld, N Tung, A Vidimos, L Vellanki, B Remzi and U Stanton-Hicks, Improving the cosmetic appearance of photoaged skin with glycolic acid, *J Am Acad Derm* 36 1011 (1997)
61. SJ Kim, JH Park, DH Kim, YH Won and HI Maibach, Increased in vivo collagen synthesis and in vitro cell proliferative effect of glycolic acid, *Dermat Surg* 24 1054 (1998)
62. BA Wolf, A Paster and SB Levy, An alpha hydroxy acid derivative suitable for sensitive skin, *Derm Surg* 22 469 (1996)
63. M Edward, Effects of retinoids on glycosaminoglycan synthesis by human skin fibroblasts grown as monolayers and within contracted collagen lattices, *Brit J Derm* 133 223 (1995)
64. B Gilchrest, Anti-sunshine vitamin A, *Nature Med* 5 376 (1999)
65. J Bhawan, Short- and long-term histologic effects of topical tretinoin on photodamaged skin, *Int J Derm* 37 286 (1998)
66. A Lundin, B Berne and G Michaeelsson, Topical retinoic acid treatment of photoaged skin: its effects on hyaluronan distribution in epidermis and on hyaluronan and retinoic acid in suction blister fluid, *Acta Dermato-Venere* 72 423 (1992)
67. Z Wang, M Boudjelal, S Kang, JJ Voorhees and GJ Fisher, Ultraviolet irradiation of human skin causes functional vitamin A deficiency, preventable by all-trans retinoic acid pre-treatment, *Nature Med* 5 418 (1999)
68. UM Agren, M Tammi and R Tammi, Hydrocortisone regulation of hyaluronan metabolism in human skin organ culture, *J Cell Phys* 164 240 (1995)
69. K Tanaka, T Nakamura, K Takagaki, M Funahashi, Y Saito and M Endo, Regulation of hyaluronate metabolism by progesterone in cultured fibroblasts from the human uterine cervix, *Febs Letters* 402 223 (1997)

Skin Whitening Agents

Keywords: pigmentation, hydroquinone, kojic acid

Review of skin-lightening agents including hydroquinone and kojic acid

Skin whitening agents have been widely used to either lighten skin (individuals who wish to change or modify their skin color) or depigment skin (treatment for abnormal hyperpigmentation skin such as melasma, freckles and actinic lentigines). Common agents include hydroquinone, kojic acid and ascorbic acid derivatives. Efficacy in treatment of hyperpigmentary disorders has been demonstrated,[1-16] but with varying success.[2,4,5,7,13] Their mechanism of action has been investigated in vitro and in vivo.[6,11,12,17-22] Their toxicology has been documented.[23-37]

Hydroquinone

Hydroquinone (1,4-dihydroxybenzene) is used in the photographic, rubber, chemical and cosmetic industries. In the late 1930s, it was observed that monobenzyl ether of hydroquinone, a chemical used in the manufacture of rubber, caused depigmented skin in some workers.[2]

The efficacy of hydroquinone as a skin-lightening agent has been established in both human and animal studies. Its structure is shown in Figure 1.

Clinically, hydroquinone is applied topically in the treatment of melasma, freckles and senile lentigines, as well as postinflammatory hyperpigmentation. In the United States, hydroquinone is available in concentrations up to 2.0% as an over the counter (OTC) drug and by prescription at higher concentrations.[2,5]

Mechanism: Hydroquinone inhibits the conversion of dopa to melanin by inhibiting the tyrosinase enzyme.[2,5,18] Other proposed mechanisms are inhibition of DNA and RNA synthesis, degradation of melanosomes and destruction of melanocytes.[5] Electron microscopic studies of black guinea pig skin treated with hydroquinone show the anatomic consequences of this action: (1) the melanosome structure is disturbed, resulting in decreased production or increased degradation of these organelles, or both; (2) hydroquinone exposure can ultimately lead to the degradation of the melanocyte; and (3) keratinocytes are spared, showing no apparent injury.[2]

Efficacy: Arndt and Fitzpatrick,[1] in a non-placebo-controlled study, compared the efficacy of 2% and 5% hydroquinone cream for treatment of pigmentary disorders in 56 patients. Hydroquinone was a moderately effective depigmenting agent in 80% of cases. There was no efficacy difference between the two concentrations, however 2% hydroquinone was less irritating than 5%.

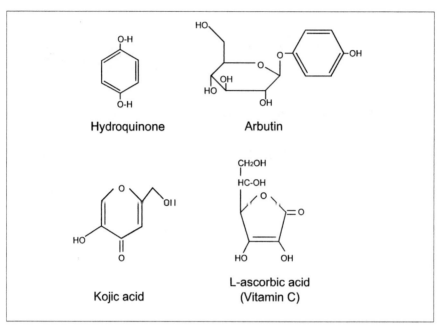

Figure 1. Chemical structures of hydroquinone, arbutin, kojic acid and L-ascorbic acid (vitamin C)

In a non-placebo-controlled study, Fitzpatrick et al.[3] evaluated the efficacy of a 2% cream of stabilized hydroquinone in 93 patients. Of those patients, 64% showed decreasing hypermelanosis.

Sanchez and Vazquez[15] treated 46 patients with melasma using two versions of a 3% hydroalcoholic solution of hydroquinone. In this non-placebo-controlled study, overall improvement was noted in 88% of the patients and moderate to marked improvement in 36%. Side effects were minimal. Sunscreen use was necessary for efficacy. The efficacy of hydroquinone may be improved when it is used in combination with other chemicals such as tretinoin, salicylic acid or corticosteroids.[2,5]

Kligman and Willis[7] noted an enhanced efficacy with 5% hydroquinone, 0.1% tretinoin and 0.1% dexamethasone in hydrophilic ointment for the treatment of melasma, ephelides and postinflammatory hyperpigmentation in a non-placebo-controlled study. In contrast, they experienced poor results with each of the aforementioned as monotherapies. However, actinic lentigines were resistant.

Gano and Garcia[4] conducted a 10-week clinical trial in 20 women with melasma. Topical applications of 0.05% tretinoin, 0.1% betamethasone valerate and 2% hydroquinone were used in a non-placebo-controlled study. There was an objective improvement rate of 65% and a subjective improvement rate of 95%. Side effects were frequent but minimal.

Caution is necessary when using potent fluorinated corticosteroids for prolonged periods on the face, because telangiectasia, atrophy or acne rosacea can develop. Pathak et al.[13] clinically tested the efficacy of hydroquinone in varying concentrations supplemented with corticosteroids or retinoic acid (tretinoin) in 300 Hispanic women with melasma in a non-placebo-controlled study, and concluded that cream or lotion formulations of 2% hydroquinone and 0.05 to 0.1% retinoic acid provided the most favorable results. In addition, avoidance of sun exposure and constant use of broad-spectrum sunscreens are requisite for efficacy.

Recently, Clarys and Barel[38] tested the efficacy of an ascorbate-phytohydroquinone complex in 14 patients with actinic lentigo in a non-placebo-controlled study. Objective skin color changes were evaluated with a chromameter. After one month of treatment, a clear depigmentation of the macules was measured.

Gellin et al.[39] established a reliable in vivo method to predict the depigmenting action of chemicals on mammalian melanocytes. These researchers used black guinea pigs and black mice as animal models to screen the depigmenting capacity of several phenols, catechols and organic antioxidants. Results showed that complete depigmentation on all test sites was achieved with mono-methyl ether of hydroquinone and p-tertiary butyl catechol in the black guinea pig. Less pronounced pigment loss was noted with these chemicals in black mice.

In some cases, higher concentrations of hydroquinone may be used. The formulations contain concentrations as high as 10% combined with nonfluorinated corticoid creams with or without the additional use of tretinoin or hydroxy acids such as glycolic acid. Extemporaneously compounded preparations are often effective in patients that have failed to respond to lower concentrations of hydroquinone. With controlled use and monitoring, side effects from these preparations have proved minimal.[5] Note, however, that hydroquinone may be quickly oxidized in such formulations.

Hydroquinone occurs in nature as the beta-glucopyranoside conjugate (arbutin). Arbutin is a mild agent for treating cutaneous hyperpigmentation including melasma and UV-induced ephelides.[20] Arbutin (Figure 1) is an active ingredient of the crude drug Uvae Ursi Folium traditionally used in Japan and contained in the leaves of pear trees and certain herbs. Maeda and Fukuda[20] determined arbutin's inhibitory action on the melanin synthetic enzyme and its effects on melanin intermediates and melanin production in cultured human melanocytes. They indicated that the depigmentation effect of arbutin works through an inhibition of the melanosomal tyrosinase activity, rather than by suppression of the expression and synthesis of tyrosinase in human melanocytes. Arbutin was less cytotoxic than hydroquinone to cultured human melanocytes.

Toxicology: Adverse reactions associated with hydroquinone use include both acute and chronic complications. Among acute reactions are irritant dermatitis, nail discoloration and postinflammatory hyperpigmentation.[2] Although generally assumed to be a common allergen, the documentation of hydroquinone allergic contact dermatitis is weak.[2] Hydroquinone use can also induce hypopigmentation and, rarely, depigmentation of treated surrounding normal skin. But, these changes are temporary and resolve on cessation of hydroquinone treatment, in contrast to monobenzone use, which can cause permanent depigmentation.[40] Hence, the only indication for monobenzone therapy is in the treatment of severe vitiligo.

A more recent concern regarding the use of hydroquinone is the occurrence of hydroquinone-induced ochronosis, a chronic disfiguring condition resulting, in general, from the prolonged use of high concentrations of hydroquinone.[40,41]

Hydroquinone's acute and chronic toxicity toward higher terrestrial organisms appears to be minimal in humans.[28,33] In an epidemiologic investigation, 478 persons who were employed as photographic processors showed no significant excess mortality, sickness/absence or cancer incidence.[28] The reported nephropathy and cell proliferation, as evidence of carcinogenicity, observed in Fischer 344/N rats,[26,27] appears to be strain- and sex-specific.[27] Hydroquinone was negative in the Ames/Salmonella and Drosophila genotoxicity assays.[30] Others suggest that carcinogenic and teratogenic potentials have been inadequately studied,[28,37] and that both hydroquinone and benzoquinone produce cytotoxic effects on human and mouse bone-marrow cells.[25] Hydroquinone in an alcoholic vehicle readily penetrates human forehead skin in vivo following a single 24 hour topical exposure; elimination was complete within five days.[24] Wester et al.[36] determined the topical bioavailability, metabolism and disposition of hydroquinone on humans in vivo and in vitro; dose recovery in urine was 45.3%, of which the majority was excreted in the first 24 hours.

Kojic Acid

Kojic acid (Figure 1), a fungal metabolic product, is increasingly being used as a skin-lightening agent in skincare products marketed in Japan since 1988. It was first isolated from Aspergillus in 1907.[32] Kojic acid suppresses free tyrosinase, mainly by chelating its copper,[17,19,32] and it has been shown to be responsible for therapy and prevention of pigmentation, both in vitro and in vivo.[17,9,42]

In Japan, it is used in non-prescription skincare products up to a concentration of 1%. To increase percutaneous absorption and thus therapeutic activity, it is usually used at the highest concentration allowed.[32]

Since it is used intensively in foods (such as bean paste, soy and sake) in some countries, particularly Japan, its oral safety has been studied.

Shibuya et al.[34] investigated the mutagenicity of kojic acid by the Ames test, by the forward mutation test in cultured Chinese hamster cells and by the dominant lethal test in mice. These researchers concluded that although kojic acid is a weak mutagen in bacteria, it is nonmutagenic in the eukaryotic system either in vivo or in vitro.

Abdel-Hafez and Shoreit[23] tested the mycotoxins using the dilution plate method; kojic acid may induce some toxins. Fujimoto et al.[29] examined the tumorigenicity of kojic acid in B6C3F1 mice. Three groups of animals were given food containing 0, 1.5 and 3.0% kojic acid for six weeks; mice in the groups ingesting kojic acid showed significantly higher frequency of induced thyroid tumors.

But true adverse effects after human oral ingestion have not been demonstrated. Nakagawa et al.[32] noted no signs of relapse of dermatitis or any other adverse effects on sensitized patients upon ingestion of foods containing kojic acid; however, these researchers also noted that topical application may induce allergic contact dermatitis with sensitized patients. They postulated that kojic acid was considered to have a high sensitizing potential, because of the comparatively high frequency of contact sensitivity in patients using one or more kojic acid-containing products.

Recently, Majmudar et al.[42] used an in vitro model to evaluate the efficacy, stability and cytotoxicity of whitening agents. They also conducted a non-placebo-controlled clinical study that indicated that kojic acid in an anhydrous base can induce more skin lightening than in an aqueous base.

Lim[9] conducted a non-placebo-controlled study to test 2% kojic acid in a gel containing 2% glycolic acid and 2% hydroquinone in 40 Chinese women who had epidermal melasma for 12 weeks. Half of the face was treated with the above formulation. The other half was treated with a formulation that was identical, except that it contained no kojic acid. Results showed similar improvement in melasma on both sides. More than half (60%) of the melasma cleared in sides receiving kojic acid, whereas less than half (48%) cleared in the side denied kojic acid; in particular, two patients had complete clearance only in the kojic acid-treated side. However, the improvement did not show a statistical difference between the formulations.

Ascorbic Acid and its Derivatives

Ascorbic acid (Figure 1) may inhibit melanin production by reducing o-quinones,[6] so that melanin cannot be formed by the action of tyrosinase until all vitamin C is oxidized. Because vitamin C is quickly oxidized and decomposes in aqueous solution, it is not generally useful as a depigmenting agent.

Recently, stable derivatives of vitamin C have been synthesized to minimize this problem.[6,11,12,21] Magnesium-L-ascorbyl-2-phosphate (VC-PMG) is a vitamin C derivative that is stable in water, especially in neutral or alkaline solutions containing boric acid or its salt.[6] VC-PMG is hydrolyzed by phosphatases of liver or skin to vitamin C and thus exhibits vitamin C-reducing activity.[6]

Kameyama et al.[6] investigated the effects of VC-PMG on melanogenesis in vitro and in vivo. Results from their non-placebo-controlled study suggested the topical application of VC-PMG was significantly effective in lightening the skin in 19 of 34 patients with chloasma or senile freckles and in three of 25 subjects with normally pigmented healthy skin.

Other Agents

Since many predisposing factors—such as pregnancy or exposure to sunlight (in the UVB and UVA ranges)—may cause hyperpigmentation, various systemic drugs and natural products have been used as protective agents. These agents include chloroquine, indomethacin, vitamin C and E, fish oil and green tea.[14]

Funasaka[43] demonstrated that oral vitamin E (alpha-tocopherol, alpha-T) supplementation can improve facial hyperpigmentation: the inhibitory effect of alpha-tocopheryl ferulate (alpha-TF) on melanogenesis was examined biochemically using human melanoma cells in culture. Alpha-TF, solubilized in ethanol or in 0.5% lecithin, inhibited melanization significantly, as did alpha-T at a concentration of 100 µg/mL, without inhibiting cell growth.

Recently, Kobayashi et al.[8] reported that neoagarobiose could be useful as a novel whitening agent because it has shown moisturizing and whitening effects with low cytotoxicity.

Ando et al.[44] evaluated the effects of unsaturated fatty acids on ultraviolet-induced hyperpigmentation of the skin in a placebo (vehicle) controlled study. Skin hyperpigmentation was induced on the backs of guinea pigs by UVB exposure. Oleic acid, linoleic acid, and α-linolenic acid (0.5% in ethanol), or ethanol alone as a control, were then topically applied daily five times weekly for three successive weeks. Results suggest that the pigment-lightening effects of linoleic acid and α-linolenic acid are, at least in part, due to suppression of melanin production by active melanocytes and to enhanced desquamation of melanin pigment from the epidermis.

A new combination product composed of 2% 4-hydroxyanisole (mequinol) and 0.01% tretinoin (all-trans-retinoic acid) in an ethanolic solution is being studied for its safety and efficacy as a topical treatment for disorders of skin hyperpigmentation.[45,46] Fleischer et al.[46] evaluated their efficacy in a controlled, double-blind trial. Subjects were randomized to treatment with the combination solution, or one of the active components (4-hydroxyanisole or tretinoin), or vehicle twice daily to all solar lentigines and related hyperpigmented lesions on the face, forearms and backs of hands for up to 24 weeks. They reported that the combination solution (2% 4-hydroxyanisole and 0.01% tretinoin) was clinically superior to each of its active components and to the vehicle in the treatment of solar lentigines. Most skin-related adverse events were mild and were similar for both the combination solution and tretinoin treatment groups.

Conclusions

In general, skin whitening agents are considered modestly effective. High concentrations are not recommended, except under a physician's supervision. The application as a combination with certain chemicals (retinoic acid and AHAs) may enhance lightening. Recently, chemical peelings with kojic acid, glycolic acid and trichloroacetic acid, either alone or in combination, have been widely introduced for treatment of hyperpigmentations.[47] However, note that few placebo (vehicle) controlled studies[44,46] are available. The real efficacy of whitening agents should be determined in a placebo-controlled study in humans. Optimal whitening agents remain a future goal.

—Hongbo Zhai, MD, and Howard I. Maibach, MD, *University of California at San Francisco School of Medicine, San Francisco, California USA*

References

1. KA Arndt and TB Fitzpatrick, Topical use of hydroquinone as a depigmenting agent, *JAMA* 194 965-967 (1965)
2. PG Engasser and HI Maibach, Cosmetics and dermatology: Bleaching creams, *J Am Acad Dermatol* 5 143-147 (1981)
3. TB Fitzpatrick, KA Arndt, AM el-Mofty and MA Pathak, Hydroquinone and psoralens in the therapy of hypermelanosis and vitiligo, *Arch Dermatol* 93 589-600 (1966)
4. SE Gano and RL Garcia, Topical tretinoin, hydroquinone, and betamethasone valerate in the therapy of melasma, *Cutis* 23 239-241 (1979)
5. PE Grimes, Melasma. Etiologic and therapeutic considerations, *Arch Dermatol* 131 1453-1457 (1995)

6. K Kameyama, C Sakai, S Kondoh, K Yonemoto, S Nishiyama, M Tagawa, T Murata, T Ohnuma, J Quigley, A Dorsky, D Bucks and K Blanock, Inhibitory effect of magnesium L-ascorbyl-2-phosphate (VC-PMG) on melanogenesis in vitro and in vivo, *J Am Acad Dermatol* 34 29-33 (1996)
7. AM Kligman and I Willis, A new formula for depigmenting human skin, *Arch Dermatol* 111 40-48 (1975)
8. R Kobayashi, M Takisada, T Suzuki, K Kirimura and S Usami, Neoagarobiose as a novel moisturizer with whitening effect, *Biosci Biotechnol Biochem* 61 162-163 (1997)
9. JTE Lim, Treatment of melasma using kojic acid in a gel containing hydroquinone and glycolic acid, *Dermatol Surg* 25 282-284 (1999)
10. H Nakayama, T Ebihara, N Satoh and T Jinnai, Depigmentation agents, in *Cosmeceuticals, Drugs vs. Cosmetics*, P Elsner and HI Maibach, (eds), New York: Marcel Dekker (2000) 123-144
11. H Nomura, T Ishiguro and S Morimoto, Studies on L-ascorbic acid derivatives. II. L-ascorbic acid 3-phosphate and 3-pyrophosphate, *Chem Pharm Bull* 17 381-386 (1969)
12. H Nomura, T Ishiguro and S Morimoto, Studies on L-ascorbic acid derivatives. 3. Bis(L-ascorbic acid 3,3')phosphate and L-ascorbic acid 2-phosphate, *Chem Pharm Bull* 17 387-393 (1969)
13. MA Pathak, TB Fitzpatrick and EW Kraus, Usefulness of retinoic acid in the treatment of melasma, *J Am Acad Dermatol* 15 894-899 (1986)
14. T Piamphongsant, Treatment of melasma: a review with personal experience, *Int J Dermatol* 37 897-903 (1998)
15. JL Sanchez and M Vazquez, A hydroquinone solution in the treatment of melasma, *Int J Dermatol* 21 55-58 (1982)
16. E Strauch, P Burke and HI Maibach, Hydroquinone, *J Derm Treatment*, Submitted for publication (2000)
17. J Cabanes, S Chazarra and F Garcia-Carmona, Kojic acid, a cosmetic skin whitening agent, is a slow-binding inhibitor of catecholase activity of tyrosinase, *J Pharm Pharmacol* 46 982-985 (1994)
18. K Jimbow, H Obata, MA Pathak and TB Fitzpatrick, Mechanism of depigmentation by hydroquinone, *J Invest Dermatol* 62 436-449 (1974)
19. V Kahn, Effect of kojic acid on the oxidation of DL-DOPA, norepinephrine, and dopamine by mushroom tyrosinase, *Pig Cell Res* 8 234-240 (1995)
20. K Maeda and M Fukuda, Arbutin: mechanism of its depigmenting action in human melanocyte culture, *J Pharm Exp Ther* 276 765-769 (1996)
21. K Morisaki and S Ozaki, Design of novel hybrid vitamin C derivatives: thermal stability and biological activity, *Chem Pharm Bull* 44 1647-1655 (1996)
22. M Nakajima, I Shinoda, Y Fukuwatari and H Hayasawa, Arbutin increases the pigmentation of cultured human melanocytes through mechanisms other than the induction of tyrosinase activity, *Pig Cell Res* 11 12-17 (1998)
23. SI Abdel-Hafez and AA Shoreit, Mycotoxins producing fungi and mycoflora of air-dust from Taif, Saudi Arabia, *Mycopathologia* 92 65-71 (1985)
24. DAW Bucks, JR McMaster, RH Guy and HI Maibach, Percutaneous absorption of hydroquinone in humans: Effect of 1-dodecylazacycloheptan-2-one (azone) and the 2-ethylhexyl ester of 4-(dimethylamino)benzoic acid (escalol 507), *J Toxicol Environ Health* 24 279-289 (1988)
25. RJ Colinas, PT Burkart and DA Lawrence, In vitro effects of hydroquinone, benzoquinone, and doxorubicin on mouse and human bone marrow cells at physiological oxygen partial pressure, *Toxicol Appl Pharmacol* 129 95-102 (1994)
26. JC English, T Hill, JL O´Donoghue and MV Reddy, Measurement of nuclear DNA modification by 32P-postlabeling in the kidneys of male and female Fischer 344 rats after multiple gavage doses of hydroquinone, *Fundam Appl Toxicol* 23 391-396 (1994)
27. JC English, LG Perry, M Vlaovic, C Moyer and JL O´Donoghue, Measurement of cell proliferation in the kidneys of Fischer 344 and Sprague-Dawley rats after multiple gavage administration of hydroquinone, *Fundam Appl Toxicol* 23 397-406 (1994)
28. BR Friedlander, FT Hearne and BJ Newman, Mortality, cancer incidence, and sickness-absence in photographic processors: an epidemiologic study, *J Occup Med* 24 605-613 (1982)
29. N Fujimoto, H Watanabe, T Nakatani, G Roy and A Ito, Induction of thyroid tumours in (C57BL/6N ¥ C3H/N)F1 mice by oral administration of kojic acid, *Food Chem Toxicol* 36 697-703 (1998)

30. E Gocke, MT King, K Eckhardt and D Wild, Mutagenicity of cosmetics ingredients licensed by the European communities. *Mutat Res* 90 91-109 (1981)
31. V Goffin, GE Pierard, F Henry, C Letawe and HI Maibach, Sodium hypochlorite, bleaching agents, and the stratum corneum, *Ecotoxicol Environ Safety* 37 199-202 (1997)
32. M Nakagawa, K Kawai and K Kawai, Contact allergy to kojic acid in skin care products, *Cont Derm* 32 9-13 (1995)
33. JW Pifer, FT Hearne, FA Swanson and JL O'Donoghue, Mortality study of employees engaged in the manufacture and use of hydroquinone, *Int Arch Occup Environ Health* 67 267-280 (1995)
34. T Shibuya, T Murota, K Sakamoto, S Iwahara and M Ikeno, Mutagenicity and dominant lethal test of kojic acid: Ames test, forward mutation test in cultured Chinese hamster cells and dominant lethal test in mice, *J Toxicol Sci* 7 255-262 (1982)
35. CI Wei, TS Huang, SY Fernando and KT Chung, Mutagenicity studies of kojic acid, *Toxicol Letters* 59 213-220 (1991)
36. RC Wester, J Melendres, X Hui, R Cox, S Serranzana, H Zhai, D Quan and HI Maibach, Human in vivo and in vitro hydroquinone topical bioavailability, metabolism, and disposition, *J Toxicol Environ Health* 54 301-317 (1998)
37. J Whysner, L Verna, JC English and GM William, Analysis of studies related to tumorigenicity induced by hydroquinone, *Regul Toxicol Pharmacol* 21 158-176 (1995)
38. P Clarys and A Barel, Efficacy of topical treatment of pigmentation skin disorders with plant hydroquinone glucosides as assessed by quantitative color analysis, *J Dermatol* 25 412-414 (1998)
39. GA Gellin, HI Maibach, MH Mislaszek and M Ring, Detection of environmental depigmenting substances, *Cont Derm* 5 201-213 (1979)
40. PE Grimes, Vitiligo. An overview of therapeutic approaches, *Dermatol Clin* 11 325-338 (1993)
41. C Levin and HI Maibach, Exogenous ochronosis: update, *Am J Clin Dermatol* Submitted for publication (2000)
42. G Majmudar, G Jacob, Y Laboy and L Fisher, An in vitro method for screening skin-whitening products, *J Cosmet Sci* 49 361-367 (1998)
43. Y Funasaka, AK Chakraborty, M Komoto, A Ohashi and M Ichihashi, The depigmenting effect of alpha-tocopheryl ferulate on human melanoma cells, *Br J Dermatol* 141 20-29 (1999)
44. H Ando, A Ryu, A Hashimoto, M Oka and M Ichihashi, Linoleic acid and α-linolenic acid lightens ultraviolet-induced hyperpigmentation of the skin, *Arch Dermatol Res* 290 375-381 (1998)
45. DW Everett, TJ Franz, TJ Chando, PJ Gale, PA Lehman, EH Schwarzel, PV Parab, CJ D'Arienzo, KJ Kripalani, Percutaneous absorption of [3H]tretinoin and systemic exposure to mequinol after dermal application of 2% mequinol/0.01% [3H]tretinoin (Solagé) solution in healthy volunteers, *Biopharmaceutics Drug Disposition* 20 301-308 (1999)
46. AB Fleischer Jr, EH Schwartzel, SI Colby and DJ Altman, The combination of 2% 4-hydroxyanisole (Mequinol) and 0.01% tretinoin is effective in improving the appearance of solar lentigines and related hyperpigmented lesions in two double-blind multicenter clinical studies, *J Am Acad Dermatol* 42 459-467 (2000)
47. C Cotellessa, K Peris, MT Onorati, MC Fargnoli and S Chimenti, The use of chemical peelings in the treatment of different cutaneous hyperpigmentations, *Dermatol Surg* 25 450-454 (1999)

The Inhibition of Metalloproteinases by *Macrocystis pyrifera* Extract

Keywords: metalloproteinases, Macrocystis pyrifera, wrinkles

An extract of an Antarctic seaweed was found to inhibit the formation of metalloproteinases and reduce wrinkles

During the aging process, complex and unavoidable physiological phenomena contribute to noticeable physical changes in skin. We know, for instance, that UVA radiation plays a major role in altering the dermis and activating a family of degradative enzymes called metalloproteinases (MMPs). These enzymes target the components of the extracellular matrix (ECM) and the dermoepidermic junction. The disorganization of the ECM leads to a slackening of the tissue, resulting in the appearance of wrinkles on the epidermis.

MMPs are endopeptidases or proteinases, enzymes that split proteins via the hydrolysis of the peptide linkage inside the protein chains. They can be considered as inductible enzymes since they are regulated at all stages in the production of a protein. This means that, to maintain homeostasis, its inhibitor is also formed whenever an MMP is formed (Figure 1). Tissue inhibitors of metalloproteinases (TIMP) control the regulation of this process. Currently, there are three TIMPs (TIMP-1, TIMP-2 and TIMP-3) known to inhibit the formation of MMPs. However, the biological equilibrium existing between MMP and TIMP can be destroyed by UV radiation, hormonal imbalances, local inflammation and normal aging processes.

Macrocystis pyrifera Seaweed

Macrocystis pyrifera (Figure 2) is a survivor of extremes. This giant brown seaweed, which reaches a length of 25 m in fields in the very cold water off the coast of the Kerguelen Islands of the Antarctic region, regularly experiences extremes of weather and other conditions. In response, it has evolved via adaptation and modifications of its metabolism, manifested in biological terms through changed compositions of its sugars, amino acids and glycopeptide polysaccharides.

The distinctive feature of this seaweed is that it grows on the surface of the water; thus, it has a most complex type of morphology, consisting of two or more fronds shooting up from the basal meristems above the tendril. *Macrocystis pyrifera* has adapted its metabolism to resist the Antarctic environmental conditions and as a result has a very long life for a seaweed: more than 10 years.

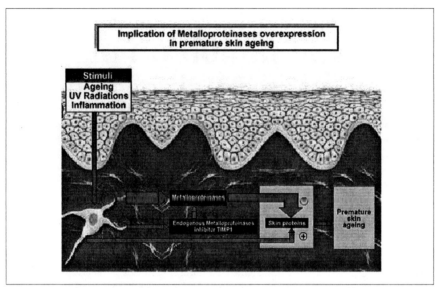

Figure 1. Mechanism of *Macrocystis pyrifera*

Because of the special traits possessed by this plant, it seemed reasonable to test the effects of an extract of the constituents of *Macrocystis pyrifera* for any palliative effects on the phenomena of physiological and actinic aging.

Screening Study: Bicellular Culture

In this co-culture model, a reconstituted epidermis based on human keratinocytes culture is treated with a *Macrocystis pyrifera* extract[a] at three different concentrations (4, 10 and 20 µg/ml), then put into contact with human fibroblasts in multiwell trays. The expression of MMP1, MMP3 and TIMP1 in each well was evaluated after UV irradiation.

The percentages of decreases of MMP1, MMP3 and TIMP1 expression, given in Table 1, indicate that the *Macrocystis pyrifera* extract can indeed inhibit the expression of the degradative metalloproteinases, even at the lower concentrations. For instance, the MMP1 expression in the control was 0.64 µg/ml; with only 4 µg/ml of the extract, that dropped to 0.48, and at 10 µg/ml, to 0.47. These results and several others led us to perform an in vivo study to determine if we could get visible antiaging results using this extract.

In Vivo Study: Wrinkle Reduction

To quantify the anti-wrinkle effect, we undertook a study using 12 female volunteers with ages 45 to 55 years. A product was developed (Formula 1) containing 0.7% of the *Macrocystis pyrifera* extract. Volunteers were to use the product twice a day (morning and evening) on the crow's foot area of the eyes for a period of two months.

---

[a]Kelpadelie, a *Macrocystis pyrifera* extract, was supplied by Secma Biotechnologies Marines, Pointrieux, France.

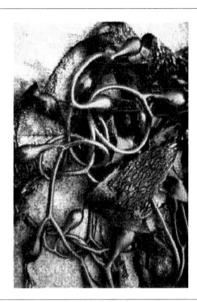

Figure 2. *Macrocystis pyrifera*

Table 1. Inhibition of MMP

Concentration (µg/ml)	MMP1 % Decrease	MMP3 % Decrease	TIMP1 % Decrease
0	-	-	-
4	10	33	35
10	20	22	27
20	-	18	12

Skin topography was measured by taking silicone imprints of the test area before and after the treatment period and subjecting them to a skin image analyzer. The parameters studied were total wrinkle surface, number and mean depth of furrows, moderate wrinkles and deep wrinkles.

The results of the in vivo study are summarized in Table 2. All of the parameters measured showed an improvement after treatment indicating the effectiveness of the *Macrocystis pyrifera* extract in helping to reverse the aging process through the inhibition of the degradative metalloproteinases. The results of in vivo testing confirm a reduction in fine lines and wrinkles and a restructuring of the skin surface.

Summary

An extract of the *Macrocystis pyrifera* seaweed showed the ability to inhibit the expression of metalloproteinases in a bicellular culture study. When the extract

Formula 1. Protective Cream with and without *Macrocystis pyrifera* extract (percents are w/w)

Product	Control	Test
A. Sorbitan stearate	3.50%	3.50%
Polysorbate 60	3.00	3.00
Dimethicone	1.00	1.00
Diethylhexyl propylene glycol	3.00	3.00
Cetearyl octanoate	5.00	5.00
Cetearyl alcohol	4.00	4.00
Ceramides III	0.10	0.10
Propylparaben	0.15	0.15
B. Water (*aqua*)	0.20	0.20
Methylparaben	3.00	3.00
C. Polyacrylamide (and) C_{13-14} isoparaffin (and) laureth-7	0.75	0.75
D. Propylene glycol	2.00	2.00
Ocymen 5-ol	0.10	0.10
Phenoxyethanol (and) methylparaben (and) butylparaben (and) ethylparaben (and) propylparaben (and) isobutylparaben	0.50	0.50
E. Water (*aqua*)	2.75	2.00
Macrocystis pyrifera	-	0.75
F. Panthenol	0.50	0.50
Retinyl palmitate	0.10	0.10
Fragrance (*parfum*)	0.30	0.30

Table 2. Results of in vivo testing

Property Measured	% Test Subjects Showing Benefit
Skin microtopography	55
Reduction of moderate wrinkles	55
Reduction of deep wrinkles	36
Reduction in wrinkle depth	64
Reduction in total wrinkled surface	46

was included in a topical formulation and applied to the skin of human volunteers, the skin showed a reduction in wrinkles, suggesting its use in formulations making an antiaging claim.

—**Anthony Ansaldi and Daniel Bosmann,** *Presperse Inc., Piscataway, New Jersey USA*

Controlling MMPs During Skin Whitening

Keywords: matrix, metalloproteinase inhibitors, UV, skin whitening

Combining matrix metalloproteinase inhibitors with skin whitening agents can help reduce their damaging side effects

Skin whitening procedures alter the skin's protective melanin layer and may subject the skin to degradation from UV-induced matrix metalloproteinases (MMPs). This article describes that degradation and suggests a formulation technique to protect against it.

Skin Whitening

In addition to being an esthetic concern, skin pigmentation–or skin tanning–holds important physiological roles in protecting skin from UV irradiation. Skin pigmentation can be triggered by UV exposure and becomes visible through melanin formation and deposition in the skin (Figure 1a).

Melanin is produced in special cells called melanocytes where the enzyme tyrosinase represents a rate-limiting step in melanin production. UV-induced pigmentation is part of a protection system that partially guards the skin from further damage by UV radiation. However, UV radiation is also known as a major environmental threat to skin homeostasis potentially leading to skin cancer and premature skin aging. Increased outdoor recreational activities and the progressive loss of the ozone layer represent major factors contributing to UV-induced biological skin damages.

A panoply of cosmetic products known as skin whiteners are aimed at lightening skin complexion. These products very often contain inhibitors of the enzyme tyrosinase as the main ingredient. Through this mechanism of action, whitening agents inhibit melanogenesis and thereby reduce the density of the melanin layer (Figure 1b). This reduction may have the unfortunate consequence of rendering the skin more susceptible to UV attack.

Side Effects of Skin Whitening

Upon exposure to sun, UV radiation energy is transferred to skin and initiates photochemical processes. Following a cascade of chemical, oxidative and biochemical reactions, UV radiation energy will modulate numerous signaling pathways, triggering various adaptive responses in skin cells. Melanogenesis is one of those responses, but so are sunburn inflammation, immunosuppression, telangiectasia and photocarcinogenesis–all of which are among the skin's acute biological responses to UV radiation.

While acute responses may be triggered after excessive exposures, other side effects of the sun, such as photoaging, can be observed after the accumulation of moderate or short UV exposures.

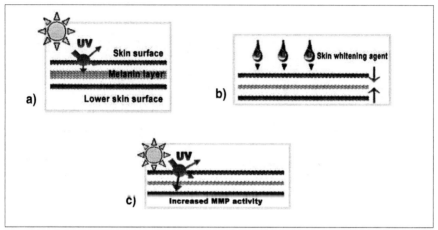

Figure 1. Skin damage side effects from skin lightening treatments
a) Skin contains a melanin layer located in the epidermis. This layer acts as a screen to reduce UV absorption in the basal layers of the skin and preserve the integrity of lower-level cells.
b) Skin whitening agents penetrate the skin to reduce the density of the protective melanin layer.
c) The thinning of the melanin layer renders the skin more vulnerable to sub-acute sun exposure. UV radiation readily penetrates the skin to induce MMP activity, resulting in accelerated photoaging of the skin.

Matrix Metalloproteinase Damage

MMPs are responsible for the degradation of several components (collagen fibers, glycosaminoglycans, fibronectin) of the skin extracellular matrix. This degradation is thought to lead to the formation of fine lines and wrinkles and ultimately skin sagging, hallmarks of skin photoaging.[2] Increased MMP activity through UV exposure is therefore closely linked to the appearance of signs of aging. Relatively low-level exposures are sufficient to initiate damage to the skin extracellular matrix. For example, an increase in the expression of MMPs can be measured in the skin after exposure to 0.1 MED (only 1/10 of the dose of UV radiation necessary to induce perceptible erythema).[1]

Skin whitening products, by their inhibitory action on the formation of the protective melanin layer, could diminish that layer's efficacy as a barrier against UV radiation. In turn, less protection leads to more UV-induced damage to the skin through MMP over-activation (Figure 1c).

Another consequence of UV exposure is the appearance of telangiectasia. It has been proposed that the appearance of telangiectasia is, in part, caused by a local degradation of the extracellular matrix surrounding capillaries,[3] a phenomenon that is potentially linked to an excessive action of selected MMPs. Therefore, concomitant inhibition of MMPs may figure as an interesting adjunct in whitening formulations.

Formulating with MMP Inhibitors

MMP inhibitors prevent the enzymatic action of MMPs. The enzymatic action of MMPs relies on the presence of a divalent metal residue (zinc). In their inactive configuration, MMPs are kept silent by the presence of a cysteine group that interacts with the zinc atom. This may explain why cysteine as an amino acid and chelators of divalent cations inhibit the action of MMPs.

Furthermore, tissue inhibitors of metalloproteinases (TIMPs) figure among the skin's natural mechanisms of defense to control the activity of MMPs. TIMPs bind to a portion of the MMP molecule responsible for substrate recognition. Undefined MMP inhibition may also arise from the direct binding of inhibitors to the enzyme or by competition with the substrate.

When selecting an MMP inhibitor, the formulation chemist must not only look at the efficacy of the ingredient in vitro but, more importantly, at the proven clinical efficacy and the safety profiles.

MMP inhibitors can be incorporated in skin whitening formulations, such as Formula 1 to maintain skin homeostasis. While skin whiteners act on melanin formation and deposition, MMP inhibitors are dedicated to interacting with the MMP pathway. The combination becomes mandatory in cosmetology especially with the exposure, whether intended or not, to UV radiation.

Such formulations present interesting features for the formulation chemist as well as for the end user. Combining skin whiteners with an effective MMP inhibitor allows the formulation chemist to provide a more complete and ideally targeted "two-in-one" product from which the customer can obtain a much wider range of action. It is important to see this combination skin whitener/MMP inhibitor not merely as the addition of ingredients–although often desirable in the case of antioxidants, for instance–but as a possibility to profit from the expected cosmetic benefits of a product without the risk of jeopardizing skin homeostasis.

Atrium's MMP Inhibitor

Atrium Biotechnologies has developed an MMP inhibitor.[a] It is a marine-derived aqueous extract that is processed through an ultrafiltration technique to focus on active fractions. Its INCI name is glycosaminoglycans, but it is unlike other cosmetic ingredients that are called glycosaminoglycans, none of which have demonstrated, in the manner of the Atrium product, an MMP inhibitory activity and efficacy in six different clinical trials.

This product inhibits the enzymatic action of various members of the MMP family: MMP-2, MMP-9 and MMP-12. The excessive activity of these MMPs is closely associated with the degradation and the disorganization of the collagen network present in the skin.[4]

Atrium's MMP inhibitor figures as a functional ingredient for skin conditions linked to potential side effects of UV exposure, especially when the skin's melanin layer has been previously impaired by the use of a skin whitening product.

This MMP inhibitor is 100% water soluble, so it is easily incorporated in formulas. Integrating it within a skin whitening formulation may help the skin to preserve an equilibrium between the enzymes and their inhibitors, allowing the skin to benefit from the lightening power of a whitening agent, without all the undesirable side effects.

---

[a] MDI Complex (Marine-Derived Innovation), Atrium Biotechnologies, Quebec City, Quebec, Canada

Formula 1. Skin whitener (with the Atrium MMP inhibitor and with magnesium ascorbyl phosphate as the whitening agent)

A	Water (*aqua*), purified	60.15%wt
	Propylene glycol (Canada Colors & Chemicals)	10.00
	Magnesium ascorbyl phosphate (MAP SL, Ikeda)	3.00
	Methylparaben (Nipagin M, Nipa)	0.40
	Sodium citrate (Charles Tennant)	0.50
	Triclosan (Irgasan DP300, Ciba Specialty Chemicals)	0.30
B	Cetearyl alcohol (and) ceteareth-20 (Promulgen D, Amerchol)	8.00
	Ethylhexyl palmitate (Tegosoft OP, Goldschmidt)	1.00
	Paraffinum liquidum (Mineral oil) (Blandol, Witco Petroleum Specialties)	3.00
	Butyrospermum parkii (Shea butter) (Fanning)	2.00
	Glyceryl stearate SE (GMS-SE, Stepan)	4.00
	Glycine soja (Soybean) oil (Lipovol Soy, Lipo Chemicals)	2.00
	Vitamin E acetate (Roche Vitamins)	0.10
	Propylparaben (Nipasol M, Nipa)	0.20
C	Fragrance (*parfum*) (Fragrance 700F47, Flavor & Fragrance Specialties)	0.35
	Glycosaminoglycans (MDI Complex, Atrium Biotechnologies)	5.00
		100.00

Procedure

Combine A in a 316SS steam-jacketed kettle. Heat A to 75-80°C and mix with a lightening mixer using a marine-type propeller at about 500 rpm. Combine B in a second 316SS steam-jacketed kettle. Heat B to 75-80°C and mix until uniform with a lightening mixer using a marine-type propeller at about 500 rpm.

When A and B are both at 75-80°C and uniform, add B to A with mixing. Insert homogenizer and homogenize for 30 min at 2000 rpm; heat to the kettle may be turned off during this step. Remove homogenizer and continue mixing with lightening mixer at about 800 rpm.

Begin cooling to 40°C with continued mixing. At approximately 30°C, add sequence C; mix until uniform. Continue mixing and cool to room temperature or 30°C.

Remove from kettle and store in polypropylene-lined drums.

—**Alain Thibodeau,** Atrium Biotechnologies, Quebec City, Quebec, Canada

References

1. GJ Fisher, ZQ Wang, SC Datta, HS Talwar, J Varani, S Kang and JJ Voorhees, Molecular basis of sun-induced premature skin ageing and retinoid antagonism, *Nature* 379 335-339 (1996)
2. A Thibodeau, Metalloproteinase inhibitors, *Cosmet Toil* 115(11) 75-82 (2000)
3. JB West and O Mathieu-Costello, Stress failure of pulmonary capillaries: role in lung and heart disease, *Lancet* 340 762 (1992)
4. Clinical efficacy of the MDI Complex in telangiectasia associated with skin redness problems, a poster presentation by Atrium Biotechnologies at the SCC Annual meeting, New York, December 2001

Ascorbic Acid and Its Derivatives in Cosmetic Formulations

Keywords: ascorbic acid, skin hydration, skin penetration

Study on effects of ascorbic acid and derivatives on epidermal tissue and skin penetration via three vehicles

Ascorbic acid (vitamin C) and its derivatives have been widely used in cosmetic formulations. Ascorbic acid is a natural antioxidant that combats the reactive oxygen species that can cause damage to the lipids, proteins and nucleic acids inside the cell, endangering tissue integrity.[6] The study of ascorbic acid in dermocosmetic formulations is important because of its suggested physiological functions in the skin. These include its inhibitory effect on melanogenesis,[3] free radical scavenger activity and antiaging properties.[8]

The relationship between reactive oxygen species and melanomas, cutaneous photoaging and other disorders is well known. These species are produced by sunlight associated with the presence of light absorbent powder components and molecular oxygen.[1,6]

Another property attributed to ascorbic acid is inhibition of the enzyme tyrosinase, which is the most important regulator of melanin pigment production. Topically applied ascorbic acid down regulates production of melanin pigment.[2, 5, 9]

Derivatives

The use of ascorbic acid in cosmetic formulations is very difficult due to its low stability in aqueous solutions; oxidation easily occurs in formulations such as gels, cream-gels and o/w emulsions.[4, 7, 8] Therefore, scientists have looked for ascorbic acid derivatives that have an action similar to that of ascorbic acid but with better chemical stability and with comparable percutaneous penetration. As a result, magnesium L-ascorbyl phosphate and magnesium ascorbate PCA have been introduced on the world cosmetic market.

Magnesium L-ascorbyl phosphate was synthesized for higher stability than ascorbic acid when added to formulations with large aqueous content and pH above or equal to 7.[8] When penetrating the skin, magnesium L-ascorbyl phosphate releases free vitamin C by the action of phosphatase enzymes, and then is able to exert its functions.[8]

Magnesium ascorbate PCA was synthesized for use in formulations with pH under 7 to permit its association with AHAs. This property promotes its percutaneous penetration and action.

The objectives of the present study were to comparatively evaluate the tissue changes and skin penetration occurring when formulations containing L-ascorbic

Formula 1. Gel cream	
Hydrogenated lecithin	0.8%
Squalane	4.0
Hydroxyethylcellulose	2.5
Propylene glycol	5.0
Methyl dibromoglutaronitrile (and) phenoxyethanol	0.02
Water (*aqua*), distilled	qs to 100 mL

acid, magnesium L-ascorbyl phosphate or magnesium ascorbate PCA are applied topically on guinea pig epidermis for a period of 15 days.

Epidermal Tissue Changes

Methods: We studied the effects of ascorbic acid and its derivatives on the guinea pig epidermis using a gel cream (Formula 1) on male guinea pigs weighing approximately 350 g.

The guinea pigs were shaved on the back. Five areas of 1.5 cm^2 each were used for the experiments. One untreated area was used for control and the others for the application of gel cream formulation alone and gel cream formulations containing 1% ascorbic acid or its derivatives. The formulations were applied once a day for 15 days and biopsies were then obtained from each area using a dermatological punch. The material collected was analyzed by using histopathological, morphometric and stereological techniques.

Results and discussion: Skin areas treated with the gel cream with or without the additives presented histologically a thickening of the basal, spinous and granulose layers of the epidermis compared to the untreated control (Figure 1 and Table 1). These layers also showed larger cells and nuclei (Table 2). This fact can be explained by the increased skin hydration promoted by these formulations, encouraging normal cell metabolism.

When we did the statistical analysis, our results concluded the addition of ascorbic acid or its derivatives to the vehicle increased the observed alterations. Ascorbic acid had the most visible action, followed by magnesium ascorbate PCA and magnesium L-ascorbyl phosphate, confirming the action of ascorbic acid and its derivatives on cell renewal. This may be due to the fact that ascorbic acid is an AHA in the lactone structure,[10] whose action on cell renewal has been extensively studied.[11] Ascorbic acid derivatives release it in the free form when it penetrates the cutaneous tissue, inducing the alterations observed.[8]

Skin Penetration

Methods: We studied the skin penetration of ascorbic acid and its derivatives by using three vehicles applied to harvested guinea pig skin in a glass Franz diffusion cell model. The three vehicles were an o/w emulsion (Formula 2), a gel cream (Formula 1) and a gel (Formula 3). All formulations were supplemented with 1% ascorbic acid, magnesium ascorbyl phosphate or magnesium ascorbate PCA.

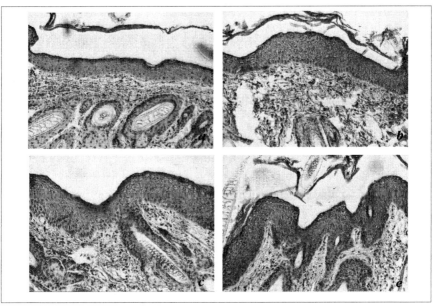

Figure 1. Photomicrograph of the skin from a guinea pig treated for 15 days. (a) control skin; (b) area treated with gel cream; (c) area treated with gel cream containing ascorbic acid; (d) area treated with gel cream containing magnesium L-ascorbyl phosphate; (e) area treated with gel cream containing magnesium ascorbate PCA. Note the thicker epithelium with larger and more numerous nuclei for all treated areas. (Magnification: x 200) (Stain: Hematoxylin and Eosin)

The receiver phase of the diffusion cell was phosphate buffer, pH 7.5. Samples of each test formulation (2 g) were applied to a second group of animals. Application was to the dorsal skin of female guinea pigs weighing 350 g. The receiver volume used for each cell was calibrated prior to use, and was constantly stirred with a Teflon-coated bar magnet. A uniform 37°C receiver temperature was maintained with a jacket connected to a circulating bath. Portions were obtained at 0, 30, 60, 90, 120, 150 and 240 min. The amounts of ascorbic acid or its derivatives in the receiver phase were obtained by ultraviolet penetration analysis at 254 nm.

Results and discussion: As seen in Figures 2, 3 and 4, the magnesium ascorbate PCA was more readily absorbed by the skin than L-ascorbic acid or magnesium ascorbyl phosphate. The best vehicle for penetration into the skin was the gel formulation, followed by the emulsion for L-ascorbic acid and the gel cream for magnesium ascorbyl phosphate.

It is well known that the polarity of a substance is important for its cutaneous penetration, and this was possibly a reason for the differences in penetration levels observed.

Some characteristics of the vehicle are necessary to facilitate the cutaneous penetration of the active principle added, such as ideal pH and viscosity. As previously mentioned, the gel was the best vehicle for penetration.

Formulating Considerations

Since L-ascorbic acid has several important properties for the prevention and treatment of skin aging, the development of formulations for skin care demands

Table 1. Mean total thickness (μm) for the different epithelial layers of the guinea pig skin treated with gel cream containing 1% ascorbic acid or its derivatives

Layers	Control	Gel Cream	AA	MAP	MAPCA
Basal	5.25	9.0	14.0	12.5	12.5
Spinous	7.25	15.0	23.5	14.0	19.5
Granulose	4.25	9.5	11.0	11.0	13.0
Total	16.75	33.5	48.5	37.5	45.0

AA = L-ascorbic acid
MAP = magnesium L-ascorbyl phosphate
MAPCA = magnesium ascorbate PCA
Statistically significant $p < 0.01$. n=5. Mann Whitney test

Table 2. Mean volume (μm^3) of the nuclei of cells in the different epithelial layers of the guinea pig skin treated with gel cream containing 1% ascorbic acid or its derivatives

Layers	Control	Gel Cream	AA	MAP	MAPCA
Basal	130.64	172.31	304.92	248.67	230.07
Spinous	162.07	276.34	349.56	310.41	257.36
Granulose	243.63	356.79	397.05	337.72	299.48

AA = L-ascorbic acid
MAP = magnesium L-ascorbyl phosphate
MAPCA = magnesium ascorbate PCA
Statistically significant $p < 0.05$. n=5. Mann Whitney test

special care, such as the selection of vehicle and agents that will favor stability for the longest possible time. Confirmation of the biological activity of a topically applied active substance at the epithelial level is an essential factor for its use.

The cosmetics formulations, regardless of the active substances contained, must undergo rigorous quality control tests, in both chemical-physical aspect and biological action.

The formulators must highlight the shelf life of formulations containing active substances subject to oxidation reactions, as in the case of vitamin C. These reactions cause product degradation, reducing its potential and resulting in skin damage through the presence of free radicals.

Conclusion

The results obtained under the present experimental conditions permit us to conclude that the magnesium ascorbate PCA was more readily absorbed by the

Formula 2. o/w emulsion

Self-emulsifying base	7.0%
Mineral oil (*paraffinum liquidum*)	4.0
Propylene glycol	5.0
Methyl dibromoglutaronitrile (and) phenoxyethanol	0.02
Water (*aqua*), distilled	qs to 100 mL

Formula 3. Gel

Hydroxyethylcellulose	2.5%
Propylene glycol	5.0
Methyl dibromoglutaronitrile (and) phenoxyethanol	0.02
Water (*aqua*), distilled	qs to 100 mL

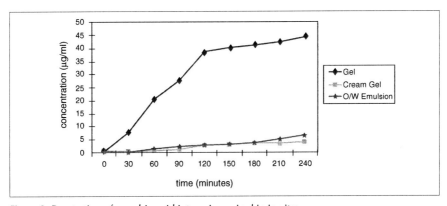

Figure 2. Penetration of ascorbic acid into guinea pig skin in vitro

skin than L-ascorbic acid and magnesium ascorbyl phosphate. The best vehicle for penetration into the skin was the gel formulation, followed by the emulsion for L-ascorbic acid and the gel cream for magnesium ascorbyl phosphate.

The gel cream formulation has a hydrating action on the epidermis of guinea pigs after 15 days of cutaneous treatment. It was also clear that the addition of vitamin C or its derivatives to this formulation intensified the intra- and extracellular hydration observed, a fact that favors normal cell metabolism.

—**Patricia M. B. G. Maia Campos and Gisele Mara Silva,** *Faculty of Pharmaceutical Sciences of Ribeirão Preto, University of São Paulo, Brazil*

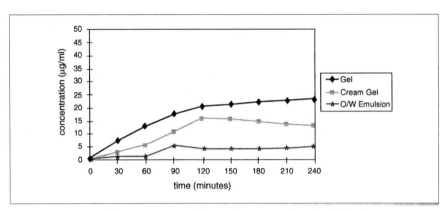

Figure 3. Penetration of magnesium ascorbyl phosphate into guinea pig skin in vitro

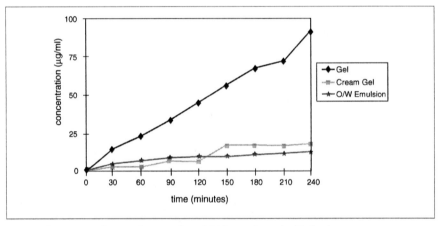

Figure 4. Penetration of magnesium ascorbate PCA into guinea pig skin in vitro

References

1. D Darr and I Fridovich, *J Inv Derm* (Review) **102**(5) 670 (1994)
2. K Iozumi, GE Hogansom, R Pennela, MA Everett and BB Fuller, *J Inv Derm* **100**(6) 806 (1993)
3. CL Phillips, SB Combs and SR Pinell, *J Inv Derm* **103**(2) 229 (1994)
4. SR Pinnell, *Rev Cosm Med Est* **3**(4) 31 (1995)
5. G Prota, *J Inv Derm* **100**(2) Sup 156-S (1993)
6. Y Shindo, E Witt and L Packer, *J Inv Derm* **100**(3) 261 (1993)
7. GM Silva and PM Maia Campos, *XIX IFSCC Congress*, Acapulco, Mexico, 3 21-225 (1997)
8. M Tagawa, K Uji and Y Tabata, *XV IFSCC International Congress,* Yokohama, Japan, 3 399 (1988)
9. H Takashima, H Nomura, Y Imai and H Mima, *Am Perf Cosm* **86**(7) 29 (1971)
10. R Hermitte, Pele envelhecida: retinóides e alfa-hidróxi ácidos, *Cosmet Toil Portugues* **5**(5) 55-58 (1993)

The Synergistic Antioxidative Effect of Ascorbyl 2-Phosphate and α-Tocopheryl Acetate

Keywords: ascorbyl 2-phosphate, tocopheryl acetate, antioxidant, UV, photoaging

Ascorbyl 2-phosphate and tocopheryl acetate show a synergistic antioxidative effect in vitro using a human keratinocyte cell line

The use of antioxidants in cosmetics has a long history. Two different modes of application can be distinguished. They are used either to stabilize the cosmetic formulation or to protect the skin.[1]

As a formulation stabilizer, they serve as a "technical antioxidant." Here they are being used to prevent the oxidative deterioration of labile ingredients such as unsaturated fatty acids or oxidation-prone active ingredients. Both synthetic compounds like butylated hydroxy toluene and natural antioxidants like α-tocopherol are frequently used for this purpose.

As a skin protectant, they serve as a "biological antioxidant." In recent years, this second application mode has become increasingly important. The prevention of oxidative processes within the skin is therefore an additional function of antioxidants in skin care cosmetics.[2,3]

Photoaging

Oxidative processes in the skin are commonly accepted as a major cause for skin aging.[4] They are involved in general processes of the intrinsic (chronological) aging of the skin; these processes resemble similar processes described for other tissues.[5] Yet, their main impact is probably on the accelerated skin aging due to environmental damage. Since the main factor of this extrinsic aging is sun exposure, it is often termed photoaging.[6-11] Within the spectrum of sunlight, UV light is the major factor leading to photodamaged skin.

The uncovered parts of the skin are permanently exposed to UV irradiation. This irradiation—especially in the UVB (280-320 nm) and UVA (320-400 nm) range—leads to the formation of reactive species.[12] These species, which are often oxygen-centered free radicals, are probably mediated via intracellular chromophores or photosensitizers. They have a detrimental effect on all molecular components of the skin cells and the extracellular matrix. Thus DNA, proteins and lipids are all affected.

- ***Direct action.*** Extracellular matrix proteins, such as collagen and elastin, are partly destroyed by the direct action of free radicals. Crosslinks between collagen fibrils occur more frequently in photoaged skin than in normal skin and the type of crosslinking is also different. Elastosis occurs when elastin fibrils are degraded to a certain extent and clump together.[13]
- ***Inflammatory reaction.*** In addition to the direct action of UV-generated free radicals, cell damage and subsequent tissue damage further involve inflammatory reactions. Mistaking UV-mediated cell damage for a mechanical insult bearing the risk of an infection, the skin cells release pro-inflammatory cytokines like interleukin-1 α or prostaglandin E_2. Subsequently, white blood cells infiltrate the skin, generating additional radical species in order to combat a non-existing infection, thus augmenting the detrimental process in a vicious circle.
- ***Antioxidant depletion.*** Furthermore, this oxidative stress leads to a depletion of both lipophilic and hydrophilic antioxidants in epidermal cells.[14-16]

Based on this model, there are three plausible ways of preventing photoaging: 1) avoid UV irradiation either by not exposing the skin to the sun (or similar light sources) or by permanently using adequate UVB- and UVA-filters on the sun-exposed skin areas,[17] 2) strengthen the antioxidative network of the skin to prevent–or at least reduce–the direct insult of free radicals, and 3) soothe the inflammatory processes, once they are generated. In this chapter, we will deal only with antioxidant reinforcement.

The Antioxidant Effect

For antioxidants to be used as actives in the skin, they have to be stabilized so they are not consumed in the cosmetic formulation. This can be achieved by specific formulation techniques or by chemical stabilization. Chemically stabilized derivatives have to be liberated in the skin, for example by enzymes present on or in the skin, in order to regain their antioxidative activity.

Typical derivatives for natural antioxidants are tocopheryl acetate and ascorbyl 2-phosphate for vitamin E and vitamin C, respectively.

Tocopheryl acetate: Tocopheryl acetate is a stable derivative of vitamin E. In this form it is not reactive towards active species like oxygen free radicals because the reactive center is protected by the derivatization. Therefore, tocopheryl acetate is not used as a "technical antioxidant" in cosmetic formulations. The preferred "technical antioxidant" form of vitamin E is the free alcohol, alpha-tocopherol, used at low concentrations.

Tocopheryl acetate is, however, used as a biological antioxidant to protect the skin. It penetrates into the skin, where it is cleaved to α-tocopherol and becomes active as an antioxidant, protecting the skin cells from the harmful effects of free radical species.[18,19] The term "biological antioxidant" is used since the compound becomes active as an antioxidant only after enzymatic conversion in a living biological environment.

Principle of the DCF Assay

The DCF assay is a method for evaluating the antioxidative capacity in vitro. In contrast to other common assays using cell-free systems, this specific method is performed in living cells. The measurement of the cellular antioxidative capacity allows not only an assessment of the antioxidative compound itself, but also takes cell biological aspects into account.

The assay is based on the use of the dichlorodihydrofluorescein diacetate (H_2DCF-DA) / dichlorofluorescein (DCF) system, an established non-fluorescent/fluorescent system for the measurement of reactive oxidative species.

The dichlorodihydrofluorescein di-(aceto-methyl ester) (H_2DCF-AM) is easily taken up from the cells because of its non-polar structure. Once inside the cell, the ester is cleaved by non-specific intracellular esterases, yielding the polar dichlorodihydrofluorescein (H_2DCF), which is trapped inside the cell. The oxidation of this non-fluorescent dye by oxidative species produced within the cell leads to DCF, which emits a characteristic fluorescence spectrum. Cells with a high content of antioxidants can prevent the dye from being oxidized. Therefore, the fluorescence measured is less in cells with a high antioxidative capacity than in cells with a low antioxidative capacity.

Ascorbyl 2-phosphate: Like tocopheryl acetate, ascorbyl 2-phosphate is also a "biological antioxidant." It is a very stable derivative of vitamin C in aqueous solution.[20] It is usually used in the form of its sodium or magnesium salt.

Studies have shown that ascorbyl 2-phosphate penetrates into mouse skin and is then cleaved to ascorbic acid.[21] However, there is little additional data on the effect of ascorbyl 2-phosphate on UV-irradiation-induced oxidation processes. Sodium ascorbyl 2-phosphate has shown a positive effect in cultured mouse skin after UVB irradiation.[21] Magnesium ascorbyl 2-phosphate was shown to protect against UVB-induced lipid peroxidation in hairless mice after intraperitoneal or intracutaneous application.[22]

A synergistic effect: Because tocopherol is lipophilic, its main site of action is the lipid membrane of the cells. The same is true for tocopheryl acetate, which is converted to tocopherol.

On the other hand, ascorbic acid is hydrophilic and therefore acts primarily in the aqueous cytosol and extracellular space. The same is true for ascorbyl 2-phosphate, which is cleaved to ascorbic acid. From this point of view, it is obvious that both antioxidants should act in a complementary or even synergistic way. A synergistic effect of vitamin E and vitamin C has, in fact, been demonstrated many times in numerous different systems, but data for a synergistic antioxidative effect of tocopheryl acetate and ascorbyl phosphate in skin is missing.

Therefore, we investigated the efficacy of alpha-tocopheryl acetate and ascorbyl-2 phosphate to inhibit hydrogen peroxide-induced oxidative stress in an in vitro model using a spontaneously immortalized human keratinocyte cell line.

Methods

Cell culture and DCF assay: The human dermal keratinocyte cell line HaCaT was used as an in vitro system. This cell line was chosen for its known ability to cleave tocopheryl and ascorbyl esters to their active components.

Principle, chemistry and further details on the DCF assay are described in the sidebar.[23] The materials used in the assay are listed in Table 1.

For the DCF assay, we seeded the cells in a 48-multiwell plate at a density of 4×10^5 cells per well. After reaching confluency, the cells were supplemented with different concentrations of alpha–tocopheryl acetate (3, 10, 30, 100 and 300 µmol/L) for 96 h or with different concentrations of sodium ascorbyl phosphate (50, 100 or 400 µmol/L) for 48 h.

Incubation periods and vitamin concentrations have been established in previous experiments and are in accordance with others.[24,25] They are sufficient for maximum uptake of the respective compounds without affecting cell viability and proliferation rate (data not shown). Solubilization of the lipophilic alpha–tocopheryl acetate in the culture media was achieved by use of ethanol as solvent. The ethanol concentration in the culture medium never exceeded 0.1% to avoid toxic effects of the solvent. Preliminary experiments proved this vehicle to be inert with respect to cell viability and the DCF assay (data not shown).

After the supplementation period, the cells were rinsed with Hank's balanced salt solution and 4-(2-hydroxyethyl)-1-piperazineethane sulfonic acid (HBSS/HEPES) and stained with 2.5 µmol/L of 6-carboxy-2´,7´-

Table 1. Materials and suppliers

Materials	Suppliers
Dulbecco's modified minimal essential medium	Biochrom[a]
Fetal calf serum	Biochrom
Hank's balanced salt solution	Biochrom
4-(2-Hydroxyethyl)-1-piperazineethane sulfonic acid	Biochrom
Ethanol	Merck[b]
Hydrogen peroxide	Sigma[c]
6-Carboxy-2´, 7´-dichlorodihydrofluorescein diacetate, di-(aceto-methyl ester)	Molecular Probes[d]
α-tocopheryl acetate	BASF[e]
Sodium ascorbyl 2-phosphate	BASF

[a] Berlin, Germany
[b] Darmstadt, Germany
[c] Deisenhofen, Germany
[d] Eugene, Oregon, USA
[e] Ludwigshafen, Germany

dichlorodihydrofluorescein diacetate, di-(aceto-methyl ester) for 45 min. After we washed the cells carefully, we covered them with 500 μL of HBSS/HEPES. Then we measured background intracellular radical flux over a period of 30 min in a fluoresence multiwell plate reader.

To evaluate the antioxidative capacity of the supplemented cells, we induced extracellular oxidative stress by adding 200 μmol/L of H_2O_2. Then we monitored subsequent intracellular radical production by measuring fluorescence emission of DCF.

Permeability studies: Penetration of sodium ascorbyl 2-phosphate and ascorbic acid was determined using a Franz diffusion chamber. The Franz diffusion chamber consists of a donor compartment on top, an acceptor compartment at the bottom and a membrane between the two compartments. In our experiments, we used dermatomized skin as the membrane. The top layer of human skin was removed with a dermatom. The dermatomized skin had a thickness of 700 μm and contained the intact stratum corneum with a thickness of approximately 17 μm.

The substances under evaluation were dissolved in the donor compartment. The compounds of interest were analyzed via HPLC in the acceptor solution at different time points to determine the rate of diffusion through the skin. Three different donor solutions were examined: 1% ascorbic acid, 1% ascorbyl 2-phosphate, and water as control. As an acceptor solution, we chose a 20% glycerol solution with 1% glucose with a pH of 4.7. Previous experiments had shown that both ascorbyl 2-phosphate and ascorbic acid are stable in this acceptor solution for at least 24.

HPLC analysis: We determined ascorbic acid by using a system consisting of an autosampler,[a,b] a pump[c] and an amperometric detector.[d] The mobile phase, consisting of 20 mM $NH_4H_2PO_4$ adjusted to pH 2.55 with H_3PO_4, was pumped at 0.8 mL/min. The stationary phase applied was a Hypersil ODS 3 μm (125 mm x 4 mm) with a 10 mm pre-column of the same material. Column oven temperature was set to 30°C ± 1°C and electrochemical detection was performed at a potential of +0.4 V. We injected 20 µL samples for analysis.

We determined ascorbyl 2-phosphate by using a system consisting of an autosampler,[e] a pump[f] and a diode array detector.[g] The mobile phase, consisting of methanol/tetrabutyl ammonium hydroxide (TBAOH) (35:65 v/v, with concentration of TBAOH at 6.48 g/L adjusted to pH 7.0), was pumped at 1 mL/min over a Merck Lichrocart 125 mm x 4 mm column filled with Lichrospher 100 RP 18.5 µm with a 4 mm pre-column of the same material. Oven temperature was set to 30°C ±1°C and ascorbyl 2-phosphate was detected at 260 nm. We injected 20 µL samples for analysis.

Results and Discussion

Permeability study: After applying 1% ascorbic acid in the donor compartment, we were unable to detect any ascorbic acid in the acceptor solution, even after several hours (Figure 1). Because ascorbic acid was found to be fairly stable in both the acceptor and the donor solution (data not shown), this indicates that essentially no ascorbic acid permeates through the dermatomized skin.

When we applied 1% ascorbyl 2-phosphate, a different picture emerged. Although we were unable to detect any ascorbyl 2-phosphate in the acceptor solution, the concentration of free ascorbic acid rose steadily over the time in the acceptor compartment (Figure 1).

This indicates that ascorbyl 2-phosphate did, indeed, permeate to a remarkable degree through the skin barrier. There was no significant lag-time detectable for the diffusion. Because no ascorbyl 2-phosphate could be detected in the acceptor solution, ascorbyl 2-phosphate must have been converted to ascorbic acid during the transport. It is likely that ascorbyl 2-phosphate was cleaved enzymatically to ascorbic acid during transport through the skin, although we did not show this. Our results indicate that, because there is practically no permeation of ascorbic acid, ascorbyl 2-phosphate is more effective in providing ascorbic acid to the inner layers of the skin than ascorbic acid itself.

These results confirm the studies of Nayama and colleagues, who measured a 1.03-, 2.17- and 6.27-fold increase of the cutaneous ascorbic acid concentration after topically applying sodium ascorbyl 2-phosphate at 2, 20, and 100 mmol/L, respectively, on cultured mouse skin.[21]

Antioxidative capacity: Tocopheryl acetate and ascorbyl 2-phosphate were tested alone and in combination over a wide concentration range.

Both compounds applied alone showed antioxidative activities in a dose-dependent manner (Figure 2). Under these specific assay conditions tocopherol acetate was able to inhibit hydrogen peroxide-induced oxidation to a maximum extent of around 10%. Ascorbyl 2-phosphate was more effective in this system, reaching values

---

[a] Model 655A-40, Merck-Hitachi HPLC system, supplied by Merck KGaA, Darmstadt, Germany
[b] Model AS 2000A, Merck-Hitachi, supplied by Merck KGaA, Darmstadt, Germany
[c] L-6220 Intelligent Pump, Merck-Hitachi, supplied by Merck KGaA, Darmstadt, Germany
[d] Model L-3500A, Merck-Hitachi, supplied by Merck KGaA, Darmstadt, Germany
[e] Autosampler Module 507, Beckman System Gold, Beckman Coulter Inc., Krefeld, Germany
[f] Solvent Module 125, Beckman System Gold, Beckman Coulter Inc., Krefeld, Germany
[g] Module 168, Beckman System Gold, Beckman Coulter Inc., Krefeld, Germany

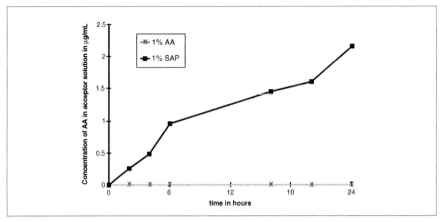

Figure 1. Amount of ascorbic acid analyzed in the acceptor compartment when ascorbic acid (AA) and ascorbyl 2-phosphate (SAP) are permeated through dermatomized human skin (700 μm thick)

of around 17% inhibition. Higher concentrations would give no further effect. This means that there is a saturation-like behavior for the antioxidative capacity of these compounds. Instead, there is evidence in the literature that at very high concentrations antioxidants could even show pro-oxidative behavior. Although this has only been shown for highly artificial, isolated models, it confirms the assumption that for a single compound, the antioxidative capacity peaks; that is, it cannot be increased continuously by simply increasing the concentration of the compound.

Use of both compounds in combination increased the antioxidative activity, compared to the activity of the compounds acting alone. In almost all cases, the values were higher than the theoretical values, shown as the dotted lines in Figure 3, obtained by calculating the sum of the individual values for the two antioxidants applied alone. This can be clearly seen for all concentrations of ascorbyl 2-phosphate combined with 3, 10 and 100 μmol/L tocopheryl acetate. A few combinations, such as for 30 and 300 μmol/L, are not over-additive, yet still higher than the respective values of the single compounds. Also, with the single compounds one could reach inhibitory effects of 10-17%. With combinations of the two antioxidants, the levels reached were in most cases clearly above that value, reaching maximum inhibitory effects close to 30%.

This indicates that the two substances did indeed act synergistically against the oxidative stress mediated by hydrogen peroxide. A combination always led to an improved antioxidative capacity in this system.

Conclusion

Photoprotection of the skin can be achieved by the application of tocopherol. This has been shown in vitro[19] or for topical application on mouse skin.[18] The combination of tocopherol with ascorbic acid was found to be photoprotective in skin after oral application.[26,27] Yet, there are few studies describing a beneficial effect of ascorbic acid or one of its derivatives after topical application,[28] and other studies failed to demonstrate any effect

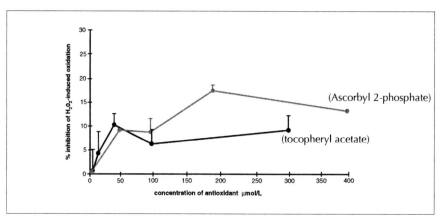

Figure 2. Antioxidative effect of alpha-tocopheryl acetate and ascorbyl 2-phosphate separately on H_2O_2-induced oxidation in HaCaT cells
Cells were pre-treated with varying concentrations of alpha-tocopheryl acetate or ascorbyl 2-phosphate for 96 or 48, respectively, prior to the stress. Stress was mediated by applying 200 μmol/L H_2O_2. Oxidation was determined by measuring the resulting fluorescence of DCF. Percent inhibition is calculated with respect to a stressed control without antioxidant. Data are expressed in means and standard errors.

at all.[29] This might be due, in part, to the instability of ascorbic acid in aqueous solutions. But ascorbic acid's low permeability through human skin may also play a role.

Currently available ascorbyl 2-phosphate compounds enable the formulator to incorporate the antioxidative activity of ascorbic acid into a cosmetic formulation. These derivatives are stable and able to permeate the stratum corneum of cultivated mouse skin[21] and excised human skin (as shown here). Both studies also showed that active ascorbic acid cleavage does in fact occur in the skin.

The combination of ascorbic acid with tocopherol is well known for its synergistic effect in many different models. This phenomenon is explained not only by their different sites of action (aqueous phase and lipid phase, respectively), but also by a direct recycling of the oxidized tocopherol by ascorbic acid.[30] The term antioxidant network is therefore often used in this context.

Ascorbyl 2-phosphate and tocopheryl acetate penetrate into the skin and are cleaved enzymatically to the biologically active antioxidants ascorbic acid and α-tocopherol, respectively. In the in vitro model used in this study, both compounds are also taken up and subsequently cleaved by the cells.

Our study showed, for the first time in an in vitro model using a human skin keratinocyte cell line, that the stable derivatives tocopheryl acetate and ascorbyl 2-phosphate act synergistically as antioxidants, having a saturation-like antioxidative activity with maximum oxidation inhibition effects of around 10% and 17%, respectively. The synergistic effect was seen over a wide concentration range and with most combinations used.

A combined use of these two antioxidants should, therefore, strengthen the antioxidative capacity of the skin and improve the antioxidant potential of cosmetics intended to prevent photoaging and other radical-mediated skin disorders.

The Synergistic Antioxidative Effect of Ascorbyl 2-Phosphate and α-Tocopheryl Acetate

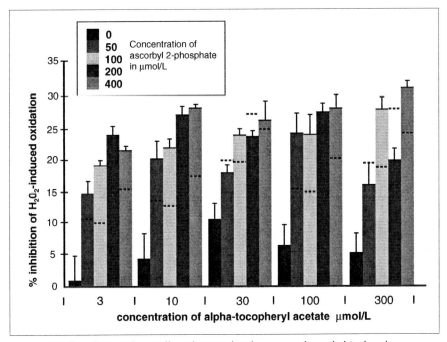

Figure 3. Combined antioxidative effect of α-tocopheryl acetate and ascorbyl 2-phosphate on H_2O_2-induced oxidation in HaCaT cells
Cells were pre-treated with 3, 10, 30, 100 or 300 µmol/L of α-tocopheryl acetate and 0, 50, 100, 200 or 400 µmol/L of ascorbyl 2-phosphate for 96 and 48 h, respectively, prior to the stress. Stress was mediated by applying 200 µmol/L H_2O_2. The dotted lines indicate the theoretical value obtained by calculating the sum of the individual values for the two antioxidants applied alone. Oxidation was determined by measuring the resulting fluorescence of DCF. Percent inhibition is calculated with respect to a stressed control without antioxidant. Data are expressed in means and standard errors.

—**Axel Jentzsch and Harald Streicher,** *BASF Aktiengesellschaft, Ludwigshafen, Germany*
—**Karin Engelhart,** *BioTeSys GmbH, Esslingen, Germany*

References

1. F Stäb, G Lanzendörfer, U Schönrock and H Wenck, Novel antioxidants: New strategies in product stabilization and skin protection, *SÖFW* 124 604-613 (1998)
2. T Förster, J Meister, H Möller, S Ortanderl and K Schlotmann, Protecting the skin against exogenous noxes, *Cosmet Toil* 114(3) 71-80 (1999)
3. *Oxidants and Antioxidants in Cutaneous Biology*, vol 29 in series *Current Problems in Dermatology*, J Thiele and P Elsner, eds, Basel: Karger (2001)
4. I Emerit, Free radicals and aging of the skin, *EXS* 62 328-341 (1992)
5. HL Dhar, Physiology of aging, *Indian J Med Sci* 52 485-497 (1998)
6. S Beissert and RD Granstein, UV-induced cutaneous photobiology, *Crit Rev Biochem Mol Biol* 31 381-404 (1996)
7. LH Kaminester, Current concepts: Photo-protection, *Arch Fam Med* 5 289-295 (1996)
8. Y Miyachi, Photoaging from an oxidative standpoint, *J Dermatol Sci* 9 79-86 (1995)

9. NH Nicol and NA Fenske, Photodamage: Cause, clinical manifestations, and prevention, *Dermatol Nurs* 5 263-275 (1993)
10. K Scharffetter-Kochanek, M Wlaschek, P Brenneisen, M Schauen, R Blaudschun and J Wenk, UV-induced reactive oxygen species in photocarcinogenesis and photoaging, *Biol Chem* 378 1247-1257 (1997)
11. CR Taylor and AJ Sober, Sun exposure and skin disease, *Annu Rev Med* 47 181-191 (1996)
12. R Kohen, Skin antioxidants: Their role in aging and in oxidative stress—new approaches for their evaluation, *Biomed Pharmacother* 53 181-192 (1999)
13. BA Gilchrest, A review of skin ageing and its medical therapy, *Brit J Dermatol* 135 867-875 (1996)
14. M Podda, MG Traber, C Weber, LJ Yan and L Packer, UV-irradiation depletes antioxidants and causes oxidative damage in a model of human skin, *Free Radic Biol Med* 24 55-65 (1998)
15. JJ Thiele, MG Traber and L Packer, Depletion of human stratum corneum vitamin E. An early and sensitive in vivo marker of UV induced photo-oxidation, *J Invest Dermatol* 110 756-761 (1998)
16. MJ Connor and LA Wheeler, Depletion of cutaneous glutathione by ultraviolet radiation, *Photochem Photobiol* 47 239-245 (1987)
17. S Seité, A Colige, P Piquemal-Vivenot, C Monastier, A Fourtanier, C Lapière and B Nusgens, A full-UV spectrum absorbing daily use cream protects human skin against biological changes occurring in photoaging, *Photodermatol Photoimmunol Photomed* 16 147-155 (2000)
18. BA Jurkiewicz, DL Bissett and GR Buettner, Effect of topically applied tocopherol on ultraviolet radiation- mediated free radical damage in skin, *J Invest Dermatol* 104 484-488 (1995)
19. P Clement-Lacroix, L Michel, A Moysan, P Morliere and L Dubertret, UVA-induced immune suppression in human skin: Protective effect of vitamin E in human epidermal cells in vitro, *Br J Dermatol* 134 77-84 (1996)
20. R Austria et al, Stability of vitamin C derivatives in solution and topical formulations, *J Pharm Biomed Anal* 15 795-801 (1997)
21. S Nayama et al, Protective effects of sodium-L-ascorbyl-2 phosphate on the development of UVB-induced damage in cultured mouse skin, *Biol Pharm Bull* 22 1301-1305 (1999)
22. S Kobayashi, M Takehana, S Itoh and E Ogata, Protective effect of magnesium-L-ascorbyl-2 phosphate against skin damage induced by UVB irradiation, *Photochem Photobiol* 64 224-228 (1996)
23. D Mayer, A Mühlhöfer and HK Biesalski, A modified system to evaluate the potency of antioxidative compounds in different cell types in vitro, *Eur J Med Res* 6 1-8 (2001)
24. AC Chan and K Tran, The uptake of (R,R,R)-alpha-tocopherol by human endothelial cells in culture, *Lipids* 25 17-21 (1990)
25. I Savini, I D'Angelo, M Ranalli, G Melino and L Avigliano, Ascorbic acid maintenance in HaCaT cells prevents radical formation and apoptosis by UV-B, *Free Radic Biol Med* 26 1172-1180 (1999)
26. B Eberlein-Konig, M Placzek and B Przybilla, Protective effect against sunburn of combined systemic ascorbic acid (vitamin C) and D-alpha-tocopherol (vitamin E), *J Am Acad Dermatol* 38 45-48 (1998)
27. J Fuchs and H Kern, Modulation of UV-light-induced skin inflammation by D-alpha-tocopherol and L-ascorbic acid: A clinical study using solar simulated radiation, *Free Radic Biol Med* 25 1006-1012 (1998)
28. D Darr, S Combs, S Dunston, T Manning and S Pinnell, Topical vitamin C protects porcine skin from ultraviolet radiation-induced damage, *Br J Dermatol* 127 247-253 (1992)
29. F Dreher, B Gabard, DA Schwindt and HI Maibach, Topical melatonin in combination with vitamins E and C protects skin from ultraviolet-induced erythema: A human study in vivo, *Br J Dermatol* 139 332-339 (1998)
30. MK Sharma and GR Buettner, Interaction of vitamin C and vitamin E during free radical stress in plasma: An ESR study, *Free Radic Biol Med* 14 649-53 (1993)

Skin Antioxidants

Keywords: antioxidants, UV, melatonin, free radicals

Combination of several antioxidants shows optimal results in photoprotection prior to UV exposure

The skin is supplied with an antioxidant defense system that includes enzymatic and nonenzymatic components.[1] Because the skin is the outermost layer of protection, it is constantly exposed to environmental oxidative stress such as ultraviolet radiation (UVR), air pollutants and chemical oxidants.

A skin antioxidant defense system is essential in protecting the epidermis from damage by free radicals generated by environmental and endogenous factors[1,2]; the antioxidants counteract free radicals by removing them from the body.

To prevent or diminish oxidative stress-induced skin damage, topical antioxidants have been widely used. This review introduces the skin antioxidant defense system and summarizes its efficacy in the photoprotection of human skin.

Reactive Oxygen Species

Reactive oxygen species (ROS) are considered a major contributor to skin aging, cancer and certain skin disorders. Although ROS normally have a short half-life, they can react with DNA, proteins and unsaturated fatty acids.[2,3] Healthy skin possesses an antioxidant defense system against oxidative stress. However, excessive free radical attack (such as overexposure to UVR) can overwhelm cutaneous antioxidant capacity, leading to oxidative damage and ultimately to skin cancer, immunosuppression and premature skin aging.[1-4]

The important role of ROS in UVR-induced skin damage is well documented.[4] UVR-induced skin damage includes acute reactions such as erythema, edema and pain followed by exfoliation, tanning and epidermal thickening. ROS and other free radicals, particularly the highly damaging hydroxyl radical, can deplete the skin antioxidants and hence damage biomolecules such as lipids, proteins and nucleic acids.[1,2]

Supplementing the endogenous antioxidant system may prevent or minimize ROS-induced photoaging. This can be accomplished by induction or transdermal delivery of various enzymatic and nonenzymatic antioxidants (see sidebar).

Enzymatic Antioxidants

Gluthathione peroxidase: Gluthathione peroxidase (GSH-Px) is considered an important antioxidant enzyme defense mechanism in skin.[3] Active against H_2O_2 and lipid peroxides, it catalyzes the reactions of GSH with these compounds. A small

> **Skin Antioxidant Defense Systems**
>
> **Enzymatic Antioxidants**
> - Gluthathione peroxidase (GSH-Px)
> - Catalase (CAT)
> - Superoxide dismutases (SOD)
> - Haem-Oxygenase (HO)
> - Thioprotein reductase
> - Metallothionein (MT)
>
> **Nonenzymatic Antioxidants**
> - Glutathione (GSH)
> - Vitamin E (α-tocopherol)
> - Vitamin C (ascorbate)
> - β-Carotene
> - Melanins
> - Ubiquinol (coenzyme Q)
> - Mannitol, xanthine, sorbate, urate

increase in GSH-Px activity may completely compensate for the total absence of catalase in fibroblasts from catalase-deficient patients.[5]

Catalase: Catalase (CAT) is a scavenger of H_2O_2 in skin. It detoxifies H_2O_2 by decomposing two H_2O_2 molecules to two molecules of water and one oxygen.[1] CAT is markedly reduced after UVR exposure.[3]

Compared to GSH-Px, CAT is thought to be less important as an antioxidant enzyme because fibroblasts from CAT-deficient patients did not show decreased survival after a single dose of UVR when compared to normal cells.[5]

Superoxide dismutase: Superoxide dismutase (SOD) catalyzes the reduction of superoxide anion, one of the ROS formed in irradiated skin, to the less reactive H_2O_2. Repeated skin exposure to UVR or other types of oxidative stress induces extra SOD activity in the long run. However, this is a minimal effect, usually not sufficient to compensate for the acute loss of activity after a single high UV dose.[3]

Haem-Oxygenase: Haem-Oxygenase (HO) breaks down haem to biliverdin, which is then converted to bilirubin; both are powerful antioxidants. Interestingly, most antioxidants are decreased when encountering oxidative stress, but HO is increased. Therefore, low dose UVA might be used to induce this enzyme to provide extra photoprotection.[3]

Other enzymatic antioxidants: The enzyme thioprotein reductase, involved in DNA synthesis, is an active scavenger of nitroxide and superoxide radicals. Metallothionein (MT), a cysteine-rich protein, has also been proven to have antioxidant properties.[3]

Nonenzymatic Antioxidants

Glutathione: Glutathione (GSH) acts as a direct free radical scavenger, and is thought to be the main protective effect at UVB wavelengths. It plays a pivotal role in the cellular defense against oxidative damage. Depletion of GSH in cultured human skin cells makes them sensitive to UVA- and UVB-induced mutations and cell death.[3]

Vitamin E: Vitamin E (α-tocopherol or α-TOC) is a lipophilic endogenous antioxidant that provides protection against UV-induced oxidative membrane damage. It is believed that the broad biological activities of vitamin E are due to its ability to inhibit lipid peroxidation and stabilize biological membranes.[3]

Vitamin C: Vitamin C (ascorbate) is an efficient scavenger of superoxide for many free radicals. The biochemical importance of vitamin C is primarily based on its reducing potential, which is required in a number of hydroxylation reactions.[1]

β-Carotene: β-Carotene is a vitamin A precursor and is an important member of the carotenoid family of antioxidants. It is capable of quenching excited triplet states and singlet oxygen and scavenging lipid peroxide radicals.[3]

Melanins: The most obvious protective property of the two cutaneous pigments, pheomelanin and eumelanin, depends on their ability to absorb and scatter light. In this process, pheomelanin and the melanin precursors 5-S-cysteinyldopa (SCN) and 5,6-dihydroxyindole (DHI) generate free radical species that can subsequently damage DNA. However, eumelanin and especially DHI and SCN also have antioxidant potential.[3]

Other nonenzymatic antioxidants: Ubiquinol (coenzyme Q) is a lipophilic chain-breaking antioxidant. It can react with ROS and, thus, prevent direct damage to bimolecules and initiation of lipid peroxidation.[1]

Mannitol, xanthine, sorbate and urate are cutaneous scavengers and quenchers of various ROS. However, these compounds may have less importance in the antioxidant system.[3]

Efficacy of Antioxidants in the Photoprotection of Human Skin

Topical application of a single antioxidant: The photoprotection effects of vitamin E have been extensively investigated.[1-6] The clinical efficacy of vitamin E alone in prevention or treatment of normal reactions of skin to UVR remains controversial.[1,6]

Topical vitamin E acetate (1%) reduced sunburn cell formation in human skin, but had no effect on xenon arc lamp UVB/A-induced erythema.[6]

In a double-blind, placebo-controlled study, an alcoholic lotion containing 2% vitamin E, applied on human skin 30 min before UVR exposure, significantly reduced erythemal responses and dermal blood flow.[7] Because this lotion lacks significant sunscreening properties, the photoprotective effect may be attributed to the antioxidant properties of vitamin E.

Vitamin E esters–particularly vitamin E acetate, succinate and linoleate–were also promising agents in reducing UVR-induced skin damage, but their photoprotective effects appear less pronounced as compared to vitamin E.[1,2]

Few studies described the photoprotective effects of vitamin C after topical application. Vitamin C can reduce UVB-induced inflammation and attenuate UVA-induced immediate pigment response in human skin,[6] but it is highly unstable and only poorly absorbed into the skin.[1] It is also strongly concentration dependent.[6]

The photoprotective effects of vitamin C were generally considered poor to modest[1,2] – 5% vitamin C alone in an alcoholic lotion failed to provide any significant protective effect when applied on human skin 30 min before UVR exposure.[7] However, more lipophilic and more stable vitamin C derivatives–such as its palmityl, succinyl or phosphoryl ester–might be promising compounds to provide increased photoprotection.[1,6,7]

Melatonin has shown beneficial effect in reducing UVR-induced erythema in human skin.[7-9] The photoprotective effect of melatonin is dose-dependent: 1% melatonin was less effective than a 2.5% melatonin formulation in preventing the induction of erythema.[7] Another double blind clinical trial evaluated erythema suppression by a topical melatonin formulation applied on human backs 15 min before and after UVR; treatment after UVR showed no erythema suppression; treatment before UVR showed that erythema was significantly suppressed by the melatonin formulation, compared to treatment with the vehicle alone.[9]

Decreases in photodamage have been reported from topical application of other antioxidants such as plant extracts (i.e., green tea and flavonoids), SOD, thiols and other derivatives.[1,2] Particularly, flavonoids were reported to reduce acute and chronic skin damage after UVR exposure.[1]

Topical application of antioxidant combinations: As already described, the antioxidant system in skin is complex and acts as network.[1] Therefore, topical application of combinations of antioxidants may enhance the photoprotective effects.[1,7,9]

A human study proved the synergistic effects of combinations of antioxidants.[7] The formulation containing melatonin in combination with 2% vitamin E and 5% vitamin C showed the greatest efficacy in preventing UVR-induced erythema in human skin.

A placebo-controlled, double blind study used combinations of the antioxidants α-glycosylrutin, ferulic acid and tocopheryl acetate applied topically on 30 patients with a history of polymorphous light eruption (PLE).[10] The test combinations were applied twice daily for one week and twice daily during the following week of photoprovocation with UVR. Some of the applied antioxidant combinations significantly reduced the development and severity of PLE in these humans.

Another study used a mixture of antioxidants and free radical scavengers. The mixture consisted of 5% vitamin E linoleate, 0.03% butylated hydroxytoluene, 0.01% nordihydroguaradinic acid and 1% magnesium ascorbyl phosphate in an o/w emulsion. When applied to human backs prior to UVR exposure, this mixture showed a significant protective effect against UVR.

Topical application of antioxidants after UVR exposure: For topical antioxidants applied prior to UVR exposure, the effects are well known; however, their efficacy after UVR exposure is less studied.[1,12]

A recent human study used vitamin E, vitamin C and melatonin. None of these antioxidants, alone or in combination, significantly reduced erythema formation when administered after UVR exposure.[12] This might be because of timing: UVR-induced ROS formation and the subsequent reaction of ROS with skin biomolecules resulting in acute skin damage is a rapid process. It may happen so quickly that antioxidants applied after irradiation do not reach the site of action (e.g., superficial skin layers) in relevant amounts during the occurrence of oxidative stress; therefore, they do not significantly reduce UVR-induced erythema formation when compared to their vehicles.[2,12]

Conclusions

The antioxidant defense system in human skin provides extremely important protection against environmental oxidative stress. The action mechanisms are complex and linked to each other. Although treatments with single components of the antioxidant system were successful against a wide variety of photodamage,[7,9] the balance between the different antioxidants in the skin is very important. Too much of a single component could even be deleterious.[3]

Supporting the skin's antioxidant defense system is a promising strategy for providing photoprotection. However, the goal should be the enhancement of the endogenous mechanism as a whole and not just supplying one or two components. The combination of several antioxidants showed optimal results in synergism of the protective effects.[7,10,11]

Topical application of antioxidants alone or in combination prior to UVR exposure demonstrated pronounced photoprotective effects on human skin. However, applying antioxidants after UVR exposure showed no substantial benefit.[12]

A recent study revealed that vitamin C alone or with vitamin E, added to a commercial sunscreen, produced an apparently greater-than-additive protection against phototoxic damage.[13] This finding suggests that sunscreens may benefit from combination with antioxidants to maximize photoprotection effects.

Today, antioxidants are found in numerous skin care products claiming to fight photoaging and reverse photodamage. Vitamin E, vitamin C, SOD, coenzyme Q10 and copper are common components in these formulations. It is true that daily applications of antioxidant-containing products may minimize oxidative stress. But there are many questions. Is the component active in the formula? Is it stable? Will it actually penetrate the skin? Does it actually get to the target organ? Scientific testing is needed before launching these products into the marketplace.

—Hongbo Zhai, MD, and Howard I. Maibach, MD, *University of California at San Francisco School of Medicine, San Fransisco, California USA*

References

1. JJ Thiele, F Dreher and L Packer, Antioxidant defense systems in skin, in *Cosmeceuticals, Drugs vs. Cosmetics*, P Elsner and HI Maibach, eds, New York: Marcel Dekker (2000) 145-187
2. F Dreher and HI Maibach, Protective effects of topical antioxidants in humans, in *Oxidants*

and *Antioxidants in Cutaneous Biology*, JJ Thiele and P Elsner, eds, Basel: Karger (2001) 157-164
3. DP Steenvoorden and GM van Henegouwen, The use of endogenous antioxidants to improve photoprotection, *J Photochem Photobiol B: Biol* 41 1-10 (1997)
4. K Scharffetter-Kochanek, P Brenneisen, J Wenk, G Herrmann, W Ma, L Kuhr, C Meewes and M Wlaschek, Photoaging of the skin from phenotype to mechanisms, *Exp Gerontol* 35 307-316 (2000)
5. Y Shindo and T Hashimoto, Antioxidant defence mechanism of the skin against UV irradiation: study of the role of catalase using acatalasaemia fibroblasts, *Arch Dermatol Res* 287 747-753 (1995)
6. J Fuchs, Potentials and limitations of the natural antioxidants RRR-alpha-tocopherol, L-ascorbic acid and beta-carotene in cutaneous photo-protection, *Free Radic Biol Med* 25 848-873 (1998)
7. F Dreher, B Gabard, DA Schwindt and HI Maibach, Topical melatonin in combination with vitamins E and C protects skin from ultraviolet-induced erythema: a human study in vivo *Br J Dermatol* 139 332-339 (1998)
8. TW Fischer and P Elsner, The antioxidative potential of melatonin in the skin, in *Oxidants and Antioxidants in Cutaneous Biology*, JJ Thiele and P Elsner, eds, Basel: Karger (2001) 165-174
9. T Fischer, E Bangha, P Elsner and GS Kistler, Suppression of UV-induced erythema by topical treatment with melatonin. Influence of the application time point, *Biol Signals Recept* 8 132-135 (1999)
10. I Hadshiew, F Stäb, S Untiedt, K Bohnsack, F Rippke and E Hölzle, Effects of topically applied antioxidants in experimentally provoked polymorphous light eruption, *Dermatol* 195 362-368 (1997)
11. N Muizzuddin, AR Shakoori and KD Marenus, Effect of antioxidants and free radical scavengers on protection of human skin against UVB, UVA and IR irradiation, *Skin Res Technol* 5 260-265 (1999)
12. F Dreher, N Denig, B Gabard, DA Schwindt and HI Maibach, Effect of topical antioxidants on UV-induced erythema formation when administered after exposure, *Dermatol* 198 52-55 (1999)
13. D Darr, S Dunston, H Faust and S Pinnell, Effectiveness of antioxidants (vitamin C and E) with and without sunscreens as topical photoprotectants, *Acta Dermatol Venereol* 76 264-268 (1996)

Efficacy Testing of a Brown Seaweed Extract

Keywords: oligosaccharide, anti-pollution, anti-inflammation, antiaging, anti-acne, hair care

Extracts from Laminaria digitata *show promising in vivo and in vitro efficacy for use in cosmetic products*

Over the past five years, there has been extreme interest in the immune stimulatory and protective properties of polysaccharides. Just working from natural sources, the cosmetic chemist has access to polysaccharides obtained from mushrooms, various terrestrial plants and crustaceans, and via biotechnology from yeast and bacteria. Many seaweeds also provide promising sources of polysaccharides. Indeed, the sea has become a leading source of cosmetic actives, as its bounty is both renewable and worthy of interesting label copy.

Using controlled enzymatic hydrolysis, an oligosaccharide composed of mannuronic and guluronic acids has been extracted from the membrane of a brown seaweed, *Laminaria digitata*. The extract has been made into four different variants: the first with a high molecular weight (20,000 daltons), the second with a molecular weight of 3500 D. The lower-molecular-weight variant is then complexed with different minerals: zinc or a combination of magnesium and manganese. The different extracts then underwent tests, both in vivo and in vitro, for cosmetic efficacy.

Protection Against Pollution

The high-molecular-weight (20,000 D) extract is a clear, viscous liquid with a pH in the range of 7 to 8 and consisting of between 5 and 8% solids.[a] The INCI name is water (and) hydrolyzed algin. Testing done on this extract using cultured human keratinocytes determined that it can protect the skin from the deleterious effects of cigarette smoke, heavy metals and pesticides.

The cultures were incubated for 30 min at 37°C under one of three conditions: without product (control) and with either 1% or 2.5% of the high-molecular-weight extract. Then the culture was exposed to the stresses: 1 h of cigarette smoke, 10 ppm of lead and cadmium, 100 ppm of lead and cadmium or a mixture of 1% DDT and 1% lindane. After exposure to the stress, the cultures were incubated for 24 h.

---

[a] Anti-Pollution, Codif International, Saint-Malo, France

Then cell viability was measured via morphological examination and by measuring the reduction of MTT. MTT is a yellow substrate that is metabolized by living cells to a purple-colored material. The development of the purple color is measured spectrophotometrically; the intensity of the purple color is directly related to the number of viable cells in a culture.

The metal chelation ability of the extract was also measured versus lead and cadmium. The high-molecular-weight extract at 5% was incubated with the metal solutions for 1 h. The solutions were then centrifuged and the amount of metal was measured in the insoluble fraction and in the supernatant.

The results for the high-molecular-weight extract are remarkable (Figures 1-3). When skin cells are exposed to cigarette smoke, their viability decreases by 40% and apoptotic cells appear in the stratum granulosum layer. However, when the cells are treated with 1% high-molecular-weight extract, there is an only 10% reduction in cell viability and histological examination shows a decrease in the number of apoptotic cells. When cells are treated with 2.5% of the high-molecular-weight extract, there is no decrease in cellular viability following exposure to cigarette smoke (Figure 4). When exposed to heavy metals, 2.5% of the high-molecular-weight extract gives 80% protection against 10 ppm of lead and cadmium (Figure 5), and 25% protection against 100 ppm of lead and cadmium. When cells are exposed to pesticides (a mixture of 1% lindane and 1% DDT), their viability decreases by 53% and apoptotic cells appear in the granulosum layer. However, when the cells are treated with 2.5% high-molecular-weight extract, there is only a 12% reduction in cell viability (Figure 6) and histological examination shows a decrease in the number of apoptotic cells.

Measurements of the chelating ability showed the high-molecular-weight extract at 5% to complex 87% of the lead and 98% of the cadmium.

Suppressing Inflammation

The second variant, with a molecular weight of 3500 D, is also a clear, viscous liquid, but with a pH in the range of 3.5 to 4.5 and between 7.5 and 8.5% solids.[b] The INCI name is water (and) hydrolyzed algin. This extract was tested for anti-inflammatory effects using three testing modalities: quantification of epidermal interleukin 1α, lactic acid sting test and protection of Langerhans cells after UVB irradiation. The low-molecular-weight extract gave excellent results in all three tests.

Interleukin 1α test: For the quantification of interleukin 1α, the ingredient was incorporated into a cosmetic base (acrylic polymer aqueous gel) at 5%. It was then applied at 12 h intervals to the forearms of 10 volunteers for one day. On the second day, the quantity of interleukin 11α was measured using ELISA (enzyme-linked immunosorbant assay). ELISA allows the measurement of selected chemicals in a system by treating the system with an enzyme specific to the chemical of interest, then quantifying the development of the enzyme-substrate. The values measured on the treatment sites were compared to the baselines and the untreated sites.

---

[b]Anti-Inflammation, Codif International, Saint-Malo, France

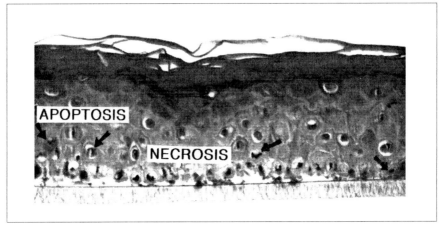

Figure 1. Epidermal tissue cross section with apoptotic and necrotic cells identified

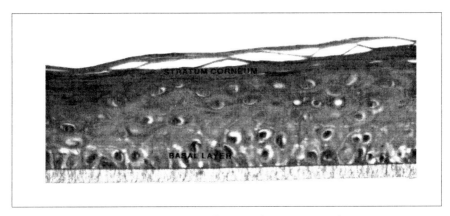

Figure 2. Epidermal tissue section showing cells exposed to cigarette smoke

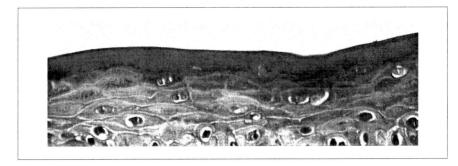

Figure 3. Epidermal tissue section showing results when cells exposed to cigarette smoke are also protected by the high-molecular-weight seaweed oligosaccharide

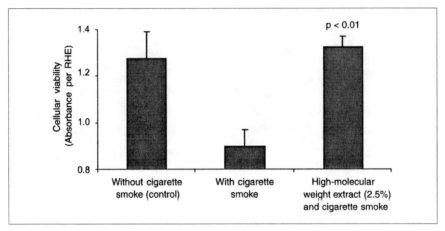

Figure 4. Changes in cellular viability following exposure to cigarette smoke, with and without protection by the high-molecular-weight seaweed oligosaccharide

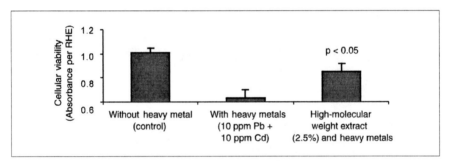

Figure 5. Changes in cellular viability following exposure to heavy metals, with and without protection by the high-molecular-weight seaweed oligosaccharide

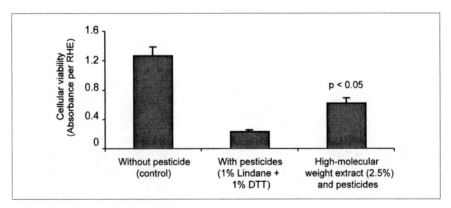

Figure 6. Changes in cellular viability following exposure to pesticides, with and without protection by the high-molecular-weight seaweed oligosaccharide

Two applications of this extract to the forearm of ten panelists gave an 11.1% increase in epidermal interleukin 1α levels over applications of the control gel. This increase of epidermal interleukin 1α indicates a soothing effect.

Lactic acid sting: The lactic acid sting test was carried out on eight panelists who had used a cosmetic base containing 5% of the hydrolyzed algin for 28 days. Treatment with the active resulted in a 36.6% reduction in the reactivity of the skin.

Langerhans cells: To determine the ability of the ingredient to protect Langerhans cells after UVB irradiation, human skin explants were treated twice daily with the low-molecular-weight product for three days and then irradiated with 1.5 J/cm^2 of UVB (Figure 7). Langerhans cells were marked using monoclonal antibodies and then counted. Treated explants were compared to untreated controls. In the in vitro immunoprotective assay, the extract was able, at concentrations of 0.5, 2.0, 3.0 and 5.0%, to give 35, 64, 75 and 100% protection of the Langerhans cells versus an untreated control.

Other anti-inflammatory testing: Further testing was done to determine the effect of the low-molecular-weight extract on other markers of inflammatory response such as cyclooxygenase (COX_2), prostaglandin E_2 (PGE_2) and 5-lipoxygenase (5-LO).

COX_2 is responsible, along with phospholipase A_2, for the enzymatic transformation of arachidonic acid to prostaglandin E_2. To test extract from test and control cultures and for inhibition of COX_2, the enzyme was isolated from human endothelial cells and then incubated with the low-molecular-weight extract at concentrations of 0.5, 1.0 and 2.0%. The radio-labeled substrate, arachidonic acid, was added and the amount of PGE_2 was measured. As an inhibitor of COX_2, the low-molecular-weight extract at 0.5, 1.0 and 5% gives 42, 61 and 64% reduction of activity.

To substantiate the results of the test done on isolated COX_2 enzyme, human keratinocytes in culture were activated with PMA to release PGE_2. In addition to control cells, the experimental design included cells treated with 1 and 5% of the low-molecular-weight extract. After incubation, the supernatant was harvested from each treatment and the amount of PGE_2 present in each was measured using radio-immunological assay. As an inhibitor of PGE_2 synthesis, the low-molecular-weight extract, tested on human keratinocytes at 1 and 5%, gives a 46 and 78% reduction in synthesis.

This same test was then done on human skin explants using UVB radiation to induce PGE_2 release. At 5% concentration, it gave a 31% decrease in PGE_2 release.

The inhibition of 5-LO, an enzyme responsible for transforming arachidonic acid to leukotriene B_4 in the inflammation pathway, was also tested. The enzyme was isolated from human granulocytes and incubated with 2.5 and 5.0% low-molecular-weight extract. The substrate, arachidonic acid, was added to the system, and then the product of the reaction, 5-HETE, was measured via radio-immunological assay. As an inhibitor of 5-lipoxygenase (5-LO), the low-molecular-weight extract at 2.5 and 5.0% gives 86 and 100% inhibition.

All of these results indicate an extract with potent anti-inflammatory capabilities.

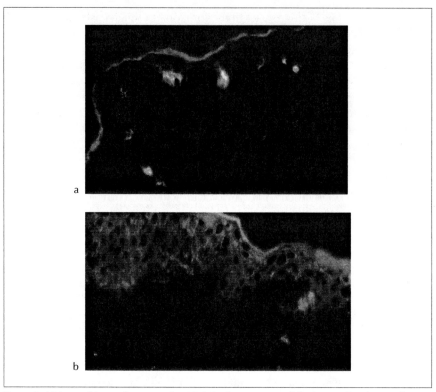

Figure 7. (a) Langerhans cells exposed to UVB in the presence of the low-molecular-weight seaweed oligosaccharide; (b) Unprotected Langerhans cells after the same UVB dose

Antiaging Complex

The low-molecular-weight extract was complexed with magnesium and manganese to give a clear viscous liquid with a pH of 5.5 to 6.0 and between 7.5 and 8.5% solids.[c] The INCI name of the Mg/Mn/oligosaccharide complex is water (and) hydrolyzed algin (and) manganese sulfate (and) magnesium sulfate. Testing showed the Mg/Mn/oligosaccharide complex to be an excellent free radical scavenger.

Free radical scavenging: Using the hypoxanthine/xanthine oxidase system, the free radical scavenging ability of this ingredient was compared to superoxide dismutase. In the hypoxanthine/xanthine oxidase model, the Mg/Mn/oligosaccharide complex reduced free radical formation by 30, 70 and 92% for 0.0001, 0.01 and 1% concentrations.

UV protection: A culture of human fibroblasts was incubated with varying concentrations of the ingredient for 24 h and then exposed to UV radiation (325 mJ/cm^2). The control was not exposed. Cellular proteins were their levels of oxidation measured by immunological determination. When tested for its ability to protect

[c]Anti-Age, Codif International, Saint-Malo, France

cellular proteins from oxidation, the Mg/Mn/oligosaccharide complex was found to have activity at 0.04, 0.2 and 1%, giving a dose dependent reduction in oxidation.

Apoptosis protection: Protection against apoptosis was determined by measuring the rupture rate in the DNA of human fibroblasts exposed to UVB irradiation. The results were compared to a reference of ascorbic acid and glutathione and also an untreated control. The Mg/Mn/oligosaccharide complex gives a protective effect to the DNA that is equal to that offered by a combination of ascorbic acid and glutathione. In the assay to measure protective effect against Langerhans cells, it was found that the Mg/Mn/oligosaccharide complex showed a 35, 64, 75 and 100% protection at concentrations of 0.5, 2, 3.5 and 5%.

Acne Reduction

The low-molecular-weight extract was complexed with zinc to maximize anti-microbial and anti-sebum effect, which gave a clear, viscous liquid with a pH in the range of 5.5 to 6 and between 7.5 and 8.5% solids.[d] Its INCI name is water (and) hydrolyzed algin (and) zinc sulfate. This zinc complex was tested for its ability to combat the development of acne.

5α Reductase inhibition: The ingredient was tested for its ability to inhibit the formation of testosterone via an inhibition of the enzyme, 5α reductase, responsible for transforming testosterone into dihydrotestosterone. Testosterone with radioactive labeling was added to samples of reconstituted epidermis.[e] The treated samples were then incubated for 24 h under one of three conditions:

1. in the presence of 10% of the extract,
2. in a 10^{-5} M solution of finasteride (positive control), or
3. in culture medium alone (negative control).

At the end of the incubation period, the samples were analyzed using thin layer chromatography and autoradiography. At a 10% concentration, the complexed extract totally inhibited the activity of 5α reductase.

Sebum reduction: The ability of the extract to reduce surface sebum levels was investigated on 12 volunteers. Using a sebumeter, casual surface sebum levels were measured. The panelists then used a product containing 5% of the extract and a placebo twice a day for 28 days. Surface sebum levels were measured again on day 28 and compared. After four weeks of twice-daily treatment, sebumeter readings documented a 67% reduction in surface sebum levels.

Bacterial inhibition: An in vitro antimicrobial assay showed that the extract has the ability to inhibit the growth of *Propionibacterium acnes*, one of the species of skin flora responsible for the development of acne. Bacteria were cultured anaerobically, either in the presence of the extract or with no extract (for the control). After 48 h of incubation, the number of organisms was determined via turbidimetric assay. The complexed assay was shown to inhibit the growth of *Propionibacterium acnes* by 45 and 75% at a 1.6 and 5.6% concentration.

---

[d]Anti-Acne, Codif International, Saint-Malo, France
[e]Skinethic 3D reconstituted epidermis, supplied by Skinethic, Nice, France

The results of these tests, taken together, clearly indicate that the zinc complex has a strong antimicrobial and anti-sebum effect.

Hair Care Applications

Three extracts (the high-molecular-weight extract and the two oligosaccharide complexes) were evaluated for substantivity on the hair. Bleached hair was used for the study, and seven treatment conditions were investigated:

1. Control: hair immersed in water and left to dry overnight at room temperature
2. Hair immersed in a solution of 5% of the high-molecular-weight extract for 15 min, then left to dry overnight at room temperature
3. Hair treated as in 2, but rinsed for 5 min before being left to dry overnight at room temperature
4. Hair immersed in a solution of 5% Mg/Mn complex for 15 min, left to dry overnight at room temperature
5. Hair treated as in 4, but rinsed for 5 min before being left to dry overnight at room temperature
6. Hair immersed in a solution of 5% Zn complex for 15 min, left to dry overnight at room temperature
7. Hair treated as in 6, but rinsed for 5 min before being left to dry overnight at room temperature

The deposition of the extracts was then evaluated using scanning electron microscopy (SEM) (results not shown). SEM micrographs clearly show the deposition of all three extracts on the surface of the hair. Furthermore, even after rinsing, the extracts can still be seen on the surface of the hair. As the extracts are substantive to the hair, it can be expected that they would also have their original effects on the hair. In other words, they would exert their pollution-protective, antiaging, and anti-sebum/anti-acne efficacies on the scalp as they do on the skin.

Conclusion

In the search for new cosmetic active ingredients, the sea provides us with many options. The four seaweed extract variants discussed here each offer a full arsenal of efficacy: They are easy to use and they have proven effectiveness.

When formulating with these products, developers should keep several points in mind. The extracts are aqueous preparations that should be incorporated at the end of processing. These extracts cannot be used in systems that contain alcohol, as alcohol causes the sugars to precipitate. However, they have been used successfully in a range of product types that includes gels, serums and both w/o and o/w emulsions. The seaweed extract actives are compatible with surfactants, so they can also be used in shampoos, body washes and face cleansers.

—**Suellen Bennett,** *Biosil Technologies, Paterson, New Jersey, USA*
—**Romuald Vallee,** *Codif International, Saint-Malo, France*

Active Substances From The Sea

Keywords: wrinkles, collagen, seaweed, moisturization, slimming, desensitization

Several seaweed extracts show effectiveness with wrinkles, moisturization, firming and sensitivity

For the past few years ingredients from the sea have been moving from myth to scientific reality. This marine vegetable, otherwise known as seaweed, was mainly used in marketing efforts up until the 1990s. Since then, its activity has evolved toward true effectiveness. Seaweed has since then left its micromarket and has taken part, thanks to its performance, in the raw materials used by the international cosmetics industry. On the whole, various algae have converged into becoming effective elements in personal care cosmetics.

Aosa Biopeptides

Aosa seaweed, or *Ulva lactuca*, lives in the rough waters of the bays facing the northeast side of the Bréhat Archipelago in France. The aosa frond is made up of only two layers of cells where the plastid forms the photoenergetic center of the seaweed. The elastic resistance of the cellular walls to the hydrostatic pressure is generated by the progression of water toward the shore or by the assault of waves which break on the rocks and drag along the aosa. This resistance is due to the protein network of elastic fibers. The weight of water which hits the rocks can sometimes create pressures superior to 25 T/m^2.

Aosa biopeptides[a] are the main protein component of the elastic fibers of the seaweed, particularly rich in proline, glycine and lysine like elastin. Because of this likeness, aosa biopeptides can play an important psychological part by maintaining the elasticity of the skin and by acting as an anti-wrinkle agent.

In the skin, the fibroblast is a differentiated mesenchymal cell, which is largely responsible for synthesis and remodeling of extracellular matrix laid down in the dermis. Elastin, the principal protein constituent of elastic fibers, is produced by human skin fibroblasts. Although elastin comprises only about 2% of the total protein in dermis, it is physiologically important in providing resiliency in the skin. During cutaneous aging, the disappearing of the elastic network is partly due to the intensification of the elastolyse thanks to the action of the elastase. This elastolyse

---

[a]Aosaïne is a registered trademark of SECMA Corporation, Pontrieux, France

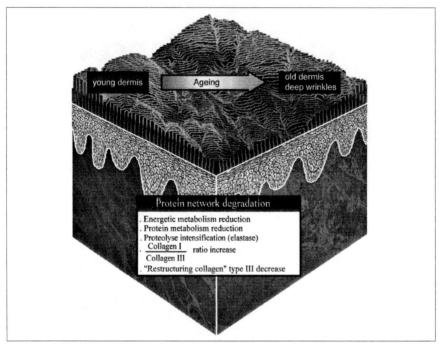

Figure 1. Protein network degradation

is also increased by the use of some detergents and by solar exposure, just like the cutaneous flora secretes particularly active bacterial elastases.

The predominant extracellular component of the dermis is the collagen. Collagen types I and III are the major interstitial, fiber-forming collagens in normal human dermis. Collagen I represents 80% of total dermis collagen of an adult's skin, while collagen III accounts for 15%. (The remaining 5% mainly correspond to type IV and type V collagens.)

Thick fibers of type I collagen are blended with a fine felting of type III collagen, which orientates big fibers during their growth. More tensile and less fibrous, the latter is predominant in fetal and postnatal skins and during the wound healing process. Thus, type III collagen has been called a "restructuring" collagen, particular to very young skin and in the process of wound healing. Collagen I is formed later and becomes predominant. Among the numerous modifications of the extra cellular matrix (ECM) during aging, collagen synthesis shows a great decrease during the aging process. Moreover, the ratio of collagen types changes throughout life. Some studies demonstrated that the content of collagen III is higher in fetal skin and in newborn infant skin than in adult skin. They found a decrease of the ratio collagen III/collagen I during aging, underlining the loss of capacity of old cells for producing collagen III.

At this level, the action of these aosa biopeptides are unique because of their numerous properties: At a cellular level, these marine biopeptides activate the energetic metabolism as well as the fibroblasts' regeneration capacity. At the dermis level, one can observe a selective amplification of the collagen III biosynthesis parallel to protein fibers protection next to the proteases. This is followed by an in vivo reduction of deep wrinkles.

Energetic metabolism: The evaluation of the activity of aosa biopeptides on the mitochondrial metabolism of human skin fibroblasts was performed by following the oxygen consumption (mitochondrial respiration) and measuring the cellular adenylic nucleotides (ATP, ADP and AMP) content. According to this study, the biopeptides stimulate the cellular respiration by a direct effect on the mitochondria. However, this stimulation of the basal cellular respiration is not due to a direct effect on the test product on the respiratory chain. So, the product acts on the energetic transducting systems (ATP synthesis). The increase in cellular ATP level and the increase in ATP/ADP ratio confirm the stimulatory effect of the aosa biopeptide on the ATP synthase.

Cellular regeneration: The cell regeneration in vitro test has been made on human fibroblasts by tracing 3H-thymidine in DNA and 14 C-uridine in RNA. Several concentrations of the aosa biopeptide have been tested and compared to a standard containing 5 to 10% of calf fetal serum which maintains the survival of the culture. A concentration of 0.4% aosa biopeptides produces an increase of cells –74% superior to cells of the standard in DNA, and 89% superior to cells in RNA.

Collagen neosynthesis: The effect of aosa biopeptides on collagen stimulation was performed on human fibroblasts. This study was conducted by determining the incorporation of hydroxyproline, an amino acid precursor of collagen. Hydroxyproline is necessary for collagen helix formation and, in its absence, collagen is improperly secreted from fibroblasts. The ascorbic acid, an essential cofactor in the hydroxylation of proline in hydroxyproline, was used as positive control. A strong stimulation (+57%) of the collagen neosynthesis was observed in the presence of the biopeptide compared with vitamin C.

Kinetics of protein synthesis: Another study has been made on a culture of human fibroblast to follow the kinetics of protein synthesis. This kinetic was realized at 4, 6 and 24 h by incorporation of a tracing substance. A concentration of 0.4% biopeptide gives a 35% improvement of the speed of protein synthesis after 24.

Restructuring collagen, type III: The effect of aosa biopeptides on the collagen type I/III ratio has been determined on human fibroblasts. The study was conducted by determining the incorporation of 3H proline, an amino acid precursor of collagen, in the macromolecular fractions extracted from the cell layer. Qualitative study of collagens was evaluated after electrophoresis. The radioactivity was then determined by liquid scintillation counting. The results show that the type I/III collagen ratio is close to 3 in the control cell layer, implying a deposit of 25% type III collagen in the cell layer with reference to type I collagen. This ratio is very similar to that observed in human dermis in vivo. Aosa biopeptides, at concentrations of 1 and 100 mg/mL, induce a decrease in the type I/III collagen ratio, implying an increase of 35 and 39% type III collagen, respectively.

Anti-wrinkle activity: The effect of aosa biopeptides skin cream on skin relief was determined after 28 days of twice-daily application. Siliconed polymer replicas

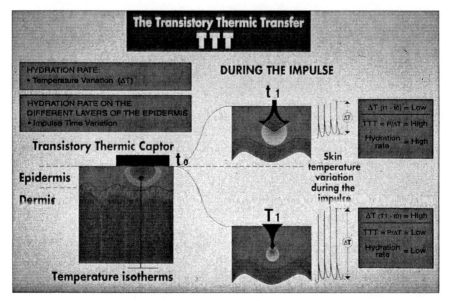

Figure 2. The transistory thermic transfer

were taken of two crow's-feet (wrinkles around the eyes) of volunteers before and after 28 days of twice-daily application. Skin relief was quantified with confocal microscopy and image analysis. Ten healthy female volunteers, between the ages of 40 and 50, were included in the study. No allergic or irritation type of intolerance reaction was observed.

Skin relief was studied as 2 micron thick microclasses between 4 and 400 microns. Three relief classes corresponding to the grouping of microclasses were calculated: microrelief, medium wrinkles and deep wrinkles. The density of these three relief classes was measured before and after utilization of the product. The product reduces deep wrinkles (-23%).

Codium tomentosum Extract

Codium tomentosum extract[b] is adapted to both low and high salt concentrations by two different mechanisms. Its adaptation to salinity is based on biosynthesis of an anionic polysaccharide contained in the cell walls of the algae and different organic substances with a low molecular weight inside the cell, compatible with the protein structures and membrane systems of the cytoplasm.

A water-soluble sulfate hetero-polysaccharide of sulfated arabinane type is isolated in *Codium tomentosum*.[4] The polymer is located outside the cell membranes forming the matrix of the cell walls. It plays a special physiological role in the alga, both mechanically and with regard to moisturizing and ionic regulation. These phycocolloids are strongly sulfated to preserve the metabolic and ionic equilibrium regardless of the variations of the environmental conditions.

---

[b]Codiavelane is a registered trademark of SECMA Corporation, Pontrieux, France

Codium tomentosum extract also stores a quantity of osmotic molecules in its cells to maintain the algae at a lower pH than its environment. This is the only way the algae maintains a hydrogen-ion activity compatible with water absorption and supply.

The cells, thus, achieve a state of equilibrium by autoregulation, whereby the osmotic effects of the marine microecosystem of the tide pools are compensated for by both the barrier and mucilage exchange effects and the osmotic effects of the organic substances in the cytoplasm.

These water-soluble fractions are collected from the fresh algae by water-glycol extraction (50% water/50% propylene glycol). The algae are harvested by hand on the foreshore at low tide by a team of special gatherers.

Codium corresponds to this hydroglycolic agent with a 1.2% dry extract for a sulfur content of 1.1 g/L, i.e. the equivalent of 3.3 g/L of SO_4. This sulfate content is characteristic of sulfated polymer that represents the most water-soluble polysaccharide fraction of the extract.

Moisturizing activity measured by corneometer: An experiment was conducted using a corneometer to measure the dielectric constant of skin moisture. The skin moisture level was evaluated on a group of 10 women. The corneometer lead was placed on the forearm on an area of approximately 4 cm^2.

The measurements were made on treated skin (reference area). The treated areas received a dose of 4 mg/cm^2 of five commercially available moisturizing creams. Then 5% *Codium tomentosum* extract was added to the same creams. The hydrating power of these compositions was measured over time at 0, 3, 15, and 30 min, and 1, 2, 4 and 6 h.

The off-the-shelf moisturizing creams, with the exception of one, caused an immediate but short-lived increase in the moisture level. The average moisture level for the five creams was higher than 28% just after application but dropped to 10% after only 3 min.

The addition of the extract to these creams reinforced their hydrating power. This effect lasted over time. Six hours after application, the moisture level remained high (an average of +15% compared to the control area), i.e. three times higher than moisturizing creams without the *Codium* extract.

Moisturizing activity measured by transitory thermic transfer: Evaluation of the moisture level by transitory thermic transfer (TTT) is based on the capability of the skin to exchange heat when in contact with another substance. The moisture level expressed as TTT is measured by a thermal device including a heating element generating a thermal wave and a temperature probe. The thermal wave generated is transitory and pulsed. The thermal pulses with a constant power and duration and is followed by a quiescent period, also with a constant duration. This pulse/quiescence cycle is the transitory characteristic of the measurement. The thermal wave propagates through the skin.

The skin temperature variation is measured during the pulse, and the device records this temperature variation as a signal. The TTT is then defined as the ratio of the power (P) of the thermal pulse applied by device to the temperature variation of the skin during the pulse: TTT = P/DT. The values are in mw/°C. The higher the skin moisture level, the smaller the temperature variations due to the thermal pulse, because of heat transfer with the water. The TTT is, thus, inversely proportional to the temperature variation and directly proportional to the moisture level. The pulse duration can be adjusted to explore the superficial, mid-layer and deep moisture levels of the epidermis successively.

The effect of a product on the skin moisture level is expressed as the percentage of variation of the TTT measurement. This percentage is equivalent to the variation in moisture level observed between time after application of the product and time before application. The experiment was conducted with a microeffusivimeter with measurement of the moisture levels in the three layers of the epidermis (superficial, mid-layer, deep layer) by TTT. The products were applied twice a day for seven days.

The hydrating power was determined as follows:

- **Measurement on day 0**: just before the first application and 1 and 3 h after.
- **Measurement on day 8:** just before application (12 h after the last application on day 7) and 1 h after.

The use of 5% *Codium tomentosum* extract in a cream has various effects on the skin moisture level:

- **Immediate hydrating action:** The *Codium* extract doubles the skin moisture level on the first application (t = 1 h).
- **Lasting hydrating action:** The cream with the extract maintains the moisture level in the skin for more than 3 h, whereas the skin moisture level drops by 25% for the placebo cream. After the 8-day application period, the moisture level in the skin remains more than 20% higher than the skin treated with the placebo cream.
- **Moisturizing of all the layers of the epidermis:** Application of the placebo cream is characterized by a systematically higher water loss in the superficial layers of the skin. By contrast, at the outcome of the treatment with the extract, regulation and control of the moisture level through the entire stratum corneum was observed. The measurement made on day 8 following application shows that *Codium* increases the action of the placebo cream by 50%.

Gelidium Extract

In the world of algae, there exist mechanisms for the storage and release of lipids. In order to meet the energy needs of the algae cell, lipids are transformed to energy and the excess is stored. This reserve is used by the algae in periods of stress, exposure to cold and weak sunshine (reduced photosynthesis). Algae agents that can transmit messages to instigate lipolysis are sterols or derivates and are particularly active in the rhodophyta, or red algae. Algae sterol[5] is obtained from a red algae of the species *Gelidium* sp, where it plays an important role in algae metabolism. *Gelidium* extract[c], which contains 1.5% of active sterol, can be used in the same way as a lipolytic and firming agent in cosmetic products.

Lipolytic activity: Lipolytic activity in cells of adipose tissue is assessed in

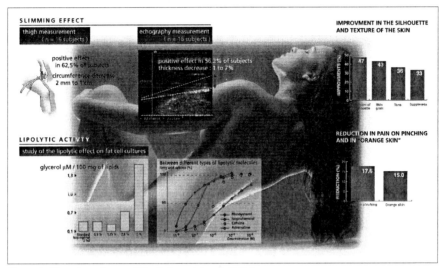

Figure 3. Slimming effect and lipolytic activity of *Gelidium* extract

culture media by measuring the concentration of liberated glycerol. Adipocytes released glycerol after application of *Gelidium*. At concentrations between 2.5 and 5%, it exerts a clear lipolytic activity compared to traditional lipolytic agents such as caffeine and theophilline.

Slimming effect: A balm containing 5% *Gelidium* extract was applied to 16 female volunteers. All presented a localized adipose overload. Each subject applied the product twice a day on the thighs for four weeks in the direction of venous circulation (from the bottom to the top of the thighs) during a period of about three minutes. The slimming effect was appraised by means of tape measurement and echography.

Centimetric measurement: The circumference of the thighs was evaluated before and after local treatment on the subject, who was standing close to a wall. We measured the diameter about midway between the knee and the hip and determined the distance from the floor and the wall. To ensure that diameters were measured parallel to the floor and perpendicular to the leg, we placed two levels on the tape measure. Thigh circumference slimming effects were observed following the treatment with *Gelidium* extract. These measurements showed a slight (-3 mm) to clear-cut (-10 mm) decrease in 10 subjects out of 16 (62.5%). Tape measurement of the thighs showed a 2 mm – 1 cm decrease in the thickness of the adipose tissue in 10 of the 16 subjects.

Echographies: These were performed on the same days as the tape measurements by means of an echograph. We scanned the thigh area in the regions of interest. Digitally stored scans were analyzed for epidermal, dermal and fat layer thickness. Echography showed between 0.2 – 1.8 mm (i.e., –1% – 7% variation) decrease in the thickness of the adipose tissue in 9 of the 16 of the subjects.

---

cRhodysterol is a registered trademark of SECMA Corporation, Pontrieux, France

The signal of lipolysis: Lipolysis, the breakdown of fat, begins with the degradation of triglycerides into glycerol and free fatty acids. Following triglycerides hydrolysis, the fatty acids are directed to the interior of the mitochondria where oxidation to CO_2 and H_2O occurs, or they are eliminated by the lymphatic system. This lipolysis can be mediated through action on the β and α adrenergic receptors, adenyl cyclase, the intercellular cyclic AMP (adenosine 3', 5' cyclic monophosphate), the phosphodiesterase and the lipase.

In this area, the sterol of *Gelidium* extract acts as a signal to the receptors of the adipocyte by stimulating lipolysis. Fields of applications of "signal molecules," now called "chemoreception," are very interesting in cosmetology. This enables a substance applied to the skin to act on physiological mechanisms of the underlying layers. It seems to be capable of breaking up the excess fats stored in the cells to allow their removal and of preventing fat from entering the normally balanced cell.

It also appears to stimulate the fibroblasts during its lipolysis action. The connective tissue appears to undergo a reorganization which gives it back its tonicity and elasticity.

Enteromorpha compressa Extract

The green algae *Enteromorpha compressa* that lives in the upper strand is one of the species that represents the transition between seaweed and land plants. It is indifferent and insensitive to land-based substances borne by fresh water, so its preferred niche is in zones directly influenced by rivers and runoff.

An active fraction of peptides[d] is extracted from *Enteromorpha compressa* and constitutes an original approach to hypersensitivity by its novel action on neuropeptides. Hypersensitivity is a global phenomenon, widespread around the world. During the past several years, dermatologists have noted a significant increase in the number of people suffering from skin reactions linked to hypersensitivity, which is characterized by skin with a lower-than-normal tolerance and involves both disorders of the stratum corneum and a disturbance of the microcirculatory system. Two clinical studies on skin with a tendency to atopy or hypersensitivity have shown the efficacy of the active ingredient toward skin sensitivity.

Desensitization as target: The desensitizing effect of a body lotion containing 2% *Enteromorpha compressa* on the sensation of pruritus (severe itching) was determined by thermography. Ten volunteers, otherwise healthy but hypersensitive to pruritus and with a propensity to atopy (skin allergy), were selected. Skin was sensitized by thermal aggression using a probe in contact with the internal side of the wrists. The 3 cm diameter probe combined a temperature sensor with 0.1°C accuracy and a heat control system that supplied heat to the skin, also with a 0.1°C accuracy. The physical method, more specifically sensorimetric, enabled the perception of pruritus to be studied and quantified by evaluating variations in sensations it caused. This technique is preferred over less precise visual methods.

Initial stimulation was applied in the absence of the product in order to determine "control" values. At first, the temperature corresponding to "very hot"–as

---

[d] Enteline is a registered trademark of SECMA Corporation, Pontrieux, France

perceived by the volunteers at the interior of the wrists–was determined for each subject. After this determination, the time required for the sensation of pruritus to appear on the symmetrical zone (control) of the wrist was noted. Each subject announced the onset and termination of the sensation of pruritus. These values were recorded by the operator. Since the results were given by the volunteers, the use of a placebo enabled the psychological effect of product application to be eliminated.

The measurement zone was again stimulated two hours later. *Enteromorpha compressa* formulated at 2% in a lotion was applied at the dose of 8 µL/cm² on one of the wrists. The placebo lotion was applied at the same dose on the other wrist. The three parameters studied were the time for the sensation of heat to appear, its duration and its intensity. The latter was expressed as a numerical scale between 0 and 4, corresponding to a gradient of increasing intensity of the sensation.

The results of this experiment showed that in comparison to the control zone, the placebo lotion tended to accentuate the problem of pruritus as judged by the three evaluation parameters. When the placebo lotion was applied, the sensation of pruritus appeared very rapidly, after 28 seconds. When *Enteromorpha compressa* was incorporated in the placebo formula, sensitisation effects were greatly retarded, since the time of appearance of the sensation of pruritus increased to 112 seconds. This improvement was obtained in 10 of the 10 volunteers when compared to the placebo, and in eight of the 10 volunteers when compared to the control.

Lotion containing *Enteromorpha compressa* also considerably reduced the duration of the sensation of pruritus: control, placebo and *Enteromorpha* extract otion values were 89, 67 and 47 seconds. This duration decreased in eight of 10 volunteers compared to the control, and in six of 10 compared to the placebo. Similarly, the *Enteromorpha compressa* containing milk also reduced the intensity of the sensation of pruritus. Thus, in the case of the control and placebo lotions, intensities were judged to be strong (2.8-2.9) in contrast to the extract lotion, where the intensity of the sensation of pruritus was genuinely attenuated, as shown by the low value (1.1). Compared to the placebo, improvement was seen in 9 of 10 volunteers.

Enteromorpha compressa, thus, has considerable soothing properties when applied to skin that is hypersensitive to pruritus. In this study, the reactivity phenomenon was delayed by a factor of four in terms of time. This reactivity was also reduced in terms of intensity (55%) and duration (30%).

Comfort as target: This study was conducted by 10 dermatologists on 45 subjects, included because of their very dry hypersensitive skin, and a sensation of pruritus.

The lotion, identical to that in the preceding study containing 2% *Enteromorpha* extract, was applied to the body once or twice daily (depending on the case) for 20 days. Clinical evaluations were conducted by the dermatologists and volunteers in order to quantify treatment efficacy as determined by the following criteria: desquamation (peeling), erythema sensation of skin straining and pruritus. Efficacy was scored from 0 (good) to 4 (poor). The improvement of well being was also evaluated according to characteristics such as dryness of the skin, its comfort and the quality of life.

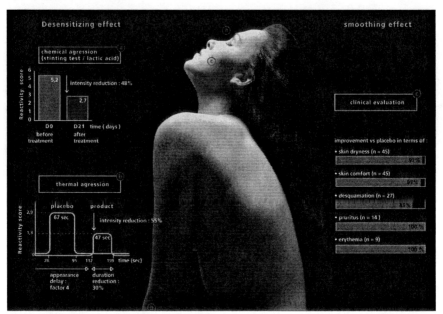

Figure 4. Effect of *Enteromorpha* extract on skin

The soothing effect of *Enteromorpha* extract was evaluated in a clinical study of a population providing a wide range of manifestations characteristic of reactive skin. Depending on the individuals, these manifestations can be characterized by problems of desquamation, erythema, pruritus and skin straining. The lotion containing *Enteromorpha compressa* was effective against all these symptoms. In the nine subjects with problems of erythema, this phenomenon disappeared following treatment. In addition, pruritus totally disappeared in the 14 subjects concerned.

At the end of treatment, the number of subjects still reporting problems of skin straining was reduced from 18 to one. At the same time, the number of individuals complaining of desquamation decreased from 27 to seven. This efficacy was also encountered in terms of the criteria of well being, since among all the subjects, 97% confirmed improvement in terms of dryness, 95% for improved skin comfort and 92% for improved general well being.

Enteromorpha *extract, the soothing and comfort response:* Faced with the extent of the phenomenon of hypersensitivity, "sensitive skin" lines of cosmetics have multiplied during the past several years, with different paths of development being undertaken. Understanding the problem of skin sensitivity has been gained not only by testing formulations with soothing action against external aggression, but also by the quest for better acceptability of the formulation itself. Thus, the cosmetics industry has, in general, become oriented toward lighter and non-occlusive textures. The products are adapted to sensitive skin by eliminating unwanted products and remaining within the pH range of the skin.

Companies have also tackled hypersensitivity by acting either on protection against deleterious environmental effects or by reconstituting the hydrolipid film (hydrating film, natural moisturizing factor (NMF), vegetable oils, etc.), or by applying venotonic agents or anti-irritants. In this setting, *Enteromorpha compressa* represents a novel approach to hypersensitivity, by initially considering it to be a neurosensorial problem before being erythymatous.

Its originality resides in the fact that it acts at the initiation stage of inflammation, pain or pruritus by blocking access of neuromediators to cell receptors. *Enteromorpha* extract modulates their activity and the attendant biological response. This leads to a reduction in the activation of epidermal cells. As a result, the extract has a positive effect on the tolerance threshold and breaks the vicious cycle of sensitive skin – the skin becomes more intolerant as aggression is more frequent and, as its reactions intensify, crises multiply. Use of *Enteromorpha* extract, thus, leads to a return to a more normal state, with the skin reacting less violently and less rapidly to external aggressions.

When *Enteromorpha compressa* is incorporated at 1-2%, it provides an overall cosmetic response to the problem of hypersensitivity. It is intended for all care or hygiene lines as well as specific "sensitive skin" lines. It is a positive addition to face, body and hair products, examples of which are makeup removers, shower gels, foaming baths, etc. While *Enteromorpha* extract provides a solution to problems of intolerant skin, it also is recommended in more specific cases of irritable and atopic skin. In this case, it acts as a genuine active ingredient by reducing the degree of skin reactivity and minimising skin reactions.

The skin is subjected to a number of aggressive factors every day. Taken individually, they may cause only minor damage, but when the entire arsenal is present over long periods of time, it may potentiate harmful effects in the epidermis and the dermis. The skin also retains a certain degree of reactivity during renewed contact with the initial factor. *Enteromorpha* extract can thus be recommended in sun or after-sun products to enable the skin to recover from prior traumas.

Enteromorpha compressa is a new response to problems of skin that is atopic, irritable or intolerant. Starting with the initiation of the sensitisation process, it leads to the inactivation of epidermal receptors involved in inflammatory reactions or in pain. Over the course of time, these reactions may cause damage which, if repeated, can lead to functional and structural changes in the skin and, even if in an indirect manner, to the formation of free radicals by lipoperoxidation of cell membranes. Hypersensitivity can lead to skin aging.

Enteromorpha compressa formulations have a several objectives: to improve tolerance to cosmetic products, improve the capacity for adaptation and the comfort of sensitive skin, and afford anti-wrinkle protection in skin aging. *Enteromorpha compressa* pushes back the frontiers of hypersensitivity.

—**Xavier Briand,** SECMA, Pontrieux, France

References

1. X Zhou and K. Mopper, Determination of photochemically produced hydroxyl radicals in seawater and freshwater, *Mar Chem* 307 88 (1990)
2. X Briand and N Mekideche, Antiradical protector from marine algae, *Cosmetics & Toiletries* 107 (8) 77-80 (1992)

3. A Moysan, P Morlière, X Briand, G Hüppe and L Dubertret, Lipid peroxidation induced by UVA light or by hypoxanthinexanthine oxidase in cultured human skin fibroblasts: Effect of exogenous SOD and of a new antioxidant extracted from seaweed, IVth Congress of European Society for Photobiology, Amsterdam (1991)
4. N Mekideche and X Briand, A marine moisturizer, *Cosm Toil* 111(6) 101-106 (1996)
5. N Mekideche and X Briand, Beauty shines in the blue, *DCI* (10) 32-40 (1996)

Phytoplankton: The New Frontier for Stress-Relieving Cosmetic Ingredients

Key words: phytoplankton, skin homeostasis, UV, protein oxidation, cell communication, barrier repair, inflammation

Three species of phytoplankton (or micro algae) help relieve the skin from various stresses and restore homeostasis

The skin suffers various stresses such as UV radiation, aging and aggressions that disrupt its normal function. Traditional cosmetic answers are found either in vegetable, biological or biotechnologically derived ingredients. A new diversified, sustainable and ecological source is now available: phytoplankton (a unicellular microscopic alga). Various phytoplankton extracts have been found to relieve skin stresses and restore skin homeostasis.

Phaeodactylum tricornutum Extract

Phaeodactylum tricornutum belongs to the Diatomophyceae order widely found in coastal waters. The lipidic fraction is obtained by a treatment with a mixture of water and alcohol under alkaline condition. The fatty acids are purified by extraction and molecular distillation. The extract offers a unique fatty acid profile. The key constituents are unsaturated fatty acid C16:1, C20:5, C22:6. *Phaeodactylum tricornutum* protects and repairs age- and UV-induced damage to proteins.

Aging and protein oxidation: Proteins are among the major targets for oxidative damage (in addition to DNA and lipids) and the build-up of potentially harmful oxidized proteins is characteristic of aging and leads to cellular dysfunction and senescence. The correlation between photodamage and protein oxidation is also well-established,[1] because there is increasing evidence for the generation of reactive oxygen species (ROS) in skin upon UV exposure.

Until now, besides using antioxidants and sunscreens for protecting the skin, no other protection was available. However a new line of defense can be established by modulating proteasome activity.

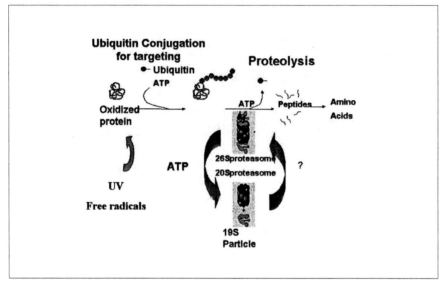

Figure 1. The ubiquitin proteasome pathway

Protein oxidation and proteasome: Each cell contains a proteolytic system called proteasome responsible for the degradation of oxidized proteins and protein turnover. Proteasome is a multicatalytic protease complex (1.5 to 2 million Daltons) located in the cytosol and in the cell nucleus. It exhibits three distinct proteolytic activities – chymotryptic-like, tryptic-like and peptidylglutamyl peptide hydrolyzing–as it cleaves proteins on the carbonyl side of hydrophobic, basic and acidic residues. Proteolysis (Figure 1) is achieved after combining the oxidized protein with ubiquitin, which unfolds the protein before it is digested by the 26S proteasome.

The Ubiquitin Proteasome Pathway (UPP) is the cell's principal mechanism for protein catabolism and has roles in both housekeeping and the turnover of many regulatory proteins (Figure 1).

Proteolysis is central to an incredible multitude of processes including the cell cycle, cell growth and differentiation, embryogenesis, apoptosis, signal transduction, DNA repair, regulation of transcription and DNA replication, antigen presentation and other aspects of the immune response, the functions of the nervous system including circadian rhythms and acquisition of memory.

Impairment of proteasome function: Proteasome function is impaired upon aging and UVA- and UVB-irradiation.[2-4] Therefore it is important to restore its activity and prevent the impairment of key enzymes for skin function such as protein kinase, catalase, p53 and proteases among others.

The *Phaeodactylum tricornutum* extract (2.5 µg/mL) prevents the increase of UV-induced oxidized proteins: 10 µg of a keratinocyte lysate is irradiated (10 J/cm^2 UVA + 0.05 J/cm^2 UVB) and treated for 7 h. The oxidized proteins are analyzed by western blot and quantified after staining of the carbonyl group (Figure 2a). A 45% reduction in the level of oxidized protein is obtained.

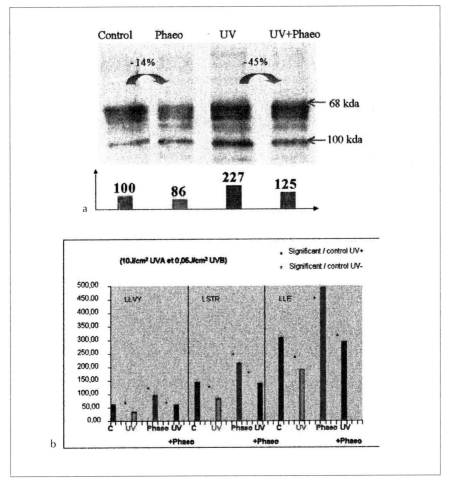

Figure 2. The repair effect of *Phaeodactylum* extract
a) Western blot
b) Restoration of the three protease activities of proteasomes in keratinocyte culture after UV irradiation

The mechanism of action of this repair effect is confirmed as the *Phaeodactylum* extract (2.5 mg/mL) restores the three protease activities of proteasome in keratinocyte culture that have been submitted to UV irradiation (Figure 2b).

Protein oxidation can be repaired: Because both chronological and photoaging increase protein damage and damage the proteosomal activity, the cells accumulate oxidized proteins that are no longer removed. The *Phaeodactylum tricornutum* extract helps to protect and repair protein oxidation and will help the skin to be purified from these wastes and reduce age- and UV-induced damages.

Furthermore, it has been proven[5] that lipofuscin–a substance composed essentially of oxidized, cross-linked proteins and a hallmark of aged non-dividing

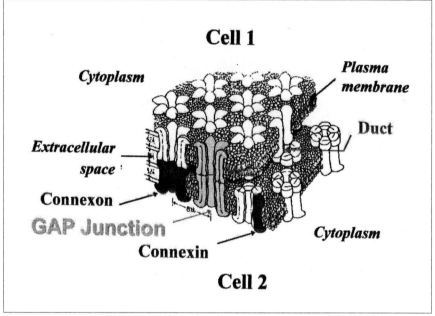

Figure 3. Gap junctions between neighboring cells

cells–present in age spots, may originate from insufficient proteasome activity. Further studies may be needed to confirm that the *Phaeodactylum tricornutum* extract may help in reducing lipofuscin content in age spots.

Skeletonema costatum Lipidic Extract

Skeletonema costatum is a mono-cellular alga from the Diatomophyceae order. It is widely found in the coastal waters. The micro algae are treated with a mixture of water and alcohol under alkaline conditions. The fatty acids are purified by extraction and molecular distillation. The extract offers a unique fatty acid profile. The key constituents are polyunsaturated fatty acids: C16:3, C18:4, C20:5 and C22:6.

Cell communication and skin homeostasis: The loss in homeostatic capacity of the aging organism depends in particular on the quality of the communication between cells through gap junctions.

Gap junctions are areas of the cell membranes that connect neighboring cells (Figure 3). These organized protein channels allow ions and small molecules to cross between the connected cells in a passive fashion. The "communicating" cells equilibrate all of their critical regulatory ions and small molecules (Ca++, C-AMP, glutathione). These protein channels consist of two "hemi-channels" or connexons that consist of six proteins called connexins. Connexin 43 is mostly expressed by keratinocytes.[6]

Physiological functions of gap junctions: The fundamental function of two or more cells coupled by gap junctions is to communicate through chemical

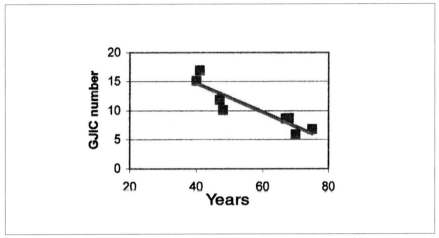

Figure 4. Aging (in years) disrupts the number of gap junctions (GJIC) in normal human keratinocytes

signals. The major physiological role of gap junctions is to synchronize "metabolic" or electronic signals between cells (regulatory ions and small molecules, e.g. Ca++, C-AMP). Cells have four basic functions: proliferate, differentiate, apoptose (or die by programmed cell death) and adaptively respond if they are already terminally differentiated. In a multi-cellular organism, a delicate coordination or orchestration of these four cellular functions must occur. So in the epidermis, growth, differentiation, apoptosis and homeostatic control of differentiated cell function must occur in a single space and this is done by coupling the keratinocytes through their gap junctions.[7,8]

Aging disrupts gap junctions: The number of gap junctions in normal human keratinocytes declines according to the age of donors (Figure 4). However, we observed a dose dependent increase in the number of gap junctions following micro-injections of confluent normal human keratinocytes (NHK) (Figure 5).

Various authors[9,10] have also proven that UV radiation disrupts gap junctional communication in human keratinocytes and contributes to skin photoaging.

Skeletonema costatum lipidic extract restores gap junctional communication: When keratinocytes from a 63-year-old donor are treated with 2.5 µg/mL of *Skeletonema costatum* lipidic extract, the number of gap junctions (GJIC) is increased by 78% and reaches the number of GAP junctions found in keratinocytes from younger donors. By this effect, a *Skeletonema costatum* lipidic extract helps to restore and maintain epidermal homeostasis and cellular harmony.

Porphyridium cruentum Extract

During its growth, *Porphyridium cruentum*, a red micro algae, synthesizes high molecular weight polysaccharides that create a protective coating around the cell membrane. Part of these polysaccharides will dissolve in seawater and increase the viscosity locally. This is really a unique self-protection mechanism!

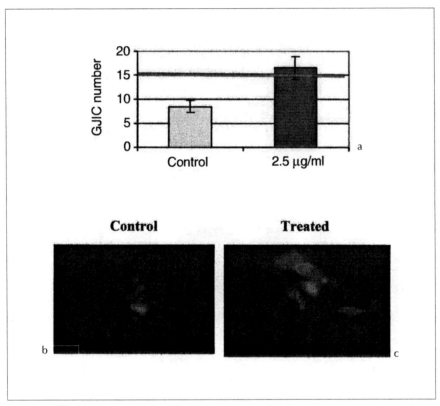

Figure 5. Effect of micro-injected *Skeletonema costatum* lipidic extract on gap junctional communication
a) Restoration of number of gap junctions (GJIC) in older donors following treatment versus untreated control, and comparison at 15 to the number of GJIC in young donors (correlation: r=0.91)
b) Control
c) Treated

Porphyridium cruentum extract is obtained from red micro algae grown in Brittany in marine spring water whose unique mineral composition boosts its duplication. The extract is therefore composed of a solution of polysaccharides in minerals.

Special polysaccharide structure: *Porphyridium cruentum* excretes a high molecular weight (around 4,000,000 D) sulfated polymer, similar to human glycosaminoglycans. Its major components[11] are listed in Table 1. It contains three disaccharides and two uronic acids. The disaccharides are 3-O(α-D-glucopyranosyluronic acid)-L-galactose, 3-O-(2-O-methyl-α-D-glucopyranosyluronic acid)-D-galactose and 3-O-(2-O-methyl-α-D-glucopyranosyluronic acid)-D-glucose. A polyanion of high molecular weight contains D- and L-galactose, xylose, D-glucose, D-glucuronic acid and 2-O-methyl-D-glucuronic acid, and sulfate in molar ratio (relative to D-glucose) of 2.12:2.42:1.00:1.22:2.61.

Table 1. Major components of the polysaccharide excreted by *Porphyridium cruentum*

Hexoses (galactose / glucose)	36.0%
Pentoses (xylose)	30.0%
Glucuronic acid	8.0%
Sulfate	9.0%
Amino acids (xylose linked)	3.8%

The *Porphyridium cruentum* extract, obtained from marine spring water, is rich is various beneficial trace elements—manganese, silicium and zinc—at levels ten times higher than in normal seawater. These trace elements are essential cofactors used by skin enzymes for its normal function.

Immuno-modulating properties: The polysaccharide of *Porphyridium cruentum* significantly increases the number of cells present in the peritoneal exudate. It also increases the activity of lysosomal acid phosphatase enzyme. These facts suggest a possible stimulation of the metabolic and functional action of the immune system.[12] Topically, the polysaccharide of *Porphyridium cruentum* limits irritant-induced cutaneous erythema, probably by inhibiting circulating immune cell recruitment.[13] Furthermore, the polysaccharide of *Porphyridium cruentum* exhibited strong antiviral activity against herpes simplex virus types 1 and 2 both in vitro and in vivo, through a strong interaction between the polysaccharide and the virus.[14]

These activities suggest anti-adhesion properties similar to chondroitin sulfate, a well-known modulator of the immune response and activator for wound healing. Further work will be needed to fully explore this field of research.

Moisturization and barrier repair: Due to its high molecular weight polysaccharide (higher than hyaluronic acid) and to its unique mineral composition, the *Porphyridium cruentum* extract offers excellent moisturizing properties in two ways: through forming a non-occlusive film on the skin surface and by enabling the skin to synthesize the epidermal lipids necessary for barrier recovery. Figure 6 shows the results of tests on a keratinocyte culture. At 100 μg/mL, the *Porphyridium cruentum* extract increases in average the barrier lipids by 45%.

Barrier repair is essential for preventing transepidermal water loss and maintaining skin homeostasis.[15] Allowing the skin to synthesize its own requirements represents an elegant way to retain its normal functioning, even under stressed conditions.

Skeletonema costatum Peptidic Extract

During an aggression (from UV or other sources), the skin defends itself by recruiting immune cells. Keratinocytes release cytokines (such as IL1a and TNFa) and chemokines (IL8). They express adhesion molecules (selectins, ICAM 1s) that allow the adhesion of leukocytes on the endothelial cell of micro-capillaries and their release

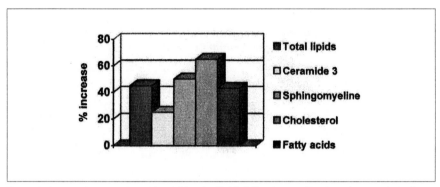

Figure 6. Average increase in epidermal barrier lipids following application of 100 µg/mL of *Porphyridium cruentum* extract

in the inflamed tissue that leads to erythema. Inhibiting leukocyte adhesion prevents the endothelial cells from acting as inflammation amplifiers.

Human leukocyte adhesion on endothelial cells from the dermal capillary is stimulated by a UV-irradiated NHK culture medium. Human keratinocytes (from plastic surgery) were cultured in the presence of tested molecules for 48 to 72 h after UVB irradiation (10 mJ/cm^2) to condition the media. Then, the leukocyte (human T cell line CEMT4) adhesion and endothelial cell (immortalized human micro vascular HSMEC 7,8) adhesion molecule expression were assayed on keratinocyte-conditioned medium-treated endothelial cells.

To evaluate leukocyte adhesion, FITC-labeled leukocytes were overlaid onto PKH26GL-labeled endothelial cell monolayer in 5:1 endothelial cell ratio and were allowed to adhere for 20 min at 4°C, under static conditions. After removing non-adhering cells, endothelial cells and adhering cells were lysed and analyzed by fluorescence. IL 8 were assayed by ELISA kits.

Adhesion molecules expression of ICAM 1s was analyzed by flow cytometry after incubation with FITC antibodies.

Skeletonema costatum peptidic extract (SKCP) was applied on keratinocyte cultures at 40 µg/mL. It inhibits IL8, ICAM 1s and leukocyte adhesion by 42, 69 and 37%, respectively.

Modulation of pro-inflammatory condition and erythema prevention are traditionally performed by inhibiting the cytokines that enhance the PLA2 activity, release arachidonic acid that is metabolized by the lipoxygenase pathway to produce leukotrienes, or by the cyclooxygenase pathway to produce prostaglandins and thromboxanes. Inhibiting leukocyte adhesion will modulate another key inflammation mechanism. Thus, *Skeletonema costatum* peptidic extract offers a new promising approach for relieving irritated skin.

Another important field of cosmetics dealing with inflammation is the acneic skin. It has been proven that the *Propionibacterium acnes* has a key role in acne-related inflammation because it induces IL-8 production.[16] Also, acne-related inflammation is mediated by CD4+T cells with a high level of ICAM-1s expression

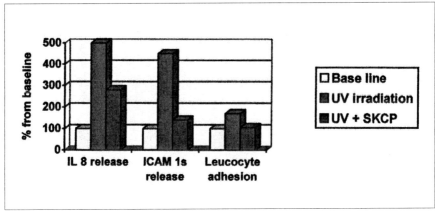

Figure 7. Anti-inflammatory effect of Skeletonema costatum peptidic extract applied to keratinic cultures at 40 µg/mL

and leukocyte infiltration.[17,18] Because *Skeletonema costatum* peptidic extract limits those phenomena, it will help improve the comfort of acneic skin.

New Perspectives for Relieving Skin Stress

By helping repair oxidized proteins, restore cell communication, promote barrier repair and limit the consequences of inflammation, various phytoplankton extracts will limit the consequences of many daily stresses and restore skin homeostasis.

—Dr. Patrice André, *LVMH Parfums & Cosmétiques, St. Jean de Braye, France*
—Dr. Anne Humeau, *SOMAIG, L'Ile Grande, France*
—Dr. Laurent Sousselier, *Naturactiva, Nogent, France*

References

1. CS Sander, H Chang, S Salzmann, CS Muller, S Ekanayake-Mudiyanselage, P Elsner and JJ Thiele, Photoaging is associated with protein oxidation in human skin in vivo, *J Invest Dermatol* 118(4) 618-625 (2002)
2. I Petropoulos, M Conconi, X Wang, B Hoenel, F Bregegere, Y Milner and B Friguet, Increase of oxidatively modified protein is associated with a decrease of proteasome activity and content in aging epidermal cells, *J Gerontol A Biol Sci Med Sci* 55(5) B220-227 (2000)
3. G Carrard, AL Bulteau, I Petropoulos and B Friguet, Impairment of proteasome structure and function in aging, Int J Biochem Cell Biol 34(11) 1461-1474 (Nov 2004)
4. AL Bulteau, M Moreau, C Nizard and B Friguet, Impairment of proteasome function upon UVA- and UVB-irradiation of human keratinocytes, *Free Radic Biol Med* 1 32(11 1157-1170 (Jun 1, 2002)
5. A Terman and S Sandberg, Proteasome inhibition enhances lipofuscin formation, *Ann N Y Acad Sci* 973 309312 (Nov 2002)
6. D Salomon, E Masgrau, S Vischer, S Ullrich, E Dupont, P Sappino, JH Saurat and P Meda, Topography of mammalian connexins in human skin, *J Invest Dermatol* 103(2) 240247 (Aug 1994)

7. DF Gibson, DD Bikle, J Harris and GS Goldberg, The expression of the gap junctional protein Cx43 is restricted to proliferating and non differentiated normal and transformed keratinocytes, *Exp Dermatol* 6(4) 167-174 (Aug 1997)
8. DJ Fitzgerald, NE Fusenig, P Boukamp, C Piccoli, M Mesnil and H Yamasaki, Expression and function of connexin in normal and transformed human keratinocytes in culture, *Carcinogenesis* 15(9) 1859-1865 (1994)
9. H Banrud, SO Mikalsen, K Berg and J Moan, Effects of ultraviolet radiation on intercellular communication in V79 Chinese hamster fibroblasts, *Carcinogenesis* 15(2) 233-239 (Feb 1994)
10. N Provost N, M Moreau M, A Leturque A and C Nizard, Ultraviolet A radiation transiently disrupts gap junctional communication in human keratinocytes, *Am J Physiol Cell Physiol* 284(1) C51-59 (Jan 2003) (Epub Sep 4, 2002)
11. J Heaney-Kieras and DJ Chapman, Structural studies on the extracellular polysaccharide of the red alga, Porphyridium, *Carbohydr Res* 52 169-177 (Dec 1976)
12. H Quevedo, C Manrique, R D'az and G Pupo, Preliminary evidence of immunomodulatory activity of the polysaccharide fraction of Porphyridium cruentum, *Rev Cubana Oncol* 16(3) 171-176 (2000)
13. MS Matsui, N Muizzuddin, S Arad and K Marenus, Sulfated polysaccharides from red microalgae have antiinflammatory properties in vitro and in vivo, *Appl Biochem Biotechnol* 104(1) 13-22 (Jan 2003)
14. M Huheihel, V Ishanu, J Tal and SM Arad, Activity of Porphyridium sp. polysaccharide against herpes simplex viruses in vitro and in vivo, *J Biochem Biophys Methods* 4 50(2-3) 189-200 (Jan 4, 2002)
15. MSM Man, KR Feingold, CR Thornfeldt and PM Elias, Optimization of physiological lipid mixtures for barrier repair, *J Invest Dermatol* 106(5) 1096-1101 (May 1996)
16. Q Chen Q, T Koga, H Uchi, H Hara, H Terao, Y Moroi, K Urabe and M Furue, Propionibacterium acnes-induced IL-8 production may be mediated by NF-kappaB activation in human monocytes, *J Dermatol Sci* 29(2) 97-103 (Aug 2002)
17. AM Layton, C Morris, WJ Cunliffe and E Ingham, Immunohistochemical investigation of evolving inflammation in lesions of acne vulgaris, *Exp Dermatol* 7(4) 191-197 (Aug 1998)
18. M Toho, Y Uchida, I Miyamoto and T Ogawa, Immunohistochemical studies of the acne-like inflammatory model, *Jikken Dobutsu* 39(4) 531-537 (Oct 1990)

Penetration of Vitamin A Palmitate Into the Skin

Keywords: vitamin A palmitate, percutaneous absorption

Study of five vehicles on the penetration of vitamin A palmitate into the skin

Among the active compounds described, prescribed and sold for cosmetic purposes, vitamins are gaining increasing notoriety and posing a challenge to many researchers.[1-6] Vitamin A has been used in cosmetics in its different forms because of its pharmacodynamic properties.[7,8] It acts on the skin by keeping it in good condition and favoring correct metabolism. It also acts on epithelialization in dry and rough skin, as well as on keratinization considered to be abnormal.

In cosmetics and dermatologic applications, Vitamin A must penetrate the skin to be effective.[9,10] Absorption can take place through stratum corneum (SC) and hair follicles (HF).[11]

Percutaneous absorption is controlled by skin pH and its integrity, by hydration condition (correlated with the individual's age), mode of application, coefficient of o/w partition, ionization state, dissolution state and active substance concentration. Some vehicle properties such as pH, solubilizing capacity of the active substance, viscosity and type of preparation are also important.[12,13]

An understanding of the microstructure of creams and their bases is important to control the manner in which creams release drugs or water into the skin.[14]

The in vivo methods developed to study skin absorption of active substances are based on the use of skin of rats, mice, rabbits, guinea pigs, pigs, hamsters, monkeys and, if possible, human skin.[15,16]

The Experiment

Vitamin A absorption by the skin is affected by the type of preparation used as a vehicle. In vivo absorption studies were carried out to investigate the influence of different dermatologic vehicles in the penetration of vitamin A palmitate into the skin. Light microscopy, differential scanning calorimetry and hot stage microscopy as well as rheological measurements were used in the physicochemical characterization of the different formulations in an attempt to correlate their microstructure with the absorption results.

Preparation of formulations: Five different formulations containing vitamin A palmitate were prepared: a hydrophilic gel, three formulations in which the

Table 1. Composition of the Five Formulations

Composition	Gel	Gels creams (% w/w)			o/w emulsion
	1	2	3	4	5
Carbomer 940	0.60	0.60	0.60	0.60	
Propylene glycol	10.00	10.00	10.00	10.00	10.00
Triethanolamine	0.50	0.50	0.50	0.50	-
Mineral oil	-	7.00	7.00	7.00	5.00
Trilaureth-4 phosphate	3.00	1.00			
Trilaneth-4 phosphate	3.00	2.00			
Self-emulsifying base	6.00				
Mixture of parabens	0.20	0.20	0.20	0.20	0.20
Butylated hydroxytoluene	0.05	0.05	0.05	0.05	0.05
Vitamin A palmitate (1.000.000 UI/g)	0.50	0.50	0.50	0.50	0.50
Distilled water qsp to	100.00	100.00	100.00	100.00	100.00

hydrophilic gel contained surfactants and mineral oil (gel creams), and an o/w emulsion (self-emulsifying base containing fatty alcohols with both anionic and non-ionic surfactants). The composition of each formulation is shown in Table 1.

In vivo absorption studies: In vivo experiments were performed using guinea pigs (average weight: 350 g) to evaluate skin absorption of vitamin A palmitate.[17,18] The dorsal area of 35 animals (seven animals for each formulation) was shaved and 100 mg of each formulation was applied, in duplicate, to four areas of 1.5 cm² each. An extra duplicate area, where no formulation was applied, was used as a blank. The animals were submitted to biopsies one, two, four and eight hours after application of the formulation. The biopsies were performed using a dermatologic punch, and 0.28 cm² of each area was removed. In each biopsy the area of one of the duplicates was washed with cotton soaked in ethanol to remove preparations still remaining on the skin surface. The other area was not washed to serve as control.

The biopsy material was ground in a mortar with a porcelain pestle. The skin homogenate was submitted to extraction with isopropyl alcohol in a bath sonicator for 24 min. After extraction, the material was filtered through qualitative filter paper, and the filtrate was brought to a defined volume and submitted to quantitative analysis of vitamin A palmitate by HPLC. The amount of vitamin A palmitate that penetrated the skin was calculated by Equation 1.

HPLC analysis: The amount of vitamin A palmitate extracted from the biopsies was determined by reverse-phase HPLC with a C-18 column (Micro Pack 5 µm particle size; 4 mm d and 125 mm long). The mobile phase consisted of pure

$$\frac{A_w}{A_{nw}}$$

Where: A_w = Amount of vitamin A palmitate in the washed area

A_{nw} = Amount of vitamin palmitate in the unwashed area

Equation 1

methanol at a flow rate of 1.5 mL/min.[19] Under these conditions, the retention time of vitamin A palmitate was 10 min and the retention time of the internal standard (vitamin K_1) was 6.5 min. The detection limit was 2.5 µg/ml.

Structure determination: Polarized light microscopy (Polyvar Optical Microscope, Rheichart - Jung, Austria), differential scanning calorimetry (DSC) (Mettler TA 4000, Germany) and rheological measurements were used in the characterization of the formulations to correlate their microstructure with the absorption results.

In the DSC analysis, used to determine phase transitions in samples between 10 and 90∞C, these samples were heated in sealed aluminum pans at a rate of 10∞C/min. The reference cell, containing water and liquid nitrogen, was used to cool to subambient temperatures and a nitrogen atmosphere was present in the furnace.

Rheological measurements were made using a Ferranti-Shirley cone and plate viscometer in conjunction with an X-Y recorder. Measurements were made at 25°C using a medium cone in a humidified chamber. The instrument was used in the automatic mode with a sweeping time of 600 seconds and a maximum shear rate of 1651 s^{-1}. Apparent viscosity values were obtained from the loop apex.

Results and Discussion

After vitamin A was extracted from the biopsy material, we calculated the percentages of vitamin A palmitate absorbed as a function of time. The results can be seen in Table 2.

Characteristics: Polarized light microscopy showed that Formulations 2, 3 and 4 (hydrophilic gels containing surfactants and mineral oil) were nearly isotropic. The presence of liquid crystalline material was detected only in Formulation 5 (o/w emulsion) (Figure 1) by the characteristic birefringence under polarized light. This result was confirmed by differential scanning calorimetry, which indicated a gel/liquid crystalline phase transition only for the o/w emulsion formulation (Figure 2).

Apparent viscosity (N_{app}) values obtained in the rheological measurements are presented in Table 3. Flow curves for all formulations are shown in Figure 3. The presence of a surfactant decreased the apparent viscosity values of the gel formulations, and the process was related to the carbon-chain length of the surfactant.

The microviscosity of the formulation is one of the factors that controls the rate of drug release, explaining why the gel creams (Formulations 2, 3 and 4) provided a higher percentage of vitamin absorption than the gel alone (Formulation 1).

Table 2. Influence of formulation on the percentage of vitamin A palmitate absorbed into guinea pig skin as a function of time

% absorption of vitamin A palmitate*				
	Time			
Formulation	1 h	2 h	4 h	8 h
1 (hydrophilic gel)	8	11	15	23
2 (hydrophilic gel trilaneth-4 phosphate)	6	15	26	35
3 (hydrophilic gel trilaureth-4 phosphate)	8	12	17	27
4 (hydrophilic gel trilaneth-4 and trilaureth-4 phosphate)	17	23	31	43
5 (O/W emulsion)	18	31	38	57

* mean of 7 determinations.

DSC heating experiments indicated a gel/liquid crystal phase transition only for Formulation 5. Therefore, both polarized microscopy and DSC analysis indicated the presence of liquid crystals in Formulation 5. Hot stage was used to confirm visually the gel-to-liquid crystalline transition temperature detected by DSC analysis.

The self-emulsifying base, the formulation that presented the best results in the percutaneous penetration experiments, was the only one in which polarized light microscopy and DSC analysis detected the presence of liquid crystals. The property of liquid crystals, increasing the drug diffusion coefficient into the skin, has been reported in the literature.[24]

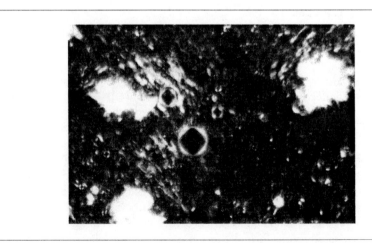

Figure 1. Photomicrograph of Formulation 5, an o/w emulsion (self-emulsifying preparation containing phosphoric esters)

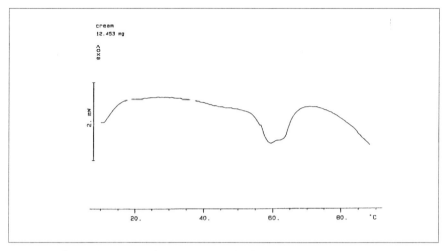

Figure 2. Differential scanning calorimetry spectra for the self-emulsifying base (Formulation 5)

Formulation 5, a self-emulsifying base containing phosphoric esters, provided the highest absorption of vitamin A in guinea pig skin, while Formulation 1, a hydrophilic gel, showed the lowest percentage of absorption. Formulations 2, 3 and 4, gel creams containing trilaneth-4 phosphate, trilaureth-4 phosphate and a combination of the two, showed intermediate percentages of absorption in relation to Formulations 1 and 5.

Chemical natures: This difference in percentage of absorption was due to the chemical nature of the vehicles. In the first situation, the vehicle contained phosphoric esters compatible with the phospholipid structure of cell membranes, while the hydrophilic gel, given its polymeric structure, probably forms a film on the epidermal surface, hindering absorption.[20-23]

The presence of a surfactant may also facilitate absorption because it increases the SC/vehicle partition coefficient.

Statistical analysis: To determine whether there were statistically significant differences among the formulations studied, analysis of variance and the Tukey test were applied to the data (these were the tests that best fitted the experimental

Table 3. Apparent viscosity (happ) of the formulations studied

Formulation	Apparent viscosity (centipoises)
1	567.5 cps
2	400.9 cps
3	173.8 cps
4	298.0 cps
5	176.7 cps

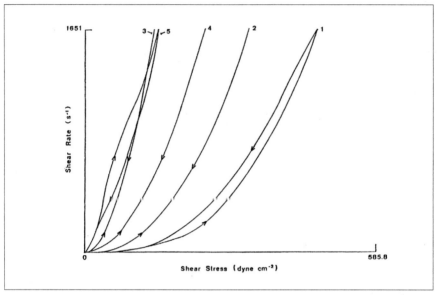

Figure 3. Ferranti-Shirley viscometer - Flow curves for hydrophilic gel (1), gel cream containing launeth-4 phosphate (2), gel cream containing laureth-4 phosphate (3), gel cream containing laurenth-4 phosphate and laureth-4 phosphate (4) and o/w emulsion (5)

Table 4. Tukey test applied to the results		
Formulation	Mean	Tukey critical value (1%)
1 (hydrophilic gel)	21.71 t	
2 (hydrophilic gel trilaneth 4 phosphate)	26.85 ?	2.02
3 (hydrophilic gel trilaureth 4 phosphate)	22.38 t	
4 (hydrophilic gel trilaneth 4 and trilaureth 4 phosphate)	31.88 X	
5 (o/w emulsion)	36.44 Y	

Note: Means followed by equal symbols did not differ significantly.

model).[25] The results are listed in Table 4 and demonstrate that, except for Formulations 1 and 3, the means obtained for each formulation were significantly different (the Tukey test critical value was 2.02 at the 1% level of significance).

Conclusion

The results obtained in the present study demonstrate that the vehicle used influences the in vivo skin absorption of vitamin A. Thus, the correct choice of a vehicle is very important for vitamin A absorption and the manifestation of its pharmacodynamic effects.

—**Patrícia Maria Berardo Gonçalves Maia Campos,** *Faculty of Pharmaceutical Sciences of Ribeirão Preto, USP, and Salete A. Benetton, USP, São Paulo, Brazil*
—**Gillian M. Eccleston,** *Faculty of Pharmacy, University of Strathclyde/Glasgow, Scotland*

References

1. R Hermitte, Aged skin, retinoids and alpha hydroxy acids, Cosm Toil 107 63-7 (1992)
2. B Idson, Vitamins in cosmetics, an update. I. overview and vitamin A., Drug Cosm Ind 146 26-8, 91-92 (1990)
3. P Mayer, The effects of vitamin E on the skin, Cosm Toil 108 99-109 (1993)
4. B Rens, Las vitaminas em productos cosméticos, Ci Cosm 2 6-9 (1991)
5. KW Strandberg, Research and development: the roots of a product, Soap Chem Spec 68 28-39 (1992)
6. PMBGM Campos, Vitaminas lipossolúveis em cosméticos, Cosm Toil 4 30-3 (1992)
7. J Franchi, MC Coutadeur, JC Archambault and G Redziniak. Effects of retinoids free or encapsulated in liposomes on human skin cells in culture, in: International Congress Yokohama 17 (1992) Yokohama. Anais. Yokohama: International Federation Societies Cosmetics Chemists, 126-39 (1992)
8. E Schwartz, FA Cruickshank, JA Mezick and LH Kligan, Topical all-trans retinoic acid stimulates collagen synthesis in vivo, J Invest Dermatol 96 975-8 (1991)
9. H Schaefer, A Zesch, Penetration of vitamin A acid into human skin, Acta Derm Venerol 74 50-5 (1975)
10. H Moeler, A Ansmann and S Wallat, Efeitos da vitamina E por aplicação tópica, Cosm Toil 3 14-25 (1991)
11. M Kaminura and T Matsuzawa, Percutaneous absorption of alpha-tocopheryl acetate, J Vitaminol 14 150-9 (1968)
12. BW Bany, Dermatological Formulations Percutaneous Absorption, Marcel Decker, New York, pp. 127-232 (1993)
13. SH Rubin, Percutaneous absorption of vitaminas, J Soc Cosm Chem 11 160-9 (1960)
14. GM Eccleston, The microstructure of semisolid creams, Pharm Intern 7 63-70 (1986)
15. BW Bany, The science of dosage form design, ME Aulton, Churchill Livingstone, 381-411(1988)
16. RL Bronaugh, RF Sterwart, Methods for in vitro percutaneous absorption studies IV: the flow through diffusion cell, J Pharm Sci 74 64-7 (1985)
17. AM Kligman and B Magnusson, Allergic contact dermatitis in the guinea pig, Charles C. Thomas, Springfield, (1970)
18. PA Viglioglia, and J Rubin, Cosmiatria II. (ed.) de Cosmiatria, Buenos Aires, 136-64,(1989)
19. MIRM Santoro, PMBGM Campos and ERM Hackmann, Determining vitamins A, D_3 and E in creams: High performance liquid chromatography, Cosm Toil 108 (1993)
20. W Skrypzak, A Reng and JM Quack, Formulating cosmetic emulsions with O-phosphoric esters, Cosm Toil 95 47-55 (1980).
21. HE Jass, The living stratum corneum: Implications for cosmetics formulation, Cosm Toil 106 47-53 (1991)
22. Z Amjad, WJ Hemker, CA Maiden, WM Rouse and CE Sauer, Carbomer resins: past, present and future, Cosm Toil 107 81-6 (1992)
23. RY Lochhead, Water-soluble polymers: solution adsorption and interaction characteristics,

Cosm Toil 107 131-56 (1992)
24. S Wahlgren, AL Lindstrom and SE Friberg, Liquid crystals as a potential ointment vehicle, J Pharm Sci 73 1484-86 (1984)
25. PMBGM Campos, GMC software version G.1, developed in the Department of Stomatology, University of São Paulo, Brazil

Defending Against Photoaging: A New Perspective for Retinol

Keywords: photoaging, matrix metallo proteinases, UV, wrinkles, antioxidants, retinol

Strategy to protect skin from photoaging includes sunscreens, antioxidants and modulating cellular processes

Photoaging of the skin is a multifactorial process with the major influence coming from UV rays. UVA and UVB both contribute to this phenomenon, but in different ways. While UVB directly affects the cellular and extracellular components to a certain extent, UVA exerts its detrimental effect predominantly via reactive oxygen species (ROS). The ROS break down lipids and proteins of the skin cells as well as the extracellular matrix of the skin. They also trigger different signal transduction cascades within the cells, thus modifying gene expression and giving rise to further detrimental processes such as inflammation and enzymatic matrix degradation through so-called matrix metalloproteinases (MMPs).

Based on this understanding of the processes of photoaging, we propose a "three-level defense strategy."

First, the amount of UV reaching the living layers of the epidermis has to be minimized. This can be achieved through a combination of UVA and UVB filters. Obviously, this can never afford full protection, and some ROS are still generated.

Therefore, at the second level, the use of antioxidants match particularly a combination of water-soluble (e.g., ascorbyl phosphate) and fat-soluble (e.g., tocopheryl acetate) compounds—reduces damage by oxygen radicals. Again, this type of defense is not absolute. Cellular responses, such as the inflammatory reaction still occur, leading to the release of matrix degrading MMPs.

The third and final level of defense modulates and, at best, normalizes these detrimental cellular processes. It comprises compounds such as bisabolol or panthenol, which help sooth the inflammatory response of the skin.

Finally, retinol has been shown to normalize the activation of MMPs after UV irradiation. This gives a new perspective on retinol. Apart from being an agent effective in treating wrinkles, it also helps to prevent the formation of wrinkles and should therefore be used at adequate levels in daily care products.

Combining all of the mentioned compounds would optimize the everyday protection of the skin against photoaging.

The Three Levels of Defense Against Photoaging

Retinol is an established antiwrinkle active ingredient for high-end, up-to-date skin care products. It increases the mitotic activity and normalizes the enzyme activity of keratinocytes, thus improving the epithelization of the skin. As a result, retinol smoothes fine lines and wrinkles, improves the structure and normalizes the physiology of the epidermis. This also leads to an improved barrier function and helps retain skin moisture and it has a normalizing effect on skin pigmentation leading to a healthier skin color.

The best known effect of retinol is its high efficiency in treating skin wrinkles. Until now it is used mostly in skin care products for "mature" skin to diminish fine lines and other signs of premature skin aging. But retinol has additional effects. Recent studies have shown that retinoids also counteract the physiological processes that actually lead to the formation of wrinkles.

It is well accepted that a major cause for skin aging is UV irradiation, which leads to certain physiological changes within the skin. This is an immediate response that is invisible to the eye but, after accumulating over the years, it leads to prematurely aged skin. Oxidative processes are commonly believed to be involved in these processes.[1] A variety of factors contribute to an increased occurrence of highly detrimental ROS. Apart from UV irradiation, these factors include various xenobiotics such as ozone, cigarette smoke or nitrogen oxides. Yet, the main factor of the extrinsic, premature aging is sun exposure. Therefore, it is often termed photoaging.[2-7]

Although a crucial part of the aging process, ROS and the oxidative stress do not represent more than a small slice of the whole process. Other physiological processes are equally important: Figure 1 depicts some crucial aspects of these events: UVA and UVB generate ROS both within the skin cells and in the intercellular space.[8] They induce different signal transduction cascades – either directly or via unknown mediators. The activation of AP-1 alters the gene expression of the skin cells. Interleukins, the mediators of inflammation, are released and initiate the inflammatory response in the skin. As this process proceeds, further ROS are generated, keeping the vicious circle going.

The altered gene expression also leads to an overexpression of matrix degrading proteins. These so-called MMPs are responsible for the degradation of collagen and elastin[9] which are important matrix proteins that give the skin its mechanical strength and elasticity. Figure 1 also shows the different targets for anti-photoaging products, which are discussed in more detail below.

Level 1: Sunscreens: Within the spectrum of sunlight, UV light is the major factor leading to skin damage. The exposed parts of the skin are permanently subjected to UV radiation. When UV filters are applied to the skin, e.g. in the form of a sunscreen, the amount of UV radiation that reaches the skin is markedly reduced.[10,11]

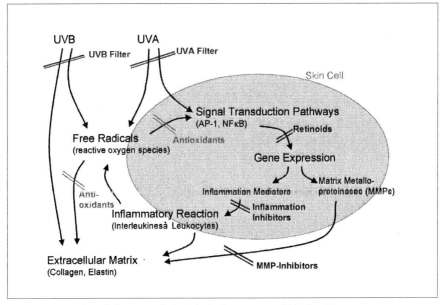

Figure 1. UV-mediated alteration of physiological processes within the skin

However, sunscreens are not usually applied on a routine basis. Only a certain percentage of the western population uses sunscreens at all,[12] and those that do only do so upon extensive sun exposure. The protection afforded is often well below the theoretical SPF value due to insufficiencies in both dose applied and mode of application.[13]

It can be estimated that for more than 95% of a year, when people are not explicitly sunbathing, they usually remain totally unprotected. This lack of protection can be overcome by the moderate use of UV filters in daily cosmetic products such as face lotions, hand creams, body milk and so forth. The SPF achieved should not be too high, so that the light-dependent synthesis of vitamin D will not be impeded.

Both UVA and UVB filters should be incorporated into daily care products. Whether this is through organic filters or inorganic pigments depends on the product to be formulated. Both UVA and UVB protective filter types are nowadays available in stable form (see Formula 1).

Level 2: Antioxidants: UV filters can never offer total protection. The remaining radiation penetrating into the skin–especially in the UVB (280-320 nm) and UVA (320-400 nm) range–leads to the formation of reactive species. These species, which are often oxygen-centered free radicals, are probably mediated via intracellular chromophores or photosensitizers.[14,15] They have a detrimental effect on all molecular components of the skin cells and on the extracellular matrix. Thus, DNA, proteins and lipids are all affected.

Damaged DNA bases could cause gene mutations,[16] leading to skin cancer in the worst case. Lipid peroxidation, as well as damage to certain enzymes, impairs

Formula 1. Sun care lotion "Walking on Sunshine" (o/w) with UVA and UVB protection, SPF 28

A. Dibutyl adipate	8.0 %wt
C12-15 alkyl benzoate	8.0
Cocoglycerides	12.0
Sodium cetearyl sulfate	1.0
Diethylamino hydroxybenzoyl hexyl benzoate (Uvinul A Plus, BASF Aktiengesellschaft)	2.0
Lauryl glucoside, polyglyceryl-2 dipolyhydroxystearate, glycerin	4.0
Cetearyl alcohol	2.0
Ethylhexyl triazone (Uvinul T 150, BASF Aktiengesellschaft)	3.0
Vitamin E acetate	1.0
B. Zinc oxide (Z-Cote, BASF Aktiengesellschaft)	4.0
C. Glycerin	3.0
Disodium EDTA (Edeta BD, BASF Aktiengesellschaft)	0.05
Allantoin	0.2
Xanthan gum	0.3
Magnesium aluminum silicate	1.5
Water (aqua)	48.45
D. Citric acid	0.5
Phenxoyethanol, methylparaben, ethylparaben, butylparaben, propylparaben, isobutylparaben	1.0
Fragrance (parfum)	qs

cellular functions, leading to decreased cell turnover. This results in dry, flaky, dull-looking skin with decreased barrier properties.

Extracellular matrix proteins, such as collagen and elastin, are partly destroyed by the direct action of free radicals. Crosslinks between collagen fibrils occur more frequently in photo-aged skin than in normal skin. Interestingly, the type of crosslinking is different in photo-aged skin. Elastosis, a typical sign of photo-aged skin, occurs when elastin fibrils are degraded to a certain extent and clump together. Accumulated damage of matrix proteins results in structural changes with impaired mechanical functionality and leads to a loss of skin elasticity with wrinkles and sagging.

The use of antioxidants in cosmetic products is therefore highly recommended.[17,18] To achieve a good antioxidant activity, two requisites must be met. First, the compounds have to be stable for the storage life of the product. Second, they have to penetrate into the skin to exert their antioxidant activity at the site where the oxidative stress occurs.

Tocopherol and ascorbic acid are well-known antioxidants, but they should only be used as technical antioxidants to prevent oxidation of the cosmetic product.

They are inadequate for protecting the skin, so more stable compounds should be chosen. Derivatives are often the choice, and both tocopheryl acetate and ascorbyl 2-phosphate salts are stable enough to survive long storage times. They are also able to penetrate into the skin and within the skin the ester bonds are cleaved liberating the respective antioxidants.

By applying a combination of these two antioxidants, a synergistic effect that gives maximum protection should be achieved. It was shown in a recent study on human keratinocytes that ascorbyl phosphate and tocopheryl acetate represent a highly efficient combination for the protection of skin cells against ROS.[19] Adequate use levels should be between 0.5% and 5% for tocopheryl acetate and around 0.2% and 2% for ascorbyl phosphate (see Formula 2).

Level 3: Modulators of cellular responses: In addition to the direct action of UV-generated free radicals, a vast variety of cellular responses occur. Mediated directly by UV irradiation or indirectly by signals coming from cell and tissue damage, the skin reacts with an inflammatory process. When skin cells mistake

Formula 2. Daily care aerosol "Care and Protect" according to the "three levels of defense" concept

A. Ethylhexyl methoxycinnamate (Uvinul MC 80, BASF Aktiengesellschaft)	4.0 %wt
Octocrylene (Uvinul N539, BASF Aktiengesellschaft)	1.5
Caprylic/capric triglyceride	9.0
Buxus chinensis (jojoba) oil	5.0
Cyclomethicone	1.5
Hydrogenated coco-glycerides	3.0
VP/hexadecene copolymer	1.0
Ceteareth-6, stearyl alcohol (Cremophor A 6, BASF Aktiengesellschaft)	1.0
B. Zinc oxide (Z-Cote, BASF Aktiengesellschaft)	5.0
C. Ceteareth-25 (Cremophor A 25, BASF Aktiengesellschaft)	2.0
Panthenol (D-Panthenol 75W, BASF Aktiengesellschaft)	1.2
Sodium ascorbyl phosphate (BASF Aktiengesellschaft)	0.2
Imidazolidinyl urea	0.3
Disodium EDTA (Edeta BD, BASF Aktiengesellschaft)	0.1
Water (*aqua*)	64.17
D. Tocopheryl acetate (Vitamin E-Acetate, BASF Aktiengesellschaft)	0.5
Bisabolol (Bisabolol rac., BASF Aktiengesellschaft)	0.2
Fragrance (*parfum*)	qs
Caprylic/capric triglyceride, retinol (Retinol 15D, BASF Aktiengesellschaft)	0.33

Procedure: Fill in suitable cans and pressurize with liquid pressure gas (e.g. 10 % Propan/Butan 25/75).

UV-mediated cell damage for a mechanical insult, bearing the risk of an infection, the skin cells release pro-inflammatory cytokines like interleukin (IL)-1α, IL-1β, and IL-6, as well as prostaglandins.[20,21] Subsequently, white blood cells infiltrate the skin, generating additional radical species in order to combat a non-existing infection, thus augmenting the detrimental process in a vicious circle.

This inflammatory reaction of the skin is an immediate response that, in extreme cases, manifests itself as an erythema or sunburn. But, even if a response is not visible, the underlying mechanisms still occur to a certain extent, and their detrimental effects accumulate over time. Few compounds are available to the cosmetic chemist seeking to alleviate these temporary processes. One such active compound is bisabolol, which is the active principle of chamomile and a traditional medical herb. A recent study has shown its activity against UV irradiation-induced erythema.[22] Another active compound for this purpose is panthenol, long known for its effectiveness at healing wounds.[23]

On the same level of cellular responses, the UV-induced gene products c-jun and c-fos form AP-1 (activator protein-1). AP-1 binds to a family of retinoid receptors, the so-called retinoic acid receptors (RAR), giving rise to an overexpression of MMPs.[24] These MMPs, in turn, are responsible for the degradation of extracellular matrix proteins, most importantly collagen and elastin.[25]

Retinoids have been shown to moderate this overexpression and, thus, normalize the cellular function.[26-27] Therefore, retinol is not only a highly potent active against existing fine lines and wrinkles, which various studies have shown previously, but also acts directly on UV-induced cellular processes, thus guarding the skin against premature aging.[28,29]

While panthenol and bisabolol are easy to handle, a few words should be said on retinol. The major drawback of this highly potent skin active for the formulator is its inherent instability, which has—until now—not been solved to satisfaction.

To achieve an adequate stability of retinol in a cosmetic product, the manufacturer has follow two prerequisites: first is the use of an inert gas atmosphere during production and packaging, and the second is the use of oxygen-impermeable packaging such as aluminum-lined tubes. These packaging types should also prevent the re-entry of air when samples are taken under use conditions.

With these two prerequisites, the stability of retinol in a typical cream formulation would be approximately 90-95% after three months at 40°C. Even with the currently available stabilized forms of retinol, one would probably have to follow these precautions to achieve this stability—and consequently the desired physiological effect in the skin.

Conclusion

Based on the previously described understanding of UV-mediated skin damage, there are three levels of defense against photoaging:

- Minimize UV irradiation either by permanently using adequate UVB and UVA filters applied to the exposed areas of the skin or by not exposing the skin to the sun (or similar light sources).
- Strengthen the skin's antioxidative network to minimize direct insult from free radicals by supplying adequate amounts of synergistic antioxidants such as

ascorbyl phosphate and tocopheryl acetate.
- Soothe the inflammatory processes with active compounds such as bisabolol and panthenol and normalize overall cellular responses by using adequate concentrations of retinol.

—Axel Jentzsch, Harald Streicher and Valérie André, *BASF Aktiengesellschaft*, Ludwigshafen, Germany

References

1. I Emerit, Free radicals and aging of the skin, *EXS* 62 328-341 (1992)
2. S Beissert and RD Granstein, UV-induced cutaneous photobiology, *Crit Rev Biochem Mol Biol* 31 381-404 (1996)
3. LH Kaminester, Current concepts: Photoprotection, *Arch Fam Med* 5 289-295 (1996)
4. Y Miyachi, Photoaging from an oxidative standpoint, *J Dermatol Sci* 9 79-86 (1995)
5. NH Nicol and NA Fenske, Photodamage: Cause, clinical manifestations, and prevention, *Dermatol Nurs* 5 263-275 (1993)
6. K Scharffetter-Kochanek, M Wlaschek, P Brenneisen, M Schauen, R Blaudschun and J Wenk, UV-induced reactive oxygen species in photocarcinogenesis and photoaging, *Biol Chem* 378 1247-1257 (1997)
7. CR Taylor and AJ Sober, Sun exposure and skin disease, *Annu Rev Med* 47 181-191 (1996)
8. R Kohen, Skin antioxidants: Their role in aging and in oxidative stress—new approaches for their evaluation, *Biomed Pharmacother* 53 181-192 (1999)
9. A Thibodeau, Metalloproteinase inhibitors, *Cosm Toil* 115 75-82 (2000)
10. H Schaefer et al., Photoprotection of skin against ultraviolet A damage, *Met. Enzymol* 319 445-465 (2000)
11. HV Debuys et al., Modern approaches to photoprotection, *Dermatol Clin*, 18 577-590 (2000)
12. B Diffey, Sun protection, Nederl. Tijdschrift Dermatol. Venerol., 9 333-334 (1999)
13. B Diffey, Personal Protection: The Way Forward, *Radiation protection dosimetry* 91 293-296 (2000)
14. K Scharffetter-Kochanek et al. UV-induced reactive oxygen species in photocarcinogenesis and photoaging, *Biol Chem* 378 1247-1257 (1997)
15. M Dalle Carbonare, MA Pathak, Skin photosensitizing agents and the role of reactive oxygen species in photoaging, *J Photochem Photobio. B: Biol*, 14 105-124 (1992)
16. J Lehmann et al., Kinetics of DNA strand breaks and protection by antioxidants in UVA- or UVB-irradiated HaCaT keratinocytes using the single cell gel electrophoresis assay, Mut. Res., 407 97-108 (1998)
17. I Savini et al., Ascorbic acid maintenance in HaCaT cells prevents radical formation and apoptosis by UV-B, *Free Radic Biol Med* 26 1172-1180 (1999)
18. L Vaillant et al., Long term topical antioxidant treatment provides protection against clinical signs of photoaging, *Trace Elem Man Anim* 10 437-439 (2000)
19. A Jentzsch, H Streicher, K Engelhart, The Synergistic Antioxidant Effect of Ascorbyl 2-Phosphate and Alpha-Tocopheryl Acetate, *Cosmetics & Toiletries* 116 55-64
20. T Schwarz, UV radiation and cytokines, In: *Skin cancer and UV radiation* edited by P Altmeyer and K Hoffmann 219-226 (1997)
21. K Isoherranen et al., Ultraviolet irradiation induces cyclooxygenase-2 expression in keratinocytes, *Brit J. Dermatol* 140 1017-1022 (1999)
22. H Streicher, A Jentzsch, Alpha-Bisabolol, a versatile active ingredient for cosmetics, PCIA Bangkok 8.-10.3.2000
23. U Wollina, Zur klinischen Wirksamkeit von DexPanthenol, *Kosmetische Medizin* 4 180-186

(2001)
24. GJ Fisher, JJ Voorhees, Molecular mechanisms of photoaging and its prevention by retinoic acid: Ultraviolet irradiation induces MAP kinase signal transduction cascades that induce Ap-1-regulated matrix metalloproteinases that degrade human skin in vivo, *J. Invest Dermatol Symp. Proc* 3 61-68 (1998)
25. AN Malak, E Perrier, TIMP-1 like: A new strategy for anti-aging cosmetic formulations, *20th IFSCC congress Cannes* 1 79-89 (1998)
26. J Varani et al., Inhibition of type I procollagen synthesis by damaged collagen in photoaged skin and by collagenase-degraded collagen in vitro, *Am J Pathol*, 158 931-942 (2001)
27. J Varani et al., Molecular mechanisms of intrinsic skin aging and retinoid-induced repair and reversal, *J. Invest Dermatol Symp Proc* 3 57-60 (1998)
28. J Varani et al., Vitamin A antagonizes decreased cell growth and elevated collagen-degrading matrix metalloproteinases and stimulates collagen accumulation in naturally aged human skin *J Invest Dermatol* 114 480-486 (2000)
29. JJ Voorhees, Retinoid repair/prevention of photoaging and natural aging in human skin in vivo, *Clin Exp Dermatol*, 25 161-162 (2000)

Moisturizing Potential of d-α-tocopherol

Keywords: d-α-tocopherol, vitamin E, moisturizing

Skin hydrating potential of topically applied vitamin E, in the form of d-α-tocopherol of natural origin is evaluated

Moisturizing products are designed with the intention of improving skin condition and appearance by increasing skin hydration. The classic approach to this problem involves the use of occlusive ingredients (to decrease the level of transepidermal water loss) and humectants (to improve the water-binding ability of the stratum corneum [SC]). More recently, different groups of moisturizing ingredients have been introduced, aiming to intervene at the level of skin physiological processes responsible for maintaining a water balance. A range of natural moisturizing factors have been revealed and used in cosmetic formulations.[1] Furthermore, the interlamellar lipids (ceramides, cholesterol and fatty acids) of the SC have been recognized as playing a major role in the barrier function of the skin,[2] and some promising results in terms of effective formulations were reported.[3]

Vitamin E (tocopherol) has been shown to provide a range of benefits to the skin. It is stated that natural vitamin E, when used in skincare products, can protect skin from UV light, reduce the appearance of fine facial lines and wrinkles and help delay the progression of aging.[4] It is also claimed that vitamin E has a skin moisturizing power,[5] attributed to its antioxidant nature. As a potent free radical scavenger, tocopherol is active in reducing phospholipid peroxidation in cell membranes. This reduction prevents the cross-linking of collagen by the oxidation-breakdown products, principally malondialdehyde.[6] Vitamin E is also known to act as an emollient.[7]

Whatever the exact mechanism of action may be, tocopherol and its esters are recognized (and marketed) as cosmetic actives with skin hydrating powers. Yet the published data which quantify these powers are surprisingly scarce. One of the rare studies on this subject has shown a significantly increased water-binding capacity of the SC after topical treatment with vitamin E acetate.[8]

The aim of this study was to generate and analyze objective data on the hydrating potential of tocopherol-enriched skin lotion compared to skin lotion containing no tocopherol. Vitamin E was used in the form of natural d-α-tocopherol, in a vehicle of an o/w emulsion type stabilized with a polymeric emulsifier; we called this the "active treatment." The concentration of d-α-tocopherol was 5% w/w. The vehicle without vitamin E was called the "placebo treatment."

The capacitance-measuring method was used to evaluate changes in skin surface hydration. Two trials were performed: a short-term study, aiming to detect immediate hydration effects of the test products, and a long-term study, designed to evaluate subtle differences in skin hydration produced by continuous use of the test products.

Before performing these trials, measurements of skin hydrating potential of individual emulsion components were carried out in order to detect the relative strength of each component in comparison with the o/w emulsion used.

Materials

The compositions of test emulsions, with and without 5% w/w vitamin E, are presented in Table 1.

The vitamin E used in the study was d-α-tocopherol[a] of natural origin. It was carefully protected from light, heat and prolonged contact with air, and used within a week after being obtained from the manufacturer. For stability reasons, we chose to produce the test emulsions at room temperature by using a polymeric emulsifier[b] that does not require an elevated temperature for emulsification; this emulsifier has the INCI name acrylates/C10-30 alkyl acrylate crosspolymer. The polyacrylic polymer was neutralized by triethanolamine.[c]

Two typical moisturizers of different types were used in the formulation: an emollient (mineral oil)[d] and a humectant (glycerol, 98% pure).[e] The preservative used was a combination of methylparaben, propylparaben and diazolidinyl urea in propylene glycol.[f] Water was of the molecular biology grade.[g]

Methods

Emulsion preparation: A "direct method"[9] was used to prepare the test products. The active emulsion was obtained by dispersing a polymeric emulsifier in the mixture of water and glycerol at room temperature, by means of an overhead stirrer. After a 20 min stirring period, the oil phase (mineral oil and tocopherol) was added, and the stirring continued for another 30 min. The acidic dispersion obtained was then neutralized with triethanolamine 10% w/w aqueous solution to the pH value of approximately 5.5. A resulting skin lotion was packed in an opaque container, with a pump dispenser for equal dosing, and labeled. A placebo emulsion was produced in the same manner, except that the level of mineral oil was increased to compensate for tocopherol.

Skin hydration testing: A new version of a skin capacitance-based instrument[h] was used to evaluate the state of skin surface hydration throughout the study. The measuring inter-digital electrode indicates changes in skin capacitance due to variations in the moisture content of the horny layer. The recorded capacitance

---

[a] Copherol F 1300, Henkel Organics, Düsseldorf, Germany
[b] Pemulen TR-1, BF Goodrich, Brecksville, Ohio USA
[c] TEA 99% pure, Merck, Darmstadt, Germany
[d] Paraffin liquid, water white grade, Fisher Scientific, Loughborough, UK
[e] Hogg Laboratory Supplies, Bilston, UK
[f] Germaben II-E, Sutton Laboratories, Chatham, New Jersey USA
[g] BDH Laboratory Suppliers, Poole, UK

Table 1. Test formulations

Ingredient	Emulsion A (placebo treatment) % w/w	Emulsion B (active treatment) % w/w
d-α-tocopherol	-	5.0
Mineral (Paraffinum liquidum) oil	15.0	10.0
Glycerol	5.0	5.0
Acrylates/C10-30 alkyl acrylate crosspolymer	0.5	0.5
Triethanolamine, 10% aqueous soln	5.0	5.0
Preservative (*propylene glycol soln*)	1.0	1.0
Purified water (*aqua*)	73.5	73.5

values were then converted in arbitrary hydration units varying from 0 to 120 rcu (relative corneometer units).

Capacitance testing on plastic: It is well documented that the product residue on the skin imparts an objective measurement of skin hydration for some time after product application.[10,11] Some published data[11] suggest that the emulsion water evaporates for less than 15 min. There are recommendations that the skin test should not start before 20 min[10] or 30 min[12] have elapsed.

To resolve this point, a short preliminary study was performed by measuring the capacitance of a thin film of different ingredients on an inert surface, as suggested by Marenus.[13] An amount of 0.01 mL each of water, mineral oil, glycerol and the test emulsions was spread over a 4 cm² area of plastic surface, at 22°C. Skin hydration measurements (rcu) were taken five, 10, 20, 30, 40, 50 and 60 min after application. By eliminating the possibility of absorption, this experiment was designed to follow the evaporation rate only.

Hydration testing on skin: In this study, the effects of d-α-tocopherol, glycerol, mineral oil and water on skin hydration were tested and compared with the effect of an active treatment emulsion. Fourteen volunteers were recruited; all were between the ages of 23 and 53 and showed no signs of skin disturbances. Three square sites (2 x 2 cm, 2 cm apart) on each volunteer's inner forearm were marked using a plastic template, allowing for the testing of five different materials and a control. The room temperature was 22°C, and the relative humidity stood at 59% throughout the trial.

After 30 min of acclimatization, the baseline values were measured, and each designated area was treated with 0.01 mL of a test substance (random and balanced distribution among the upper, middle and lower test sites). The test materials were applied with a 1 mL plastic syringe and spread homogeneously using the flattened tip of a glass rod.

---

[h] Corneometer CM 825, Courage + Khazaka, Cologne, Germany

On the basis of the preliminary study, a period of 45 min was allowed to elapse before the first measurement was taken. Three skin hydration readings of each test site were recorded, with at least 5 sec between the readings. The change in skin hydration was followed for 3.5 h.

Short-term emulsion trial: Ten female volunteers were recruited for this study; all were between the ages of 17 and 27 and had healthy skin of normal to dry type. The treatments were labeled as follows: treatment A is the placebo treatment (emulsion without vitamin E applied); treatment B is the active treatment (emulsion with 5% vitamin E applied) and treatment C is the control (no product applied). The $4 cm^2$ sites were marked on both the left and right arm of each volunteer, using the same template as described above, resulting in 20 readings at each time interval for each treatment. The same test procedure as above was employed. Skin hydration readings were taken over a period of 2.5 h.

Long-term emulsion trial: Nine females and three males, all between the ages of 23 and 56, were recruited and asked to sign an informed consent statement. Six volunteers received treatment A. The other six received treatment B. One volunteer receiving treatment B withdrew from the study due to unrelated health problems. The left inner forearm of each subject was used as a treatment site, while the contralateral site served as a control.

The trial was spread over a period of 10 weeks, with the volunteers applying a designated treatment twice daily for nine weeks. The quantity applied was standardized to one portion obtained by a pump dispenser. The skin hydration tests were performed on the second day of the study, and then on the first, second, fourth and ninth weeks. Measurements were obtained 12 to 16 h after the last product application, as suggested by Marenus.[13] One week after treatment ceased, the last measurement was taken. No other products were used on each forearm for the duration of the trial. The trial was performed over the summer months (July through September), with variations in the external temperature of 18 – 28°C and the external relative humidity of 55 – 85%. This was taken into account in the result analysis.

Results

Capacitance test on plastic: Each of the five materials tested on plastic showed the highest surface moisture values occurring 5 min after application (Figure 1).

The largest initial increase was detected in water, as expected, followed by the o/w emulsions. A sharp drop in the moisture was evident after 10 min, with water returning to baseline levels after 20 min (Figure 1). The water bound in the form of o/w emulsion evaporated at a lower rate, with a significant difference to the baseline being lost after 30 min. This finding corresponds well with the data obtained on human skin by Blichmann et al.,[11] who used different instrumentation in their study. The proposed waiting time of 30 min was, therefore, acceptable. Only glycerol, being a potent humectant, showed a measurable surface hydration after 60 min.

Hydration tests on skin: For the second preliminary study, an even safer option of a 45 min delay was used (Figure 2). This time interval was still not enough for

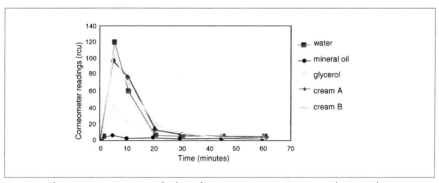

Figure 1. Changes in capacitance of selected moisturizers over time on a plastic surface

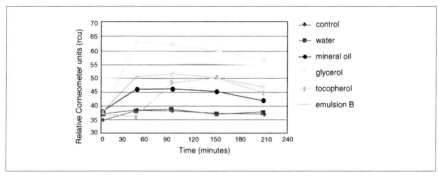

Figure 2. Mean values of the changes in skin hydration on inner forearms of 14 volunteers over time following single applications of selected moisturizers and a vitamin E emulsion

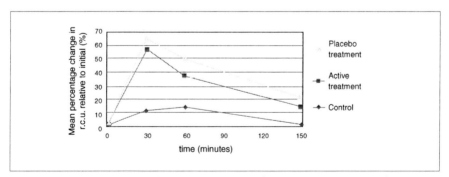

Figure 3. Short-term percentage change in mean values of skin hydration at 20 inner forearm sites over time following single application of a vitamin E emulsion or a placebo emulsion

Table 2. Mean values of skin hydration measurements over time at 20 test sites for each of three conditions: placebo treatment, vitamin E emulsion treatment and no treatment

Treatment	Mean skin hydration measurements (rcu*)			
	0 min	30 min	60 min	150 min
Placebo (emulsion A)	33.7±6.8[a]	54.2±10.5	49.7±7.7	40.1±6.6
Active (emulsion B)	35.3±8.7	53.7±9.5	47.2±7.1	39.8±8.1
Control (no treatment)	34.1±7.1	37.5±5.7	38.3±5.7	34.4±7.5

* rcu = relative corneometer units (arbitrary units from the skin hydration measuring device)
[a] Standard deviation

the absorption of neat d-α-tocopherol into the upper skin layers. Because d-α-tocopherol is an extremely viscous liquid, it impaired the capacitance measurement, giving a false negative value after 45 min in all volunteers. Following the absorption of vitamin E, the rcu values increased, with the peak at 150 min. The highest overall skin hydration was achieved by glycerol, with the tocopherol-enriched emulsion (treatment B) being the second best.

The coefficient of variation between the three replicate instrument readings taken at each site was less than 5%, with large inter-individual variations, as expected. The analysis of standard variations of the treatment means, however, showed no significant difference and allowed for the application of a one-way analysis of variants test to the results. The moisturizing performance of glycerol was significantly higher than the rest of the tested ingredients at the $p < 0.01$ level and better than the test emulsion at $p < 0.05$. Skin hydration at the water-treated site was not significantly different from skin hydration at an untreated site. The increase in hydration produced by the test emulsion was not significantly different from the effect of either neat mineral oil or d-α-tocopherol.

Short-term emulsion test: Having established the good moisturizing performance of the tocopherol-enriched emulsion, we next set out to test the hypothesis that this effect was due to the presence of tocopherol. The results of a short-term moisturizing trial with emulsion A (placebo) and emulsion B (active treatment) are presented in Table 2 and Figure 3.

Table 2 shows mean skin hydration readings for each treatment obtained from 20 test sites. Analysis of variance of each time point has shown that after 30, 60 and 150 min the mean readings for the sites treated with emulsions A and B were statistically significantly higher ($p < 0.05$) than the mean readings for the control site.

To compensate for the initial and inter-individual differences, a further analysis was carried out on the percentage change in readings for each point in time (Figure 3). The comparison of the two treatments, however, was made less straightforward by the existence of significant difference in the standard deviations of the treatment means at 30 and 60 min. This necessitated the use of the Kruskall Wallis non-parametric analysis of variance. Treatments A and B were both significantly better (p< 0.05) than the control for the whole period of study. There was no significant difference between placebo and active treatment.

Long-term emulsion test: After obtaining no increase in skin hydration with 5% vitamin E in the short-term study, we designed and carried out a long-term trial. The aim of this study was to explore the possibility of cumulative effects of d-α-tocopherol on the skin hydration. Figure 4 shows the results in the form of percentage change over time.

The mean readings for treatments A and B were significantly higher than the mean control readings from week one to nine of the trial. There were no direct-paired comparisons of treatments A and B, and large differences in standard deviation of the three treatments made it inappropriate to compare the mean readings directly. The readings were standardized by converting to percentage change from baseline and subtracting the percentage change for the control. These values, as shown in Figure 4, confirmed that the percentage change in treatment B (active), relative to the control, was significantly higher (p< 0.01) than the relative percentage change for treatment A (placebo).

Discussion

Vitamin E: Vitamin E for cosmetic use is available in the form of tocopherol of natural origin (mostly d-α-tocopherol), a synthetic analogue (dl-α-tocopherol) and as an ester (acetate, linoleate, nicotinate).[5] The ester is known to be more stable, but generally less effective than the alcohol form of vitamin E. Tocopheryl linoleate is claimed to be the most effective moisturizing agent within the group,[5] with both tocopherol and linoleic acid playing their respective roles. It would be difficult to distinguish the effect of tocopherol from the fatty acid in such a combination, hence a pure alcoholic form was chosen for this study. A natural isomer was preferred over a synthetic, less effective form, with the stability aspect being considered throughout the study.

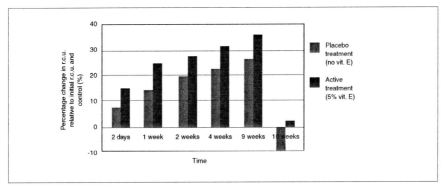

Figure 4. Long-term percentage change in mean values of skin hydration during twice daily applications of a vitamin E emulsion (five inner forearm sites) or a placebo (six inner forearm sites), standardized by subtracting the percentage change for the control

Although legally unlimited,[14] the concentration of tocopherol in cosmetic products is very much cost-limited. A realistically high concentration of 5% w/w in a skin lotion was chosen for this study. Gehring et al. studied tocopheryl acetate concentration in both o/w and w/o emulsions and found that within the 2.5 to 7.5% range of concentrations studied, tocopheryl acetate was most effective at a concentration of 5.0%.[8]

The emulsion: The formulation of the emulsion vehicle used in our study had to meet certain requirements: It had to be fairly simple in composition, but appealing to volunteers; it had to allow a room-temperature preparation method; and it had to be effective as a cosmetic active carrier. We used the polymeric emulsifier called acrylates/C10-30 alkyl acrylate crosspolymer. This emulsifier reportedly provides an immediate availability of the oil phase upon application.[15] This availability is due to the process of deswelling of the polymer emulsifier hydrogel upon contact with the skin electrolytes.[15]

It has been shown that the absorption of vitamin E into the skin is higher from a w/o emulsion system than from either an o/w emulsion or an oil solution.[16] The rational behind the choice of an o/w system was largely our aim to maximize the volunteers' compliance.

To achieve a homogeneous release of active substance out of the oil phase and to accelerate the formation of an occlusive film, an isotropic oil phase must be formed following the evaporation of water.[17] It has been suggested that the use of polymeric emulsifiers, whose network structure is being destroyed during the rubbing-in process, offers the fastest release of an active.[17]

The results of testing the hydration potential of neat ingredients in comparison with the test emulsion validated the use of the emulsion. As shown in Figure 2, at least in the moisturizing domain, there is no advantage in using neat vitamin E, because an even better hydrating effect may be achieved with an aesthetically pleasing and cost-effective emulsion product.

Short-term effects: The immediate effects on the skin hydration by the two products were evaluated in the short-term study (Figure 3). The fact that no significant difference was found in the performance of the placebo and active emulsion leads to the conclusion that, in the short-term, the effect of d-α-tocopherol is predominantly occlusive. It was shown (Figure 2) to be equal to the moisturizing effect of the same quantity of mineral oil.

Long-term effects: A long-term study has the advantage of being able to detect the more subtle benefits that can be derived from moisturizers due to repeated long-term use. However, it also has a major drawback, and that is the possibility that during the long treatment period the subjects could be influenced by environmental factors, such as changes of seasons or climate.[18] Also, it is often difficult to discriminate between different products having similar moisturizing potential.

When analyzing the results of our long-term trial, we found it was, indeed, impossible to detect any significant difference using paired comparison at the same time points. This was due to several facts: the small number of volunteers, the large inter-individual variations and climate changes. However, after subtracting the control means from the respective treatment means, and expressing the effects in the form of the percentage change to baseline and control, it became evident (Figure 4) that the vitamin E-enriched emulsion performed consistently better than the same emulsion with no vitamin E. It appears that d-α-tocopherol in the concentration of 5% w/w does increase the moisturizing potential of a topically applied product when used on a long-term basis.

The moisturizing effect: It is reasonable to assume that vitamin E has to be absorbed and maintained in certain concentration in the skin in order to generate its moisturizing effect. The exact hydrating mechanism is still unclear. It could be speculated that it is a consequence of a potent antioxidant nature of vitamin E. As an effective free-radical scavenger, it is able to prevent and repair skin perturbations induced by aggressive situations. This, in turn, enables the skin to exercise its own powerful moisture-control system through the natural moisturizing factors and barrier lipids.

Conclusion

The aim of this study was to investigate whether vitamin E (in the form of d-α-tocopherol) could improve the moisturizing potential of a cosmetic emulsion when present in the concentration of 5% w/w. The short-term trial revealed no significant difference between the hydrating effect of vitamin E-enriched o/w emulsion and the placebo treatment.

The long-term study (nine weeks) detected a significant increase in the performance of vitamin E treatment, compared with the placebo treatment. This indicates that a measurable difference in the skin hydration status depends on the absorption of vitamin E and its subsequent physiological effects.

—**Slobodanka Tamburic and Grace Abamba** , *The London Institute, London College of Fashion, London, UK*
—**Joyce Ryan,** *Joyce Ryan Consultancy, Bolton, UK*

References

1. AV Rawlings, IR Scott, CR Harding and PA Bowser, Stratum corneum moisturization at the molecular level, *J Invest Dermatol* 103 731-740 (1994)
2. G Imokawa and M Hattori, A possible function of structural lipids in the water-holding properties of the stratum corneum, *J Invest Dermatol* 84 282-284 (1985)
3. K De Paepe, Barrier Lipids in Dermato-Cosmetic Products, in *Intensive Course in Dermato-Cosmetic Sciences,* Brussels: Vrije Universiteit Brussel (1998) 68-94
4. P Mayer, W Pittermann and S Wallat, The effects of vitamin E on the skin, *Cosmet Toil* 108 (2) 99-109 (1993)
5. T Sayama, The E factor, *SPC* 66 33 (1993)
6. RC Wester and HI Maibach, Absorption of tocopherol into and through human skin, *Cosmet Toil* 112 (4) 53-57 (1997)
7. JP Marty, Vitamins and skin aging, in *Intensive Course in Dermato-Cosmetic Sciences*, Brussels: Vrije Universiteit Brussel (1998) 115-140
8. W Gehring, J Fluhr and M Gloor, Influence of vitamin E acetate on stratum corneum hydration, *Arzneimittelforschung* 48 772-775 (1998)
9. BF Goodrich, Bulletin No 13 (1994)
10. V Rogiers, MP Derde, G Verleye and D Roseeuw, Standardized conditions needed for skin

surface hydration measurements, *Cosmet Toil* 105 (10) 73-82 (1990)
11. CW Blichmann, J Serup and A Winther, Effects of single application of a moisturizer: evaporation of emulsion water, skin surface temperature, electrical conductance, electrical capacitance, and skin surface (emulsion) lipids, *Acta Derm Venerol* (Stockh) 69 327-330 (1989)
12. K-P Wilhelm, Skin hydration measurements: general considerations and possible pitfalls, in *Proceedings from In-Cosmetics 1997*, Verlag für Chemische Industrie, H Ziolkowsky GmbH, Augsburg, Germany (1998) 175-187
13. KD Marenus, Skin conditioning benefits of moisturizing products, in *Cosmetic Claims Substantiation*, LB Aust, ed, New York: Marcel Dekker (1998) 97-113
14. Merck, Cosmetic Ingredient Dossier, DL-alpha-tocopherol (1996)
15. BF Goodrich, Polymers for personal care, TDS-114 (1994)
16. Th Foster, B Jackwerth, W Pittermann, W von Rybinski and M Schmitt, Properties of emulsions: Structure and skin penetration, *Cosmet Toil* 112 (10) 73-82 (1997)
17. GH Dahms, The impact of the emulsion structure on adsorption and release of actives on skin, in *Emulsions: Technology, Structures, Ingredients, Formulations*, Verlag für Chemische Industrie, H Ziolkowsky GmbH Augsburg, Germany (1998) 15-24
18. E Berardesca, Measurement of skin hydration, in *Intensive Course in Dermato-Cosmetic Sciences*, Brussels: Vrije Universiteit Brussel (1998) 258-267

Polyethoxylated Retinamide as an Anti-Wrinkle Agent

Keywords: polyethoxylated retinamide, retinol, wrinkles

Tests of collagen synthesis, skin permeation, stability, toxicity and wrinkle reduction suggest cosmetic possibilities for PERA

In this chapter we describe a novel retinol derivative that we believe has cosmetic potential as an anti-wrinkle agent.

Retinol is required for normal epidermal cell growth and differentiation, and it is an important regulator of keratinocyte terminal differentiation.[1] It is also known that retinol can alter or modulate total collagen synthesis.[2-4] Retinoic acid has demonstrated the ability to alter the type of collagen synthesized.[5] Retinol has the potential to alter the expression of protein molecules in both the epidermis and the dermis.

All these effects of topically applied retinol tend to oppose changes that occur with aging, and they maintain the skin in a more youthful condition, suggesting the use of this material in antiaging products.[6-8] The antiaging effect on human skin after treatment with retinoic acid is convincingly documented in a number of open and controlled studies.[9-12]

Although it has potential as an antiaging agent, retinol has shown some problems when used in cosmetic formulations. Among these problems are its stability, permeability and toxicity. Side effects of the dermatitis type are a major problem when cosmetics containing retinoids receive prolonged use.[10] The known toxicity of this class of compounds[13] has prompted a continuing search for retinoids with decreased undesirable side effects. It has been reported that retinoids are unstable in light, oxygen, temperature, lipid peroxide and water[14-16] and retinol's instability makes it difficult to incorporate this ingredient in cosmetic products without using stabilization techniques.[17]

To increase the stability and decrease the toxicity of retinol, retinyl esters (such as acetate, propionate and palmitate) are employed in cosmetic formulations as precursors of retinoic acid. However, such retinol derivatives are so lipophilic that it is difficult for them to penetrate the skin.[18] They can, therefore, not be expected to have clinical efficiency in treatment of wrinkles and other symptoms of skin ag-

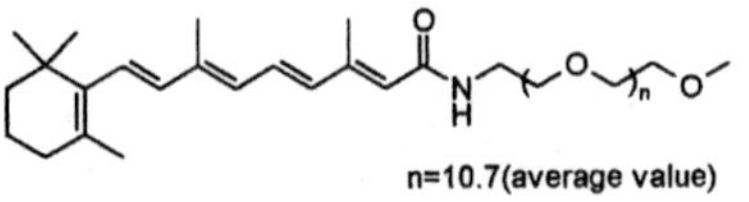

Figure 1. Chemical structure of polyethoxylated retinamide

ing. It has been reported that suitable substances, such as complex polysaccharides, have been used to increase the skin permeability of highly lipophilic substances.[19,20] Conjugation with a complex polysaccharide or another suitable substance will enhance the skin permeability of lipophilic molecules and, at the same time, protect the substance against oxidation (degradation).

In the present study, a polyethoxylated retinamide[a] (PERA) with an average molecular weight of 831 has been developed by coupling retinoic acid with polyethylene glycol (PEG) to enhance skin permeability and stability (Figure 1). Collagen synthesis, skin permeability, stability, wrinkle reduction effect and toxicity of PERA have been evaluated and compared with those of retinol and retinyl palmitate, which are widely used in cosmetics.

Methods and Materials

Materials: PERA was synthesized by LG Chemical Ltd., Korea. Other materials[b] were obtained from reputable sources.

Caprylic capryl triglyceride solution containing 35 mM of retinol, retinyl palmitate or PERA was prepared for in vitro skin permeation experiments. Solubilized states (Formula 1) and o/w emulsions (Formula 2) containing 0.1% of retinol, retinyl palmitate or PERA were prepared for stability studies. O/W emulsions containing 0.5% and 0.2% of PERA were prepared for toxicity and clinical tests, respectively.

Preparation of PERA: The key intermediate, 1-amino-polyethyleneglycol monomethyl ether ($MPEG_{550}$-NH_2) was prepared through four steps from the starting material, polyethyleneglycol monomethyl ether ($MPEG_{550}$-OH).

First, chlorination of $MPEG_{550}$-OH with thionyl chloride, followed by substitution reaction with sodium azide in dimethylformamide gave $MPEG_{550}$-N_3 in quantitative yield. Reductive amination of $MPEG_{550}$-N_3 was carried out with triphenyl phosphine in tetrahydrofurane. Finally, the coupling reaction of retinoic acid with $MPEG_{550}$-NH_2 by use of dicyclohexylcarbodiimide and 4-dimethylaminopyridine in dry dichloro-methane under mild conditions gave a pale yellow product, PERA, in 85% yield starting from $MPEG_{550}$-N_3. PERA was purified by column chroma-

---

[a] Medimin A, LG Chemical Ltd, Teajon, Korea
[b] Retinol was purchased from Aldrich Chem. Co. (Milwaukee, Wisconsin, USA). Retinyl palmitate was purchased from Roche (Basel, Switzerland). DMEM (Dulbecco's Modified Eagle's Medium), fetal bovine serum (FBS), streptomycin and penicillin were all purchased from GibcoBRL (Grand Island, New York, USA). L-[2,3-^{3}H]-proline was from Amersham (USA). Potassium phosphate, sodium chloride and sodium phosphate were obtained from Sigma Co. (St. Louis, Missouri, USA). HPLC grade methanol, methylene chloride and acetonitrile were all purchased from J. T. Baker Chemical Co. (Muskegon, Michigan, USA). Polyoxyethylene oleyl ether (Volpo 20) was from Croda Inc. (Parsippany, New Jersey, USA). Caprylic capryl triglyceride was obtained from Inolex Chemical Co. (Philadelphia, Pennsylvania, USA). All other ingredients were reagent or cosmetic grade.

tography (SiO$_2$, mesh size 270-400, dichloromethane/methanol 15:1, v/v). The structure and purity of PERA were characterized by ^{1}H NMR, HPLC and FAB mass spectroscopy.

Collagen synthesis: Determination of collagen synthesis was performed according to the method of Webster and Harvey.[21] Human fibroblasts (WI-38 fibroblast) were cultured in a T-75 flask[c] with Dulbecco's Modified Eagle's Medium, supplemented with 10% fetal bovine serum, streptomycin (100 µg/mL) and penicillin (100 µg/mL) in a humidified atmosphere of 5% CO$_2$, 95% air. The cells were initially seeded at 5×10^4 cells per well in microplates[d] and incubated for 24 h. The cells were treated with retinoic acid or PERA in a medium containing 3 µCi L-[2,3-^{3}H]-proline and the culture plates were incubated for an additional 24 h. At the end of incubation, culture supernatants were collected and treated with purified collagenase. Radioactivity incorporated into collagenase-sensitive protein was measured over a 2 min period in a Beckman liquid scintillation counter.

[c] Falcon, Franklin Lakes, New Jersey, USA
[d] Costar, Acton, Massachusetts, USA

Formula 1. Retinoids in solubilized state for stability studies

POE(25) Octyldodecyl ether	3.0%
POE(40) Hydrogenated castor (*Ricinus communis*) oil	1.0
Polyethoxylated retinamide (or *Retinyl palmitate*)	0.1
Ethanol	10.0
Water (*aqua*)	qs 100.0

Formula 2. Retinoids in o/w emulsion for stability studies

Cetearyl alcohol	0.7%
Glyceryl stearate SE	1.5
Glyceryl stearate (and) PEG-100 stearate	1.0
Mineral (*paraffinum liquidum*) oil	10.0
Trioctanoin	2.0
Polysorbate 60	1.5
Sorbitan stearate	0.3
Polyethoxylated retinamide (or Retinyl palmitate or Retinol)	0.1
Propylene glycol	5.0
TEA	0.15
Carbomer	0.12
Water (*aqua*)	qs 100.00

Skin permeation studies: Vertically assembled Franz-type diffusion cells[e] were used for in vitro skin permeation experiments. The system consisted of Franz type diffusion cells with an effective diffusion area of 1.776 cm^2, receptor volume of 7.0 mL, autosampler and cell drive system with rpm controller.

The fundamental experiments were performed according to the method given in our previous report.[22] Briefly, the excised skin of the female hairless mouse was obtained from animals 8-9 weeks old weighing 27-33 g. The skin was mounted on a diffusion cell, and the receiver compartment was filled with 7 mL of 50 mM phosphate buffer saline pH 7.4 with 2% polyoxyethylene oleyl ether[f] (a nonionic surfactant, HLB=16) and maintained at 32°C by circulating water within a jacket around the lower chamber. Polyoxyethylene oleyl ether was used to insure that solubility in the receptor solution would not limit penetration through skin. Diffusion characteristics of skin are unaffected by exposure to polyoxyethylene oleyl ether.

Samples containing retinoids were applied in the donor compartment. Caprylic capryl triglyceride solution was uniformly distributed with a micropipette (20 µL) on the skin surface. The receptor fluid was mixed by a magnetic stirrer throughout the experiment. The receptor fluid was collected from the receiver compartment every 6 h after sample application and replaced by fresh fluid.

At the end of the experiment 24 h after sample application, receptor fluid was collected, and the donor compartment was washed three times with 500 µL of methanol each time. Twenty-four hours later skin samples were taken out of diffusion cells. The skin was homogenized by 4 mL of methylene chloride to extract retinoids. After filtration on Millex filter FG (pore size: 0.2 µm, millipore), solutions were assessed by HPLC. Five-hundred µL of receptor fluid withdrawn from the receiver compartment at predetermined times was treated with 1 mL of methylene chloride and shaken by a vortex mixer. Following centrifugation (13,000 rpm), the supernatant was subjected to HPLC.

Stability studies: To measure the temperature stabilities, retinol, retinyl palmitate and PERA were prepared in solubilized state and in o/w emulsion. Samples were stored at 25°C and 40°C for as long as two months. Periodically, 50 µL aliquots of each sample were pipetted out and diluted with methanol. The amount of residual retinoids was measured by HPLC.

To measure the UV stabilities of retinol and PERA, 1 g aliquots of o/w emulsion containing retinol or PERA were added to each test tube. The samples were illuminated with UV from an artificial light generator[g] for 30 min. The amount of residual retinoids was measured by HPLC.

Toxicity tests: We tested o/w emulsions containing 0.5% PERA on animals in vivo using toxicity tests based on the Organization for Economic Cooperation and Development's 1960 guidelines for testing of chemicals. To estimate the toxicity, we performed tests on acute oral toxicity, acute transdermal toxicity, primary skin irritation, ocular irritation, skin sensitization and human patch test according to the methods of Seo et al.[23]

---

[e] Microette transdermal diffusion system, Hanson Research Corp., Chatsworth, California, USA
[f] Volpo 20, Croda Inc., Parsippany, New Jersey, USA
[g] Sun Test CPS, Heraeus Co., Postfach, Hanau, Germany

Wrinkle reduction tests: We studied wrinkles in the faces of 29 Korean female volunteers between the ages of 22 and 55. All volunteers gave informed consent before participation. Each volunteer used either an emulsion containing 0.2% PERA or a placebo. The emulsions were applied twice daily for 10 weeks. Wrinkle measurements were taken before and after the 10-week period.

On the day of the wrinkle measurements, each volunteer was first acclimated to room temperature for 30 min. We used Provil[h] to take replicas of the crow's feet and under-eye region. The Provil, mixed with a catalyst, was poured on the test site and then peeled off after hardening. To correct for wrinkle measurement errors resulting from curvature of the face, we positioned the replica so its bottom side was on a flat pane of glass. Then we poured Xantopren[j] with a catalyst over the replica and pressed it down for a few minutes under a heavy pane of glass. This curve-corrected replica was used to measure the wrinkles.

For measuring wrinkles, we used the method of Corcuff et al.,[24] taking before and after photographs of wrinkle shadows in the main deep wrinkled area (WA) in the captured image of the replica, and then calculating the ratio. We set the replica on the sample stand so that the measurement surface would be horizontal and produced wrinkle shadows by illuminating the replica with parallel light of a fixed intensity from a fiber optic illuminator[k]. The wrinkle image was transferred to image capture software[m] through a color digital camera.[n] The shadow area of the main deep wrinkle was measured by image analyzation software.[p]

Results and Discussion

Collagen synthesis: Figure 2 represents the increase in collagen synthesis by retinoic acid or PERA treatment. There were 23, 46 and 46% increases in collagen synthesis after retinoic acid treatment at 10^{-8} M, 10^{-9} M and 10^{-10} M, respectively. There were 25, 31 and 40% increases after PERA treatment at 10^{-8} M, 10^{-9} M and 10^{-10} M, respectively. There was a maximum 46% increase at 10^{-10} M retinoic acid and 40% increase at 10^{-10} M PERA. This result indicates that the effect of PERA on collagen synthesis is similar to that of retinoic acid, and the function of retinoic acid is retained.

Skin permeation: Table 1 shows the permeated amounts of retinol, retinyl palmitate and PERA in oil (caprylic capryl triglyceride) 24 h after application to the excised hairless mouse skin. The skin permeability of PERA was higher than that of retinol or retinyl palmitate. The permeated amount of PERA was three times greater than that of retinol, and six times greater than that of retinyl palmitate. We suggest that this result was due to the increase in partition of PERA to skin by PEG conjugation. It has already been reported that conjugation with complex polysaccharides or another suitable substance enhances the water solubility and the skin permeability of lipophilic molecules. At the same time it protects the substance against oxidation (degradation).[18-19]

[h]Provil is a proprietary name registered to Bayer, Leverkusen, Germany
[j]Xantopren is a proprietary name registered to Bayer, Leverkusen, Germany
[k]GLS-V100B, Halla Optical Eng., Seoul, Korea
[m]On Air TV, Sasem, Seoul, Korea
[n]CCS-212, Samsung, Seoul, Korea
[p]Ultimage, Graftek, Miramande, France

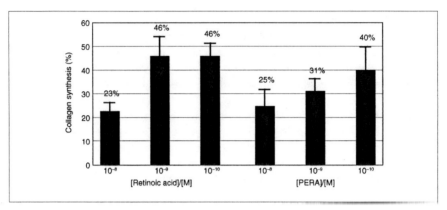

Figure 2. Increased collagen synthesis after treatment with retinoic acid or PERA

Table 1. Total amount (nmole) of retinoids permeated across excised hairless mouse skin 24 h after application

	Retinol	Retinyl palmitate	PERA
Skin	23.38	18.73	54.17
Receptor solution	11.48	0.48	62.72
Total permeated amount	34.86	19.21	116.87

Stability: Figures 3 and 4 demonstrate the stabilities of retinyl palmitate and PERA in a solubilized state and in o/w emulsion during two months storage at 25°C and 40°C. PERA was more stable than retinyl palmitate, a retinol derivative widely known to be stable. Both retinyl palmitate and PERA were more stable in o/w emulsion than in solubilized solution. This indicates that retinoids are less stable in water.

Figure 5 shows stabilities of retinol, retinyl palmitate and PERA in o/w emulsion exposed to UV for 30 min. These results show that PERA was more stable than retinol and retinyl palmitate when exposed to UV.

Toxicity: To estimate the safety of an o/w emulsion containing 0.5% PERA, various tests were carried out.

- **Acute oral toxicity.** When we administered the emulsion orally in doses of 10,000 or 5,000 or 2,500 mg/kg of body weight in healthy Sprague-Dawley rats and New Zealand white rabbits, we observed neither unusual symptoms nor death.
- **Acute transdermal toxicity.** When we applied the emulsion in doses of 10,000 or 5,000 or 2,500 or 1,250 or 625 mg/kg of body weight on the skin of selected healthy Sprague-Dawley rats and New Zealand white rabbits, we observed no changes. Autopsies revealed no visual pathological symptoms.

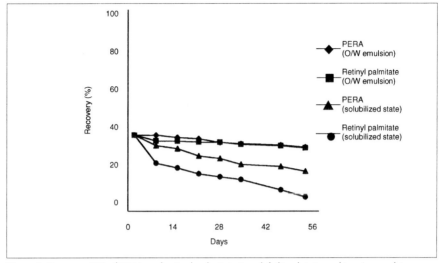

Figure 3. Recovery (%) of PERA and retinyl palmitate in solubilized state and in o/w emulsion stored at 25°C

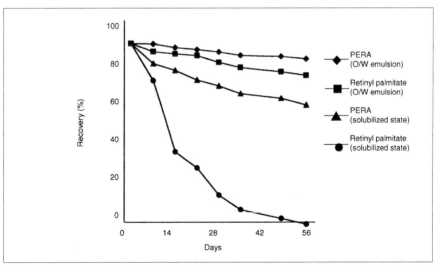

Figure 4. Recovery (%) of PERA and retinyl palmitate in solubilized state and in o/w emulsion stored at 40°C

- **Dermal primary irritation in rabbits.** We observed neither general symptoms nor weight change when the emulsion was applied to the skin. Nor did we see any erythema or formation of scale or edema at the application site. The P.I.I. (Primary Irritation Index by Draize) value was zero. Thus, there is no skin irritancy associated with an o/w emulsion containing 0.5% PERA.

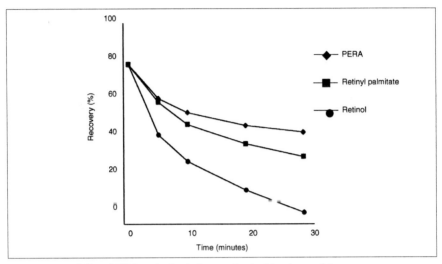

Figure 5. UV stability of PERA, retinyl palmitate and retinol in o/w emulsion

Table 2. Wrinkle area values from two skin treatments, each lasting 10 weeks, resulting from a computer-aided image profile analysis

	Variable	Mean	SD	N
Placebo (emulsion)	WA before	9.68	4.14	29
	WA after	8.24	3.60	29
	ΔWA	1.44	2.30	29
Emulsion (PERA 0.2%)	WA before	9.48	3.92	29
	WA after	5.58	2.36	29
	ΔWA	3.89	0.73	29

N = number of sites evaluated

- **Ocular irritation in rabbits.** Neither general symptoms nor weight change were observed. There was no observable turbidity of cornea, abnormality of the iris, redness of the conjunctiva, edema or secretion. We concluded that the o/w emulsion containing 0.5% PERA does not induce eye irritancy in the New Zealand white rabbits.
- **Skin sensitization in guinea pigs.** According to Magnusson and Kligman's evaluating standard,[25] the hypersensitivity score was zero and the hypersensitivity induction rate was 0%. Thus, the o/w emulsion containing 0.5% PERA caused no sensitized hypersensitivity in guinea pigs.
- **Human patch test.** Draize score was zero.

In summary, the emulsion with 0.5% PERA was nontoxic in various toxicological tests, proving that it can be safely introduced into skin care formulations.

Wrinkle reduction: Wrinkle area (WA) value is a measure of the roughness of a surface profile. Because the measurement is contact-free, we used it to evaluate the anti-wrinkle effectiveness of either a skin treatment emulsion containing 0.2% PERA or a placebo emulsion without the PERA.

The change in wrinkle area values before and after the 10-week treatment is significant for both treatment groups. Significant differences were found between the two groups regarding mean wrinkle area (WA) after treatment and the change in wrinkle area (ΔWA = WA before treatment – WA after treatment). The results are shown in Table 2. ΔWA values and WA values before treatment were used to calculate the percentages of skin WA reduction, which correlate to reduced skin roughness. After treatment the placebo emulsion reduced WA by an average of 14.9%. The anti-wrinkle emulsion containing 0.2% PERA reduced WA by an average of 45.0%.

Conclusion

The effect of PERA on collagen synthesis was similar to that of retinol. The skin permeability of PERA was higher than that of retinol or retinyl palmitate. The PERA was more stable than retinol and retinyl palmitate. An o/w emulsion containing 0.5% PERA was nontoxic in various toxicological tests. The emulsion containing 0.2% PERA reduced wrinkles by an average of 41.0%.

These results suggest that PERA would be a good anti-wrinkle agent for enhancing bioavailability and stability.

—Young Sook Song, Bong Yul Chung, Sun Gyoo Park, Mun Eok Park, Sung Jun Lee, Wan Goo Cho and She Hoon Kang, *LG Chemical Ltd., Taejon, Korea*

References

1. J Kubilus, J Invest Dermatol 81 55s-58s (1983)
2. H Oikarinen et al, J Clin Invest 75 1545-1553 (1985)
3. R Beach and C Kenny, Biochem Biophys Res Commun 114 395-402 (1983)
4. M Kenny et al, Biochem Biophys Acta 889 156-162 (1986)
5. P Benya and S Padilla, Develop Biol 118 296-305 (1986)
6. M Pfahl, Skin Pharmacol 6 suppl 1 24-34 (1993)
7. JH Saurat et al, J Invest Dermatol 103 203 (1994)
8. GJ Gendimenico and JA Mezick, Skin Pharmacol 6 suppl 1 24-34 (1993)
9. AM Kligman et al, J Am Acad Dermatol 15 836-859 (1986)
10. KS Weiss et al, JAMA 259 527-532 (1988)
11. J Bhawan et al, Arch Dermatol 127 666-672 (1991)
12. FS Rafal et al, The New England J Med 6 368-374 (1992)
13. L Packer et al, Methods in Enzymology, vol 190, San Diego, California: Academic Press (1990)
14. C Kwasaki and M Hida, Vitamins 15 383-386 (1958)
15. T Tabata, Vitamins 18 164-167 (1961)
16. S Hayashi and Y Nishii, Vitamins 28 269-273 (1971)
17. A Semenzato et al, Il farmaco 47 1407-1417 (1992)

18. T Erling, J Appl Cosmetol 11 71-76 (1993)
19. R Sperman and A Jarrett, Br J Dermatol 90 553-560 (1974)
20. R Cecchi et al, J Med Chem 24 622-625 (1981)
21. D Webster and W Harvey, Anal Biochem 96 220-224 (1979)
22. YS Song et al, Evaluation of in vitro skin permeation of UV filters, 20th IFSCC Congress Cannes (1998)
23. DS Seo et al, Antimicrobial inorganic pigments for cosmetics, Cosmet Toil 112(5) 83-90 (1997)
24. P Corcuff et al, J Soc Cosmet Chem 34 177-190 (1983)
25. B Magnusson and AM Kligman, J Invest Dermatol 52 268-276 (1969)

A Collagen III Amplifier System

Keywords: liposome, collagen III, keratinocytes, fibroblast activity, cytokines

Unique collagen-containing liposomal system provides benefits in skin care

In this chapter[a], we discuss the design and application of a liposomal system for age-defying skin care products. This unique system modulates epidermal cytokine production to selectively increase collagen III. A mechanism for this activity is proposed and supported by in vitro and in vivo data. The proper incorporation of this liposome into skin rejuvenating products is summarized and the practical benefits are noted.

Changes in the characteristics of human skin during aging are caused by intrinsic factors resulting in alterations of the connective tissue in the dermis. The connective tissue, which is composed of fibroblasts embedded into the extracellular matrix, is also susceptible to alterations by extrinsic factors, such as ultraviolet radiation and environmental pollution.[1]

A major component of the extracellular matrix is collagen. Other constituents found in significant amounts include elastin, proteoglycans, glycoproteins and fibronectin. During the aging process, a thinning of the dermis occurs due to a decrease of both collagen[2] and glycosaminoglycan[3]. The junction between the epidermis and dermis also changes during the aging process, affecting skin integrity. This weak link leaves the skin more vulnerable to mechanical trauma.[4]

Collagen, the most abundant protein in mammals, accounts for about 30% of all proteins. Collagen molecules secreted from fibroblast cells assemble into characteristic fibers to provide functional integrity of such tissues as bone, cartilage, skin and tendon. They constitute a structural framework for other tissues including blood vessels and most organs. This microscopic collagenous mesh that permeates every corner of the human body is responsible for contributing to our physical appearance. This mesh turns out to be extraordinarily important to the way we look and feel. It is the main actor in the extracellular matrix, a support system for the survival of every cell in the body.[5]

---

[a] This chapter is adapted from a paper presented by R. Chaudhuri at the HBA Global Expo in New York in 1999.

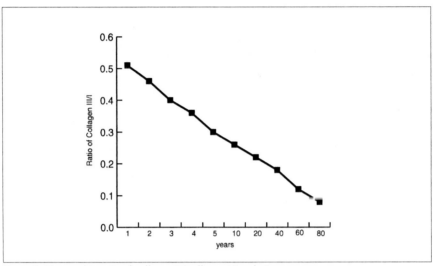

Figure 1. Human skin ratio of collagen III/collagen I in relation to age

In human tissues there are at least 10 genetically and chemically distinct collagen types that have been well-characterized.[6] Several others are currently under study. Collagen I represents about 80% of the dermal collagen in an adult's skin, while collagen III accounts for about 15%.[7] The remaining 5% is made up of collagens IV and V. Collagen III is predominant in young skin and during the wound healing process. Thus, collagen III is also known as a "restructuring" collagen.

One of the numerous modifications of the extracellular matrix during aging is the significant decrease in collagen synthesis. The ratio of collagen types changes throughout life. Specifically, the age-related decrease in the ratio of collagen III/collagen I is a dramatic one (Figure 1).[8]

The reduction and alteration of the natural collagen support layer that lies just beneath the skin causes facial lines and wrinkles. Creams and lotions can moisturize or exfoliate the surface of the skin, but they can't diminish the lines or wrinkles that are caused by the reduction and alteration of the underlying collagen support.

In summary, three major changes in skin structure involving collagen occur during the aging process:

- Decrease in collagen biosynthesis by fibroblast cells
- Relative thinning of extracellular matrix, which becomes more pronounced with reduction in collagen III than with collagen I
- Insolubilization of fibrous collagen leading to loss of skin's biomechanical properties

Design of a Collagen III Amplifier System

Product concept: Keratinocytes are the skin's "antenna" for the reception of external signals. Fibroblasts are the "machines" that synthesize most of the skin's

Figure 2. Scanning electron microscopy of CIIIAS liposome

supportive structures. When these supportive structures are damaged, the fibroblasts start to resynthesize the damaged structures. This requires the transduction of a biological signal from the keratinocytes to the fibroblast. These biological signals are called "cytokines."

Our product concept stems from the basic understanding of "biological signals." The question posed during development of the collagen III amplifier system was: could the biological system be fooled in such a way that keratinocytes send signals to fibroblasts, which in turn will start synthesizing collagen III? The answer is yes.

We have developed a unique liposomal system[b] that can mimic the biological system to modulate the fibroblast phenotype via keratinocytes. For the purposes of this article, we'll call this development a collagen III amplifier system (CIIIAS). When CIIIAS liposome is added to human keratinocytes, it appears to express mediators that in human fibroblasts specifically induce the synthesis of collagen III.

Product description: CIIIAS liposome is a suspension of phospholipidic vesicular carriers, in which the external lipophilic wall contains the amphiphilic dipalmitoyl hydroxyproline (DPHP). Its INCI name is water, lecithin, dipalmitoyl hydroxyproline, phenoxyethanol, tall oil sterol, linoleic acid, tocopherol, sodium ascorbate, methylparaben, butylparaben, ethylparaben, propylparaben, mannitol.

CIIIAS liposome selectively amplifies the biosynthesis of collagen III in human skin. This unique liposome is shaped like a hexagonal pyramid. Its membrane always shows angled structures (Figure 2).

Mode of action: Figure 3 schematically shows the method we have used to demonstrate the activity of CIIIAS liposome. This involves transduction of biological

---

[b] ASC III is a trade mark product of Merck KGaA, Darmstadt, Germany. Rona/EM Industries is an affiliate of Merck KGaA.

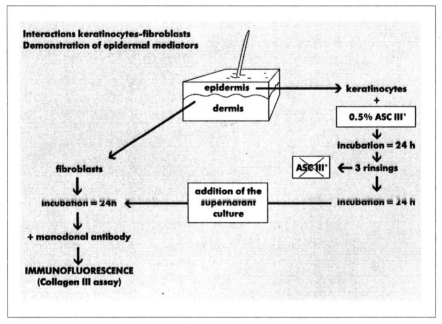

Figure 3. Mechanism of action of CIIIAS liposome

signals from keratinocytes to fibroblasts resulting in selective stimulation of collagen III expression in human skin fibroblasts.

In this method, the epidermis is separated from the dermis of a normal human skin sample. The epidermal cells (consisting mainly of keratinocytes) are then incubated in the culture media with CIIIAS liposome, stimulating the production of cytokines. The supernatant of the keratinocyte containing the cytokines is collected, filtered and then added to the fibroblast cultures. If the keratinocytes were stimulated successfully, they would synthesize cytokines, concomitantly enabling the fibroblasts to produce collagen.

Our data (immunofluorescence staining using a solution of murine anti-immunoglobulin antibodies from rabbits, coupled to fluorescein) shows that CIIIAS liposome appears capable of increasing collagen III synthesis selectively in elderly fibroblasts by similar degrees as in newborn cells, even though basal levels of collagen synthesis are age-dependent.

In Vitro Studies

Immunostaining of collagen III: The selective increase of collagen III in elderly fibroblasts was observed via selective immunostaining of collagen III.[9] Qualitative identification of the antibody/collagen III precipitant was achieved using a solution of murine anti-immunoglobulin antibodies from rabbits coupled to fluorescein. The intensity of the fluorescence is proportional to the content of collagen III in human fibroblasts.

Figure 4a shows young fibroblasts (four years old) with a high collagen III content. Figure 4b shows lower content in older fibroblasts (66 years old). Interestingly, when the same 66-year-old fibroblast culture was treated with CIIIAS liposome, an intense staining was observed, similar to the intensity of four-year-old fibroblasts, as shown in Figure 4c.

Cell proliferation of fibroblasts: We studied cell proliferation in human fibroblasts.[10] Optical density directly correlates to the number of cells present in the human fibroblast culture. This was recorded over time (Figure 5) using CIIIAS liposome having different levels of DPHP and a control (containing all the ingredients of CIIIAS liposome without the three-dimensional structure). The results for both the control and the CIIIAS liposome do not show cell proliferation. Therefore, the increase in collagen III biosynthesis in the dermis (after application of CIIIAS liposome) is due to the induction of fibroblast activity and not due to fibroblast proliferation.

Selective amplification of collagen III: We studied selective amplification of collagen III after induction of fibroblasts.[11] Collagen I and III can be simultaneously quantified in human fibroblasts using radioimmunoassay. Figure 6 summarizes the selectivity of results obtained after 48 hrs of incubation using CIIIAS liposome (with two different levels of DPHP) and a mixture of all the ingredients in CIIIAS liposome (reference solution). The results show a ratio of about 60/40 (Collagen III/I) with CIIIAS liposome irrespective of DPHP concentration. The total increase of collagen I and III was found to be well over 800 mcg/L in a non-dose dependent manner. Meanwhile, the reference solution showed a ratio of about 27/73 (collagen III/I) with a total increase of collagen III and I well below 100 mcg/L.

Effect of three-dimensional structure: A series of in vitro tests (selective immunostaining of collagen III in human fibroblasts) was done to determine the effect of the three-dimensional structure of CIIIAS liposome and its effect on fibroblasts. Results of this work are summarized in Table 1.

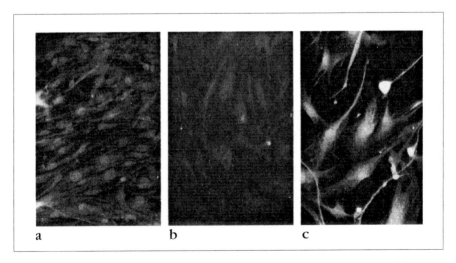

Figure 4. Selective immunostaining of Collagen III in human fibroblasts: a) four years old; b) 66 years old; c) 66 years old treated with 0.5% CIIIAS liposome

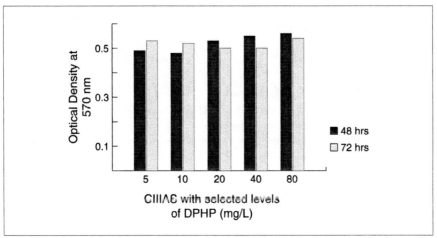

Figure 5. Cell proliferation study in human fibroblasts

Table 1. Active compound in CIIIAS liposome

Composition	Structure	Effect on fibroblasts
CIIIAS without DPHP	spherical liposome	none
DPHP in ethanol or in a formulation	-	none
Short chain peptides from collagen hydrolysis (PC) in ethanol	-	none
CIIIAS with PC without DPHP	spherical liposome	none
CIIIAS	bipyramidal/angled	increase of collagen III decrease of collagenase
CIIIAS with PC without DPHP	bipyramidal/angled	decrease of collagen III increase of collagenase

This work clearly shows that both DPHP and the three-dimensional structure in CIIIAS liposome are necessary to show selective increase of collagen III and decrease in collagenase activity.

Comparative studies: In an in vitro comparative study, we incubated cultured human dermal fibroblasts (obtained from a 54-year-old woman, cells at 4th passage) in Dulbecco's modified eagle medium[c] (DMEM) in the presence of CIIIAS liposome (0.2% or vitamin C (50 mg/mL) or retinoic acid (1 mg/mL).[10,12-14] As a control, we used human fibroblast culture in DMEM supplemented with 10% fetal calf serum, penicillin and streptomycin. At the start of the experiments, the number of cells was about 50,000/well.

In the cell proliferation assay (Coulter Counter method), no effect was observed compared to the control when cells were treated with CIIIAS liposome. However,

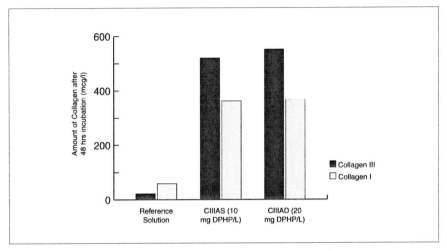

Figure 6. Quantitative determination of collagen III and I in fibroblast culture

Table 2. Summary of comparative studies of CIIIAS liposome, retinoic acid and vitamin C*

Effects on fibroblast cells[d]	None	Reduction	Increase
Collagen I amplification[e]	75.0%	67.0%	82.0%
Collagen III amplification[e]	83.0	42.0	67.0
Collagen III/I	1.2	0.7	0.9
Collagenase formation[e]	4.0	63.0	43.0
Net collagen III increase vs collagenase[f]	44.0	1.4	3.3

*Cell type: Human dermal fibroblasts; 54 year-old woman; incubation time = 72 h
[d] Coulter Counter method
[e] Selective Immunostaining method
[f] These are calculated by normalizing the data for % amplification of collagen III and % formation of collagenase and dividing accordingly

retinoic acid causes reduction in cell counts, whereas vitamin C causes an increase in cell counts.

The results (based on fluorescence intensity) showed that both collagen I expression and collagen III expression are strongly increased in the presence of each of the three products (Table 2). In the case of retinoic acid and ascorbic acid, collagen I was altered to a larger extent than collagen III. However, in the case of the CIIIAS liposome, collagen III increased more than collagen I.

[c] Purchased from Life Technologies Inc., Rockville, Maryland

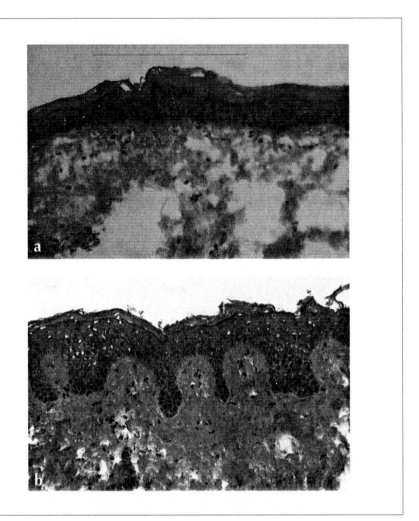

Figure 7. Study of human skin after biopsy: a) before treatment; b) after three weeks of treatment with CIIIAS liposome.

Regarding collagenase synthesis, our in vitro studies showed an increase of the enzyme expression in the cultured human dermal fibroblasts with retinoic acid or vitamin C. In contrast, when cells were treated with CIIIAS liposome, the collagenase synthesis was not at all affected resulting in positive net production of collagen III.

In vivo Studies

Cutometer study for skin elasticity: This analysis quantifies the degree of elasticity in the upper layers of a subject's skin. In this procedure, skin is sucked into the orifice of a probe using constant vacuum pressure for a set time. Two optical lenses located at the probe orifice measure the depth to which the skin penetrates into the probe.

Two gels—one containing 5% CIIIAS liposome and one without—were applied daily for 27 days on each half of the face (panel size 6, average age 50 years, females with fine wrinkles and dry skin). Measurements were carried out before and after the 27 days of treatment. For a gel containing the CIIIAS liposome, we observed a 35% increase in skin elasticity over the 27 days. Another gel containing no CIIIAS liposome showed only a 10% increase in skin elasticity during the treatment period.

Biopsy studies for CIIIAS efficacy: We used human biopsy studies to evaluate the efficacy of CIIIAS liposome on the skin.[13,14] A study lasting five weeks was conducted. Under local anesthesia, samples of both dermis and epidermis were taken from the face of a 44-year-old male volunteer at the end of weeks two, three

Formula 1. Age-defying lotion

A. Water (*aqua*), demineralized	qs 79.90% w/w
Propylene glycol	2.00
Glycerin	3.00
Allantoin (Allantoin, Rona)	0.20
Methylparaben	0.15
B. Carbomer (Carbopol Ultrez 10, BFGoodrich)	0.20
C. Caprylic capric triglyceride (Myritol 318, Henkel)	3.00
Isopropyl myristate (Emerest 2314, Henkel)	3.00
Cetyl alcohol (and) glyceryl stearate (and) PEG-75 stearate (and) ceteth 20 (and) steareth 20 (Emulium Delta, Gattefosse)	3.50
PEG-8 (and) tocopherol (and) ascorbyl palmitate (and) ascorbic acid (and) citric acid (Oxynex K Liquid, Rona)	0.10
D. Triethanolamine, 99%	0.35
E. Water (*aqua*) (and) lecithin (and) dipalmitoyl hydroxyproline (and) phenoxyethanol (and) tall oil sterol (and) linoleic acid (and) tocopherol (and) sodium ascorbate (and) methylparaben (and) butylparaben (and) ethylparaben (and) propylparaben (and) mannitol (ASC III, Rona)	4.00
DMDM hydantoin (Mackstat DM, McIntyre)	<u>0.60</u>
	100.00

Procedure: Combine A; heat to 50°C while stirring until all solids are dissolved. Disperse B in A with a sifter. Heat AB to 65°C. Combine C; heat to 65-70°C while stirring. Add C to AB while stirring. Homogenize ABC. Add D at 55-60°C. Continue homogenizing allowing mixture to cool to 35-40°C. Adjust pH with TEA to 6.8-7.2. When mixture temperature reaches 30-33°C add E and stir gently until mixture is homogeneous. Note: Viscosity 12,000 cps (Brookfield RV 5, 20 rpm) at 23°C

Formula 2. Rejuvenating Aqueous Skin Gel

A. Glycerin	2.00% w/w
Propylene glycol	3.00
Propylene glycol (and) diazolidinyl urea (and) methyl-paraben (and) propylparaben	0.80
Allantoin (Allantoin, Rona)	0.20
Water (*aqua*), demineralized	74.72
B. Carbomer (Carbopol Ultrez 10, BFGoodrich)	0.50
C. Tromethamine	0.78
Water (*aqua*), demineralized	14.00
D. Water (*aqua*) (and) lecithin (and) dipalmitoyl hydroxyproline (and) phenoxyethanol (and) tall oil sterol (and) linoleic acid (and) tocopherol (and) sodium ascorbate (and) methylparaben (and) butylparaben (and) ethylparaben (and) propylparaben (and) mannitol (ASC III, Rona)	4.00
	100.00

Procedure: Combine A and stir. Disperse B in A. Dissolve C and add to AB, stir until mixture is homogeneous. Add D and stir gently until mixture is homogeneous.
Note: Yellowish-opaque gel, viscosity 120,000 cps (Brookfield RV6, 5 rpm) at 23°C, pH at 23°C = 7.3

and five. Treatments occurred on a daily basis. Between two and five weeks, the left side of the volunteer's face was treated daily with a cream containing 5% CIIIAS liposome, while the right side was treated with an identical cream without CIIIAS liposome (control). In order to avoid possible cytokine-mediated carry-over effects from the CIIIAS liposome treated left side, skin biopsies were done on the right side (control) first.

Target parameters for measurement of CIIIAS liposome efficacy on skin were the thickness of epidermal keratinocyte layers and relative content of pro-collagen III. Pro-collagen III is a precursor of collagen III and allows one to distinguish between newly formed and pre-existing collagen III.

Results of immunofluorescence staining of skin show that the untreated skin contained practically no pro-ollagen III at day 1 (control) and showed a very weak response for the two-week control (without CIIIAS treatment). That was done with treatment of the right side with an identical cream without CIIIAS and evaluation of facial skin after 1 day and two-week intervals. In contrast, the application of CIIIAS resulted in a very significant increase in pro-collagen III content. This increase in pro-collagen III content was accompanied by an equally pronounced increase in thickness of epidermal keratinocyte layers, and improvement in dermal/epidermal junction area with reformation of collagen fibers in the dermis (Figure 7).

Formula 3. Rejuvenating o/w lotion

A. Polyglyceryl-3 methyl glucose distearate (Tego Care 450, Goldschmidt)	3.50% w/w
Glyceryl stearate, PEG-100 stearate (Arlacel 165, ICI)	2.50
Dicapryl ether (Cetiol OE, Henkel)	5.00
Coco-caprylate/caprate (Cetiol LC, Henkel)	5.00
Propylene glycol dicaprylate/dicaprate (Myritol PC, Henkel)	3.00
Almond (*Prunus amygdalus*) oil (Cropure Almond Oil, Croda)	2.00
Cetyl alcohol (Crodacol C-70, Croda)	1.50
PEG 8 (and) tocopherol (and) ascorbyl palmitate (and) ascorbic acid (and) citric acid (Oxynex K Liquid, Rona)	0.30
B. Glycerin	3.00
Propylene glycol	3.00
Allantoin (Allantoin, Rona)	0.20
Methylparaben	0.15
Water (*aqua*), deionized	66.35
C. Phenoxyethanol (and) isopropylparaben (and) isobutyl-paraben (and) butylparaben	0.50
D. Water (*aqua*) (and) lecithin (and) dipalmitoyl hydroxyproline (and) phenoxyethanol (and) tall oil sterol (and) linoleic acid (and) tocopherol (and) sodium ascorbate (and) methylparaben (and) butylparaben(and) ethylparaben propylparaben (and) mannitol (ASC III, Rona)	4.00
	100.00

Procedure: Combine A, stir and heat to 65°C. Combine B, stir and heat to 65°C. Add A to B while stirring. Homogenize at moderate speeds to avoid foaming, while allowing mixture temperature to cool to 40°C. Add C; homogenize. When mixture temperature reaches 30-33°C add D and stir gently until mixture is homogeneous.
Note: Viscosity 14,000 cps (Brookfield RV 5, 10 rpm) at 23°C. pH at 23°C = 6.4

Formulation Guidelines

CIIIAS liposome is stable at least for 18 months, if unopened and stored at 4 to 35°C. Three-dimensional structure is lost on prolonged heating at temperatures above 35°C or below freezing. However, in formulated products the liposome is stable over a broader temperature range from 40°C to –10°C.

CIIIAS liposome can easily be incorporated into lotions, creams and gels. Ionic surfactants and bivalent cations must be avoided because they could disrupt the liposome structure over time. Water-soluble film formers and silicone polymers should also be avoided because they hinder the reception of external signals by keratinocytes, resulting in very little production of collagen by fibroblasts. Formulation pH requirements are neutral to slightly acidic. CIIIAS liposome should be added to the formulation after cooling to 35°C.

Formula 4. Rejuvenating skin (o/w) cream

A. Glyceryl stearate, stearth-25, ceteth-20, stearyl alcohol (Tego Care 150, Goldschmidt)	9.00% w/w
Isopropyl myristate	7.00
Propylene glycol dicaprylate/dicaprate (Myritol PC, Henkel)	7.00
C_{12-15} alcohols benzoate (Finsolv TN, Finetex)	7.00
Hydrogenated castor flakes	1.50
Cetyl alcohol (Crodacol C-70, Croda)	1.50
B. Allantoin (Allantoin, Rona)	0.20
Glycerin	3.00
Propylene gycol	3.00
Methylparaben	0.15
Triethanolamine	0.05
Water (*aqua*), demineralized	55.80
C. Phenoxyethanol (and) isopropylparabem (and) isobuty-paraben (and) butylparaben	0.80
D. Water (*aqua*) (and) lecithin (and) dipalmitoyl hydroxyproline (and) phenoxyethanol (and) tall oil sterol (and) linoleic acid (and) tocopherol (and) sodium ascorbate (and) methylparaben (and) butylparaben (and) ethylparaben (and) propylparaben (and) mannitol (ASC III, Rona)	4.00
	100.00

Procedure: Combine A, stir and heat to 70-75°C. Combine B, stir and heat to 65-70°C. Add B to A while stirring. Homogenize at moderate speeds allowing mixture to cool to 40°C. Add C, continue homogenization. Add D when mixture temperature reaches 30°C. Stir with anchor mixer thoroughly at slow rpm until mixture is homogeneous. *Note:* Viscosity 40,000 cps (Brookfield RV 6, 5 rpm) at 23°C. pH at 23°C = 6.9

Four formulations are included to demonstrate the ease of incorporation into different skin care formulations.

Summary

We have described a system that modulates cytokine production to selectively increase collagen III. The method consists of selecting epidermal cells (mainly keratinocytes) from normal human skin, incubation in culture media with CIIIAS liposome, collection of keratinocyte supernatant culture containing cytokines followed by addition to the fibroblast culture. Data shows CIIIAS liposome is capable of increasing collagen III synthesis selectively in elderly fibroblasts by similar degrees as in newborn cells, even though basal levels of collagen synthesis are age-dependent.

We performed several in vitro studies, namely, selective immunostaining of collagen III, cell proliferation study in human fibroblasts and selective amplification of collagen III after induction of human fibroblasts. These studies clearly demonstrate selective increase in collagen III production by the CIIIAS liposome. It is noteworthy that this amplification is the result of fibroblast activity and is not due to fibroblast proliferation. The role of the three-dimensional structure in CIIIAS was also substantiated using an in vitro test method (selective immunostaining of collagen III in fibroblasts).

From comparative studies of CIIIAS liposome with retinoic acid and vitamin C, we concluded that the net production of collagen III, the desired collagen type, is higher with CIIIAS liposome than from either retinoic acid or vitamin C.

A biopsy study of human facial skin showed the effects of CIIIAS liposome resulting in a very significant increase in pro-collagen III content accompanied by a pronounced increase of the epidermal keratinocyte layer thickness. It also showed improvement in dermal/epidermal junction area and reformation of collagen fibers in the dermis. Additionally, an in vivo Cutometer study for skin elasticity showed a 35% increase in skin elasticity.

CIIIAS liposome can easily be incorporated at temperatures below 35°C into lotions, creams and gels having a slightly acidic to neutral pH. Ionic surfactants, water-insoluble film formers and silicone polymers should be avoided as they can hinder efficacy.

In summary, skin care products containing CIIIAS liposome can repair the natural collagen support layer that lies just beneath the skin, resulting in increased skin elasticity and smoothness and reduction of facial lines and wrinkles.

>—**Ratan K. Chaudhuri and George Majewski,** *Rona/EM Industries, Hawthorne, New York, USA*
>
>—**Gilles Guttierez and Mustafa Serrar,** *Texinfine, Lyon Cedex, France*

References

1. NA Fenske and CW Lober, Structural and functional changes of normal aging skin, *J Am Acad Dermatol* 15 571 (1986)
2. S Shuster, MM Black and E McVitte, The influence of age and sex on skin thickness, skin collagen and density, *Br J Dermatol* 93 639 (1975)
3. R Fleischmajer, JS Perlish and RI Bashey, Aging of human dermis, *Front Matrix Biol* 1 90 (1973)
4. JL Cook and LM Dzubow, Aging of the skin: Implication for cutaneous surgery, *Arch Dermatol* 133 1273 (1997)
5. J Uitto, DR Olsen and MJ Fazio, Extracellular matrix of the skin: 50 years of progress, *J Soc of Invest Dermato* 92 61S-77S (1989)
6. P Bornstein and H Sage, Structurally distinct collagen types, *Ann Rev Biochem* 49 957 (1980)
7. EH Epstein Jr and NH Munderlob, Human skin collagen. Presence of type I and type III at all levels of the dermis, *J Biol Chem* 253 1336 (1978)
8. PK Mays, JE Bishop and GJ Laurent, Age related changes in the proportion of type I and III collagen, *Mechanisms of Ageing and Development*, 45 203-212 (1988)

9. K Fukai, M Ishii, M Chanoki, H Kobayashi, T Hamada, Y Muragaki and A Ooshima, Immunofluorescent localization of type I and III collagens in normal human skin with polyclonal and monoclonal antibodies, *Acta Dermato-Venereologica* 68 196-201 (1988)
10. J Varani, RS Mitra, D Gibbs, SH Phan, VM Dixit, R Mitra, T Wang, KJ Siebert, BJ Nickoloff and JJ Voorhes, All-trans retinoic acid stimulates growth and extracellular matrix production in growth-inhibited cultured human skin fibroblasts, *J Invest Dematol* 94 717-723 (1990)
11. H Magloire, A Calle, D Hartmann, A Joffre, JA Grimaud and F Schue, Type I collagen production by human odontoblast – like cells in explants cultured on cyano-acrylate films, *Cell and Tissue Research* 244 133-140 (1986)
12. JC Geesin, LJ Hendricks, JS Gordon and RA Berg, Modulation of collagen synthesis by growth factors: the role of ascorbate-stimulated lipid peroxidation, *Archives of Biochem and Biophys* 1 6-11 (1991)
13. J Varani, SEG Fligiel, L Schuger, P Perone, D Imman, CEM Griffiths and JJ Voorhees, Effects of all-trans retinoic acid and Ca ++ on human skin organ culture, *Am J Pathol* 142 189-198 (1993)
14. J Varani, P Perone, DR Inman, W Burmeister, SB Schollenberger, SEG Fligiel, RG Sitrin and KJ Johnson, *Am J Pathol* 146 210-217 (1995)

Patents

1. French Pat 2 649 335, Procede et dispositif de production directe de liposomes, G Guttierez, September 20, 1991
2. French Pat 2 683 038, Vecteurs vesicularis intracellulaires et leurs applications en cosmetique, G Guttierez, J-A Grimaud, M Andujar and M Serrar May 24, 1995

Effects of Gelatin-Glycine on Oxidative Stress

Keywords: antioxidant, diet supplement, free radicals

Study of the effects of gelatin-glycine applied topically and orally in the skin and blood of heavy smokers.

Oxidative damage can occur by direct chemical oxidation. It can also result from a more insidious chain reaction involving molecules (such as unsaturated fatty acids) that are oxidatively reactive because of the presence of a single unpaired electron. These molecules are known as free radicals.[1]

It is also well established that free radicals and other reactive oxygen species (ROS) continuously produced in vivo are the most important intermediates in the mechanism of toxicity.

Although most ROS produced during cell metabolism are metabolized to non-reactive species, cellular injury occurs in some situations. Thus, decreased activity of cellular defenses against radicals and/or increased production of ROS creates oxidative stress in the cell.[1]

Oxidative stress appears therefore to be associated with an increased production of oxygen free radicals, and this increased production alters the natural antioxidant defense mechanism present in most tissues.[2,3] These antioxidant processes work by direct scavenging of the initiating pro-oxidant species, by sequestering of heavy metals (thereby preventing initiation or propagation of free radical reactions), by scavenging secondary free radical species (which terminates on-going chain reactions), or by returning oxidized groups to their reduced state.

The natural antioxidant systems of the skin are represented by detoxification enzymes (such as superoxide dismutases or glutatione peroxides), water-soluble molecules (such as ascorbate or glutatione) and lipid-soluble molecules (such as tocopherols, ubiquinone or carotenoids).

In addition to these approaches, cells also have mechanisms for repairing damaged macromolecules, for enhancing cell renewal and for minimizing oxidative injury.[4,5]

The balance between cellular production and catabolism of these oxidants is critical for the maintenance of tissue homeostasis.[6-8]

Moreover, there is increasing evidence that ROS may be involved in a variety of skin disorders, including carcinogenesis, cutaneous inflammation and photosensitization. This damage increases with age, and is speeded up and enhanced by UV rays, environmental pollutants, alcohol abuse and smoking.[9,10]

Gelatin-Glycine

Gelatin-glycine is a patented mixture. When these two well-known raw materials are mixed together and fused at 50°C, the resulting mixture shows unique characteristics unlike those that might be expected from a simple mixture of the two powdered raw materials. Probably, the temperature and fusing conditions modify the order of the hydrogen bonds able to link different amino acids with glycine, giving the final product its unique characteristics.

Goals of the Study

Antioxidants are the body's natural defense mechanism against oxidative stress. To combat oxidative stress in the skin, antioxidants can be delivered topically or systemically. In our lab, we studied both delivery systems using a patented material called gelatin-glycine.[11]

Having previously obtained interesting results using gelatin-glycine as a skin-hydrating and antiaging compound,[12-16] we wanted to determine its potential antioxidant activity, topically and systemically, by including this mixture in a cosmetic emulsion and in a diet supplement.

In addition, we wanted to determine if including in the formulation well known antioxidants (such as vitamins C and E and ubiquinone) would strengthen this potential activity.[17-23]

ROS Measurement

It is known that free radicals are extremely reactive and have a very short life. Because of its unopposed electron, a radical is slightly attracted to a magnetic field; it is paramagnetic. This unique physical property allows for its detection and analysis by electron paramagnetic resonance (EPR) spectroscopy. But only a small number of radicals are stable enough to be detected by such spectroscopy in aqueous solution at room temperature.

Fast flow techniques are used to detect short-lived radicals. The most popular methodology is called spin trapping. In this technique, a diamagnetic (capable of being repelled by both poles of a magnet) organic molecule, called the spin trap, reacts with the radical to be detected, producing a secondary but more stable radical called a spin adduct, which is more readily detectable by EPR.[24]

Recently a simpler method was developed. The D-Roms test[25] has already been used in our lab[26] and others.[27] This method is based on a property of transition metals. In the presence of peroxides, transition metals will catalyze the formation of free radicals, which are then trapped by an alkylamine. The alkylamine reacts, forming a colored radical detectable at 505 nm through a kinetic reaction, which is linear up to 500 Carratelli Units (UCarr). One UCarr is equal to a hydrogen peroxide concentration of 0.08% mg.

For our study of the free radicals in blood serum, we use the D-Roms test and we detect free radicals by using a spectrophotometer[a]. Our reagents are the chromogens R1 (an alkylamine) and R2 (a pH 4.8 buffer). We add 10 µL of free serum to 1 mL of R2, and then immediately add that mixture to 10 µL of R1.

We mix the sample gently and incubate it in a cuvette for 1 min at 37°C. The resulting deep red coloring can be photometrically detected at a wavelength of 505 nm by the spectrophotometer. The color changes, but after another minute the sample is red again. We calculate the difference (delta A) and multiply by a K-factor (9000). Thus:

$$UCarr = Delta\ A \times K$$

Normally, the level of free radicals in the blood is in the range of 250-300 UCarr.[25-27] Carratelli has suggested ranges of UCarr values that correspond to levels of oxidative stress (Table 1).

Oxidative Stress in Smokers

We devised an experiment to measure the potential antioxidant activity of gelatin-glycine applied topically and systemically to persons known to have oxidative stress. The test material was formulated in a cosmetic emulsion and a diet supplement and administered to smokers, whose blood serum free radicals, skin lipid peroxide, skin hydration and skin surface lipids were monitored over a 60-day period.

The smokers: As already shown by other authors, the excessive use of cigarettes increases the level of free radicals in blood serum.[28] For our study of the effect of gelatin-glycine on oxidative stress, we selected 60 female volunteers aged 35 to 45. All the subjects were inveterate smokers. They had typically smoked 20-30 cigarettes per day for at least two years and suffered from dryness and clear skin dehydration. All gave their written informed consent in accordance with the ethic of cosmetic experimentation. All showed UCarr values in excess of 400, which surely classified them as having "heavy oxidative stress," according to Carratelli's table.[29]

The materials: The gelatin-glycine was incorporated into capsules for daily ingestion as dietary supplements and also into a cream for topical application.

We used three soya oil-based supplement formulations (Table 2) to test for any anti-free-radical activity by gelatin-glycine alone and with known anti-free-radical vitamins. One formulation contained gelatin-glycine. Another contained gelatin-glycine and vitamins C and E and ubiquinone. A control formulation contained only soya oil and starch.

We also used three glycolic acid cream formulations (Table 3) to verify whether gelatin-glycine, when used as an agent partially neutralizing glycolic acid (to pH 4.5), was able to enhance both the superficial moisturizing activity and the antiaging activity

[a] ROS-meter System, Rome, Italy

Table 1. Carratelli's baseline UCarr values, indicating relative levels of oxidative stress

UCarr value	Oxidative stress
300-320	borderline oxidative stress
320-340	slight oxidative stress
340-400	oxidative stress
400-500	heavy oxidative stress
above 500	very heavy oxidative stress

Table 2. Composition of the dietary supplement formulations C, D and E

Ingredient	C	D	E
Soya oil	0.2%	0.2%	0.2%
Gelatin-glycine	35.8	35.8	-
Vitamin C	5	5	-
Vitamin E	1.6	1.6	-
Ubiquinone	1.6	1.6	-
Starch	-	-	25%
Fe, Zn, Mn, Cu, vitamins B_6, PP, folic acid	4.5	4.5	-

Table 3. Composition of the topical cream formulations A, A1 and B

Ingredient	A	A1	B
Glycolic acid (10%)	10%	10%	10%(w/w)
Gelatin-glycine	-	2%	-
Gelatin-glycine /Arginine	1.6%	-	-
Arginine	-	-	40%

Base Cream: Water (aqua), caprylic-capric triglyceride, octyl stearate, cyclodimethicone, glycerin, cetyl dimethicone, copolyol, cetearyl octanoate, isopropyl lanolate, squalane, sodium PCA, sodium chloride, tocopheryl acetate, BHT, soluble collagen, linoleic acid, linolenic acid, petynil palmitate, imidazolidinyl urea, methylparaben, propylparaben, disodium EDTA, fragrance (parfum), magnesium sulfate

Table 4. Treatment plan (using formulation codes from Tables 2 and 3)

Treatment	Diet supplement	Topical cream
Active 1	C	A
Active 2	D	A
Active 3	E	A1
Control	E	B

of the glycolic acid on the skin. One formulation was partially neutralized by gelatin-glycine and arginine. Another was partially neutralized by gelatin-glycine alone. A control formulation, containing no gelatin-glycine, was partially neutralized by arginine.

Using these six formulations, we devised the four treatment plans shown in Table 4.

The procedures: The cosmetic and diet supplement treatment lasted eight weeks between December 1997 and January 1998. Tests were made always by the same operator at day 0 and then every 10 days for 60 days, always after completing that day's treatments.

During the treatment period, no other cosmetic products or diet supplements were used except the products provided by us.

The subjects were randomly divided into four groups of 15 people each, receiving cream A, A1 or B and diet supplement C, D or E, according to Table 4. Neither operator nor subject was able to identify the assigned product. The cream was applied twice daily over the entire face and on both forearms. Four capsules were taken each day (two in the morning and two in the evening). The skin was always cleaned with a supplied cleansing lotion.

Thirty days before starting the study all systemic drugs or diet supplements were discontinued.

Measuring Oxidative Stress

The determination of antioxidant activity was done directly on the skin and blood serum of the volunteer smokers.

Measuring ROS: Blood samples (0.2 mL) were taken from fingertips at 8 a.m. on the day before beginning administration of the diet supplement (day 0) and at days 10, 20, 30, 40, 50 and 60 of treatment. The blood samples were immediately evaluated[a] by spectrophotometry.

Measuring skin surface lipids and hydration: Quantitative measurements[b] were taken before treatment (day 0) and at days 10, 20, 30, 40, 50 and 60 of treatment, always between 8 and 10 a.m., on skin cleaned the night before. According to instructions, no cream was applied the night before the measurement. The measuring instrument collects 10-15 measurements over a 25 sec sampling period and automatically reports the mean values, standardizing the environmental conditions at 22°C and 50% RH.[30]

The surface lipids determination is based on photometric measurement of light transmission through a skin surface imprint obtained by applying a frosted plastic foil to designated skin areas on the cheeks, forehead, chin and nose. Skin lipids adhere to a 1 cm² area of the foil.

Hydration of the skin horny layer is assessed by a probe that measures electrical capacitance of the skin surface. When the probe is placed on the skin for 0.5 sec, the measuring instrument displays capacitance digitally in arbitrary units. The results are expressed as mean values of the measurements performed on the four sites.

Measuring skin peroxides: Skin lipids were extracted from the forearm skin according to the method of Pugliese.[31] A glass cylinder measuring 5 cm in diameter was placed on the labeled area of the forearm skin and held snugly, while the lipids were extracted by two different aliquots of 5 mL portions of acetone.

The two portions of acetone were dried under a nitrogen stream. The lipid residue was emulsified with 0.2 mL of 8% sodium dodecyl sulfate, 1.5 mL of 20% acetic acid and 0.5% of thiobarbituric acid solution, with enough water to make the final volume 4 mL. Finally, the concentration of peroxides, determined as malonyl aldehyde (MDA) precursors on supernatant extracted with 4 mL of n-butanol, was read at 531 nm.

Results and Discussion

Skin performs its protective function as a barrier to environmental insult in an oxygenated atmosphere where peroxidants, such as ROS, may be generated from a wide spectrum of exogenous and endogenous reactions. If there is a significant increase in generation of these radicals, or if antioxidant defenses are depleted, the delicate balance between insult and defense is disrupted, and a state of oxidative stress occurs. As matter of fact, the skin biological membranes are rich in polyunsaturated fatty acids (PUFA) and are therefore vulnerable targets for free radical attack.

The primary targets of free radical attack and resulting oxidation damage are the unsaturated lipids in the stratum corneum of the skin.[32] In fact, 70% of these lipids on a dry weight basis consist of ceramides with unsaturated fatty acyl side chains and cholesterol, which also has unsaturated double bonds.[33] The stratum corneum also contains squalene, an unsaturated hydrocarbon that is readily peroxidized upon irradiation with UV light in the presence of oxygen. Therefore, those free-radical-mediated reactions resulting in lipid peroxidation[34] are of particular importance.

---

[b] 3C System, Dermotech, Rome, Italy

In this case, the normal antioxidants that serve as free-radical quenchers are no longer sufficient to protect the skin. Therefore, it is necessary to supply the skin with protective substances able to compensate its needs. It seems possible to do that with diet supplements administered orally, or by applying topical cosmetic emulsions,[29,35] or by adding fruit and vegetables to the daily diet.[36]

Results on ROS: As shown in Figure 1 through 4, gelatin-glycine seems to perform an interesting antioxidant activity, if taken simultaneously orally and topically. The oxidative stress present in the examined subjects at the beginning of the treatment decreases by approximately 27% ($p< 0.01$) after the first ten days of treatment, and reaches values less than 47% ($p< 0.01$) after the second month.

As expected, enriching the diet supplements with vitamins C and E and ubiquinone produces a further reduction of approximately 25% ($p< 0.01$) of the free radicals present in the blood of the subjects tested. This is due to the well known antioxidant activity of those actives.[37-39] This is especially interesting in light of the fact that the volunteers continued to smoke their habitual 20-30 cigarettes per day throughout the treatment period.

It's clear, then, how this diet supplement seems to be able to fight the oxidative stress caused by cigarettes, and probably also by other external and internal factors. Surprisingly, the activity of gelatin-glycine, used either topically or systemically, is not achieved by simply using gelatin or glycine alone or mixed, even when they are at concentrations like those in our formulations.[11,40]

Results on skin lipid peroxides: Figure 4 shows skin lipid peroxides in the groups that simultaneously received gelatin-glycine both orally and topically (Active 1, 2 and 3). If compared to the control group (B-E treatment), which received a normal cream with 10% glycolic acid partially neutralized by arginine and a diet supplement containing starch only, the group treated with the gelatin-glycine diet supplement and the glycolic acid cream, enriched and partially neutralized with gelatin-glycine (Active 2), showed a decrease in skin lipid peroxides of 25% ($p< 0.01$) after only 20 days of treatment and 41% at 60 days ($p< 0.01$).

The group receiving both the cream and the diet supplement further enriched with vitamins C and E and ubiquinone (Active 1) showed more marked results. Cutaneous lipid peroxides decreased by 40% after the first 20 days of treatment and 56% at 60 days ($p< 0.01$).

The 10% glycolic acid cream partially neutralized by arginine (Control) shows no activity on skin peroxides, which remain almost unchanged.

Results on skin hydration: Figure 3 shows that skin hydration is enhanced by cream containing glycolic acid partially neutralized by arginine (Control) or by gelatin-glycine alone (Active 3), as widely shown by our group and by many other authors.[41-48] Compared to treatments with that cream, the treatments combining topical glycolic acid partially neutralized by gelatin-glycine and those based also on systemic gelatin-glycine reveal a considerably greater increase in the skin's moisture level. The increase was 10-30% greater after the first ten days of gelatin-glycine treatment ($p< 0.01$).

Results on lipids: Gelatin-glycine also seems to be better at rebalancing surface lipids, which are crucial to keep the skin soft and healthy (Figure 2).

Moreover, it is interesting to note that at the same pH, with the same quantity of free glycolic acid present in bioavailable forms in the emulsion, gelatin-glycine

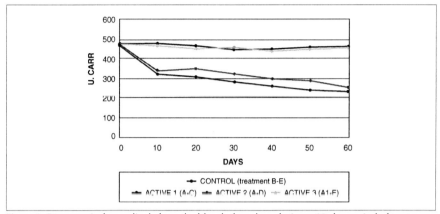

Figure 1. Decrease in free radicals from the blood of smokers during a 60-day period of treatment with gelatin-glycine (topical and oral) or with a control. (n=60; all Active 1 and Active 2 p values are highly significant (p< 0.01) vs. Control at different days; Active 3 vs. Control is not significant)

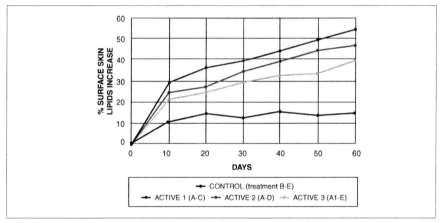

Figure 2. Increase of skin surface lipids on the face of smokers during a 60-day period of treatment with gelatin-glycine (topical and oral) or with a control. (n=60; all p values are highly significant (p< 0.01) vs. Control at different days)

seems also to make the formulation more active both in rehydrating and in rebuilding the skin surface lipidic film (Figures 2 and 3). This mixture acts by itself to promote rehydrating and antiaging, independent of glycolic acid, which does not seem able to reduce the free radicals found at the cutaneous level in the subjects examined (Figure 4). In fact, the same emulsion, partially neutralized at pH 4.5 using the arginine only, did not produce the values obtained by adding gelatin-glycine instead (Figures 2, 3, 4).

In this case, however, gelatin-glycine seems to have a unique function. It is not a simple buffering system to control the release of glycolic acid into the skin, as seems to be the case when arginine, glycine or gelatin is used alone.[40]

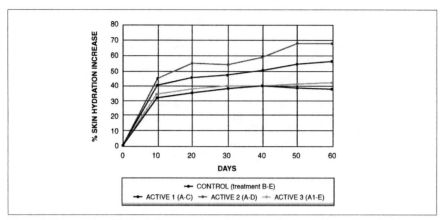

Figure 3. Increase in skin hydration at selected sites on the head of smokers during a 60 day period of treatment with gelatin-glycine (topical and oral) or with a control. (n=60; Active 1 and Active 2 values are highly significant ($p< 0.01$) vs. Control and vs. Active 3 at different days)

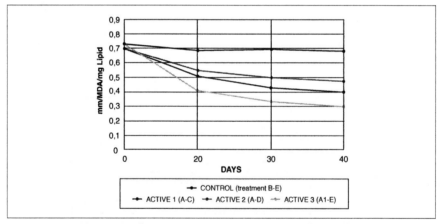

Figure 4. Decrease in acetone-extracted lipid peroxides from forearm of smokers during a 60-day period of treatment with gelatin-glycine (topical and oral) or with a control. (n=60; all p values are highly significant ($p< 0.01$) vs. Control at different days)

A comparison of the data in Figures 2 and 3 shows that topical and oral gelatin-glycine causes basic surface lipids to increase by 25-55%, and skin hydration to increase by 40-68%. These values are approximately three times greater than values obtained by using only the cream with 10% glycolic acid ($p< 0.01$) partially neutralized by arginine alone.

Summary

These new data seem to confirm the results previously obtained by our team: Gelatin-glycine increases the skin's moisture level when used both topically and systemically. It is interesting to note that gelatin-glycine also seems able to reduce free radicals in the serum and at the skin level, which paves the way for other future uses of this patented mixture.[11]

—**P. Morganti,** *Mavi Sud S.r.l., Aprilia, Italy*
—**G. Fabrizi,** *Department of Dermatology, Catholic University of Rome, Italy*

References

Address correspondence to Dr. P. Morganti, c/o Editor, *Cosmetics & Toiletries* magazine, 362 South Schmale Road, Carol Stream, IL 60188-2787 USA.

1. J Euchs and L Packer, *Oxidative Stress in Dermatology*, New York: Marcel Dekker (1987)
2. B Halliwell, Free radicals, antioxidants and human disease: curiosity, cause or consequence, *Lancet* 344 721-734 (1988)
3. AC Bowling and M Flint Belal, Bioenergetic and oxidative stress in neurogenerative diseases, *Life Sciences* 56 1151-1171 (1995)
4. M Ruberfroid and PB Calderon, *Free Radicals and Oxidation Phenomena in Biological Systems*, New York: Marcel Dekker (1994)
5. AW Girotti, Photodynamic lipid peroxidation in biological systems, *Photochem Photobiol* 51 497 (1990)
6. I Fridorich, Superoxide radicals: an endogenous toxicant, *Ann Rev Pharmacol Toxicol* 23 239 (1983)
7. R Zimmerman and PA Cerutti, Active oxygen acts as a promoter of transformation in mouse C3H/10T1/2/C18 fibroblasts, *Proc Natl Acad Sci USA* 81 2085 (1984)
8. CE Cross et al, Oxygen radicals and human disease, *Ann Intern Med* 107 526 (1987)
9. M Athar, R Agarwal, DR Bickers and A Mukatar, Role of reactive oxygen species in skin, in *Pharmacology of the Skin*, A Mukatar, ed, Boca Raton, Florida: CRC Press (1992) 269-279
10. CH Hennekens, Antioxidant vitamins and cardiovascular disease: current knowledge and future directions, *Nutrition* 14 50-51 (1998)
11. US Pat 4,806,525, P Morganti (Feb 21, 1989); see also US Pat 4,863,950 and EC Pat 77430 (Mar 1986)
12. P Morganti and SD Randazzo, Enriched gelatin as skin hydration enhancer, *J Appl Cosmetol* 5 105-120 (1987)
13. P Morganti, The future of cosmetics dermatology, *J Appl Cosmetol* 5 145-158 (1987)
14. P Morganti, SD Randazzo and C Bruno, Oral treatment of skin dryness, *Cosmet Toil* 103(4) 77-80 (1988)
15. P Morganti and B James, Gelatin-glycine: improved cutaneous water retention capacity, *J Appl Cosmetol* 7 103-109 (1989)
16. P Morganti, B James and SD Randazzo, The effect of gelatin-glycine on skin hydration, *J Appl Cosmetol* 8 81-88 (1990)
17. S Kaiser, P DiMascio, ME Murphy and H Sies, Quenching of single molecular oxygen by tocopherols, *Adv Exp Med Biol* 246 117-24 (1990)
18. N Kheltab et al, Photo-protective effects of vitamin A and E on poyamine and oxygenated free radical metabolism in hairless mouse epidermis, *Biochimie* 70 1709-1713 (1988)
19. IR Record, IE Dreosti, N Kostantinopoulos and RA Buckley, The influence of topical and systemic vitamin E on ultra-violet induced skin damage in hairless mice, *Nutr Cancer* 162 219-225 (1991)
20. DI Roshchupkin, MY Pistsov and AY Potapenko, Inhibition of ultra-violet light induced erythema by anti-oxidants, *Arch Dermatol Res* 226 91-94 (1979)
21. M Hiramatsu and L Packer, Anti-oxidant activity of retinoids, in *Methods in Enzymology*, L Packer, ed, Academic Press (1990) p 273-280
22. P Mayer, W Pittmann and S Wallat, The effects of vitamin E on skin, *Cosmet Toil* 108(2) 99-109 (1993)
23. A Mellors and AL Tappel, The inhibition of mitochondrial peroxidation by ubiquinone and

ubiquinol, *J Biol Chem* 241 4353 (1966)
24. CF Babbs and DW Griffin, Scatchard analysis of metan sulfinic acid production from dimethylsulfoxide. A method to quantify hydroxyl radical formation in physiologic system, *Free Radical Biol Med* 6 493-503 (1989)
25. A Alberti, I Bolognini and M Carratelli, The radical cation of N,N-diethyl-paraphenylendiamine: A possible indicator of oxidative stress in biological samples, *Res Chem Intermed*, 26(3) 253-267 (2000)
26. P Morganti, G Fabrizi, C Bruno and A Cardillo, Fotoprotezione e stress ossidativo, presented at 37th Congresso Nazionale ADOI, 23-26 Sept 1998, Ischia, Italy, *Cosmet Technology* 2 39-43 (1999)
27. MR Cesarone et al, A single test to monitor oxidative stress, *International Angiology* 18(2) 127-130 (1999)
28. P Morganti, The radical protection factor for innovative nutriceuticals, J Appl Cosmetol 18 29-35 (1999)
29. P Morganti, G Fabrizi, B James and C Bruno, Effect of gelatin-cystine and serenoa repens extract on free radicals level and hair growth, J Appl Cosmetol 16 57-64 (1998)
30. A Cardillo and P Morganti, Fast and non-invasive method for assessing skin hydration, J Appl Cosmetol 12 11-16 (1994)
31. PT Pugliese, Assessment of anti-aging products, in *Clinical Safety and Efficacy Testing of Cosmetics*, WC Waggoner, ed, New York: M Mekker (1990) 306
32. M Okkido, K Yoshiro and I Matsuo, Lipid peroxide of human skin, *Curr Problems in Derm* 10 269-278 (1982)
33. HG Yardley, Epidermal lipids, in *Biochemistry and Physiology of the Skin,* LA Goldsmith, ed, New York: Oxford University Press (1983) 363-381
34. A Girotti, Mechanism of lipid peroxidation, *J Free Radical Biol Med* 1 87 95 (1985)
35. FJ Wright, Benefical effects of topical application of free radicals scavengers, *J Appl Cosmetol* 13 41-50 (1995)
36. SC Bolton, M Woodward and H Tunstall-Pedoc, The Scottish heart health study. Dietary intake by food frequency questionnaire and odds ratios for coronary hearth disease risk II. The antioxidant vitamin and fibre, *Eur J Clin Nutr* 46 85-93 (1992)
37. B Fei, R Stoker and BN Ames. Ascorbate is an outstanding antioxidant in human blood plasma, *Proc Natl Acad Sci USA* 86 6377 (1989)
38. PB McCay, EK Lai, G Bruggemann and SR Powell, A biological antioxidant function for vitamin E: electron shuttling for a membrane-bound "free radical reductase", in *Fat Production and Consumption Technologies and Nutritional Implications*, C Galli and E Fedeli, eds, New York: Plenum Press (1987) 145-156
39. H Nohol, W Jordan and RJ Youngmann, Quinones in biology: functions in electron transfer and oxygen activation, *Adv Free Rad Biol Med* 108 710 (1986)
40. P Morganti, unpublished data (1999)
41. EJ Van Scott and RJ Yu, Substances that modify the stratum corneum by modulating its formation, in *Principles of Cosmetics for the Dermatologists*, P Frost and SN Horwitz, eds, St. Louis: CV Mosby Co (1984) 70-74
42. EJ Van Scott and RJ Yu, Alpha hydroxy acids: therapeutic potentials, *Canadian J Dermatol* 15 1253-1258 (1989)
43. WF Dial, Preparations prescribed in anti-wrinkling therapy, *Cosmet Dermatol* 3 32-34 (1990)
44. JM Ridge, RJ Siegle and J Zuckerman, Use of alpha hydroxy acids in therapy for photo-aged skin, *J Am Acad Dermatol* 23 932 (1990)
45. P Morganti, S Persechino and C Bruno, Effects of topical AHAs on skin xerosis, *J Appl Cosmetol* 12 85-90 (1994)
46. WP Smith and W Smith, Hydroxy acids and skin aging, *Cosmet Toil* 109(9) 41-48 (1994)
47. P Morganti, SD Randazzo and C Bruno, Alpha hydroxy acids in the cosmetic treatment of photo-induced skin aging, *J Appl Cosmetol* 13 1 (1996)
48. P Morganti, SD Randazzo and C Bruno, Stratum corneum turnover time in aged skin, *Cosmet Toil* 112(7) 61 (1997)

Potassium Azeloyl Diglycinate: A Multifunctional Skin Lightener

Keywords: potassium ayeloyl diglycinate, melanin, skin whitening, sebum

Skin lightening and sebum normalization are among the useful cosmetic functions of a soluble derivative of azelaic acid

The dermatological use of azelaic acid is of interest because of its lightening, anti-seborrheic, anti-mycotic and anti-acne properties. However, its limits in cosmetic formulations are well known, even when it is used at lower concentrations than in pharmacological applications.

A new molecule, potassium azeloyl diglycinate, has been developed to overcome these limitations. This chapter describes the new ingredient, and reports preliminary results of its efficacy as a skin lightening agent and as a sebum-normalizing agent.

Azelaic Acid

The use of azelaic acid (in its free-acid form) has long been known by dermatologists, who consider azelaic acid as a topical drug. This ingredient has found use as an important skin lightener and is also useful in treatment of seborrheic skin. Note that in Europe azelaic acid is not considered a drug and can be used as a cosmetic ingredient in skin care applications.[1]

Azelaic acid is produced by the microorganism *Pityrosporum ovale* (and other species). This organism is responsible for the cutaneous disease known as pitiryasis vescicolor, which causes leucodermic spots in which melanin is not present. Its mechanism of action has been shown to be a competitive inhibition of tyrosinase, the main enzyme involved in the formation of melanin. From this evidence, azelaic acid has been topically used in the dermatological treatment of hypermelanic spots (see sidebar).

Another important application of azelaic acid in dermatology is explained by its bacteriostatic activity. Azelaic acid has bacteriostatic properties against aerobic microorganisms such as *Staphylococcus epidermidis, Staphylococcus aureus, Proteus*

Melanin and Skin Lighteners

Melanin is the substance responsible for the color of skin. Its main function is the protection of the deepest layers of epidermis, protecting them from damage by UV radiation. In fact, the exposure to UV rays, especially UVB, promotes the synthesis of new melanin that protects the genetic molecules inside keratinocytes from damaging radiation.[2]

The synthesis of melanin takes place in specialized cells called melanocytes, where the amino acid tyrosine, in the presence of UVB radiation, is converted into dopa and dopaquinone due to the action of the enzyme tyrosinase. Tyrosinase also requires the presence of oxygen and copper.[3]

From the dermatological perspective, the treatment of hyperchromic spots involves several substances active at different stages:

- Use of sunscreens to reduce the stimulating effect of UV radiation on melanogenesis
- Promotion of cellular turnover in order to replace corneocytes containing melanin granules
- Inhibition of the synthesis of new melanin

Compared to other skin lighteners, azelaic acid is not a photosensitizer, and skin shows moderate tolerance. Azelaic acid has been used for a long time, even at high levels (20%).[4]

mirabilis, Escherichia coli, Pseudomonas aeruginosa and *Candida albicans*, and against anaerobic ones such as *Propionibacterium acnes*. This activity is probably due to azelaic acid's ability to inhibit the cells' protein synthesis.

Additionally, azelaic acid may also cause a reduction of free fatty acids in cutaneous sebum due to a competitive inhibition of the enzyme 5-α reductase, thus inhibiting the conversion of testosterone to 5-dehydrotestosterone.

These two properties also make azelaic acid especially effective in the treatment of acne, which involves sebum production and excretion, microbial colonization of the pilosebaceous unit and inflammatory reaction of the perifollicular area. Azelaic acid possesses activity against all these factors.[5]

Potassium Azeloyl Diglycinate

Azelaic acid, in spite of its valuable properties, presents some technical and formulating problems. First of all, it must be present in high concentrations to be effective; however, it is not soluble at high concentrations, and gives poor cosmetic properties to formulations, particularly thick systems difficult to spread.

Furthermore, it has a melting point around 105-106°C, which is quite high for a cosmetic ingredient and makes azelaic acid difficult to handle under standard conditions.

Finally, the solubilization of azelaic acid through unusual chemical methods results in the loss of azelaic acid content over time because it undergoes decarboxylation.

By reacting the acid chloride of azelaic acid with two moles of glycine and one mole of potassium hydroxide (Figure 1), we obtain potassium azeloyl diglycinate,[a] a new molecule with better technical performance than its precursor, azelaic acid. Because this derivative of azelaic acid also contains the amino acid glycine, it exhibits very high water solubility, amphiphilicity, highly specific activities at low concentration and low toxicity, and excellent chemical stability and compatibilities.

Potassium azeloyl diglycinate is a soluble azelaic acid derivative that maintains all the cosmetic properties of the original molecule but improves upon its technical characteristics. Completely water soluble, potassium azeloyl diglycinate has increased bioavailability compared to azelaic acid, so its required use levels are much lower. Furthermore, skin tolerance is much enhanced.

The chemical modification of azelaic acid to potassium azeloyl diglycinate has therefore led to an equally active ingredient, but much improved from a technical point of view. Its physical and chemical properties are shown in Table 1.

[a]Azeloglicina is a trade name of Sinerga Srl, Milan, Italy. The INCI name is Potassium Azeloyl Diglycinate.

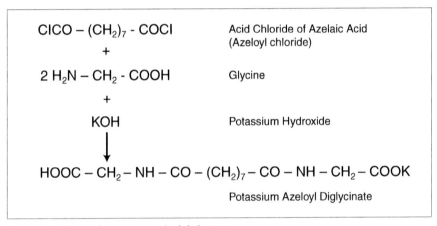

Figure 1. Synthesis of potassium azeloyl diglycinate

Table 1. Physical and chemical properties of potassium azeloyl diglycinate

Property	Description
Appearance	clear liquid
Color	colorless to light yellow
Odor	odorless
pH	6.5-7.5
Active ingredient	28-32%
Specific gravity (at 25°C)	1.160-1.170 g/mL
Molecular weight	341.42 g/mole
Water solubility	complete

Autoradiography techniques[6] have identified the mode of action of both azelaic acid and potassium azeloyl diglycinate to be the inhibition of tyrosinase. On melanocytes irradiated by radiotagged potassium azeloyl diglycinate (^{3}H dodecanoic acid), the localization of radioactivity in mitochondrions and cell nucleus was observed. Azelaic acid is similarly incorporated in the cell nucleus.

Efficacy Evaluations

In order to prove the effectiveness of the product, several efficacy evaluations were carried out on human volunteers.

Whitening: The purpose of this test[7] was to evaluate the whitening efficacy of potassium azeloyl diglycinate on both hypermelanic and normal (unspotted) skin.

The product, in the form of a 3% aqueous solution, was applied to five volunteers having hypermelanic spots on the back of their hands. The spots were carefully selected and their location noted.

Each subject applied the product on the back of one hand (chosen randomly for each subject) twice a day for three weeks.

At the beginning and after three weeks of treatment, the skin color was measured using a colorimeter[b] on the following skin areas:

- A hypermelanic spot on the treated hand (treated spot)
- An area without hypermelanic spots on the treated hand (treated skin)

[b] Minolta Chroma Meter CR 300, Minolta, Osaka, Japan

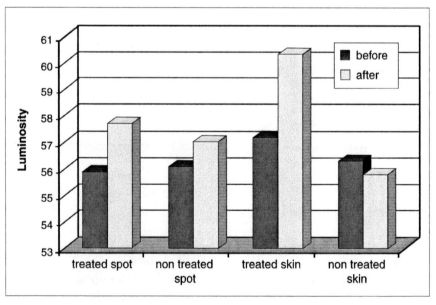

Figure 2. Whitening efficacy of potassium azeloyl diglycinate in 3% aqueous solution on human hands (N = 5)

- A hypermelanic spot on the untreated hand (untreated spot)
- An area without hypermelanic spots on the untreated hand (untreated skin)

The parameters evaluated were "L°" (luminosity), "a°" (red-green axis), "b°" (yellow-blue axis); together they define skin color. As an internal reference, untreated sites were measured as controls at the beginning and at the end of treatment.

The results of measuring the "L°" parameter (Figure 2) reveal a significant increase in skin brightness on both areas treated with the product. The whitening efficacy of the product was also confirmed by a decrease in skin color as measured by the "a°" parameter, which showed a decrease of 10.1% on treated spots and decrease of 12.2% on treated skin.

Sebum normalization: The purpose of this study[8] was to evaluate the sebum-normalizing efficacy of potassium azeloyl diglycinate. The product was applied as a 3% aqueous solution on the face of five volunteers with oily and acne-affected skin twice a day for three weeks.

Measurements of the resultant levels of cutaneous lipids were obtained through the use of a sebometer[c] on the forehead, nose and chin.

At the end of the treatment, the measured levels of cutaneous lipids were lower than initial levels by 29.4%, 27.0% and 31.5% for forehead, nose and chin, respectively. We conclude that potassium azeloyl diglycinate is effective in the treatment of oily and acne-affected skin, effectively reducing the excess of cutaneous lipids.

Hydration and elasticity: The purpose of this test[9] was to evaluate the efficacy of the product in improving skin elasticity, moisture and smoothness after long-term use. Five volunteers applied a 3% aqueous solution to specific facial areas twice a day for three weeks.

At the beginning and at the end of the treatments, instrumental measurements of moisture[d] and elasticity[e] were taken.

Measurements showed that the basal value of skin moisture increased by 12.7% and 8.2% for forehead and cheek, respectively. Skin viscoelasticity on the forehead increased by 2.5%.

We conclude that potassium azeloyl diglycinate maintains all of the valuable properties of azelaic acid, but shows additional characteristics as a multifunctional ingredient.

Toxicological Evaluations

Innocuousness and skin tolerance of cosmetic products and cosmetic raw materials are important concerns for suppliers, formulators and consumers. Here we report our evaluations of the major toxicological properties of potassium azeloyl diglycinate.

Skin irritation: This test[10] aimed to evaluate the skin irritation potential of potassium azeloyl diglycinate. The product, as received (30% active), was applied in an occlusive patch on the backs or forearms of 20 selected subjects for 48 h. The

---

[c] Sebometer SM 810, Courage & Khazaka, Cologne, Germany
[d] Corneometer CM 820PC, Courage & Khazaka, Cologne, Germany
[e] Cutometer SEM 474, Courage & Khazaka, Cologne, Germany

Formula 1. Protective and lightening day cream

Ingredient	Amount
Arachidyl alcohol (and) behenyl alcohol (and) arachidyl glucoside	5.0%wt
Triticum vulgare (wheat germ) oil	5.0
Olea europaea (olive) oil unsaponifiables	5.0
Persea gratissima (avocado) oil	3.0
Oryzanol (Gamma Oryzanol)	0.5
Dimethicone	0.48
Tocopheryl acetate	0.5
Zinc oxide	1.0
Tocopherol (and) lecithin (and) ascorbyl palmitate (and) citric acid	0.02
Phenoxyethanol (and) methylparaben (and) ethylparaben (and) propylparaben (and) butylparaben	0.5
Fragrance *(parfum)*	0.2
Water *(aqua)*	qs
Glycerin	2.0
Potassium azeloyl diglycinate (Azeloglicina, Sinerga)	5.0

Properties
Appearance: homogeneous viscous emulsion
Color: ivory
pH: 6.35
Viscosity: 10000 mPa.s (Viscotester Haak, spindle 1.25 rpm)
Stability: centrifuge (30' at 4000 rpm)

irritating activity was clinically evaluated at 30 min after application (immediate irritating effect) and again 48 h later.

None of the subjects showed any signs related either to immediate irritation or to long-term irritation. Therefore, potassium azeloyl diglycinate can be considered as a non-irritant.

Hypoallergenicity: The aim of this test[11] was to determine if potassium azeloyl diglycinate contains either a single ingredient or mixtures that behave as common allergens. The product, as received, was applied as an occlusive application on the backs or forearms of 20 subjects for 48 h. After removing the occlusive patch, the cutaneous reactions induced by the product were evaluated at 24 and 48 h.

No allergic reactions were observed, allowing us to conclude that potassium azeloyl diglycinate is hypoallergenic.

Eye irritation: Irritation potential on mucous membrane was evaluated by observing the adverse changes occurring to egg chorioallantoic membrane after exposure to the product being tested.[12] Results enabled us to conclude that potassium azeloyl diglycinate can be viewed as a non-irritant to mucous membranes.

Formula 2. Low viscosity lightening cream

Glyceryl stearate (and) cetearth-20 (and) ceteareth-12 (and) cetearyl alcohol (and) cetyl palmitate	4.5%wt
Ceteareth-20	1.2
Squalane	5.0
Coco caprylate/caprate	5.0
Phenoxyethanol (and) methylparaben (and) ethylparaben (and) propylparaben (and) butylparaben	0.5
Water *(aqua)*	qs
Potassium azeloyl diglycinate (Azeloglicina, Sinerga)	5.0
Fragrance *(parfum)*	0.5

Properties

Appearance: low viscosity emulsion
Color: white, (Tyndall effect)
pH: 6.55
Viscosity: <100 mPa.s (Viscotester Haak, spindle 3.25 rpm)
Stability: centrifuge (30' at 4000 rpm) - stable

Formulations

Formulas 1 and 2 are examples of lightening products. Note that because it is not sensitive to temperature, potassium azeloyl diglycinate is very easily added to emulsions at any step during processing.

Conclusion

Potassium azeloyl diglycinate is a new derivative of azelaic acid. In preliminary results from efficacy tests on humans, this new material proved very effective as a skin lightening agent and as a sebum-normalizing agent. In addition, skin moisturization and viscoelasticity parameters showed remarkable improvements.

Potassium azeloyl diglycinate is a multifunctional ingredient that also provides other beneficial characteristics, yet presents no toxicity in cosmetic application. Potassium azeloyl diglycinate enriches, through innovation, the panorama of lightening ingredients on the market today.

—**G. Maramaldi and M. A. Esposito,** *Sinerga Srl, Pero, Milan, Italy*

References

1. M Cucchiara, G Proserpio and H Sedghi, Dall' acido azelaico insolubile all' acido azelaico solubile, a technical paper from Sinerga R&D (Mar 1998)
2. R Caputo and M Monti, *Manuale di Dermocosmetologia Medica*, Milano: Raffaello Cortina (1995) 319-378

3. G Prota, Melanine e melanogenesi, *Cosmet Toil*, Italian ed, 18(2) 9-22 (1997)
4. G Penazzi and H Sedghi, Pigmentazione cutanea e depigmentanti, *Cosm Tech* 3(1) 30-33 (2000)
5. S Passi, M Picardo, C de Luca and M Nazzaro Porro, Mechanism of azelaic acid action in acne, *Ital Dermatol Venereol* 124(10) 455-463 (1989)
6. A Fitton and KL Goa, Azelaic acid – a review of its pharmacological properties and therapeutic efficacy in acne and hyperpigmentary skin disorders, *Drugs* 41(5) 780-798 (1991)
7. Institute of Skin and Product Evaluation (ISPE) Study 104/97/00 (Oct 8, 1997)
8. ISPE Study 102/97/00 (Oct 8, 1997)
9. ISPE Study 103/97/00 (Oct 8, 1997)
10. ISPE Study 31/98/00 (Feb 23, 1998)
11. ISPE Study 52/01/01 (Mar 21, 2001)
12. ISPE Study 26/98/00 (Mar 6, 1998)

Dietary/Nutritional Supplements: The New Ally to Topical Cosmetic Formulations?

Keywords: wrinkles, sallowness, hydration, elasticity, diet supplement

Combining the efficacy of a topical cream with a dietary/nutritional supplement improved the skin's appearance. In the trial conducted, parameters associated with aged skin appearance.

The skin is an organ that creates a boundary between our internal physiological system and the environment, thereby protecting against external insults such as UV and cosmic radiation, reactive oxygen species, air pollutants and mechanical damages. Skin is also an important part of the immune system[1] and prevents excessive water loss through the epidermis.

Located externally, the skin bears the signs of its daily contacts with environmental stresses, which translates into the appearance of fine lines, wrinkles, uneven skin tone and decreased skin hydration. Ultimately, the deleterious consequence of UV radiation–actinic aging–conjugates with the inevitable chronological aging in the appearance of signs of aging.[2]

At the molecular level, constant exposure of skin to the environment may trigger the formation of free radicals, depletion of endogenous antioxidants, suppression of immune functions and the activation of specific matrix metalloproteinases (MMPs).[3] From a physiological standpoint, these molecular reactions may be associated with oxidative stress, membrane lipid peroxidation, weakened immune surveillance, inflammation and the progressive dismantlement of the extra cellular matrix (ECM) 3-D structure. The collapse of the ECM is believed by many to represent a key event in the loss of skin elasticity and hydration and in the appearance of signs of aging.[4]

Biotechnology-based cosmetic companies have developed numerous topical formulations aimed at alleviating the appearance of signs of aging by acting directly on the underlying molecular mechanisms. In some cases, these topical formulations have proven to be efficacious in improving the skin appearance of consumers.

However, applying active ingredients on the surface of the skin may not be the only mode to provide benefits to the skin.

Dietary/Nutritional Supplements

Dietary/nutritional supplements can be used to make active nutrients available to all organs of the body. As mentioned earlier, skin is an organ and may therefore benefit from active nutrients conveyed by dietary/nutritional supplements. The repercussions of nutrition on skin health are well exemplified by the fact that some skin disorders are directly linked to nutritional deficiencies.[5]

Conversely, skin plays a major role in maintaining bone health through the synthesis of vitamin D.[6] The interrelation between skin and the nutritional homeostasis has been recently highlighted[7] and calls upon the understanding of the cellular and molecular processes in play.

We have performed a clinical trial in which a topical cream formulation and a dietary/nutritional supplement were concomitantly administered. The dietary/nutritional supplement provided proteoglycans, collagen, glucosamine, carotenoid pigment (astaxanthin esters) and omega-3 essential fatty acids (EPA and DHA). The efficacy of this regimen was demonstrated on the visual appearance of signs of aging as well as by the amelioration of functional properties of the skin.

Study design: This study was an evaluator-blinded, parallel group design clinical trial consisting of three evaluation visits over a period of 12 weeks. Three regimens were randomized among approximately 100 subjects in order to complete the study with about 30 subjects per cohort. Efficacy of regimen was assessed through visual evaluations for fine lines, unevenness of skin tone, sallowness and tactile roughness. In addition, instrumental measurements using a Dermal Torque Meter[a] and Corneometer[b] tested for skin elasticity and skin hydration, respectively. Self-evaluation questionnaires were administered to assess sensory and acceptance parameters.

Visual and instrumental evaluations were used to assess the efficacy of the test articles in supporting specific claims: diminution of fine lines was assessed visually, by the subjects and by the evaluator; improved skin elasticity was measured with the Dermal Torque Meter[a]; skin tone was assessed visually by the evaluator and by the subjects; improvements in skin tone (sallowness) were assessed visually by the evaluator and by the subjects, skin hydration was assessed visually by the evaluator, measured through Corneometer readings and by the subjects.

Subject selection: Inclusion criteria for subjects were as follows: female, age 35 to 60, inclusive; completion of a minimum 3-day conditioning period immediately prior to study enrollment with the presence of fine lines on the face as assessed by a trained evaluator; motivated and willing to comply with test procedures; willing to refrain from changing cosmetic use habits for the duration of the study; and the signature of an agreed informed consent document after the features of the study had been fully explained. The use of oral retinoid medications or topical retinoids on the face within the last three months (e.g. Retin-A or Retin-A type preparations, Accutane) stood as exclusion criteria.

---

[a] Dermal Torque Meter is a product of Dia-Stron Ltd., Hampshire, UK
[b] Corneometer is a product of Courage&Khazaka, Cologne, Germany

Methodology: The regimen for cohort A consisted of 12 weeks home use of a topical active cream and a dietary/nutritional supplement. Efficacy parameters monitored for cohort A were: fine lines, unevenness of skin tone, sallowness, tactile roughness, skin elasticity and skin hydration. Cohort A subjects completed a self-evaluation questionnaire at week 0 and 12 as well.

The regimen for cohort B and cohort C consisted of 12 weeks home use of the dietary/nutritional supplement on top of a placebo cream (cohort B) and sole use of the topical active cream (cohort C). Efficacy parameters monitored for cohort B and C were: skin elasticity and skin hydration.

The topical active cream[c] provided a highly potent MMP inhibitor (INCI: glycosaminoglycans) and was included in a cream base. The placebo cream consisted of the exact same cream base in which the cosmetic active ingredient was omitted.

As for the dietary/nutritional supplement[d], each 250 mg #1 clear gelatin capsule supplied 125 mg of a patented glycosaminoglycan extract as well as 125 mg of hydrolyzed extract powder providing, among others, highly potent MMP inhibitor, complex proteoglycans, collagen, glucosamine, carotenoid pigment (astaxanthin esters) and omega-3 essential fatty acids (EPA and DHA).

Procedure: Three to five days preceding the baseline visit (week 0), the candidates were required to engage in a conditioning period. Candidates replaced their facial cleanser with a regular consumer soap[e] and refrained from the use of moisturizers. Use of facial cosmetics such as foundation, powder, eye makeup and remover and mascara was permitted.

Candidates arrived at to the test locations with faces free of facial makeup, with the exception of lipstick and eye makeup. Candidates were screened for entrance into the study through completion of an inclusion/exclusion form and through visual examination by the evaluator. Baseline visual evaluations were conducted, and Dermal Torque Meter[a] and Corneometer[b] readings were performed. Subjects were randomly assigned to either cohort A, B or C. Subjects applied the topical active cream (cohorts A and C) or placebo cream (cohort B), twice daily both morning and evening, for the entire period. Subjects of cohorts A and B, in addition to applying their respective topical procedure, took the dietary/nutritional supplement once daily in the morning (e.g., two 250 mg capsules with breakfast). Subjects returned after approximately 1 and 12 weeks of using the assigned regimen.

Visual assessments and compliance checks were completed at week 1. Test areas were examined for irritation, which in due case was noted and followed up. All test products were collected, weighed and reissued. At week 12, visual assessments and instrumental readings were taken from the same locations as at the baseline. Subjects of cohort A completed a self-assessment questionnaire. All products of the regimen were collected.

Visual evaluation: The facial area of each subject was visually evaluated by a trained evaluator using a standard light source and, if necessary, a magnifying glass to view the area.

---

[c]MRT$_{EX}$ is a product of Atrium Biotechnologies Inc., Quebec City, Canada
[d]MRT$_{IN}$ is a product of Atrium Biotechnologies Inc., Quebec City, Canada
[e]In this study, Neutrogena was used. Neutrogena is a product of Neutrogena Corporation, Los Angeles, California, USA

- Fine lines were defined as shallow indentations or superficial wrinkling. Generally, these lines were eliminated by pulling the skin taut. The grades were as follows: 0 = no evidence of facial lines; 2 = occasional number of fine lines widely spaced; 4 = few number of discreet fine lines; 6 = moderate number of fine lines in close proximity; 8 = many fine lines densely packed.
- Unevenness of skin tone grades were as follows: 0 = even skin tone; 2 = slight differences in skin tone over small area(s); 4 = slight differences in skin tone involving moderate areas or moderate differences in skin tone involving small area(s); 6 = moderate differences in tone over moderate-sized area, slight differences over large areas, extreme differences in tone involving small area(s); 8 = extreme differences in skin tone over large areas, small areas of hyperpigmentation.
- Sallowness refers to the color of the skin tone. Sallowness grades were as follows: 0 = skin has very pink color; 2 = skin is pale; 4 = skin has slight suggestion of yellowness; 6 = skin is pale with moderate suggestion of yellowness; 8 = skin is quite pale with distinct suggestion of yellowness.
- Tactile roughness refers to the texture of cheek skin when gently palpated. Tactile roughness grades were as follows: 0 = skin is very smooth; 2 = skin is smooth with occasional rough area; 4 = mild roughness; 6 = moderate roughness; 8 = severe roughness.

Grades of 0, 2, 4, 6 or 8 reflect a generalized condition. Grades of 1, 3, 5, or 7 may be used to represent an intermediate condition or less than 50% of the test area having the next highest scoring condition. A score of 8 is the maximum grade assigned.

Instrumental evaluation: The subjects equilibrated in a room maintained at $20 \pm 2°C$ and $35 \pm 5\%$ relative humidity for at least 30 min prior to Corneometer[b] readings. Triplicate readings were taken for each site. These readings were averaged for statistical analysis. Dermal Torque Meter[a] measurements were conducted under

Table 1. Summary of cohort A visual evaluations

Visual Evaluation	Mean score Baseline n=32	Responders[a] Week 1 n=32	Responders[a] Week 12 n=29
Fine lines	4.63(4-6)	16/32	15/29
Unevenness of skin tone	2.22(1-4)	10/32	17/29
Sallowness	3.13(1-4)	26/32	26/29
Tactile roughness	1.00(0-3)	12/17	10/16

[a] Number of subjects that improved by at least one grade.
[b] Number of subjects that improved by one grade from those who improved after 1 and 12 weeks of use
[c] Number of subjects that improved by two grade from those who improved after 1 and 12 weeks of use
[d] Number of subjects that improved by three grade from those who improved after 1 and 12 weeks of use
* $p < 0.05$ ** $p < 0.0005$

ambient conditions. Subjects were in a prone position with the head approximately perpendicular to the supporting surface.

Statistical Analyses: Statistical analyses were conducted on all data collected, except the self-perception questionnaire. Within-regimen analyses were conducted evaluating the changes from baseline for each regimen. The method used for the visual evaluations was the signed rank test. The method used for the instrumental readings was the paired t-test. Between-regimen analyses were conducted using repeated measures analysis of variance technique comparing the changes from baseline among regimen. The model was a one-way comparison of cohorts with subject within cohort as the error stem. Significance-testing was performed at the $\alpha=0.05$ level.

Results

Results reported in this study were obtained from 93 subjects: 29 in cohort A, 32 in cohort B and 32 in cohort C. The subjects' demographics of this clinical study ranged from 35 to 55. The mean and range of values for baseline visual evaluation parameters of cohort A (topical active cream and dietary/nutritional supplement) are reported in Table 1.

Subjects were diverse as shown by the range of values for each visual evaluation. Mean scores at baseline demonstrated that subjects had relatively moderate visible signs of skin aging. For each visual evaluation, the changes were unidirectional as per the grading scale.

A substantial number of subjects displayed improvements as soon as one week of use. After 12 weeks of use and for all visual evaluations, more than half of all subjects in cohort A had at least improved by one grade with a more pronounced response in the case of sallowness. We observed a noticeable progression between week 1 and 12 in the actual number of grades gained by the subjects who improved (Table 1).

-1G [b]		-2G [c]		-3G [d]		Efficacy Week 1	Efficacy Week 12
w1	w12	w1	w12	w1	w12		
11	9	4	6	1	-	34%*	29%*
9	8	1	7	-	2	45%	58%*
14	9	10	10	2	7	43%**	60%**
7	5	5	4	-	1	89%	78%

Of particular interest was the widespread range of improvements surveyed. All visual evaluations had three grade improvements after 12 weeks of use except for fine lines. In the case of sallowness, striking improvements were noticed with seven improvements of three grades out of 26 subjects who experienced positive changes in this specific visual evaluation performed by investigators. The subjects who were given a grade 0 at baseline and also after 12 weeks of use for tactile roughness were not considered in the total number of subjects when calculating the ratio of subjects who improved.

In order to better quantify the effectiveness of the cohort A regimen within the subjects who improved, an efficacy measure was designed discerning between the various improvements as per the grading scale (shift of 1, 2 or 3 grades). The efficacy value after one and 12 weeks of use was tested against baseline for statistical difference (Table 1). Sallowness demonstrated a clear significant statistical difference ($p < 0.005$) as soon as one week after use and after 12 weeks of regimen use.

Unevenness of skin tone and fine lines both significantly improved with regimen use ($p < 0.05$), as soon as one week after use in the case of fine lines. Regimen use did not produce any statistical difference against baseline in the case of tactile roughness. This is probably explained by the presence of subjects with grade 0 evaluation at baseline and therefore the lower number of subjects that could be entered in the statistical calculation.

Table 2. Self-assessment questionnaire cohort A*	
Fine lines	48%
Wrinkles	35%
Firmness	31%
Elasticity	41%
Global quality of skin	59%
Nutritive effect	79%
Telangiectasia	38%
Tone homogeneity	52%
Imperfections improvements	41%
Silky aspect	83%
Skin's overall comfort	76%
Signs of fatigue	31%
Tone	45%
Hydration	83%
Brown spots	45%
Did the treatment meet your expectations?	86%
Global evaluation of regimen: Excellent	17%

* percentage of cohort A subjects who selected among notes excellent, very effective and effective for each parameter after twelve weeks of regimen use

The results from self-assessment questionnaires filled out by subjects of cohort A are shown in Table 2. Self-assessment was evaluated using a questionnaire that included 17 questions regarding skin quality and appearance parameters at baseline and after 12 weeks of use. For each question, subjects had to select between excellent, very effective, effective, average, mediocre and not applicable in appreciating their skin.

The regimen met the expectations of 86% of cohort A subjects. Generally the subjects noted that the regimen was effective in regards to all parameters, especially the hydrating and nutritive effects as well as the skin silky aspect. They were all clearly eminent with 83%, 79% and 83% of subjects from cohort A reporting the regimen was effective, very effective or excellent, respectively, in these particular cases. In some instances improvement of the texture of hair and nails were reported.

Skin hydration measurements were performed after 12 weeks of use by subjects of cohorts A, B and C. As shown in Figure 1 and demonstrated by Corneometer[b] instrumental readings, the combined regimen (cohort A) was the most successful in improving the skin hydration by a mean of 7.9% ($p < 0.01$). Notice the greater proportion of cohort A subjects who improved their skin surface hydration (75%). The use of the dietary/nutritional supplement alone (cohort B) still statistically improved the skin surface hydration opposite to baseline but to a lesser extent of 6.5% ($p < 0.05$). The use of topical active cream (cohort C) was unsuccessful at distinguishing itself from baseline value nonetheless exhibiting a hydrating action. Between-regimen analysis showed a significant statistical difference between cohorts A and C ($p < 0.05$).

Skin elasticity was evaluated based on average improvement for skin extensibility (U_e), viscoelastic component (U_v) and tonicity (U_r) parameters. As shown in Figure 2, U_e was best improved after 12 weeks of use by the combined regimen of cohort A with a global increase of 38%. This result was significantly statistically different against the baseline ($p < 0.01$). Cohort C also exhibited significant improvement ($p < 0.05$) but to a lesser extent (19%).

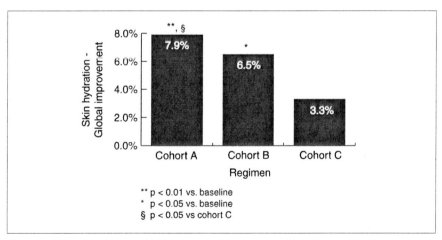

Figure 1. Skin hydration at week 12

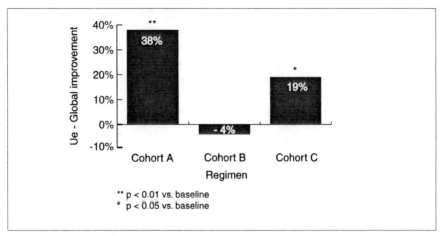

Figure 2. Skin extensibility (Ue) at week 12

Cohort A revealed a noticeable but non-statistically significant 33% global improvement between baseline and 12 weeks of use values for Uv (Figure 3). Cohort A once again differentiated itself by displaying a statistical difference in the case of Ur ($p < 0.05$) by improving 32% after twelve weeks of use (Figure 4). Cohort C had a tendency toward improvement (14%) but not sufficient to be statistically significantly different. Cohort B only produced moderate efficacy in every skin elasticity parameter monitored when compared to baseline values.

Discussion

We have demonstrated that combining the efficacy of a topical cream with that of a dietary/nutritional supplement resulted in the improvement of skin appearance. In the clinical trial conducted, parameters associated with an aged skin appearance were improved: fine lines, sallowness, unevenness of skin tone and tactile roughness. Improvement of skin appearance was observed and judged by trained investigators, assessed directly by the subjects through a questionnaire and quantified through skin hydration and skin elasticity measurements using appropriate devices. Statistical significance was reached according to the parameters and the time point tested. In some instances (sallowness and tactile roughness), the extent of improvement attained three grades on a 0-8 scale.

The regimen used in this clinical study consisted of a topical cream containing an active ingredient endowed with a potent MMP inhibitory activity and a dietary/nutritional supplement providing proteoglycans, collagen, glucosamine, carotenoid pigment (astaxanthin esters) and omega-3 essential fatty acids (EPA and DHA). A MMP inhibitory activity is also present in the dietary/nutritional supplement. Based on the data reported by the hydration and elasticity measurements, the concurrent use of the topical cream and the dietary/nutritional supplement brings additive or synergistic benefits.

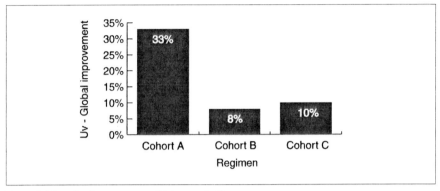

Figure 3. Viscoelastic component (Uv) at week 12

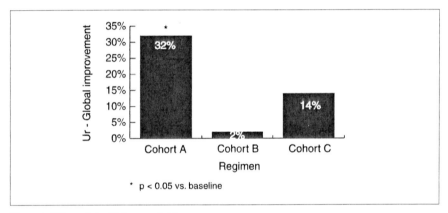

Figure 4. Skin tonicity (Ur) at week 12

When used alone, the dietary/nutritional supplement (cohort B) created a significant hydration effect (Figure 1), however, its effect on skin elasticity when assayed by the Dermal Torque Meter was moderate. Conversely, the topical active cream scored better in the extensibility parameter (Figure 2) and less for hydration (Figure 1). This demonstrates the complementary effect between the oral and topical regimen in acting on different features of skin functionality.

Conclusion

Despite its important functions, skin is too often seen as an inert envelope. Indeed, skin is a living organ just as much as the heart, the kidneys and the liver. Therefore, skin highly relies on what we use topically as well as what we ingest orally to ensure its physiological homeostasis.

We have shown that combining a dietary/nutritional supplement and an active topical cream both formulated with selected ingredients provides benefits to the skin's appearance by reducing visual signs of aging. To elucidate the precise mechanisms underlying the clinical improvement of each parameter tested in this study would require further investigation. It is, however, tempting to suggest that all potential pathways involved: inhibition of MMP enzymatic action, chelation of reactive oxygen species and anti-inflammatory action, team up to prevent excessive oxidative damages and degradation activities to ultimately preserve the multi-functionality of the skin ECM leading to a better appearance.

Working from the "inside out" represents a new and exciting global cosmeceutical approach to supply the skin with biologically active ingredients that can act at the surface of the skin and through nutrition. In this way, synergistic actions can be expected at the molecular level of the skin's ECM.

—**Alain Thibodeau and Édouard Lauzier,** *Atrium Biotechnologies, Inc., Quebec, Canada*

References

1. C Debenedictis, S Joubeh, G Zhang, M Barria and RF Ghohestani, Immune functions of the skin, *Clin Dermatol* 19(5) 573-85 (2001)
2. M Yaar and BA Gilchrest, Aging versus photoaging: Postulated mechanisms and effectors, *J Investig Dermatol Symp Proc* 3(1) 47-51 (1998)
3. VM Kahari and U Saarialho-Kere, Matrix metalloproteinases in skin, *Exp Dermatol* 6(5) 199-213 (1997)
4. M Kulozik and T Krieg, Changes in collagen connective tissue and fibroblasts in aging, *Z Hautkr* 64(11) 1003-4, 1007-9 (1989)
5. B Bunch, The family encyclopedia of disease: A complete and concise guide to illnesses and symptoms, Scientific Publishing Inc. Published by W. H. Freeman and Co., New York (1999)
6. MF Holick, Environmental factors that influence the cutaneous production of vitamin D, *Am J Clin Nutr* 61(3 Suppl) 638S-645S (1995)
7. E Boelsma, HF Hendriks and L Roza, Nutritional skin care: Health effects of micronutrients and fatty acids, *Am J Clin Nutr* 73(5) 853-64 (2001)

New Laminin Peptide for Innovative Skin Care Cosmetics

Keywords: peptide, laminin, extracellular matrix, basement membrane

A new peptide stimulates extracellular matrix biosynthesis and promotes cell adhesion, both of great interest in tissue regeneration.

Epidermal basement membrane is a specialized cutaneous zone of extracellular matrix (ECM) that separates the epidermis from the dermis. The membrane, however, serves more than simple structural and filtering roles.

The basement membrane contains specific structures that ensure the stability of the connection and communication between the two major skin compartments, the epidermis and the dermis. Among these structures we find different molecules, such as collagen IV and laminin.

Laminin represents a family of diverse critical multifunctional molecules, and is a very important component of the extracellular matrix. Laminin[1-4] plays a critical role in cell behavior, is a potent cell adhesion molecule and its defects are involved in some skin pathologies.[5]

Laminin is an essential constituent of the basement membrane and plays a primordial role together with the basement membrane in cell communication, adhesion and cutaneous regeneration. Moreover, recent studies have confirmed the involvement of the basement membrane in skin aging due to the early alteration of the membrane's constituents during the aging process.[6,7]

Therefore, we were interested in developing a laminin-like peptide for topical skin care and cosmetics. In this chapter, we describe in vitro studies concerning the properties of the new laminin peptide and its effects on human skin cells.

Laminin

Laminin is made of multidomain glycoproteins and is thus directly involved in many biological functions.

To date, many laminin isoforms have been identified. They are large disulfide-bonded heterotrimers composed of three genetically distinct polypeptide chains: alpha, beta and gamma.

Because laminin plays a very important role in cell biology, especially as a powerful cell adhesion molecule, it has recently gained recognition as an essential key to basement membrane care and skin aging.

Based on all of the biological functions of laminin in the skin, we developed a synthetic laminin-like peptide. The peptide is made of 15 amino acids, and will be identified simply as "the peptide" in this chapter. In our studies, we investigated the possibility of a general effect of an ECM peptide. We compared the effect of this peptide with other specific ECM peptides, and we also included inactive peptides as controls. These studies, too numerous to report here, enabled us to select this peptide because it showed particularly interesting effects.

This chapter presents studies concerning the selected peptide's stability and degradation, as well as its effect on cell adhesion, migration and synthesis of ECM molecules.

Materials and Methods

Stability studies: Analytical reverse-phase HPLC was carried out in order to determine the peptide's stability under different conditions, such as when it is exposed to water or to cell culture media, and at different time points.

Cell culture and H&E staining: Human fibroblasts and epidermal HaCaT and A431 cells were cultured in 8-well culture chambers and treated with 10^{-6} M of the peptide for 24-72 h, or left untreated as control cells. (In these studies we used a weak concentration of 10^{-6} M to show that the peptide is very active at even such small amounts, which is expected from peptides and not, in general, from the addition of ECM proteins.) The cells were then submitted to standard hematoxylin and eosin (H&E) staining for morphological studies.

MTT and protein dosage assay: Cultured human fibroblasts, HaCaT and A431 cells were treated or not treated with 10^{-6} M of the peptide for 24 h, and then submitted to the viability test of MTT colorimetric assay,[8] or to standard BCA kit protein dosage assay.

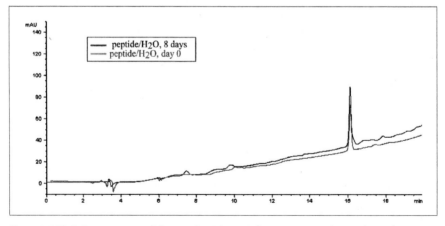

Figure 1. HPLC chromatogram of the new laminin peptide in water at 8 days and at 0 days

Cell adhesion assays: To study cell adhesion,[9] HaCaT cells were grown in microplates in the presence or absence of 10^{-6} M of the peptide. The percentage of cells that adhered to the microplate was determined using MTT colorimetric assay.[8]

Cell migration assay: Cells from the three different human cell lines (HaCaT, A431 and fibroblasts) were cultured in dishes around an insert that kept a 2 cm diameter circle free from cells. At the confluence of cells around the insert, the insert was taken out and the peptide was added to the cultured cells. Then, cell migration into the inner circles was monitored for up to seven days.

Immunostaining studies: We performed direct immunofluorescence[10] (DIF) microscopy following standard methods, using rabbit polyclonal antibody for human collagen I, III and fibronectin. We also used mouse monoclonal antibody for human laminin V, integrin β 1, integrin α 6 and pan-cytokeratin.

Results and Discussion

Stability studies: Analytical reverse-phase HPLC showed that the peptide exhibited high long-term stability in water. Figure 1 shows the new peptide's stability after eight days of incubation in water.

We were also interested in investigating the stability of this peptide under different biological conditions, such as when exposed directly to human skin cells and added into their culture media.

Protease inhibitors were added or not added into epidermal A431 cell culture media, and the peptide was incubated with the cells for 0.5, 1, 2, 6 and 24 h. At these time points, HPLC of the peptide in these conditions was performed and the results are shown in Figures 2 and 3.

These studies confirmed the peptide's stability under different conditions, especially when administered to epidermal cells, which supports the idea of its application into the skin.

Cell morphological studies by H&E staining: When the cells were cultured in the presence of the peptide, they exhibited an excellent morphology, their adhesion was enhanced and the cells spread uniformly in the wells making a nice homogenous sheet of cells within 2-3 days of culture.

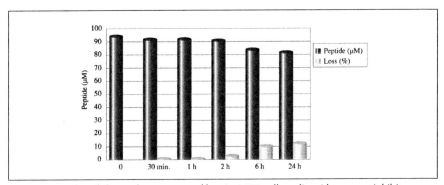

Figure 2. Peptide stability and percentage of loss in A431 cell media with enzyme inhibitors

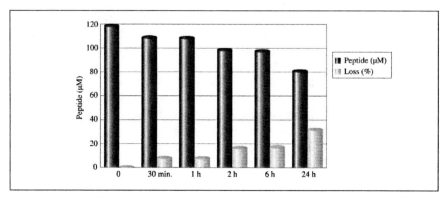

Figure 3. Peptide stability and percentage of loss in A431 cell media without enzyme inhibitors

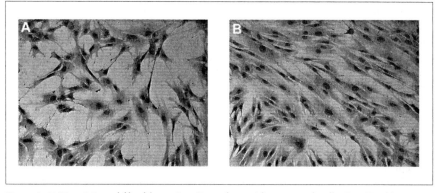

Figure 4. H&E staining of fibroblasts: A = Control, peptide-untreated cells; B = Peptide-treated cells

In Figure 4 we see the nice cell sheet in the peptide-treated cells compared to the control, peptide-untreated cells after three days of culture.

Moreover, MTT and protein dosage assays showed a very small (insignificant) and late increase in the number of cells when cultured with the peptide, which confirms that the cellular sheet aspect that we see is due more to an excellent adhesion followed by homogenous spreading of the cells, than to increased cell proliferation.

Cell adhesion assay: Because laminin is principally and directly involved in cell adhesion, after the results obtained by the observation of cultured cells following H&E staining, we were interested in confirming the peptide's effect on epidermal cell adhesion.

Cell adhesion assay using HaCaT cells showed that when the cells were put in culture in the presence of 10^{-6} M of peptide, cell adhesion onto the microplate was clearly enhanced, as shown in Figure 5. This result confirms laminin's role in enhancing cell adhesion[11,12] and presents the action of the peptide as being similar

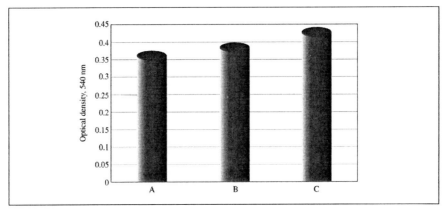

Figure 5. Adhesion assay of HaCaT: A = Control cells, adhesion onto plastic; B = Positive control, adhesion to an ECM peptide; C = Peptide-treated cells

to that of the physiological laminin molecule in the skin, which serves as a potent adhesion molecule for the cells.

Cell migration assay: It has been hypothesized that because laminin V promotes epithelial cell adhesion, it may promote the motility of keratinocytes[13,14] under different circumstances. Thus, in this study, we investigated whether the new peptide enhances epidermal cell migration after it modulates cell adhesion.

Migration of cultured cells from the three different cell lines was evaluated by monitoring the migration of the confluent cells into the inside of the inner circle, which was kept free from cells by an insert. The results showed that the presence of the peptide did not significantly increase cell migration into the free circle. Surrounding confluent cells migrated similarly in both conditions, with and without peptide treatment.

This result confirmed the peptide's primary role in improving cell adhesion (Figure 6) but not migration, which is of interest in case of the presence of pathological cells.

Immunostaining studies: The result of improved cell morphology after treatment with the peptide, and the improved adhesion and homogenous spreading of the cells in culture, suggested that after adhering so properly, the cells probably differentiate and function better. Therefore, immunofluorescence studies were performed on subconfluent fibroblasts, HaCaT and A431 cells, in order to evaluate the expression of different proteins synthesized by these cells, after the administration of the peptide.

These studies revealed that treatment of the cells with 10^{-6} M of peptide, for 24-36 h, induced, in the related cells, an interesting increased expression of cytokeratin and ECM molecules. These ECM molecules included collagen I and III, fibronectin and laminin V (Figures 7 and 8).

Among laminins, laminin V is a particularly key molecule.[15] Laminin V initiates hemidesmosome formation and provides stable attachment of the epidermis to the

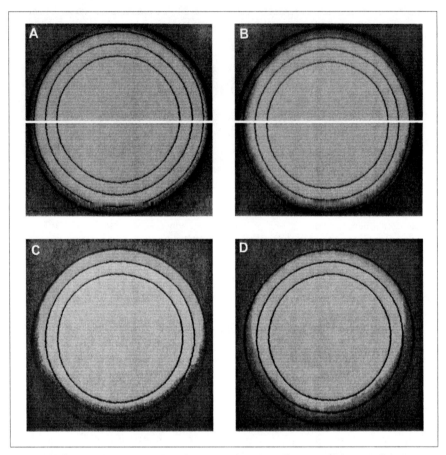

Figure 6. Cell migration assay: A = Peptide-untreated HaCaT cells; B = Peptide-treated HaCaT cells; C = Peptide-untreated A431 cells; D = Peptide-treated A431 cells

dermis. Laminin V also accelerates the assembly of basement membrane and may enhance the recovery of damaged skin.

Laminin V is one of the most important molecules of cell extracellular matrix, is an important ligand for β 1 and α 6 integrins, and its defects have been shown to be implicated in pathologies and some skin cancers.[16-18] Thus, we studied its expression in HaCaT and A431 cells treated with the peptide. Immunofluorescence studies showed that the peptide specifically stimulated laminin V synthesis in these cells (Figure 9).

In the skin, laminin V interacts with integrins (transmembrane receptor proteins) to anchor epidermal basal cells to underlying dermis.

Integrins are heterodimeric proteins made of alpha and beta subunits; they play critical roles in both cell-to-cell and cell-to-matrix adhesion. Their transmembrane location allows them to link the cell cytoskeleton to extra-cellular molecules, thereby firmly anchoring the cells to the matrix. They are involved in intercellular communication, cell differentiation, migration, proliferation and basement membrane assembly.

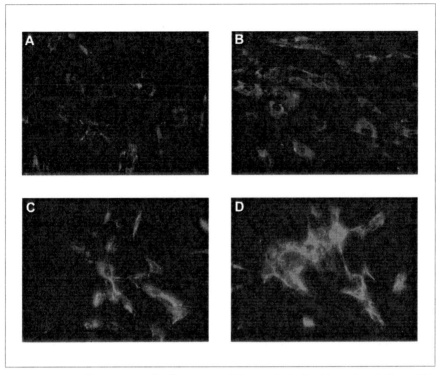

Figure 7. Immunofluorescence of collagen III and fibronectin in fibroblasts: A = Collagen III in peptide-untreated fibroblasts; B = Collagen III in peptide-treated fibroblasts; C = Fibronectin in peptide-untreated fibroblasts; D = Fibronectin in peptide-treated fibroblasts

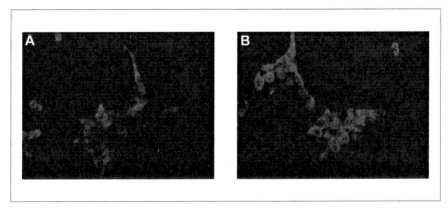

Figure 8. Immunofluorescence of cytokeratin in A431 cells: A = Cytokeratin in peptide-untreated A431 cells; B = Cytokeratin in peptide-treated A431 cells

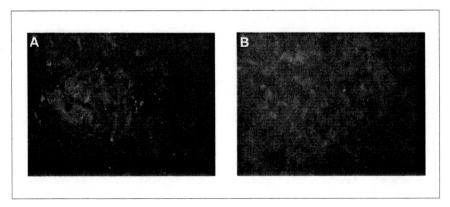

Figure 9. Laminin V immunofluorescence after HaCaT cells extraction with 2 M urea: A = Control, peptide-untreated; B = Peptide-treated

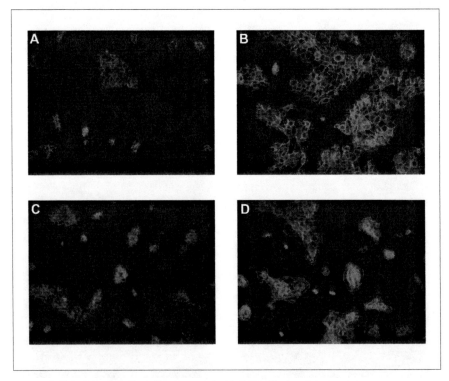

Figure 10. β 1 and α 6 immunofluorescence in HaCaT cells: A = β 1 expression in peptide-untreated HaCaT cells; B = β 1 expression in peptide-treated HaCaT cells; C = α 6 expression in peptide-untreated HaCaT cells; D = α 6 expression in peptide-treated HaCaT cells

Because β 1 integrins[19] include receptors implicated in the adhesion of keratinocytes to ECM components such as collagen, laminin and fibronectin, and because we found that laminin V expression is increased by peptide treatment, we studied β1 integrin expression in subconfluent HaCaT keratinocytes cultured in the presence of 10^{-6} M of peptide. These studies showed that treatment of the cells with the peptide enhanced β 1 expression in the cells (Figure 10).

Moreover, alpha 6 integrin is known to play a key role together with laminin V in the healthy adhesion and maintenance of the basement membrane. Because we found laminin V synthesis increased after the treatment of cells with the peptide, we examined alpha 6 expression in epidermal cultured cells (Figure 10), and found that alpha 6 expression increased with laminin V level. These results show that administration of the peptide enhances the "adhesion" property of epidermal cells.

Conclusion

Like biological laminin, this new laminin peptide demonstrates very interesting properties and high specificity shown by its obvious efficacy at such a low concentration (10^{-6} M). It promotes the synthesis of extracellular matrix and some basement membrane components, and improves cell adhesion by enhancing the different structures involved. These are essential features in preventing skin aging.

Our findings also suggest a very likely role for the peptide in improving tissue regeneration.

Taken together, these studies show that this laminin peptide is of great interest for future innovative skin care cosmetics.

—E. Bauza, C. Dal Farra, F. Portolan and N. Domloge, *Vincience Research Center, Sophia Antipolis, France*

References

1. M Aumailley and P Rousselle, Laminins of the dermo-epidermal junction, *Matrix Biol* 18(1) 19-28 (1999)
2. M Aumailley and N Smyth, The role of laminins in basement membrane function, *J Anat* 193 (Pt 1) 1-2 (1998)
3. KM Malinda and HK Kleinman, The laminins, *Int J Biochem Cell Biol* 28(9) 957-959 (1996)
4. M Aumailley and T Krieg, Laminins: a family of diverse multifunctional molecules of basement membranes, *J Invest Dermatol* 106(2) 209-214 (1996)
5. KA McGowan and MP Marinkovich, Laminins and human disease, *Microsc Res Tech* 51(3) 262-279 (2000)
6. B Le Varlet, C Chaudagne, A Saunois, P Barre, C Sauvage, B Berthouloux, A Meybeck, M Dumas and F Bonte, Age-related functional and structural changes in human dermo-epidermal junction components, *J Investig Dermatol Symp Proc* 3(2) 172-179 (1998)
7. F Vazquez, S Palacios, N Aleman and F Guerrero, Changes of the basement membrane and type IV collagen in human skin during aging, *Maturitas* 25(3) 209-215 (1996)
8. E Borenfreund, H Habich and N Martin-Alguacil, Comparison of two in vitro cytotoxicity assays: the neutral red and tetrazolium MTT tests, *Toxic in vitro* 2(1) 1-6 (1988)
9. MR Mariappan, EA Alas, JG Williams and MD Prager, Chitosan and chitosan sulfate have opposing effects on collagen-fibroblast interactions, *Wound Repair Regen* 7(5) 400-406 (1999)

10. N Domloge-Hultsch, P Bisalbutra, R Gammon and KB Yancey, Direct immunofluorescence microscopy of 1 mol/L sodium chloride-treated patient skin, *J Am Acad Dermatol* 24(6) 946-951 (1991)
11. KM Malinda and HK Kleinman, The laminins, *Int J Biochem Cell Biol* 28(9) 957-959 (1996)
12. AM Belkin and MA Stepp, Integrins as receptors for laminins, *Microsc Res Tech* 51 280-301 (2000)
13. M Shang, N Koshikawa, S Schenk and V Quaranta, The LG3 module of laminin-5 harbors a binding site for integrin alpha3 beta1 that promotes cell adhesion, spreading, and migration, *J Biol Chem* 276(35) 33045-33053 (2001)
14. F Decline and P Rousselle, Keratinocyte migration requires alpha2 beta1 integrin-mediated interaction with the laminin 5 gamma2 chain, *J Cell Sci* 114(Pt 4) 811-823 (2001)
15. T Nishiyama, S Amano, M Tsunenaga, K Kadoya, A Takeda, E Adachi and RE Burgeson, The importance of laminin 5 in the dermal-epidermal basement membrane, *J Dermatol Sci* 24(suppl 1) S51-S59 (2000)
16. N Domloge-Hultsch, JG Anhalt, WR Gammon, Z Lazarova, R Briggaman, M Welch, DA Jabs, C Huff and KB Yancey, Antiepiligrin cicatricial pemphigoid, *Arch Dermatol* 130 1521-1529 (1994)
17. N Domloge-Hultsch, WR Gammon, RA Briggaman, SG Gil, WG Carter and KB Yancey, Epiligrin, the major human keratinocyte integrin ligand, is a target in both an acquired autoimmune and an inherited subepidermal blistering skin disease, *J Clin Invest* 901628-901633 (1992)
18. G Giannelli and S Antonaci, Biological and clinical relevance of laminin-5 in cancer, *Clin Exp Metastasis* 18(6) 439-443 (2000)
19. CM Di Persio, KM Hodivala-Dlike, R Jaenish, JA Kreidberg and RO Hynes, Alpha 3 beta 1 integrin is required for normal development of the epidermal basement membrane, *J Cell Biol* 137(3) 729-742 (1997)

Copperceuticals and the Skin

Keywords: copper peptides, wound healing

Newer copper peptides may be useful in higher potency skin renewal products

> During the past 15 years, the role of growth factors—both in skin disease (e.g., wound healing and psoriasis) and in the development of skin care products—has continued to evolve at lightning speed. Factors such as vascular endothelial growth factor (VEGF), transforming growth factor beta (TFG-beta), keratinocyte growth factor (KGF) and platelet-derived growth factor (PDGF) play critical intrinsic roles in maintaining healthy and young-looking skin.
>   Biotechnology has provided recombinant bacteria that can produce large quantities of these purified factors for formulating into products for skin care and hair care. In standalone and multiregimen skin care product lines, the continuing consumer desire for antiaging products suggests an important role for growth factors, including mineral actives.
>   In this chapter, Dr. Loren Pickart of Skin Biology Inc. discusses the unique role for copper peptides as a key growth (i.e., "tissue remodeling") factor in future products for skin care and hair care.
>
> – Larry Rheins

Reversing the effects of aging on human skin was a primary goal of ancient alchemists and their successors, the modern cosmetic chemists. During human aging, skin becomes thinner and accumulates various blemishes, lesions and imperfections. The structural proteins are progressively damaged, causing collagen and elastin to lose their resiliency. The skin's water-holding proteins and sugars start to diminish, the dermis and epidermis become thin, the capillary network becomes disorganized and the subcutaneous fat cells diminish in number. These effects are further intensified by decades of exposure to ultraviolet rays, irritants, allergens and various environmental toxins. The result is dry, wrinkled, inelastic skin populated by unsightly lesions.

Restoration to a biologically younger skin morphology requires two linked processes: the removal of damaged proteins and aberrant skin lesions and their replacement with normal, blemish-free skin. This process is similar to the remodeling phase of wound healing in which scar tissue is removed over several years to slowly restore the skin to its original state. In young children this process functions efficiently and skin damage is rapidly removed. But in adults this process slows drastically, and various skin lesions may persist for years or decades.

The various skin renewal methods that have been developed produce a limited type of skin restoration, but all have drawbacks. Retinoic acid slowly remodels skin but at the price of chronic irritation and redness. Collagen-inducing peptides, melatonin and vitamin C increase skin collagen, but skin also needs to increase its microcirculation and rebuild elastin and water-holding proteoglycans. Collagen-stimulatory molecules, such a TGF-β-1 and fibronectin-binding peptides, were tested in wound-healing studies but produced unacceptable skin thickening and scarring. Controlled skin damage (i.e., peels, dermabrasion, lasers) works well only if there is a vigorous post-therapy regenerative response by the damaged skin.

The good news today is that certain types of copper peptides possess all the necessary biochemical actions that can, in a morphological sense, restore skin to a younger state without causing skin irritation. Such types of copper peptides are increasingly used in cosmetic skin care products such as Neutrogena's Active Copper line. These copper peptides also are used to improve post-treatment skin recovery after dermatological skin renewal procedures, such as chemical peels, dermabrasion and laser resurfacing. Unlike most purported skin improvement therapies, the skin regenerative actions of such types of copper peptides are documented by numerous well-controlled, published studies from leading laboratories and universities.

GHK-Copper for Skin Renewal

A human copper peptide complex forms the basis for these developments. The complex is glycyl-l-histidyl-l-lysine:copper(II), or GHK-Cu. The tripeptide, which this researcher discovered while searching for methods to reverse human aging, is generated by proteolysis after tissue injury or skin turnover. Its high affinity for copper(II) allows it to obtain copper from carrier molecules such as albumin and form GHK-Cu. The complex is a normal constituent of human plasma, saliva and urine.[1-5]

When injected into skin or applied to the skin's surface, GHK-Cu activates the processes that remove scarred or damaged tissue and deposit new tissue. Researchers at the Universite de Reims have published several reports demonstrating the role of GHK-Cu in wound healing.

- GHK-Cu acts as an inducer of the second phase of healing when skin remodeling processes remove scars and tissue debris while rebuilding healthy skin.[6-8]
- GHK-Cu concomitantly stimulates the degradation of existing collagen and synthesis of new collagen.[9-11]
- At the molecular level, GHK-Cu aids the rebuilding of new skin by increasing angiogenesis, the production of m-RNA for collagen, elastin, proteoglycans, glycosaminoglycans and decorin, while simultaneously stimulating the m-RNA production of, and synthesis of, certain metalloproteinases and anti-proteases that clear damaged protein and remove scars.[10-12]
- In addition, GHK-Cu suppresses secretion of scar-forming TGF-β-1 by fibroblasts.[13]

- GHK-Cu also acts indirectly as a chemoattractant for cells that stimulate repair, such as macrophages and mast cells, which release protein growth factor proteins that stimulate tissue repair.[6]

GHK-Cu also appears to function in humans as a circulating non-steroidal anti-inflammatory.[14] After episodes of tissue damage, ferric ion is released from ferritin and catalyzes damaging tissue oxidations. GHK-Cu counters this action by blocking ferritin channels and stopping the release of oxidizing iron ions.[15] It also blocks the oxidation of low density lipoproteins by loosely bound copper.[16] Interleukin-1β is also released after tissue injury producing cellular damage. At hormonal levels, GHK-Cu prevents damage to pancreatic cells by interleukin-1.[17]

Dermatological Actions of Copper Peptides

In 1985, a company called ProCyte was formed to develop products based on GHK-Cu. We found that the application of GHK-Cu-containing creams to the human skin increased the thickness of the epidermis and dermis, increased skin elasticity, reduced wrinkles and resulted in the removal of skin imperfections such as blotchiness and sun damage marks while producing a significant increase in subcutaneous fat cells.

Recently, more extensive and statistically significant human studies solidified these observations. Abulghani et al. reported[18] GHK-Cu was more effective in stimulating new collagen development than vitamin C, retinoic acid or melatonin. Appa et al. reported[19] that in eight weeks, a GHK-containing liquid foundation improved epidermal thickness, increased skin elasticity and improved skin appearance. Leyden et al. found[20] such creams to reduce visible signs of photodamage and increased skin density in eight weeks on facial skin. A second placebo-controlled study[21] (71 females, 12 weeks) by this group found that GHK-Cu-containing face creams reduced wrinkles and fine lines while increasing skin elasticity, density and thickness. A third placebo-controlled study[22] (41 females, 12 weeks) found that a GHK-Cu-containing eye cream reduced wrinkles and fine lines and improved eye appearance.

However, despite these promising properties, GHK-Cu failed in clinical trials conducted by the U.S. Food and Drug Administration (FDA) on the healing of indolent wounds such as venous stasis and diabetic ulcers. The actions of GHK-Cu are limited by its fragility and tendency toward breakdown as well as its lack of adhesion to the skin surface.

Therefore, in 1994, Skin Biology Inc. was formed to develop an improved second generation of skin regenerative copper peptides with enhanced potency, breakdown resistance and high adherence to skin. We isolated a fraction of peptide fragments from soy protein that possessed the desired qualities when chelated to copper(II). Such soy peptides have a long history of safe use in cosmetic products. Howard Maibach and colleagues (University of California at San Francisco) tested these new copper peptides in four small, placebo-controlled human studies. They found that creams made from these new copper complexes produced significantly faster skin healing and reduced redness and inflammation after mild skin injuries induced by nickel allergy inflammation,[23] tape stripping,[24] 24-hour detergent irritation[25] and acetone burns (removal of skin lipids).[26]

Scarless healing: Copperceuticals like those studied by Maibach et al. are opening an approach to scarless or nearly scarless healing of surgical incisions. GHK-Cu markedly reduces scarring after surgery if injected (an aqueous solution of 1 mg/mL) down the incision line. However, the second-generation copper peptides appear to be superior for such uses.

In veterinary studies, the application of second-generation copper peptides in creams immediately after surgery produced rapid and scarless healing in dogs after spaying operations and in young horses after leg-straightening operations. This allowed the dogs to be returned to their owners in four days instead of the usual five, while the foals were returned in five days instead of seven. For such uses, the incisions should be left open to air or only lightly covered with gauze. Occlusive dressing and wet bandages nullify the positive effect.

Post-procedure treatments: Both GHK-Cu and the second generation copper peptides function well to improve the recovery of skin after procedures such as deep peels, laser resurfacing and dermabrasion. If the skin is treated with copper peptides within two hours after the procedure, post-treatment redness and inflammation are avoided without the use of corticosteroids. Skin rebuilding is hastened while scars and infections are greatly lessened.

We developed a water-based product, called CP Serum, which works especially well. This is often followed by either emu oil or squalane, which act as moisturizers and as penetrating agents that push more copper peptide into the skin.

Removal of skin lesions: The second generation copper peptides work well for the removal of skin lesions. Our research finds that most skin imperfections (hypertrophic scars, pitted scars, skin tags, moles, actinic keratoses) can be removed or greatly reduced by repeated daily use of hydroxy acids to loosen and dissolve the lesions followed by strong copper peptides to aid skin regeneration.

Hydroxy acids such as 2% salicylic acid (pH=3.2) or 14% glycolic acid (pH=3.9) are effective with copper peptides, but obtaining a cosmetically satisfying result may require several months of applications.

The use of stronger hydroxy acids followed by the application of copper peptides greatly speeds the removal of lesions and scars, but can be very irritating if not expertly applied. For example, to remove moles, 70% glycolic acid is applied for six minutes, washed off, then the strong copper peptides are applied to the moles. Two to three days of this procedure often removes moles permanently.

Safety of Copper Peptides

Numerous safety tests of skin regenerative copper peptides have failed to find any toxicity problem. Extremely small amounts penetrate the skin and no rise in blood copper has ever been found in animals or humans treated with copper peptides.

Further reassurance as to the safety of copperceuticals is that in nutritional studies, increased supplementation with copper has been reported[27] to:

- Increase DHEA levels
- Raise brain enkephalin
- Reduce carcinogenesis and cancer growth

- Reduce protein glycation and deleterious peroxidation of fats
- Reduce brain developmental defects in offspring
- Increase antioxidant defenses by activation of superoxide dismutase

Formulating with Copper Peptides

When formulating copper peptide products, great care must be taken to minimize interactions with the ionic copper. Other ingredients of creams, lotions and solutions may interact with the ionic copper and neutralize the positive copper peptide actions and, in some cases, generate copper complexes that inhibit cell replication.

All products should be carefully tested for their effects on skin repair. Not all copper peptide complexes are skin regenerative; some may even inhibit skin repair. The existing products based on GHK-Cu or the newer complexes are supported by credible, published evidence of efficacy.[27]

—**Loren Pickart, PhD,** *Skin Biology, Bellevue, Washington USA*

References

1. Pickart, PhD thesis, Univ of California, San Francisco (1973)
2. Pickart and Thaler, *Nature New Biol* 243 85 (1973)
3. Pickart, In Vitro 17 459 (1981)
4. Pickart, Freedman, Loker, Peisach, Perkins, Stenkamp and Weinstein, *Nature* 288 (1980)
5. Pickart and Lovejoy, *Methods Enzymol* 147 314 (1987)
6. Maquart, Gillery, Monboisse, Pickart, Laurent and Borel, Ann NY Acad Sci 580 573 (1990)
7. Sage and Vernon, *J Hypertens Suppl* 12 S145 (1994)
8. Maquart, Simeon, Pasco and Monboisse, *J Soc Biol* 193 423 (1999)
9. Oddos, Jumeau-Lafond and Ries, *Amer Acad Derm* Abstract P72 (Feb 2002)
10. Simeon, Monier, Emonard, Gillery; Birembaut, Hornebeck and Maquart, *J Invest Dermatol* 112 957 (1999)
11. Simeon, Emonard, Hornebeck and Maquart, *Life Sci* 67 2257 (2000)
12. Simeon, Wegrowski, Bontemps and Maquart, *J Invest Dermatol* 115 962 (2000)
13. McCormack, Nowak and Koch, *Arch Facial Plast Surg* 3 28 (2001)
14. Pickart, *Speciality Chemicals* 29 (Oct 2002)
15. DeSilva, Pickart and Aust, *Adv Exp Med Biol* 264 79 (1990)
16. Thomas, Biochem Biophys Acta 1128 50 (1992)
17. Vinci, Caltabiano, Santoro, Rabuazzo, Buscema, Rizzarelli, Vigneri and Purrello, *Diabetologia* 38 39-45 (1995)
18. Abulghani, Shirin, Morales-Tapia, Sherr, Solodkina, Robertson and Gottlieb, *J Invest Dermatol* 110 686 (1998)
19. Appa, Barkovic, Finkey and Stephens, Abstract P66, Amer Acad Derm Meeting (Feb 2002)
20. Leyden, Grove, Barkovic and Appa, Abstract P67, Amer Acad Derm Meeting (Feb 2002)
21. Leyden, Stephens, Finkey and Barkovic, Abstract P68, Amer Acad Derm Meeting (Feb 2002)
22. Leyden, Stephens, Finkey and Barkovic, Abstract P69, Amer Acad Derm Meeting (Feb 2002)
23. Zhai, Chang, Singh and Maibach, *Contact Derm* 40 205 (1999)
24. Zhai, Poblete and Maibach, *Int J Dermat* 37 386 (1998)
25. Zhai, Leow and Maibach, *Skin Res Tech* 4 24 (1998)
26. Zhai, Leow and Maibach, *Clin Exp Derm* 23 11 (1998)

27. More extended references to copper peptide actions on tissue regeneration are posted at www.skinbiology.com/copperpeptideregeneration.html and www.skinbiology.com/copperhealth.html.
28. US Pat 5,348,943, Loren Pickart, assigned to ProCyte Corporation (Sep 20, 1994 as a "continuation in part")

Targeting the Cutaneous Nervous Network

Keywords: neuroprotection, nerve growth factor, UV

A synthetic agent's neurotrophic and antioxidant properties protect the skin from environmental stress and neurodegenerescence

The coordinating role of the brain and its network of nerves, previously assigned by Aristotle to the heart and the vessels, wasn't admitted until the 18th century. The "leadership" of the brain was definitely established by Gall and Broca during the 19th century, but during the 20th century another organ that doesn't "think" but "senses" has emerged as a "brilliant second," leading to the concept of neuronal skin.

An important neuroendocrine activity was discovered in this organ, with the local production of many neurotransmitters.[1] While recent progress in neurobiology has generated new pharmaceutical approaches for the treatment of neuropathies, neuromuscular diseases and neuropsychiatric disorders, we can now consider the opportunities for cosmetic products.

The Skin's Nervous System

Neurons are fragile cells with low regenerative potential. Therefore neuroprotection qualifies as a special application for skin care. Benefits of neuroprotection are not limited to improved or preserved sensory functions. Maintenance of the skin's neuroendocrine activity, which is involved in many biological processes, is needed for global preservation of skin homeostasis. Effective protection requires specific approaches that take into account the particular survival mechanisms of neurons.

Consistent with its sensory function, skin is highly innervated. Many sensitive fibers reach the most superficial living layers of the epidermis. Therefore, cosmetic ingredients can easily access this exposed nervous system (no blood/brain barrier here). They may have psycho sensorial effects by activating superficial nervous endings, but the generation of pleasant feelings is difficult to achieve due to extreme variability in individual perception.

Alternatively, the skin's anatomical considerations suggest that its nervous system is highly exposed to environmental challenges. It is therefore important to protect this peripheral nervous system. This "neuroprotection" is even more important considering the low regenerative potential of the nervous tissue, responsible for partial recovery of sensitive functions after injury, for instance. Neuroprotection also applies to anti-aging, fighting against the well-known neuron loss associated with age progression.

Preservation of the anatomical link between the central nervous system and the cutaneous peripheral sensitive neurons enables preservation of good sensitivity. Besides, preservation of the link with the cutaneous neurovegetative neurons maintains "thermoregulation systems" and sebaceous secretion.

In addition, recent studies on the relationships between the skin and the nervous system open many other opportunities for cosmetic applications. Cutaneous nerve endings release various neuromediators capable of activating specific cutaneous targets, such as keratinocytes, Langerhans cells, melanocytes, endothelial cells and fibroblasts. On the other hand, metabolic functions of cutaneous neurons are regulated by mediators produced by the cutaneous cells.

This two-way cell communication participates in many biological processes such as inflammation,[2] immune response,[3] wound healing,[4] pigmentation[5] and hair growth.[6] Modulation of these cross talks could lead to valuable cosmetic responses. However, the skin's neuroendocrine activity is complex, involving a large number of mediators (such as cytokines, growth factors, kinins and neuropeptides), with multiple activities.

In this regard, the maintenance of the cutaneous neuroendocrine system (homeostasis preservation) appears as an accessible and valuable approach, in good agreement with the definition of cosmetic activity. Nevertheless, the particularities of the nervous cells require some specific approaches for effective neuroprotection.

Neuronal Loss and Survival in the Skin

Mechanisms responsible for neuronal loss in the skin: Peripheral and central neuronal death is the consequence, as for other cell types, of extrinsic factors (stress and diseases) or intrinsic factors (aging). Due to the low regenerative potential of the nervous tissue, neuronal death leads to an irreversible neuronal loss that is mostly counterbalanced by the formation of new extensions from surviving neurons, creating new connections (synapses) to rebuild the network.

Oxidative stress is a major challenge for cutaneous neurons because it is generated by several environmental sources (UV radiation, pollution) and by some neurotransmitters' auto-oxidation (catecholamines). In addition, it is one consequence of aging.[7] Neurons happen to be very sensitive to oxidative stress due to relatively low levels of antioxidant enzymes and high dependence on mitochondrial respiration.[8] Modifications of the cellular environment (such as glutamate accumulation and a reduced supply of growth factor), generated by acute cellular injury, aging and diseases, can also induce neuronal death.

However, a particular feature of neurons to be considered for the design of specific protective agents is the particular occurrence of apoptosis, the programmed cell death process. Apoptosis is the main issue in many injuries, and neuronal death is apoptotic in several neurodegenerative diseases and (at least in part) in aging.

Opposing the inappropriate triggering of the apoptotic cascade is therefore necessary for the preservation of nervous tissue in the adult and the aged. This has already originated a new class of anti-apoptotic therapeutic agents, which are under current clinical evaluation.[9]

Neurotrophic factors and properties: A number of growth factors termed "neurotrophic factors" are now considered key players in neuronal survival. These factors, addressed to neurons, induce a number of positive biological effects, such as proliferations, differentiation and survival. They govern neuron death by initiating or repressing the apoptotic cascade.[10] Furthermore, neurotrophic factors are endowed with some cytostimulating activity, inducing neurite outgrowth and arborization, promoting neurotransmitter synthesis and improving synaptic function. Neurotrophins, a family of structurally-related small proteins including nerve growth factor (NGF), neurotrophin-3 (NT-3), NT-4 and brain-derived neurotrophic factor (BDNF), and other neurotrophic factors such as basic fibroblast growth factor (b-FGF), are expected to be able to oppose neurodegenerescence, and thus hold promise for therapeutic uses. Animal testings have already demonstrated the memory-restoring properties of NGF.[11]

Nerve growth factor in the skin: Neurotrophic factors – chiefly NGF – regulate the sensory innervation of the skin.[12] Experiments on transgenic mice overexpressing NGF have shown that this growth factor determines the number of cutaneous sensory neurons, thus modulating the innervation density and thereby the sensitivity of the skin to several stimuli.[13] Within the epidermis, keratinocytes of the basal layer produce NGFs that probably constitute a vital "local" supply for neuron survival and function in the adult.[14] Decreased endocrine activity of cutaneous cells may be one mechanism of sensory function decline during aging.

In addition, NGF has many effects on non-neural cells and is involved in the regulation of local skin metabolic pathways.[15] It participates in the inflammation process by regulating the synthesis of the pro-inflammatory neuropeptides released by sensory terminals. It also has direct effects on inflammatory cells. NGF-induced keratinocyte proliferation probably plays a role during wound healing. It has been proposed that NGF is a major "survival factor" in the skin, able to protect keratinocytes and melanocytes from UV-induced apoptosis.[16,17]

A Synthetic Neuroprotective Agent

Neurotrophic factors such as NGF are not suitable for skin care due to their polypeptidic nature and possible side-effects.[18] Small synthetic peptides, agonistic analogues of NGF developed for therapeutical purposes, have very low in vivo efficacy when topically applied. The skin needs tailored compounds with suitable bioavailability.

Researchers at Exsymol have designed an original neuroprotective agent[a] endowed with some neurotrophic and antioxidant properties (Figure 1). Its INCI name is glutamylamidoethyl indole. This synthetic neuroprotective agent, which we will call SNA, has some structural characteristics that give it good potential for cutaneous innervation. It is a small amphiphilic molecule resistant to enzymatic deactivation by proteolytic enzymes that can readily penetrate the epidermis. In addition, it has structural analogies with neuromediators expressed in the skin. These structural analogies enable it to target some specific receptors on the neuron cell membrane.

Properties of SNA

Antioxidant properties: Free radical scavenging properties of SNA were investigated by measuring its hydroxyl radical (OH°) scavenging rate constant, according to the procedure of Halliwell.[19] As anticipated, the presence of an unsaturated

---
[a] Commercialized by Exsymol under the name of Glistin, a registered trade name.

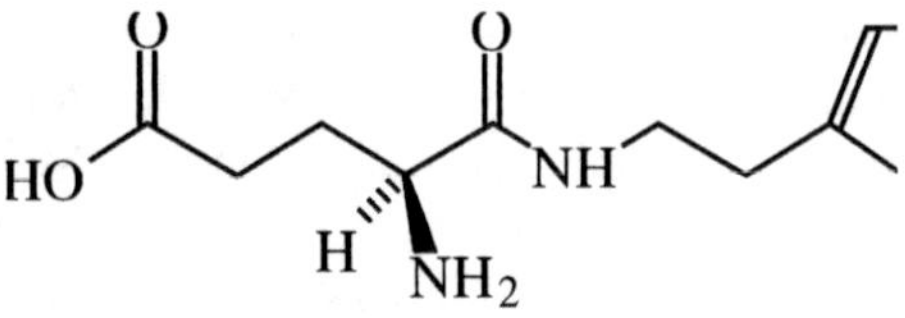

Figure 1. Chemical structure of glutamylamidoethyl indole, a neuroprotective agent

heterocycle (the indole ring) confers a strong antioxidant effect to the molecule, compared (Table 1) to the antioxidant effects of mannitol, polyphenols[b] or indolethylamine. This antioxidant effect was further demonstrated with an experimental model generating a set of ROS reported to be neurotoxic (Table 2).

Neuroprotection against UVB: The neuroprotective properties of SNA were demonstrated using the NGF-responsive murine cell line PC12, a well-accepted model of peripheral neurons.

In this assay, PC12 cells are differentiated by addition of NGF until optimum differentiation is reached (Figure 2), and then exposed to UVB (285 ± 5 nm; 50 mJ/cm^2) in the presence or the absence of a protective agent (SNA or vitamin E). Cell protection was evaluated by the measurement of lactate dehydrogenase (LDH) released in the culture medium. In this experimental design SNA shows very effective neuroprotection in the millimolar-micromolar range (Table 3). The reference antioxidant vitamin E provides moderate protection in this model.

The morphological study shows that UVB induces a reversion of the neuronal morphology together with cell death (apoptosis, necrosis). In line with previous reports, cell protection correlates well with retained differentiated morphology.[20] Comparison of Table 2 and Table 3 suggests that neuroprotection obtained with micromolar concentrations of SNA mostly results from a mechanism other than free radical scavenging.

[b] Silymarin is a mixture of polyphenols marketed under the name "Silymarin group" (Catalog reference 25,492-4) by Sigma-Aldrich, Milwaukee, Wisconsin USA. Silymarin is registered as a hepatoprotective drug and is marketed in a pharmaceutical product whose trade name is Légalon.

Table 1. Free radical scavenging rate constant (KsOH°) for SNA, compared with the rate for reference antioxidants mannitol, indolethylamine, and a mixture of polyphenols

Tested substance	Scavenging rate constant (10^9 M^{-1} s^{-1})
Ref. antioxidants mannitol	1.9 ± 0.1
Polyphenols (Silymarin)	10.0
Indolethylamine	23.5 ± 3.4
SNA	32.1 ± 1.2

Table 2. Dose-related protective effect of SNA against ROS generated by the enzymatic system xanthine oxidase / hypoxanthine (+EDTA). Control contained no SNA.

SNA concentration (mM)	Protection relative to the control (%)
0.125	4
0.25	10
1	32
2	52

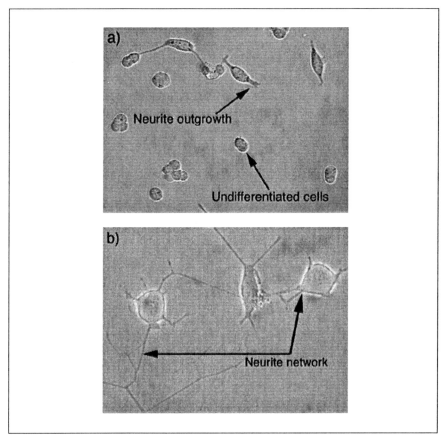

Figure 2. NGF-promoted differentiation of PC12 cells:
a) after two days of treatment (a few cells initiate differentiation, with cell flattening and neurite sprouting), and
b) after five days of treatment (fully differentiated state with extensive network of connected neurites)

Table 3. Dose-dependent neuroprotective effect of SNA on neuronal PC12 cells irradiated by UVB (50 mJ/cm^2), compared to the neuroprotective effect of vitamin E. Protection percentage is the ratio between LDH released in the presence of SNA and LDH released in the absence of SNA

Neuroprotective ingredient	Dose (mM)	Protection (%)
SNA	1.72	69
SNA	0.86	61
SNA	0.43	48
SNA	0.1	39
SNA	0.05	26
Vitamin E	2	34

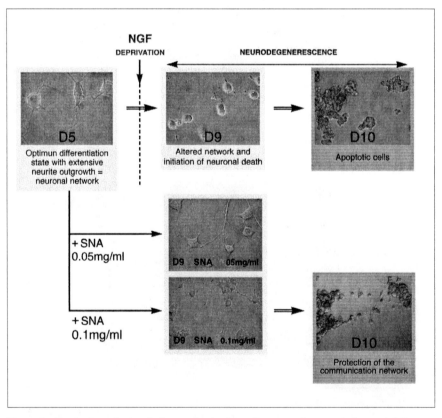

Figure 3. Monitoring of growth factor deprivation effects on the morphology of fully differentiated PC12 cells (at day 5). Comparison between treated (0.1 mg/mL or 0.05 mg/mL) and nontreated cells after four days (day 9) and five days (day 10) of deprivation

Prevention of neurodegeneration: The neurotrophic effect of SNA was demonstrated when differentiated PC12 cells were submitted to growth factor deprivation. In this experimental design, fully differentiated cells (at day 5) experience a prolonged complete deprivation of serum and NGF. Withdrawal of growth factors, essential for maintenance of neurons of the peripheral system, induces a progressive regression of the neurite network and de-differentiation of PC12 cells (Figure 3; untreated at day 9). It is followed by the initiation of the apoptotic process and extensive cell death (at day 10). Such progression can be likened, to a certain extent, to the neurodegenerative process associated with aging.

Neurotrophic/neuroprotective effects of SNA were monitored qualitatively (cell morphology study by optical microscopy) and quantitatively (LDH release assay).

Qualitative evaluation of SNA properties (Figure 3) shows that sub-millimolar concentrations of SNA counteracts deprivation-triggered neurodegenerescence; both concentrations tested at day 9 showed retention of neuronal morphology and preservation of the network. Some protection is still observed at the highest tested concentration at day 10 (five days of deprivation).

In control experiments where growth factor-deprived PC12 cells were treated with indolethylamine (proved to have an antioxidant activity in Table 1), no neuroprotective/neurotrophic effect was observed (data not shown).

By use of the LDH release assay, these morphological data were correlated with cell viability improvement and compared to nontreated PC12 cells (Table 4). Both tested treatment concentrations are neuroprotective after four or five days of deprivation. This test reveals a striking change in cell viability between day 9 and day 10.

Opposition to neurodegenerescence-driven apoptosis: The rescue of PC12 cells from apoptosis by SNA was studied in the growth factor deprivation model at day 9 and day 10, when most cells are committing programmed suicide, as revealed by the morphological study. Apoptosis activity was monitored by the dosage of the pro-apoptotic Bax protein, an early effector of the apoptotic program.[21] Time course of Bax expression during the deprivation process was monitored by the reverse transcription-polymerase chain reaction (RT-PCR) technique and biophotometry (Figure 4). Control experiments show that NGF withdrawal results in an important increase of Bax mRNA expression. Addition of SNA just before

Table 4. LDH release assay on growth factor-deprived PC12 cells

Treatment ingredient	LDH Activity (nmol NADH min^{-1}ml^{-1})		
	Before deprivation	After four days deprivation (Day 9)	After five days deprivation (Day 10)
Control PC12 cells (nontreated)	2.45	4.1	8.7
PC12 + SNA (0.05 mg/ml)	not determined	3.45	6.95
PC12 + SNA (0.1 mg/ml)	not determined	2.55	5.9

deprivation induces a strong "rescue signal" that markedly reduces the expression of the pro-apoptotic Bax protein.

Stimulation of neuroprotective endocrine activity: Further understanding of the mechanism of action of SNA was provided by a neurochemical study.

PC12 cells exposed to growth factor withdrawal produce interleukin-6 (IL-6), which was reported to correspond to an autocrine protection against the effects of serum deprivation[22] and oxidative stress.[23] IL-6 behaves as a "cooperative" neurotrophic factor believed to function primarily as an "SOS signal."

It was found that SNA induces a strong overexpression of IL-6 by neuronal cells (Figure 5). This may account in part for the neuroprotective properties of SNA.

As a control experiment, human reconstructed epidermis was exposed to SNA, and the production of the inflammatory mediator interleukin-1 (IL-1) was monitored by a standard ELISA assay. No increase of interleukin-1 or interleukin-8 was detected upon addition of millimolar-micromolar concentrations of SNA (data not shown). These data

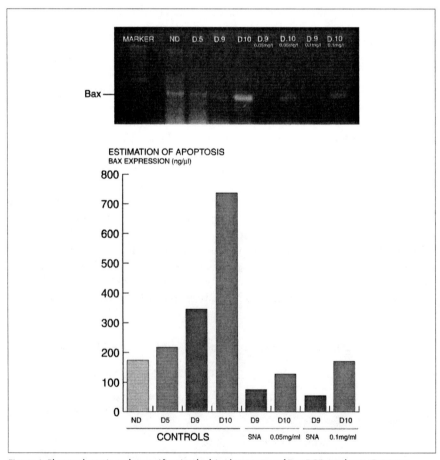

Figure 4. Electrophoresis and quantification by biophotometry of Bax PCR products. Comparative expression by growth factor-deprived nontreated PC12 cells (control) and cells treated with SNA at 0.05 mg/mL or 0.1 mg/mL

are in line with a clinical study showing the absence of cutaneous irritation, sensitization, phototoxicity and photoallergy after application of SNA on healthy volunteers.

Conclusions

A newly designed active ingredient is capable of protecting sensory neurons from environmental stress and neurodegenerescence, thanks to its neurotrophic and antioxidant properties.

Prevention of neuronal loss will preserve a number of physiological responses under the control of the nervous system (such as sebum secretion and sweat production). This active ingredient may also oppose the age-related decline of the sensory function.

Other benefits are expected from neuroprotection due to the participation of neuromediators in hair growth, skin pigmentation or immune response, and more generally in the proliferative activities within the skin. Such properties are under current evaluation.

—Jean-François Nicolaÿ and Isabelle Imbert, *Exsymol S.A.M., Monaco, Principality of Monaco*

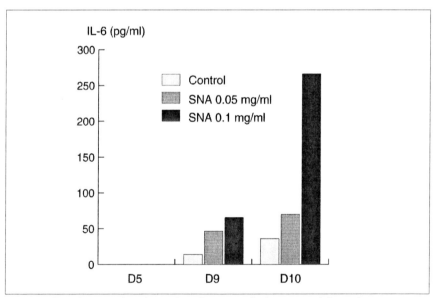

Figure 5. Effect of SNA on the release of IL-6 by PC12 cells during growth factor deprivation. Dosage by the ELISA technique. The control is non-treated cells.

References

1. A Slominsky and J Wortsman, Neuroendocrinology of the skin, *Endocrine Review* 21 457-487 (2000)
2. R Rossi and O Johansson, Cutaneous innervation and the role of neuronal peptides in cutaneous inflammation: a mini-review, *European J Dermatol* 8 299-306 (1998)

3. L Misery, Skin, immunity and the nervous system, *Brit J Dermatol* 137(6) 843-850 (1997)
4. PG Smith and M Liu, Impaired cutaneous wound healing after sensory denervation in developing rats: effects on cell proliferation and apoptosis, *Cellular and Tissue Research* 307(3) 281-291 (2002)
5. M Hara, M Toyoda, M Yaar, J Bhavan, EM Avila, IR Penner and BA Gilchrest, Innervation of melanocytes in human skin, *J Experimental Med* 184 1385-1395 (1996)
6. R Paus, EMJ Peters, S Eichmüller and VA Botchkarev, Neural mechanisms of hair growth control, *J Invest Dermatol*, Symposium Proceedings 2(1) 61-68 (1997)
7. ME Göetz, G Künig, P Riederer and MBH Youdim, Oxidative stress: free radical production in neural degeneration, *Pharmacology & Therapeutics* 63 37-122 (1994)
8. TD Buckman, MS Sutphin and B Mitrovic, Oxidative stress in a clonal cell line of neuronal origin: effects of antioxidant enzyme modulation, *J Neurochemistry* 60(1) 2046-2058 (1993)
9. DW Nicholson, From bench to clinic with apoptosis-based therapeutic agents, *Nature* 407 810-816 (2000)
10. F François, MJ Godinho, M Dragunow and ML Grimes, A population of PC12 cells that is initiating apoptosis can be rescued by nerve growth factor, *Molecular and Cellular Neuroscience* 18(4) 347-362 (2001)
11. KS Chen, E Masliah, M Mallory and FH Gage, Synaptic loss in cognitively impaired aged rats is ameliorated by chronic human nerve growth factor infusion, *Neuroscience* 68(1) 19-27 (1995)
12. M Bothwell, Neurotrophin function in skin, *J Invest Dermatol, Symposium Proceedings* 2(1) 27-30 (1997)
13. BM Davies, BT Fundin, KM Albers, TP Goodness, KM Cronk and FL Rice, Overexpression of nerve growth factor in the skin causes preferential increases among innervation to specific sensory target, *J Comprehensive Neurology* 387(4) 489-506 (1997)
14. FL Rice, KM Albers, BM Davies, I Silos-Santiago, GA Wilkinson, AM LeMaster, P Ernfors, RJ Smeyne, H Aldskogius, HS Philips, M Barbacid, TM DeChiara, GD Yancopulos, CE Dunne and BT Fundin, Differential dependency of unmyelinated and A delta epidermal and upper dermal innervation on neurotrophins, trk receptors, and p75 LNGFR, *Developmental Biology* 198 57-81 (1998)
15. C Pincelli and M Yaar, Nerve growth factor: its significance in cutaneous biology, *J Invest Dermatol*, Symposium Proceedings 2(1) 31-36 (1997)
16. A Marconi, C Vaschieri, S Zanoli, A Giannetti and C Pincelli, Nerve growth factor protects human keratinocytes from ultraviolet-B-induced apoptosis, *J Invest Dermatology* 113 920-927 (1999)
17. S Zhai, M Yaar, SM Doyle and BA Gilchrest, Nerve growth factor rescues pigment cells from ultraviolet-induced apoptosis by upregulating CBL-2 levels, *Experimental Cell Research* 224 335-343 (1996)
18. C Pincelli, Nerve growth factor and keratinocytes: a role in psoriasis, *European J Dermatol* 10 85-90 (2000)
19. B Halliwell, Antioxidant characterization. Methodology and mechanism, *Biochem Pharmacol* 49 1341-1348 (1995)
20. I Charon, G Zuin-Kornmann, S Bataille and M Schorderet, Protective effect of neurotrophic factors, neuropoietic cytokines and dibutyryl cyclic AMP on hydrogen peroxide-induced cytotoxicity on PC12 cells: a possible link with the state of differentiation, *Neurochemistry International* 33 503-511 (1999)
21. M Crompton, Bax bid and the permeabilization of the mitochondrial outer membrane in apoptosis, *Current Opinion in Cell Biology* 12 414-419 (2000)
22. H Umegaki, K Yamada, M Naito, T Kameyama, A Iguch and T Nabeshima, Protective effect of interleukin-6 against the death of PC12 cells caused by serum deprivation or by the addition of a calcium ionophore, *Biochem Pharmacol* 52(6) 911-916 (1996)
23. E Sterneck, DR Kaplan and PF Johnson, Interleukin-6 induces expression of peripherin and cooperates with Trk receptor signaling to promote neuronal differentiation in PC12 cells, *J Neurochem* 67(4) 1365-1374 (1996)

Reflecting on Soft Focus

Keywords: soft focused/scattered reflection, sphericals, wrinkles

Particles with rounded surfaces provide some light scattering, and improved feel and spreadability

In the battle against fine lines and wrinkles, there is a new wrinkle: soft focus, sometimes called optical blurring or light diffusing. But none of these terms gets directly to the point, which is scattered reflection.

Wrinkles are perceptible because of the contrast of light and dark that accentuates even very subtle creases in the skin. One could fill that crease with a thick material, but that looks unnatural. One could level that crease with a chemical treatment, but that's a slow process. Instead, cosmetic chemists have found ways to throw light into the crease and immediately eliminate the dark that makes the wrinkle so obvious.

Tiny spherical particles 2-20 microns in diameter reflect incoming parallel light rays and scatter it into and around the crevice, as if a portrait photographer had moved the light source to eliminate unwanted shadows on a subject's face.

These spheres have many functions besides their scattering effect. They roll, so they can help deliver a slick feel to the product. Spheres with a porous surface can entrap chemicals that you want to protect in the product or release on the skin. Light scattering was not their first function, but it plays a role today.

For example, in September 2003 Avon launched a product called Line and Wrinkle Corrector as the first product in its Anew Clinical line. This product, with Derma 3X Technology, is reportedly designed to treat lines and wrinkles by addressing the three key structural components of skin: collagen, elastin and hydroproteins (glycosaminoglycans including hyaluronic acid). According to Robert Kalafsky, vice president, skin care R&D, "In addition to the proven biological benefits, our scientists also employed state-of-the-art silica beads that provide the immediate, consumer-perceivable soft focus benefits through modification of skin surface optical properties."

Achieving Soft Focus

In 1996, when a feature article[1] about soft focus appeared in *Cosmetics & Toiletries* magazine, not many people were talking about soft focus. Spheres were not mentioned in the article, which discussed measuring the light diffusing characteristics of cosmetic pigments and powders such as TiO_2 and mica. Since then, procedures have been developed to manufacture these spheres and their soft focus functionality has been explored.

Qinyun Peng is a technical manager at Rona, a business unit of EMD Chemicals in Hawthorne, New York. She explains the properties an ideal material should have to achieve the soft focus effect:

- Light absorption should be minimal. If the light is absorbed, then no light is reflected, which will result in a dark or grayish appearance.
- Specular reflection (observed as gloss or luster) should be minimal. Gloss or luster makes the wrinkles more visible.
- Scattered reflection should be maximal. Light is reflected from all directions to provide an even light distribution.
- Direct transmission should be minimal. Very transparent materials do not have enough optical effect.
- Diffuse transmission should be maximal. Diffuse transmission can provide even light distribution.
- Refractive index of the material should be close to that of skin, (1.5-1.6).
- The color of the material is usually beige, skin-tone or white.
- The preferred particle shape is spherical. This shape allows maximal scattered reflection and diffuse transmission. In addition, it gives a good skin feel, due to the ball bearing effect of spherical particles.
- Particle size should not be too big (it feels grainy) or too small (it agglomerates with similar particles). Optimally, particle size should be 2-20 microns.
- The ingredient(s) have to be acceptable for cosmetic use globally.

Soft Focus Pigments and Fillers

One veteran technical person at a cosmetic pigment company recalls the evolution of sphericals during the last 25 years. The early sphericals were nylon (6 and 12) beads. Within 10 years, the trend was from nylon to polymethylmethacrylate (PMMA) to silica. Each became popular when improved manufacturing techniques made it cheaper than its predecessors.

Today both the shapes and sizes of these particles can be carefully controlled to contribute to their function. Some are spherical, some are football-shaped, and some look like little balls fused together. They can be solid or hollow. Their surfaces can be smooth or porous. But as long they have rounded surfaces, there will be a certain amount of light scattering as well as improved feel and spreadability.

Table 1 shows a sampling of pigments and fillers used to achieve soft focus. Pigments, of course, are intended to supply color, to opacify and to hide skin's discoloration.

Cardre Inc.[*], a South Plainfield, New Jersey-based pigment technologies company, has described functional fillers as a broad category of new materials ranging from platy minerals to spherical polymers and other microspherical particles whose functionality depends on the intended application. In

---

[*]Candre Inc. was purchased in 2001 by Sensient, South Plainfield, New Jersey USA

Table 1. Selected techniques and products (pigment or filler) to achieve soft focus

Technique	Ingredients	P/F	Product
Plates	(No scattering, except from edges. Very little soft focus)		
Coated Plates	Mica platelets coated with transparent TiO2 layer	P	Extender W (Rona)
	Talc platelets and spherical polyethylene coated with TiO2 and iron oxide	P	Ultracolor Soft Focus (Ultra Chemical)
	Mica platelets coated with spherical silica	F	SM-1000 (Presperse)
	Mica platelets coated with methylmethacrylate crosspolymer	F	Sphericite (Presperse)
	Mica platelets coated with polymethyl methacrylate	F	MCP-45 (Presperse)
	Mica platelets and bismuth oxychloride coated with spherical silica	F	SP-29 UVS (Presperse)
	Bismuth oxychloride platelets coated with spherical calcium aluminum borosilicate	F	Pearlsil (Presperse)
Spheres (untreated)	Calcium aluminum borosilicate hollow beads	F	Luxsil (Presperse)
	Crosslinked polystyrene spherical beads	F	Ganzpearl GS-0605 (Presperse)
	Polymethylmethacrylate spheres	F	73233 Cardre pmma II (Cardre)
	Nylon 12 microspheres	P	SP-501 (Kobo)
	Silica spheres	P	79688 Cardre LDP 1500 (Cardre)
	Silica spheres coated with TiO_2 and iron oxides	F	017166 Ronasphere LDP (Rona)
	Polyethylene microspheres	P	CL-2080 (Kobo)
	Ethylene/Acrylic acid copolymer microspheres	P	EA-209 (Kobo)
	Polymethylsilsesquioxane microspheres	P	Tospearl 145A (Kobo)
	Cellulose microspheres	P	Cellulo Beads D-10 (Kobo)
Spheres (treated)	PMMA beads coated with UF TiO2 and surface treated with alkyl silane	P	79686 Cardre LDP 1000 (Cardre)
	Polymethylsilsesquioxane microspheres surface treated with perfluoroalcohol phosphates	P	PF-5 Tospearl 145A (Kobo)

powder formulations, fillers provide slip and coverage as well as anti-caking properties. They contribute to payoff, evenness of color and stick strength of lipsticks. The viscosity of a mascara formulation is balanced by the filler. Film gloss and hardness in nail enamel results from the choice of filler. In emulsions the final finish and evenness of color is achieved from the filler used, according to Cardre.

Pigments and fillers used for delivering a soft focus effect should maximize both scattered reflection and diffuse transmission. Thus, both spheres and plates can play a role. For example, mica's high transmissivity offsets its poor performance as scatterer of reflected light; but when mica platelets are coated with spherical particles, both soft focus priorities are achieved.

One example of a coated plate is Ultracolor Soft Focus from Ultra Chemical Inc. in Red Bank, New Jersey. It is suggested for use in blushes, lip and eye products, and foundation makeup. In this product, a layer of TiO_2 (approximately 20%) and iron oxide (~2%) is coated around a core of very platy talc (~70%) and polyethylene spheres (~5%). Some of the incident light approaching the TiO_2 layer is reflected, but most penetrates that layer and scatters off the core. Some of that scattered light diffuses out through the TiO_2 layer, but some is internally reflected back to the core. "Basically, the trapped light is released a little at a time, producing the soft focus effect," explained Art Lynch, president of Ultra Chemicals.

The spheres typically are colorless or pale yellow or skin color (although Kobo Products' Cellulo Beads with iron oxides come in a range of colors). Rona's Ronasphere LDP is a spherical silica coated with one layer of TiO_2 and another layer of iron oxide. "The refractive index of this material is very close to that of skin. It does not appear to be opaque, hence, no whitish appearance, and it is not completely transparent due to the coating of TiO_2 and iron oxide. This product has all the properties of an ideal light diffuser," Peng said.

Silica spheres can be coated with other inorganic materials such as titanium dioxide or iron oxides. The bare spheres (i.e., PMMA, silica) and coated spheres are sometimes surface treated. In most cases, the surface treatment involves some types of chemical bonding between the treatment material and the substrate surface. Surface treatment modifies the substrate so it behaves differently in formulations. This may have some effects on the final formulation's ability to meet claims such as long-wearing, waterproofing, age-defying, defining/volumizing and transfer-resistant.

Cardre does soft focus surface treatments with alkyl silane because the resulting colorants are extremely hydrophobic, they can be wet easily by oils and silicones, and they are ideal for wet or dry application. Kobo uses C9-15 fluoroalcohol phosphates to give its pigments and powders repellency to both water and oil; these treated colorants reportedly facilitate "long-lasting makeup resilient to perspiration and sebum, promoting a more natural appearance."[2]

Formulating for Soft Focus

The basic principle of soft focus is scattered reflection and the best way to accomplish this effect is with a sphere. But what explains the considerable variety of substrates, coatings, treatments and combinations with platy materials that is only suggested by Table 1? Formulations are complex systems and whatever you put in them affects performance.

One consideration in a water-in-oil emulsion is gelling. When you put tiny beads into a liquid material, the material adsorbs onto the bead surface. With a load of beads (it can be up to 10% of the formulation), there is a huge surface area, so gellation occurs. In this case formulators might want a bead whose surface was treated to preadsorb on it a material that is much more soluble in the cosmetic phase so the viscosity build-up is avoided. On the other hand, the final formula has to have high enough viscosity so you don't get settling. "It's really an interesting balancing act to make sure that everything lays down nicely and stabilizes itself out so you're not finding some formulary stability issues," said a spokesman at one pigment company.

Another consideration is that the applied sphere has to be accessible to incoming light. In a skin care product that leaves a very thin film on the skin, the film has to be thinner than the spheres themselves, or the film has to be transparent; otherwise, incoming light will never reach the spheres and no scattered reflection can occur. Thus, particle size is sometimes critical.

"Each individual material presents a number of options," says Tony Ansaldi, marketing director at Presperse Inc. in Somerset, New Jersey. "They can differ in oil absorption and particle size, and the formulator may want to choose one based on whether the product is an emulsion or an anhydrous system.

"They differ in their feel properties on the skin; so depending on what the other components in the formulation are, you may want to use a particular spherical material to improve the feel, improve the slip, improve the rubout or get a little better transparency.

"What a formulator would want to look at would be the overall formulation desired and the effect desired, and then start playing around with the different spherical materials that can give you those particular effects in the final formulation. Realistically, everything is geared toward the final formulation in terms of how it presents itself to the consumer." Ansaldi said.

The consumer sees a wrinkle and wants it to disappear. "The question is how, cosmetically, you make yourself look younger by using makeup alone or by using makeup in combination with skin care products," Peng said.

"Skin care products claiming to be 'antiaging' or 'age-defying' prevent and/or diminish wrinkles over time. Meanwhile and as complement, the soft focus materials provide a more instantaneous effect. I think the cosmetic companies are realizing that. One smart choice is to provide both long-term effect (chemical) and instantaneous effect (optical) at the same time."

—**Bud Brewster,** *Allured Publishing, Carol Stream, Illinois, USA*

References
1. R Emmert, Quantification of the soft-focus effect, *Cosmet Toil* 111(7) 57-61 (1996)
3. www.koboproducts.com (Jul 16, 2003)

Building a Better Barrier from the Inside Out

Keywords: ursolic acid, ceramides, barrier function, ceratinocytes

Liposomal ursolic acid lotions increased ceremide sythesis thereby strengthening theraby strengthening the barrier function

Barrier function in the skin depends on ceramide lipids. Without them, the barrier degrades and signs of aging appear. This chapter reports on a botanical ingredient that has the unexpected property of stimulating ceramide synthesis in keratinocytes.

The Skin Barrier and Aging

As we age, our skin changes. Some of the changes are due to the accumulated exposure to the sun's ultraviolet rays, while other changes are intrinsic to skin.

A readily identifiable change is the decline of the barrier function of skin, as measured by increases in transepidermal water loss (TEWL). People of all ages regulate water loss from the skin; for example, on the inner forearm the rate is approximately 5 g/m²/h. Differences arise, however, when the barrier is challenged, such as when strips of stratum corneum are removed from the forearm skin by cellophane tape (Figure 1). The number of tape strips required to elevate TEWL to 20 g/m²/h (barrier strength) decreases with increasing age in a nearly linear fashion. The loss of barrier strength is about 30% in 30 years.

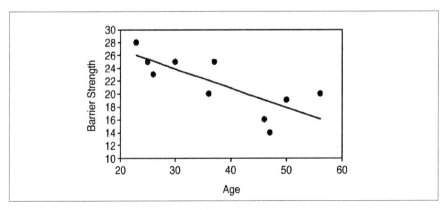

Figure 1. The effect of age on barrier strength (indicated by the number of tape strips required to elevate TEWL to 20 g/m²/h)

Key elements in the maintenance of the barrier function of skin are the ceramide lipids.[1] Ceramides are produced by a chain of enzymatic transformations of phospholipids. Ultimately, the ceramides are hydroxylated at two sites on the molecule, which serve as linkers to adjacent proteins such as keratin and involucrin. This interlocking of lipid and protein is central to the formation of the skin barrier. During aging, the ceramide content of the skin of the face, arms and legs declines, coincidently, by approximately 30% over 30 years.[2] This decline in ceramides is likely the cause of the decline in barrier function, because the linking proteins involucrin and keratin do not change with age.[3,4]

The production of ceramides begins with the keratinocytes at the basal layer of the epidermis. They synthesize phospholipids. These basal cells divide and differentiate as they move up the epidermis, and the profile of their lipid metabolism changes. Above the basal layer (stratum spinosum) the keratinocytes convert the polar lipids into neutral lipids, like triglycerides and free fatty acids. Farther up in the epidermis, in the stratum granulosum, keratinocytes use these neutral lipids to make glucosylceramides. Finally, the cells destined to become the stratum corneum produce ceramides and hydroxylate them in order to link them to the keratin and involucrin proteins they are also making. This covalent linkage of ceramides to proteins is responsible for the strength of the barrier.[5]

The progressive change in the metabolic profile is reflected in the lipid composition of the various layers: polar lipids are 44% of the lipids in the basal layer, while they comprise only 5% of the lipids in the stratum corneum; ceramides are 4% of the lipids in the basal layer, while they comprise 18% in the stratum corneum.[6] The dependence of each keratinocyte layer on the lipids manufactured by the layer below it means that it takes the whole epidermis to make a strong barrier.

How to Build a Better Barrier

To improve the ceramide content of the skin and thereby attempt to strengthen the barrier function, three approaches are available:

- Apply supplemental ceramides topically on the skin surface
- Accelerate keratinocyte differentiation so more of them are in the late stage of development, where the lipids they produce have a higher percentage of ceramides
- Stimulate all keratinocytes, regardless of stage of development, to increase the percentage of ceramides in their lipid production

We'll discuss each of these approaches in detail.

Applying supplemental ceramides: The first approach is to simply apply ceramides in a cosmetic formulation to the outside of the stratum corneum.

Ceramides are readily miscible with stratum corneum lipids and much of the delivered ceramides may not reach the lower layers of keratinocytes that link them with proteins to form the barrier.

These topically applied ceramides are not produced in the physiological location where the natural ceramides are made. This may be of some concern because ceramides, particularly short-chained ceramides, are a double-edged sword: outside the cell they are beneficial components of the barrier, but intracellular ceramides are a signal of stress. Cell membranes damaged by UV, heat or pollution are cleaved by sphingomyelinase to produce intracellular ceramides.[7] This triggers various biological pathways, such as growth inhibition, differentiation and apoptosis.[8] Inappropriate delivery of ceramides may over the long run trigger undesirable stress responses.

Accelerating keratinocyte differentiation: The second approach is to speed up maturation of keratinocytes, so that more are in the later stages of differentiation where ceramides are made in higher percentages. Several agents, such as calcium, are well known to drive keratinocytes into differentiation.

Over the short term this accelerated differentiation will also increase ceramide production in the skin, as more cells are devoted to metabolizing the free fatty acids and di- and triglycerides. However, there are fewer cells making these crucial substrates and, eventually, the pool of precursor lipids will be exhausted. At this point the improvement in ceramide levels will be lost. This is not a long-term strategy for improved barrier function.

Stimulating ceramide production: The third approach is to stimulate the keratinocytes to produce more late-stage lipids, particularly ceramides, as they did when the skin was 30 years younger. We have identified a botanical component of well-known herbs that produces such a response in skin.

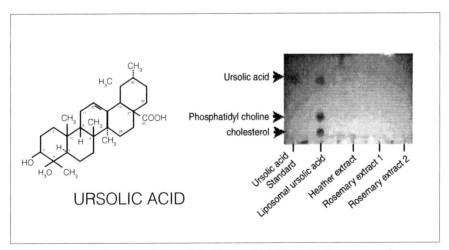

Figure 2. Ursolic acid structure (left) and identification (right) by TLC in liposomes and commercial plant extracts

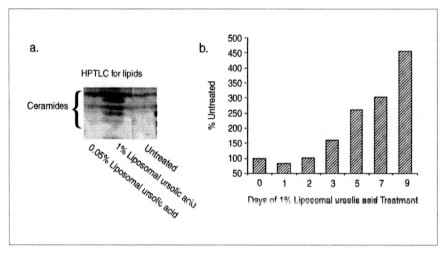

Figure 3. Lipids extracted from treated and untreated human keratinocytes in culture and detected by HPTLC
a) Ceramides for selected concentrations of liposomal ursolic acid
b) Daily percentage of total lipids in treated keratinocytes, compared to untreated keratinocytes

An Overlooked Botanical

Ursolic acid is a pentacyclic triterpenoid (Figure 2, left) widely found in the waxy coating of leaves such as rosemary and heather and in the waxy coating of apples. These plants and fruits are well known in folklore for their anti-ulcer activity and safety.[9] In fact, ursolic acid is not unknown among cosmetic chemists, but it has not been used in formulations because of its low solubility in most functional solvents. However, the flat, planar structure of ursolic acid can be stacked within the lipid membrane of liposomes, thereby assuming the water suspension characteristics of the charged liposomes. When we tested commercial extracts of rosemary and heather (10% w/v) by thin layer chromatography we did not find ursolic acid, most likely because the solvent extraction process failed to solubilize it. But, we did find ursolic acid in samples where it had been charged in liposomes (Figure 2, right).

These ursolic acid liposomes[a] allow easy formulation, delivery to the skin, and biological activity that is not characteristic of the free acid.

Turning Back the Lipid Clock

We have found that liposomal ursolic acid increases the production of ceramides in human skin.[10] We first observed the increase in ceramides extracted from treated human keratinocytes in culture and analyzed by thin layer chromatography (Figure

---

[a] Merospheres is a pending trademark of AGI Dermatics, Freeport, New York, USA. The INCI name is Ursolic acid and lecithin.
[b] Carbopol-981, Noveon Corp., Cleveland, Ohio, USA

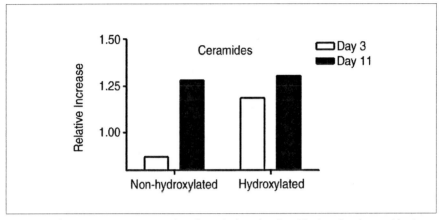

Figure 4. Relative increase in TLC-analyzed non-hydroxylated and hydroxylated ceramides in living human skin treated with 0.3% liposomal ursolic acid

3, left). Not only were ceramides increased, but free fatty acids and triglycerides (products of the stratum spinosum) were also increased.

However, phospholipid synthesis (a marker of proliferation) was not. Levels of keratin-1 and involucrin were also unchanged, indicating that the liposomal ursolic acid was not inducing differentiation.

The increase in ceramides was observed as early as three days after treatment began, and by nine days the lipids had increased more than fourfold (Figure 3, right).

These results were confirmed in clinical studies using three volunteers (45-55 years old) and a lotion applied twice daily. The lotion was a 1% carbomer[b] hydrogel with 0%, 0.3% or 1% liposomal ursolic acid. Lipids were extracted from the test skin on the third and 11th day of treatment and analyzed by thin layer chromatography.[10] The non-hydroxylated ceramides, precursors to the barrier lipids, showed an increase of more than 25% by the eleventh day of treatment with 0.3% liposomal ursolic acid (Figure 4). The fully hydroxylated ceramides, which are crosslinked to proteins to form the barrier, were already increased by the third day, and were also increased by more than 25% by the 11th day.

This 30% increase in both the hydroxylated and non-hydroxylated ceramides by day 11 represents a reversal of three decades of decline in ceramide production.

A Better Barrier from the Inside Out

The increase in ceramides reflects increased production of components of the skin barrier, and we expected to find a strengthened barrier function. This was tested in two ways: by challenging the barrier with tape stripping and by monitoring the

[c] Scotch 810 Magic Tape, 3M Companies, St. Paul, Minnesota USA. Magic is a trademark of 3M Companies.
[d] DermaLab TEWL meter, Cortex Technology, Media, Pennsylvania USA, interfaced with an IBM PC computer and Dasylab software. Dasylab is a registered trade name of Cortex Technology ApS, Hadsund, Denmark.

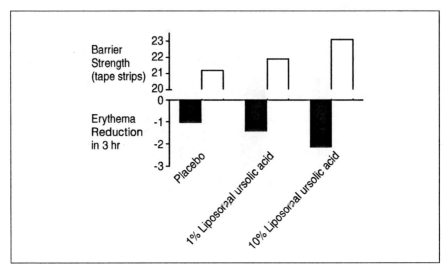

Figure 5. Strengthened barrier function in living human skin treated with liposomal ursolic acid
a) Number of tape strips required to increase the TEWL to 20 g/m²/h
b) Reduction in average erythema scores from hour 0 (when TEWL is 20 g/m²/h) to hour 3

recovery from the redness produced from the tape stripping.

For the test we prepared three lotions. The lotions were 1% carbomer[b] hydrogels, to which 0%, 1% or 10% liposomal ursolic acid[a] was added. After giving informed consent, 11 volunteers aged 25-55 applied one of the three lotions to a 3-cm² site on the dominant volar forearm every day for 11 days.

On day 12, the treated sites were tape stripped. The tape[c] was applied with rubbing, and then rapidly pulled from the skin for each strip. TEWL was measured[d] and tape stripping was repeated until the TEWL increased to 20 g/m²/h. Three TEWL measurements were made after a 15 min equilibration of the volar forearm with ambient air, which reduced the coefficient of variation to less than 10% for each measurement. The three measurements were averaged to determine when the level of 20 g/m²/h had been reached. Since the number of tape strips required to reach this level prior to treatment was linearly age-dependent, the treatment data were age-adjusted by normalizing to 23-years-old using the slope of the regression line.

The results demonstrated that the sites treated with liposomal ursolic acid required more tape strips on average to reach this increased level, indicating a stronger barrier and resistance to disruption (Figure 5, top). The increase of approximately 30% in the barrier resistance restored approximately 30 years of decline in barrier function.

Once this level of TEWL was reached by tape stripping, the slight pink color of the test site was measured by a digital spectrophotometer[e]. After 3 h, the sites

---

[e] DermaSpectrophotometer, Cortex Technology, Media, Pennsylvania USA

were again measured in triplicate by the spectrophotometer. Each set of three readings was averaged. The recovery was reported as the difference between the two averages.

These measurements of recovery from erythema were not age-dependent, and the recovery data for all subjects was averaged without adjustment. The results showed that sites treated with liposomal ursolic acid showed a greater reduction in the irritation measure than the untreated sites showed (Figure 5, bottom). This finding is consistent with the medical folklore that ursolic acid is an anti-inflammatory botanical.[9]

Conclusion

The perception of aging skin is perhaps most directly experienced as dry and easily irritated skin, a consequence of the decline of the strength of the stratum corneum. Prior approaches to improving the barrier have included adding back barrier lipids or stimulating differentiation to increase lipid synthesis.

A more attractive pathway is the stimulation of endogenous lipid synthesis by liposomal ursolic acid. The mechanism of action may be that ursolic acid triggers cell receptors, such as the glucocorticoid receptor.[11] This results in the production of ceramides at the physiologically appropriate time and space.

We found that liposomal ursolic acid lotions increased ceramide synthesis in less than two weeks and resulted in a strengthened barrier reminiscent of skin 30 years younger.

—**Daniel Yarosh, Ph.D., and David Brown, PhD,** *AGI Dermatics, Freeport, New York, USA*

References

1. M Schmuth, M-Q Man, F Weber, W Gao, K Feingold, P Fritsch, P Elias and W Holleran, Permeability barrier disorder in Niemann-Pick Disease: Sphingomyelin-ceramide processing required for normal barrier homeostasis, *J Invest Dermatol* 115 459-466 (2000)
2. J Rogers, C Harding, A Mayo, J Banks and A Rawlings, Stratum corneum lipids: The effect of ageing and the seasons, *Arch Dermatol Res* 288 765-770 (1996)
3. M Engelke, J Jensen, S Ekanayake-Mudiyanselage and E Proksch, Effects of xerosis and ageing on epidermal proliferation and differentiation, *Br J Dermatol* 137 219-225 (1997)
4. T Tezuka, J Qing, M Saheki, S Kusuda and M Takahashi, Terminal differentiation of facial epidermis of the aged: Immunohistochemical studies, *Dermatol* 188 21-24 (1994)
5. M Behne, Y Uchida, T Seki, P Ortiz de Montellano, P Elias and W Holleran, Omega-hydroxyceramides are required for corneocyte lipid envelope (CLE) formation and normal epidermal permeability barrier function, *J Invest Dermatol* 114 185-192 (2000)
6. P Elias, Epidermal lipids, barrier function, and desquamation, *J Invest Dermatol* 80 (suppl) 44s-49s (1983)
7. M Krönke, The mode of ceramide action: The alkyl chain protrusion model, *Cytok & Growth Factor Reviews* 8 103-107 (1997)
8. F Bonté, Skin lipids: Their origin and function, *Recent Res Devel Lipids Res* 3 43-62 (1999)
9. K Lintner, Purified plant extracts, *Cosmet Toil* 113(3) 67-73 (1998)
10. D Both, K Goodtzova, D Yarosh and D Brown, *Arch Dermatol Res* 293 569-575 (2002)

11. H-J Cha, M-T Park, H-Y Chung, N Kim, H Sato, M Seiki and K-W Kim, Ursolic acid-induced down-regulation of MMP gene is mediated through the nuclear translocation of glucocorticoid receptor in HT 1080 fibrosarcoma cells, *Oncogene* 16 771-778 (1998)

Applications of Essential Fatty Acids in Skin Care, Cosmetics and Cosmeceuticals

Keywords: essential fatty acid, lipid, barrier function, homeostasis, inflammation, moisturization

Physiologic lipids benefit skin from diet and even more so by application of essential fatty acid-enriched formulations

The essential fatty acids (EFAs) include fatty acids of the omega-6 and omega-3 families. The omega-6 family is comprised of linoleic acid (LA) and its longer chain derivatives such as gamma-linolenic acid (GLA) and arachidonic acid (AA); the omega-3 fatty acid family is comprised of alpha-linolenic acid (ALA), and its longer chain derivatives eicosapentaenoic acid (EPA) and docosahexaenoic acid (DHA).

The beneficial effects obtained from these physiologic lipids may be achieved by increasing the dietary intake of essential fatty acids, and perhaps more so, by topical application of essential fatty acid-enriched formulations. Topical products incorporating essential fatty acids include moisturizing creams and lotions, bar soaps, lipsticks, shampoos, and therapeutic skin preparations for dryness, eczema and psoriasis.

The recent trend towards producing natural products that utilize a more physiologic mix of lipids in skin and personal care products has spurred interest in the essential fatty acids linoleic acid, gamma-linolenic acid, eicosapentaenoic acid and docosahexaenoic acid. Oils providing these useful physiologic lipids include borage oil, evening primrose oil and refined fish oil; these oils may serve as vehicles, adjuncts or active ingredients in skin and personal care products, cosmetics, cosmeceuticals and topically administered pharmaceutical agents.

Importance of EFAs

The nutritional and health benefits of essential fatty acids (EFAs) are well documented: they are required for normal health and well-being, growth and development; they are also of benefit in the prevention and treatment of numerous disorders and diseases including cardiovascular disease, neurological disorders, diabetes, inflammatory conditions, immune dysfunction and skin disorders. The physiological changes associated with EFA consumption have been attributed to modifications in cell membrane structure and function, modulation of endogenous eicosanoid production, and effects on regulatory processes such as inflammation.

Physiologic effects apparent during essential fatty acid deficiency include a) epidermal hyperproliferation leading to a characteristic scaly skin disorder, b) abnormal barrier lipid structure and function with increased permeability of the skin to water and concomitant increased trans-epidermal water loss, and c) altered production of eicosanoids–bioactive substances derived from specific fatty acids. The result is a myriad of skin symptoms, notably dryness, erythema, poor wound healing and inflammation.

These severe skin symptoms are reversible by the addition of EFAs in the diet or by topical administration of EFAs. Increasing the levels of key essential fatty acids a) increases fluidity of the cell membrane, b) enhances barrier function and repair, c) decreases trans-epidermal water loss and improves moisturization, d) provides protection against environmental toxins, e) increases production of beneficial anti-inflammatory and anti-proliferative eicosanoids, and f) reduces or reverses the symptoms of skin disorders.

In addition to being good moisturizers, protectives and anti-inflammatory ingredients, fatty acids are used as carriers or vehicles in skin care formulations and may also be used as penetration enhancers to increase the absorption of other bioactives.

EFAs also exert their beneficial effects by increasing cell membrane fluidity, flexibility and permeability and modulation of regulatory processes through effects such as cell signaling and cell messenger systems. Not surprisingly, physiologic lipids such as GLA and EPA may be useful active cosmeceutical agents.

Metabolism of EFAs

The omega-3 and omega-6 essential fatty acids are metabolized in the body by a series of competing enzymes in alternating desaturation and elongation reactions. The parent fatty acids–ALA in the omega-3 series and LA in the omega-6 series–are metabolized to longer chain PUFAs (polyunsaturated fatty acids) and eicosanoids–bioactive substances that are produced locally at the site where they are needed on a minute to minute basis. Eicosanoids include prostaglandins, leukotrienes and hydroxy fatty acids. The key steps in the metabolism of LA and ALA and the production of important eicosanoids are shown in Figures 1 and 2, respectively.

The delta-6-desaturase (D6D) enzyme preferentially acts on ALA over LA but is the slowest and rate-limiting step in both pathways. The activity of this enzyme diminishes with age and in the presence of certain diseases such as inflammation and eczema. Abnormalities in EFA metabolism such as poor functioning of the D6D enzyme may be a causative factor in the development of skin disorders including xerosis (dryness), eczema, inflammation and poor wound healing.

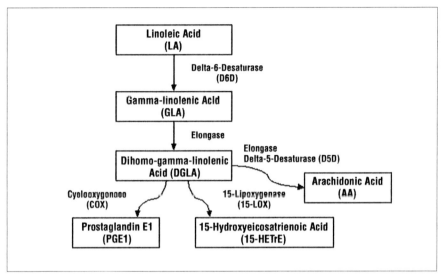

Figure 1. Metabolism of linoleic acid (LA)

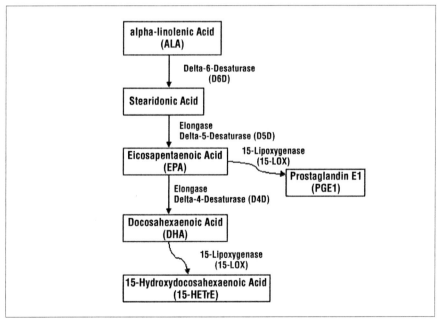

Figure 2. Metabolism of alpha-linolenic acid (ALA)

Indeed, individuals suffering from inflammatory skin disorders such as atopic dermatitis, eczema, psoriasis and xerosis have abnormal EFA levels such as low GLA levels.[1] As shown in Figure 1, GLA is the precursor for prostaglandin E1 (PGE1) – a key mediator for maintaining healthy skin. PGE1 exhibits potent anti-inflammatory properties, regulates water loss and protects skin from injury and damage.[2] The balance between the different eicoisanoids is crucial for maintenance of healthy skin.

Functional Role of EFAs in the Skin

The metabolism of essential fatty acids is highly active in the skin epidermis. The skin, in contrast to the brain and liver, does not have the ability to desaturate LA into GLA, or DGLA into AA, although it does have the ability to convert GLA into DGLA.[2] Thus the concentration of AA in the epidermis is not synthesized from epidermal LA, and depends upon endogenous sources.

Metabolites of AA, produced in a process known as the arachidonic acid cascade, are implicated in a number of pathological skin conditions. The prostaglandins and leukotrienes derived from AA are potent mediators in cutaneous inflammatory and proliferative disorders and are formed in response to injury and exposure to ultraviolet radiation. The metabolism of AA by the cyclooxygenase and lipoxygenase enzymes into bioactive eicosanoids is shown in Figure 3.

A certain amount of AA is required for maintaining healthy skin. However, an overabundance of AA and a deficiency of GLA is associated with inflammatory skin disease, hyperproliferation of epidermal cells and impaired barrier function of the skin causing increased transepidermal water loss (TEWL). Leukotrienes (LTs) derived from AA, such as LTB4, are significant mediators in cutaneous inflammatory and proliferative disorders, and these metabolites accumulate in skin disorders such as psoriasis. The metabolism of LA and its metabolite GLA into anti-inflammatory and anti-hyperproliferative substances is important to counteract the inflammatory effects of eicosanoids derived from AA.

The active lipoxygenase enzyme, 15-lipoxygenase, converts LA into predominantly 13-hydroxyoctadecadienoic acid, a fatty acid with antihyperproliferative properties. A high level of LA does not, however, ensure healthy skin; abnormal

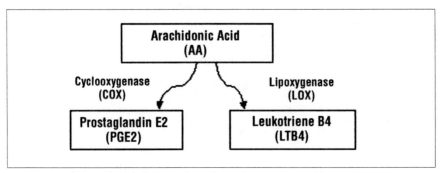

Figure 3. Metabolism of arachidonic acid (AA) by cyclooxygenase and lipoxygenase

or low conversion of LA into GLA is apparent in inflammatory skin disorders that may be characterized by high levels of LA, but low levels of GLA.[1]

GLA is easily converted to DGLA by active elongase enzymes in the epidermis, and then subsequently converted by cyclooxygenase enzymes into prostaglandins of series 1 (PGE1), and by lipoxygenase into 15-hydroxyeicosatrienoic acid (15-HETrE). 15-HETrE is a potent inhibitor of LTB4 synthesis, counteracting its inflammatory effects. Also, the lipoxygenase derivatives from EPA and DHA are 15-hydroxyeicosapentaenoic acid (15-HEPE) and 17-hydroxydocosahexaenoic acid (17-HDHE). These derivatives inhibit the formation of LTB4, an inflammatory eicosanoid derived from AA.

Thus, the anti-inflammatory and anti-proliferative effects of the eicosanoids produced from LA, GLA, EPA and DHA counteract the inflammatory and hyper-proliferative eicosanoids produced from arachidonic acid. The balance between the various eicosanoids is critical for maintaining healthy skin barrier structure and functions and skin homeostasis.

Effects of EFAs on Barrier Function and Homeostasis

Improving the barrier function of the skin will reduce TEWL, hydrate and moisturize the skin, protect the skin against environmental insults and impart softness and suppleness to the skin.

While all moisturizers temporarily improve skin dryness and roughness, they usually offer little or no improvement to the integrity of the skin barrier. In fact, long-term use of some occlusive agents containing non-physiologic lipids such as lanolin and petrolatum has a disruptive effect on the skin barrier, contributing to the development of skin disorders.[3] While non-physiologic lipids such as various hydrogenated oils, waxes and butters may effectively moisturize the skin and decrease water loss, they have to be reapplied frequently and are not naturally-occurring in the skin, often imparting an unnatural, greasy feel to the skin.

Increased interest in the development of a more physiologic mix of lipids for topical application to skin has led to the use of EFAs such as GLA in skin care formulations, cosmetics and cosmeceuticals.

In addition to moisturization and reduced TEWL, both topical and oral administration of EFAs improves the structure and function of cell membranes. They also maintain skin barrier homeostasis and enhance barrier recovery after injury or stress. By preventing undesirable moisture loss and maintaining skin barrier homeostasis, the skin remains hydrated, supple, pliable, soft and smooth. An increase in skin smoothness and softness has been demonstrated with both topical and oral supplementation of GLA.[4] However, topical administration may offer improved absorption over oral administration, supplying EFAs to the site of action earlier and more effectively.

Physiologic lipids are useful protective agents to protect against environmental insults such as decreased humidity, detergents and solvents, and aid the skin in the recovery from environmental insults and stress.

Due to their beneficial effects on skin barrier recovery and homeostasis, and their ability to hydrate the skin, physiologic EFAs may be used as moisturizers, protectants or occlusive agents in many formulations. Moreover, increasing skin surface lipids also generates a certain skin gloss and imparts mild antimicrobial effects.

The usefulness of EFAs in antiaging formulations is also apparent. Physiological changes in aged skin reflect abnormalities in skin barrier function and integrity that alter skin permeability. The deficit of EFAs that is often apparent may be related to abnormal EFA metabolism (such as low D6D activity). The result— decreased ability to withstand disturbances, slow barrier recovery after injury and altered TEWL – can limit normal function and cause severe xerosis, altered drug permeability and increased susceptibility to irritant contact dermatitis.[5] The effects are further accentuated in photoaged skin. Through their beneficial influences on the skin barrier, EFAs help normalize the function and improve the appearance of aged skin.

The fluidizing effect of fatty acids on cell membranes may be responsible for their skin penetration enhancement effects. Enhanced skin permeability of non-steroidal anti-inflammatory drugs (NSAIDS) and hormones, among other pharmaceutical agents, has been observed with EFAs such as linoleic acid. The combination of their own physiological effects, along with their ability to increase the transport of bioactive agents across biological membranes such as the skin, allows for synergistic activity with other bioactives, such as anti-inflammatory agents.

Anti-Irritant/Anti-Inflammatory Effects of EFAs

The anti-irritant and anti-inflammatory effects of EFAs on the skin may have application in protection against and treatment of various skin dysfunctions, including damage induced by ultraviolet (UV). Exposure to UV radiation causes sunburn and photosensitivity in the short term and photoaging and skin cancer over the long term. UV radiation induces changes in fatty acids and activates epidermal enzymes to increase the production of prostaglandins (PGs) derived from AA, causing inflammation.[6]

EPA and GLA act competitively with cyclooxygenase substrates to reduce PG synthesis from AA and produce anti-inflammatory PGs and leukotrienes (LTs). Oils high in GLA (such as borage or evening primrose oils) and EPA (such as fish oil) modify the cutaneous response to UV, reduce UV-induced erythema and provide protection against sunburn. Reductions in erythemal response with oral doses of fish oil and evening primrose oil occur over periods of one to six months.[7-9] Further study found that GLA supplementation significantly reduced ultraviolet damage to a group of arthritic patients.[10]

The moisturization, protective and barrier recovery effects of physiologic lipids makes them ideal as carriers, moisturizers, occlusive agents, protectants and antiaging agents in a wide range of personal care products including skin creams and lotions, lip care products, soaps and cosmetics. The anti-irritant/anti-inflammatory effects can make them ideal as ingredients in various skin care products (especially sunscreens and after-sun moisturizers) for the prevention and treatment of erythema damage.

A summary of the effects and applications of EFAs in skin care products, cosmetics and cosmeceuticals is found in Table 1.

Table 1. Summary of effects and applications of EFAs in skin care, cosmetics and cosmeceuticals

Activity	Biochemical and Clinical Effects	Use
• Anti-inflammatory effects	• Increase production of anti-inflammatory and anti-proliferative eicosanoids • Reduce UV-induced erythema • Reduce inflammation and erythema	• Sunscreens, After-sun moisturizers, Moisturizers • Bioactive
• Cell membrane structure and function; • Skin barrier function	• Enhance barrier function repair • Maintain barrier homeostasis • Decrease TEWL • Increase skin hydration and moisturization • Maintain optimal cell membrane structure and function • Impart skin softness, suppleness, flexibility, smoothness	• Carrier or vehicle • Emollient • Moisturizer • Occlusive agent • Protectant • Anti-aging agent • Skin barrier recovery agent
• Penetration enhancer	• Increase skin absorption • Increase absorption of other bioactives • Synergistic effect with anti-inflammatory agents	• Carrier or vehicle • Synergistic bio-active

Summary

Lipids play a significant role in cosmetics and cosmeceuticals because of their beneficial effects. They impart emolliency, moisturization, softness, flexibility, suppleness and smoothness to the skin. The use of physiologic lipids, the EFAs, imparts additional benefits to the skin in terms of enhancement of cell membrane structure and function, barrier maintenance and recovery, anti-irritant and anti-inflammatory effects and penetration enhancement. Lipids also deliver other cosmetic benefits such as occlusivity, cleansing, emulsification, gloss, adhesion and lubricity.

The administration of EFAs has been shown to improve and manage skin conditions–perhaps especially in topical dosage form, but also in oral dosage form. EFAs are increasingly found in skin care products, cosmetics and cosmeceuticals, as the benefits of these nutrients become more well-known.

—Janice Brenner, BSP, MSc, MH, WT, *Bioriginal Food & Science Corp., Saskatoon, Saskatchewan, Canada*

References

1. DF Horrobin, Essential fatty acid metabolism and its modification in atopic eczema, *Am J Clin Nutrition* 71(1) 367S-372S (2000)
2. V Ziboh and C Miller, Essential fatty acids and polyunsaturated fatty acids. Significance in cutaneous biology, *Ann Rev Nutr* 10 433-450 (1990)
3. MM Man, BE Brown, S Wu-Pong, KR Feingold and PM Elias, Exogenous non-physiologic vs physiologic lipids, *Arch Dermatol* 131 809-816S (1995)
4. US Pat 4,997,657, Method for improving skin smoothness, DF Horrobin and JCM Stewart, assigned to Efamol Holdings (Mar 5, 1991)
5. R Ghadially, BE Brown, SM Sequeira-Martin, KR Feingold and PM Elias, The aged epidermal permeability barrier – Structural, functional, and lipid biochemical abnormalities in humans and a senescent murine model, *J Clin Invest* 95 2281-2290 (1995)
6. K Punnonen, T Puustinen and CT Jansen, Ultraviolet B irradiation induces changes in the distribution and release of arachidonic acid, dihomo-gamma-linolenic acid, and eicosapentaenoic acid in human keratinocytes in culture, *J Invest Dermatol* 88(5) 611-614 (1987)
7. IF Orengo, HS Black, AH Kettler and JE Wolf, Influence of fish oil supplementation on the minimal erythema dose in humans, *Arch Dermatol Res* 284 219-221 (1992)
8. LE Rhodes, BH Durham, WD Fraser and PS Friedmann, Dietary fish oil reduces basal and ultraviolet B-generated PGE2 levels in skin and increases the threshold to provocation of polymorphic light eruption, *J Invest Dermatol* 105 532-535 (1995)
9. LE Rhodes, S O'Farrell, MJ Jackson and PS Friedmann, Dietary fish oil supplementation in humans reduces UVB-erythemal sensitivity but increases epidermal lipid peroxidation, *J Invest Dermatol* 103 151-154 (1994)
10. TM Hansen, A Lerche, V Kassis, I Lorenzen and J Sondergaard, Treatment of rheumatoid arthritis with prostaglandin E1 precursors cis-linoleic acid and gamma-linolenic acid, *Scand J Rheum* 12(2) 85-88 (1983)

Scratching the Skin Surface

Keywords: profilometry, product claim support, long-term trial

Considerations for using profilometry in a long-term trial to provide claim support for anti-aging products

It is well-established clinically and intuitively that skin relief, or skin surface topography, is affected by dryness and aging. Both influences affect the surface, texture and appearance of lines and wrinkles. Cosmetic benefits have traditionally been moisture-driven, giving short-term aid to skin texture. More recently, however, new products have provided substantive benefit to aging features, such as fine lines and wrinkles. While profilometry can measure all of these effects, the technique does not readily distinguish between short-term moisturization and long-term treatment effects.

Background: Profilometry is a widely used means of quantifying surface topography. During the last 50 years, researchers have developed various techniques for quantifying the microrelief of metallic surfaces. By the late 1970s, scientists were using methods of mechanical profilometry to describe the surface texture of human skin.

However, because the normal profilometry scanning process is painstakingly slow, it prevents the ability to take direct in vivo measurement on human skin. Therefore, human skin profilometry has always required the use of high-quality skin replicas.

The techniques and issues involving the use of replicas are well-summarized.[12] The work of Cook,[1,2] Makki[8] and Hoppe[7] has demonstrated that the methods can be used to effectively describe skin surface texture and the smoothing effects of cosmetic preparations. While the introduction of laser profilometry has helped overcome some of the contact issues of mechanical profilometry, the principles and output are essentially the same.

More recently, optical profilometric methods have become popular due to the availability of low-cost, high-speed personal computers and image-analysis software. Optical methods are based on the principle of incident light from a fixed angle producing a shadow pattern on the surface. Unlike the classic profilometry methods, which laboriously scan the surface, image analysis instantly captures the entire surface area for computer processing. While image analysis techniques have some disadvantages compared to mechanical profilometry, the work of Grove,[5] Corcuff[3] and Schrader[11] has established optical methods as valid alternative means of describing skin surface.

Application to Cosmetics

Profilometry can be used in cosmetic applications to show surface changes in skin due to dryness and textural effects from photoaging. One of the earliest uses of skin profilometry was to describe the effects of moisturizers on microrelief of skin.

Photodamage also contributes to changes in the skin surface with the development of fine lines, wrinkles, furrows and skin grain. The most common in vivo approach to objectively assess fine lines and age-related texture changes is clinical, visual evaluation, such as the Packman SFL counting method.[10] Using skin replicas, ex vivo instrumental analysis such as profilometry, has been utilized to describe changes in age-related appearance based on texture patterns.[4] Moreover, optical profilometry has shown to be an effective method for demonstrating the antiaging effects of treatment products on photoaged skin.[6]

An important issue in measuring skin changes is that the profilometry process fails to distinguish between dryness and aging factors. This poses a particular concern for antiaging treatment studies. The use of any emollient cream will affect the surface, which may cause a reduction in profilometry results. Commercial lotions and creams, such as products containing AHAs, provide both a short-term moisturization benefit and long-term correction of photodamage. If not careful, this moisturization effect can contribute to or even dominate the measured reductions in profilometry, and thus mask or confound the antiaging treatment. However, with understanding and careful study design, these different levels of effect can be controlled.

The cumulative benefit of moisturizing even very dry skin will plateau after a few weeks. Therefore, to measure the true reduction in features like lines and wrinkles, a clinical trial enduring from six weeks to six months is required. However, such long-term studies have particular requirements for profilometry due to the critical need for consistent and reproducible skin replicas.

Long-Term Trial Considerations

Sample population: In a limited-population clinical trial, it is important to have a homogenous group with appropriate skin condition. It is axiomatic that an antiaging study requires subjects with visibly photoaged skin. Females in the 45-60 age range are a good starting point, but selected subjects must also visibly exhibit a measurable amount of photodamage to demonstrate subsequent improvement with treatment. Similarly, subjects with an extremely severe condition should be avoided. The excessive amount of photodamage on such individuals may appear as outliers in your data set, making it difficult to draw clear conclusions.

Control group: To serve as reference for normal biological and environmental changes, a control group should be included in any long-term test. If a researcher wishes to examine the effect of a particular ingredient, a placebo product (base lotion without active) should be run. When a base vehicle itself is used over many weeks, it can have a marked effect on skin.[12]

An untreated control group (complete elimination of product) exerts a different influence on the trial, since it alters the routine of the subjects. A "no-treatment" control group that continues its normal habit of product use may be the more appropriate control to assess overall efficacy. Note that such a group is not the same as an untreated control, which stops all treatment. The "no-treatment" control group effectively maintains its "normal" state for the course of the study.

Replica site reproducibility: In making repeated skin replicas over many months, particularly on the face, reproducibility is critical. Small changes in facial expression or deviations in repositioning can have a large influence on contours and lines. For the subjects, the clinical environment should be consistent and free of distraction.

Before each evaluation, subjects should lie quietly, eyes closed, in the clinician's chair for at least 15 min. This will help them to reach a relaxed physiological state and to equilibrate fully to the room conditions. Detailed measurements should be taken from fixed points on the face to precisely reposition the replica site. If possible, take a reference photo of each subject at baseline with the exact test site clearly marked. This can be extremely valuable later. Use of replica rings with a tab is a simple yet helpful means of maintaining precise orientation of the replica after removal.

Skin state: To measure the true state of the skin, avoid transient or superficial effects following product application. When making skin replicas, the site must be clean and free of any makeup or residual product. (However, subjects need to remember that washing their face just prior to measurement can also influence the skin's state.) It is best to have subjects suspend all product application on observation days until they have completed the measurement.

Timing: In a trial running over several months, environmental effects such as changes in humidity and temperature can significantly influence skin condition. Therefore, avoid testing during winter or summer extremes.

Additional measurements: In any clinical trial, other measures such as visual clinical grading, self-assessment evaluations, photography and other bio-instrumental measures, are usually included. It is extremely important to consider these measures in relation to replicas. Because replicas can leave trace material on the skin, self-assessment and clinical visual grading should precede the replica process. Likewise, because instrumental probes can leave telltale indentations on the skin, replicas should not be made on a skin site immediately following instruments with contact probes. If both contact instruments and replicas are needed, separate sites should be considered.

Antiaging Cream Test Results

Over a six-month period, we successfully obtained replica results from a study of creams containing 4% and 8% glycolic acid.[a] We used the considerations stated previously as well as double blinding, encoded products and detailed panelist instructions.

A total of 150 subjects participated in this trial with clinical observations scheduled monthly. Subjects were assigned to either a treatment population or a no-treatment control group. Results of the visual assessment of aged features were previously described; the features showed significant improvement in the appearance of fine lines over the course of the trial.[9]

In addition to visual grading and subject self-assessment, three replicas were made for fine line and texture analysis: side of the eye, upper cheek and corner

[a]Pond's Age Defying Complex Cream, Chesebrough-Pond's, Greenwich, Connecticut, USA

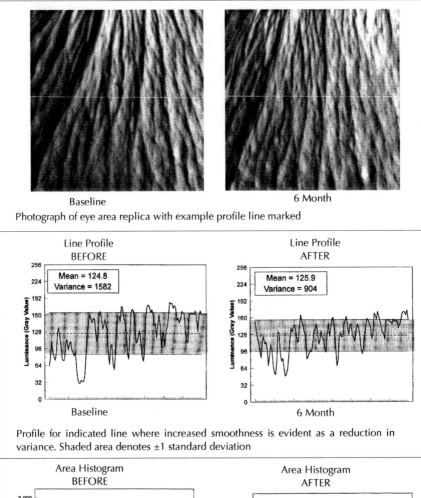

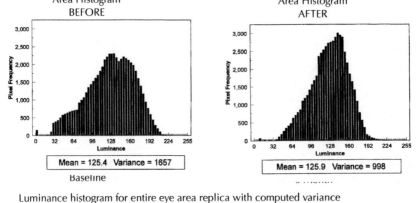

Figure 1. Surface texture

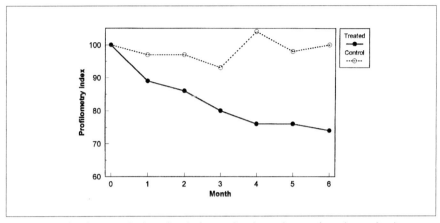

Figure 2. Results of six month clinical trial of a 4% glycolic acid cream for reducing fine lines on the face

of the mouth. Replica images were processed with commercially available image analysis software. A 1 cm² area of each replica was imaged using a fixed fiber optic light source incident to the surface and with a high resolution receiving camera set perpendicular to the surface of the replica. Image luminance was converted to a 256 level gray scale, and the values for each image presented as a histogram with the variance of the histogram being computed. Variance in luminance is a direct function of the uniformity in the surface (Figure 1).

Based on the percentage change in the luminance variance, fine lines on the side of the eye are quantified over the six-month trial. Treatment with 4% glycolic acid provided a consistent reduction in variance (effacement of fine lines), but there was no discernible change in the control group (Figure 2).

This instrumental data is corroborated by both the clinical visual findings and the subjects' self-assessment results (Figure 3).

Claim Opportunities

The successful use of profilometry provides opportunity for a very strong claim support position, particularly when reinforced by clinical grading and self-perceptions of benefits. While there is active debate on the relative objectivity of clinical visual grading in comparison to instrumental measures, the combined presentation allows for very direct statements of benefit. For example, an impactful claim of "Significantly reduces fine lines" can be strongly supported compared to a softer claim of "Helps the appearance of fine lines," which would be based only on consumer perceptions. As such, using profilometry as an objective instrumental measure is invaluable in product claim support as it provides factual evidence of the product benefit.

—Gregory Nole, Anthony Johnson, Raymond Gomes, Paul Baker, Alex Znaiden,
Unilever Home & Personal Care, USA

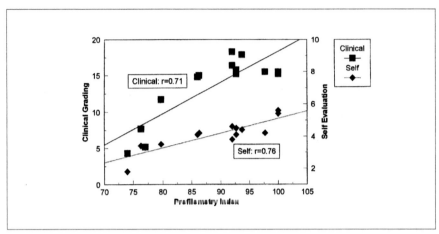

Figure 3. Correlation of profilometery to clinical and panelist self assessment

References

1. TH Cook, Profilometry of skin –A useful tool for the substantiation of cosmetic efficacy, *J Soc Cosmet Chem* 31 339-359 (1980)
2. TH Cook, TJ Craft, RL Brunelle, F Norris and WA Griffin, Quantification of the skin's topography by skin profilometry, *Int J Cosmet Sci* 4 195-205 (1982)
3. P Corcuff, F Chatenay and J-L Lévêque, A fully automated system to study skin surface patterns, *Int J Cosmet Sci* 6 167-176 (1984)
4. P Corcuff, J de Rigal, J-L Lévêque, S Makki and P Agache, Skin relief and aging, *J Soc Cosmet Chem* 34 177-190 (1983)
5. GL Grove, MJ Grove and JJ Leyden, Optical profilometry: An objective method for quantification of facial wrinkles, *J Am Acad Dermatol* 21 631-637 (1989)
6. GL Grove, MJ Grove, JJ Leyden, L Lufrano, B Schwab, BH Perry and EG Thorne, Skin replica analysis of photodamaged skin after therapy with tretinoin emollient cream, *J Am Acad Dermatol* 25 231-237 (1991)
7. U Hoppe, G Sauermann and R Lunderstädt, Quantitative analysis of the skin's surface by means of digital signal processing, *J Soc Cosmet Chem* 36 105-123 (1985)
8. S Makki, P Agache, J Mignot and H Zahouani, Statistical analysis and three-dimensional representation of the human skin surface, *J Soc Cosmet Chem* 35 311-325 (1984)
9. G Nole, S Edgerly, AW Johnson and A Znaiden, Global face assessment, A clinical evaluation method, *Cosmet & Toilet* 109 (7) 69-72 (1994)
10. EW Packman and EH Gans, Topical moisturizers: Quantification of their effect on superficial facial lines, *J Soc Cosmet Chem* 29 79-90 (1978)
11. K Schrader and S Bielfeldt, Comparative studies of skin roughness measurements by image analysis and several in vivo skin testing methods, *J Soc Cosmet Chem* 42 385-391 (1991)
12. J Serup and GBE Jemec, *Handbook of Non-Invasive Methods and the Skin*, Chapter 5: Skin Surface Contour Evaluation in CRC Press, Boca Raton, FL (1995) pp 83-131
13. MJ Stiller, J Bartolone, R Stern, S Smith, N Kollias, R Gillies and LA Drake, Topical 8% glycolic acid and 8% L-lactic acid creams for the treatment of photodamaged skin, *Arch Dermatol* 132 631-636 (1996)

Photoaging and Photodocumentation

Keywords: wrinkles, clinical trial grading, photographic evaluation, light imaging techniques

Sophisticated imaging and photographic tools for evaluation in photoaging trials are discussed

An image conveys an immediate and powerful message, both to the "untrained" consumer and to the expert grader. A comparison of two images, such as before and after treatment, creates an even stronger statement. Therefore, significant attention has been given to the standardization of imaging and photographic tools in order to use images as part of scientific evaluation methods in acne and photoaging.[1-7]

"Skin photodamage" refers to the changes in the skin tissue caused by acute or chronic ultraviolet (UV) exposure. In photoaging, the photodamage is superimposed on the changes occurring with intrinsic aging.[8] Clinical characteristics of cutaneous photoaging include fine lines, coarse wrinkles, crinkles, roughness, laxity, sallowness, dullness, telangiectasia, pebbly appearance, mottled pigmentation, disruption of microtopography, actinic keratoses and skin cancers (Table 1).[1,9-14]

Antiaging studies usually rely on the clinician to score the severity of photodamage parameters by visual and tactile inspection using a descriptive scale.[3,4,15] In acne studies, however, it has become routine to score acne severity by comparing the patients to standard photographic grades.[5] On this premise and to design a more reproducible evaluation system in photoaging, a few photonumeric scales have been published with the aim of anchoring the evaluator to fixed visual grades.[1,2]

Additionally, thanks to the recent advances and excellent standardization in photographic systems, clinicians have used sequential photographs to evaluate retrospectively the improvement of photodamage.[6,7] Finally, state-of-the-art technologies in the optical arena are offering new methodologies, such as PRIMOS, which merge three-dimensional imaging of the skin surface with quantification of macrostructures such as wrinkles.[16]

The aim of this review is to discuss various evaluation approaches and imaging methods based on the specific needs of the photoaging trial.

Clinical Grading of Photoaging

When designing a photonumeric scale for aging it is important to recognize that the face can age differently depending on the phototype, genetic background and sun exposure history of the patient.

In Caucasoid subjects, severe photodamage on the face can appear with severe hyperpigmentation, actinic keratoses and skin cancers on an atrophic, telangiectatic and pinkish skin.[14] Alternatively, severe photodamage may present with only deep and coarse wrinkles all over the face on a yellowish, pebbly and thickened skin, while hyperpigmentation and keratoses are rare.

Additionally, it is important to note that different photoexposed body sites may present different photoaging characteristics. Wrinkles are, of course, studied on the face because it is in this location that they are of most concern and most prominent. On the other hand, evaluation of crinkling and microtopography is best

Table 1. Typical parameters evaluated in photoaging studies

Graded Parameter*	Measurement Site	Description
overall severity[4]	face, dorsal forearm	overall general appearance
fine lines[1,21]	face	fine shallow wrinkles that usually disappear when the skin is slightly pulled
coarse wrinkles[1,21]	face	deeper and wider furrows than fine lines
mottled pigmentation[15,17,25]	face, forearms, hands	solar lentigines, hypopigmented and hyperpigmented spots
sallowness or yellowing[1]	face	appearance of the skin tone from pinkish to yellowish
brightness or clarity[15]	face	skin appearance is scored from bright to dull
roughness[17,21]	face, forearms	assessed by gently touching the skin
laxity[15,17]	face, forearms	inability of the skin to return to normal after being pulled. It is sometimes indicated by the "pinch recoil time" where the skin lateral to the zygomatic arch is pulled and suddenly released and the time for the skin to return to normal is measured in hundredths of a second.[15]

*Other parameters such as lentigines[4] (independent from pigmentation) and telangiectasia are less frequently graded.

achieved on the forearm.[11] The dorsum of hands is also often studied for mottled pigmentation.

Wrinkles and hyperpigmentation (also called mottled pigmentation) are the two major parameters studied in photoaging.

Wrinkles: Kligman has described five types of wrinkles: crinkles, linear facial wrinkles, glyphic facial wrinkles, facial creases and naso-labial folds.[13] Of these, usually the first two are the focus of topical treatments.

Linear facial wrinkles are the typical facial furrows such as "crow's-feet" in the eye region and "rhytids" on the upper lip. Their distribution is dictated by the insertion on the skin of the muscles of mimicry.

Crinkles, instead, are redundant fine folds of skin caused by dermal degradation and loss of skin elasticity. They are most noticeable on the arms and can be displaced by massaging the skin or by movements.

Mottled pigmentation: The second parameter, mottled pigmentation, refers to irregular pigmentation. Nevi are not included. Some authors have included only patchy hyperpigmentation, solar freckling and melasma[1] while others have also included lentigines and hypopigmentation.[17]

The mottling is caused by the heterogeneity in the activity of epidermal melanocytes, with some being very active while others are no longer producing melanin.[12] Since areas of decreased melanin (hypopigmentation) are also due to the damaging effects of solar radiation and contribute to the mottled appearance, they are usually included as part of the overall mottled pigmentation score.

Lentigines and melasma: At times, parameters such as lentigines and melasma are graded separately. Solar lentigines are sun-induced, well-defined, hyperpigmented lesions, which, unlike ephelides or freckles, do not fade during absence of sun exposure.

"Melasma" refers to an acquired hyperpigmentation that involves larger areas and it is usually related to hormonal variations. Melasma is graded using a specifically designed clinical index (MASI = Melasma Area and Severity Index).[18,19] Here the total face area is divided into four parts: forehead (F, 30%); right malar (MR, 30%); left malar (ML, 30%); and chin (C, 10%). The melasma in each part is graded according to the following:
- Percentage of site involvement (A) is graded 0 to 6, representing 0 to 90-100% site involvement
- Darkness (D) is graded 0 to 4, representing none to severe darkening
- Homogeneity (H) is graded 0 to 4, representing minimal to maximum homogeneity

MASI is then calculated as follows: MASI = 0.3(DF+HF)AF + 0.3(DMR+HMR)AMR + 0.3(DML+HML)AML + 0.1(DC+HC)AC. Using standardized photography the MASI score could also be calculated from images.

Clinical photoaging scales: Several clinical photoaging scales are reported in the literature.[1,3,4,17,20] A peculiar scale designed by Glogau is sometimes used for a rough categorization of the overall facial photodamage.[20] The distinctiveness of this scale is the analysis of wrinkle appearance during facial movements, which denotes a problem difficult or impossible to manage with only topical products.

In Glogau's scale the face is categorized into four types. Type I patients (mild) have "no wrinkles." Type II patients (moderate) have "wrinkles in motion." Type III patients (advanced) present "wrinkles at rest." Type IV patients (severe) appear with "only wrinkles." Because of the parameters evaluated and of their limited range of severity, the Glogau scale is not usually used to study topical formulations.

To detect improvement in at least some of the photoaging features, grading should involve evaluation of various photodamage parameters, both visual and tactile (see Table 1). The grading is often designed on a 10-point descriptive scale (0-9), where 0 equals absence, 1-3 is mild, 4-6 is moderate and 7-9 is severe.[1,2,4]

Typical antiaging studies usually enroll subjects who fit into mild to moderate overall photodamage (4-6 or 3-7) (Figure 1). Extreme grades would not be able to show changes because either the damage is too little to be quantified or it is too severe to improve with topicals. Clinical evaluations can be conducted unanchored or anchored to baseline photographs or scores.[3,4]

Photographic Evaluation of Photodamage

Anchored evaluation and photonumeric scales: Anchored evaluation may indicate that the subject's baseline photograph is used by the evaluator at each grading visit as a reference to assess the subject's progression.[3,7] On the other hand, it may also refer to the use of standard images from a photonumeric scale to match the subject severity to that image grade.

Griffith et al. have published a photonumeric scale to anchor the clinical visual assessment to a photographic grade.[1] They have presented five anchor severity grades (0, 2, 4, 6 and 8), frontal and side view, of a nine-point scale. This scale, however, is based only on wrinkle severity while hyperpigmentation is not evaluated.

Since severity of photodamage is based both on wrinkles and dyspigmentation, Larnier et al. addressed this issue by publishing a second photonumeric scale.[2] Here, they depicted three photodamage variations within each grade. Although this scale is more refined than the previous one, its six-point range may be restrictive when used to detect subtle improvement in topical antiaging studies.

Retrospective evaluation of photographs: Another photographic approach consists in the use of standardized sequential photography to evaluate the improvement of photodamage retrospectively. This method is believed to detect the more subtle within-patient improvements during antiaging treatments.[6,7]

Maddin et al. have used a system in which baseline and final photos were projected side-by-side in a random right/left manner, assuring a blinded analysis.[7] A 13-point balanced categorical scale was used by each evaluator to indicate that the photograph projected on the corresponding side was better. The center of the

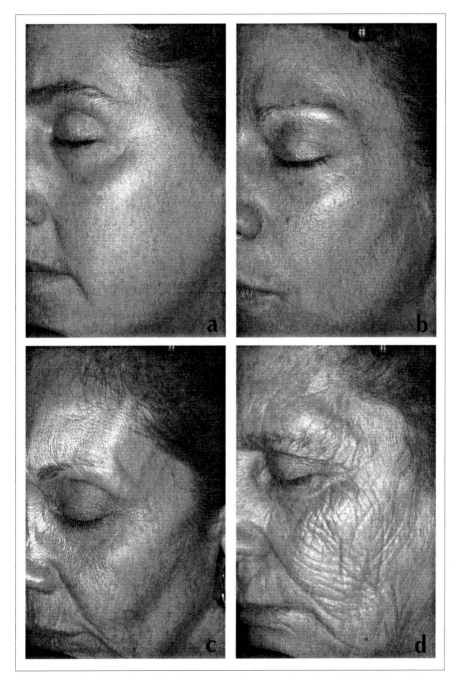

Figure 1. Examples of photoaging grades classified using a 0-9 descriptive scale for Overall Photodamage (OP), Coarse Wrinkles (CW) and Fine Lines (FL) a) OP=3, CW=0, FL=4; b) OP=4, CW=2, FL=5; c) OP=6, CW=4, FL=6; d) OP=8, CW=8, FL=0

scale corresponded to "no changes" between the two images, while six categories to the right and six to the left were used to quantify the change (much better <6>—better—slightly better—no change—<0>—slightly better—better—much better <6>).

To aid the retrospective analysis of photographs, attention should be paid to the type of photography used. For example, to study the changes in wrinkles, a "parallel polarized" photography may be better than unfiltered photography.[22-24] A discussion of the different imaging methods may help in deciding which technique would be better for a particular study.

Light Imaging Techniques

Photography is becoming an important part of testing methods, owing to recent advances in photographic systems. Various standardized photographic and lighting techniques are available that can best highlight specific skin characteristics.

Flash light versus polarized light: We define *flash photography* as an unfiltered system. It is used to document the general skin appearance (Figure 2a).

Polarized photography, on the other hand, is usually a filtered system.[22-24] It was developed to satisfy the need to enhance the visualization of either surface or sub-surface details. To achieve polarization, most photographic systems place a linear polaroid filter on each of the light sources (usually two) and a rotatable third linear polaroid filter on the camera lens.

In *parallel polarization* (Figure 2b), the light is reflected by the stratum corneum-air interface, allowing a better discrimination of the skin superficial structures (such as pores, wrinkles, fine lines and glyphics). In this case the filter on the lens is aligned parallel with the flash filters.

In *cross (perpendicular) polarization* (Figure 2c), the light is instead remitted after traversing the epidermis and papillary dermis, bringing with it information about the deeper skin components (information such as erythema, telangiectasia, hyper- or hypopigmentation). This lighting is achieved by aligning the filter on the lens perpendicularly to the flash filters.

Black-and-white UV/fluorescence photography: Fluorescence photography employs light emitting in the UVA range. As melanin strongly absorbs in the UVA, UVA photography can better detect hyperpigmentations such as freckles and melasma as well as hypopigmented lesions such as vitiligo or idiopathic guttate hypomelanosis. Sites with higher epidermal melanin appear darker compared to the surrounding area.

Particular care should be taken when photographing skin with both hyperpigmentation and inflammation. In fact, hemoglobin also attenuates UVA, and erythema may appear as dark as hyperpigmentation.[26] In subjects with inflammatory conditions, it may be impossible to distinguish on photographs what is inflammation (e.g., acne lesion) and what is hyperpigmentation.

Since UVA photography is monochromatic, its images are more suitable for quantification by image analysis than the color cross-polarized ones. Image

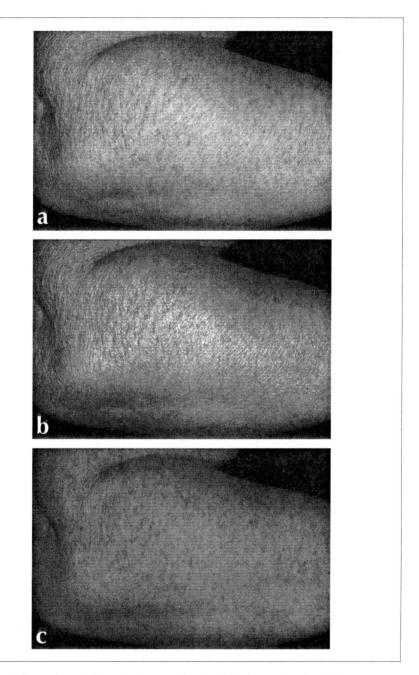

Figure 2. Examples of filtered photography (polarization) of a dorsal forearm. a) Unfiltered photograph; b) Parallel polarization enhances surface information such as wrinkling and microtopography; c) Cross polarization highlights subsurface details such as erythema and pigmented spots.

analysis may be used to quantify both the darkness of the pigmented spot, as well as its size. [25,26]

Usually, two filtered light sources (Schott UG1 filter[25]) emitting at a center wavelength of 365 nm (UVA) are positioned symmetrically at the side of the camera body. These can be substituted with Wood's light. The camera lens can be kept unfiltered or it can be fitted with a UVA-cutting filter. If UVA light reflected from the skin is allowed to pass into the camera (unfiltered), wrinkles, pores and other surface details are enhanced together with hyperpigmentation.[25] However, the intensity and delineation of hyperpigmentation is compromised by the surface reflection. Therefore, for capturing pure dyspigmentations such as lentigines, freckles and melasma, a UVA-cutting filter is placed in front of the camera lens.

Digital camera versus 35mm camera: A question often arises concerning the best type of camera to use for a study: digital or 35 mm. Although professional 35mm cameras still have overall a better resolution compared to professional digital cameras, the quality and the "risk/benefit ratio" of the latter is such that more and more studies are documented using digital photography. The resolution of two million pixels or higher in professional digital cameras is enough to achieve excellent information.

Advantages of a high-resolution digital camera system are multiple.

- Images are captured, downloaded and viewed within a couple of minutes, increasing significantly the efficiency and turnaround time. Bad images can be immediately retaken.
- Pictures are consistent in color because the development and differences in film lots are avoided.
- Repositioning of subjects, especially during high magnification, is easier because baseline images are pulled up on the computer screen to guide the subject set-up.
- Retrospective evaluation of images is easier and faster.

When using digital cameras, attention must also be paid to the type of software used for storage and database. More sophisticated software, such as the Mirror DPS,[a] can have great features, one of which is the tracking of any manipulation that occurred to the stored original picture (image authentication).

Videomicroscopy: High-resolution videomicroscopy is used to record ultra-fine in vivo changes in experimental and clinical dermatology.

An example of a UVA-videomicroscope is the Visioscan.[b] Visioscan is a hand-held black-and-white videocamera, which uses UVA light (350-400 nm, peak at 375 nm) to produce a sharp definition of surface morphology. The strong reflection of the UVA at the skin surface allows a clear definition of glyphics (Figure 3) and dry skin. Its best application in aging is in the documentation of changes in microtopography (Figure 3). The camera is easy to operate and to standardize since it works under fixed magnification on a 6 x 8 mm field of view.

---

[a] Mirror is a registered trademark of Canfield Clinical Systems, Fairfield, New Jersey, USA.
[b] Visioscan is a registered trademark of Courage+Khazaka Electronic GmbH, Cologne, Germany.

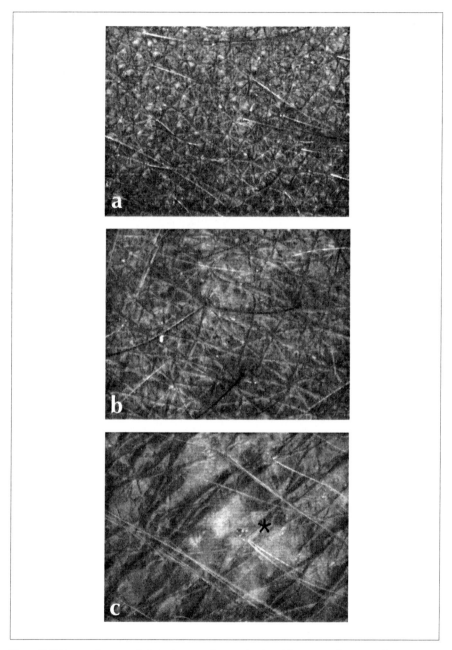

Figure 3. Visioscan images of photodamage changes in dorsal forearms of persons six years old (a), 30 years old (b) and 50 years old (c) under the same magnification.

In the young adult (b) only minor changes in glyphics are visible. In the photodamaged person (c) the microtopography is very disorganized, with some glyphics disappearing while others become more pronounced giving a "micro-wrinkled" look to the arms. Also, dyspigmentation (*) is visible.

The use of color videomicroscopy has brought new insights in the study of the skin under magnification in vivo.[27] Several types of videomicroscopes are available with magnifications ranging from 1x to approximately 700x or more.

In selecting a videomicroscope it is important to consider the resolution, the polarization system and the clarity of the images. The most useful range of magnification is 20x-80x. In photoaging, videomicroscopy is used to study hyperpigmented lesions under cross polarization and microtopography under parallel polarization mode.

Three-dimensional in vivo measuring system: Among methodologies for the measurement of skin topography, a new 3-D in vivo scanning system seems very promising.[16] Quantification of fine lines and wrinkles has usually relied upon the silicon replica technique and its evaluation by "profilometry." The need for a quantification of surface structures in vivo and real-time led to the development of PRIMOS[c] (Phaseshift Rapid In vivo Measurement Of Skin).

PRIMOS allows "contact-free" analysis of different measuring areas of the skin with a resolution of 800 x 600 or 1024 x 768 micro mirrors. It is based on the principle of the stripe projection technique, where an optical parallel stripe pattern is projected from a digital micromirror projector onto the skin area to be measured and then displayed on the chip of a CCD-matrix camera. If the skin surface presents even small differences in height, this will cause deflection of the parallel stripes, which can be quantified for the respective height profile. Quantitative assessment of roughness is usually obtained from the 2D-profile cuts using dedicated software. Additionally, by enlargement or reduction of the projection area, it is possible to make 3-D-measurements of smaller or larger areas (from 4 x 4 to 100 x 100 mm), for wounds, scars, nevi, wrinkles, limbs or even larger body parts.[16]

Conclusion

Documenting photodamage changes by photographic or imaging means has reached new levels of sophistication. These have greatly reduced the subjectiveness of photodamage grading systems and helped to identify true skin amelioration with topical treatments. A future goal will be the decrease in cost of these state-of-the-art imaging systems so that they will become standard methodology in antiaging trials.

—**A. Pagnoni,** *Hill Top Research, Inc., Milltown, New Jersey, USA*

References

1. CEM Griffith, TS Wang, TA Hamilton, JJ Voorhees and CN Ellis, A photonumeric scale for the assessment of cutaneous photodamage, *Arch Dermatol* 128 347-351 (1992)
2. C Larnier, JP Ortonne, A Venot, B Faivre, JC Beani, P Thomas, TC Brown and E Sendagorta, Evaluation of cutaneous photodamage using a photographic scale, *Br J Dermatol* 130 167-173 (1994)
3. D Piacquadio, M Dobry, S Hunt, C Andree, G Grove, KA Hollenbach, Short contact 70% glycolic acid peels as a treatment for photodamage skin. A pilot study, *Dermatol Surg* 22 449-452 (1996)

[c] PRIMOS is a product of GF Messtechnik GmbH, Teltow, Germany.

4. CEM Griffiths, S Kang, CN Ellis, KJ Kim, LJ Finkel, LC Ortiz-Ferrer, GM White, TA Hamilton and JJ Voorhees, Two concentrations of topical tretinoin (retinoic acid) cause similar improvement of photoaging but different degrees of irritation, *Arch Dermatol* 131 1037-1044 (1995)
5. BM Burke and WJ Cunliffe, The assessment of acne vulgaris: the Leeds technique, *Br J Dermatol* 111 83-92 (1983)
6. RB Armstrong et al, Clinical panel assessment of photodamage skin treated with isotretinoin using photographs, *Arch Dermatol*, 128 352-356 (1992)
7. S Maddin, J Lauharanta, P Agache, L Burrows, M Zultak, L Bulger, Isotretinoin improves the appearance of photodamaged skin: Results of a 36-week, multicenter, double-blind, placebo-controlled trial, *J Am Acad Dermatol* 42 56-63 (2000)
8. BA Gilchrest, A review of skin aging and its medical therapy, *Br J Dermatol* 135 867-975 (1996)
9. CR Taylor, RS Stern, JJ Leyden and BA Gilchrest, Photoaging/photodamage and photoprotection, *J Am Acad Dermatol* 22 1-15 (1990)
10. BA Gilchrest and M Yaar, Ageing and photoageing of the skin: observation at the cellular and molecular level, *Br J Dermatol* 127 25-30 (1992)
11. A Pagnoni, AM Kligman, I Sadiq and T Stoudemayer, Hypopigmented macules of photodamaged skin and their treatment with topical tretinoin, *Acta Dermato-Venereologica* 79 305-310 (1999)
12. JP Ortonne, Pigmentary disorders associated with sun exposure, *J Dermatol Treatment* Suppl 2 S7-S8 (1996)
13. AM Kligman, The classification and treatment of wrinkles, in *Cutaneous Aging*, AM Kligman and Y Takase, eds, Tokyo: University of Tokyo Press (1988) 547-555
14. R Marks, Aging and Photodamage, in *Sun-Damaged Skin*, R Marks, ed, London: Martin Dunitz (1992) 13-34
15. A six-month clinical study to evaluate the long-term efficacy and safety of an AHA lotion, *Cosmet Derm* 9 33-40 (1996)
16. S Jaspers, H Hopermann, G Sauermann, U Hoppe, R Lunderstadt and J Ennen, Rapid in vivo measurement of the topography of human skin by active image triangulation using a digital micromirror device, *Skin Res Technol* 5 195-207 (1999)
17. J Sefton, AK Kligman, SC Kopper, JC Lue and JR Gibson, Photodamage pilot study: A double-blind, vehicle-controlled study to assess the efficacy and safety of tazarotene 0.1% gel, *J Am Acad Dermatol* 43 656-63 (2000)
18. CK Kimbrough-Green et al, Topical retinoic acid (Tretinoin) for melasma in black patients, *Arch Dermatol* 130 727-733 (1994)
19. N Lawrence, SE Cox and HJ Brody, Treatment of melasma with Jessner's solution versus glycolic acid: a comparison of clinical efficacy and evaluation of the predictive ability of Wood's light examination, *J Am Acad Dermatol* 36 589-593 (1997)
20. RG Glogau, Aesthetic and anatomic analysis of the aging skin, *Semin Cutan Med Surg* 15(3) 134-138 (1996)
21. LM Harnisch, MK Raheja, LK Lockhart, A Lopez and A Gabbianelli, Substantiating antiaging product claims, *Cosmet Toil* 114(10) 33-47 (1999)
22. JA Muccini, N Kollias, SB Phillips, RR Anderson, AJ Sober, MJ Stiller and LA Drake, Polarized light photography in the evaluation of photoaging, *J Am Acad Dermatol* 33 765-769 (1995)
23. J Philp, NJ Carter and CP Lenn, Improved optical discrimination of skin with polarized light, *J Soc Cosmet Chem* 39 121-132 (1988)
24. RR Anderson, Polarized light examination and photography of the skin, *Arch Dermatol* 127(7) 1000-1005 (1991)
25. N Kollias, R Gillies, C Cohen-Goihman, SB Phillips, JA Muccini, MJ Stiller and LA Drake, Fluorescence photography in the evaluation of hyperpigmentation in photodamaged skin, *J Am Acad Dermatol* 36 226-230 (1997)
26. A Pagnoni, AM Kligman and T Stoudemayer, UVA photography to monitor the progression

of freckles in young children, 55th meeting, American Academy of Dermatology, P 362, Mar 21-26, 1997
27. PL Dorogi and EM Jackson, In vivo video microscopy of human skin using polarized light, *J Toxicol - Cut & Ocular Toxicol* 13 97-107 (1994)

Influence of a Formulation's pH on Cutaneous Absorption of Ascorbic Acid

Keywords: ascorbic acid, pH, penetration

The level of ascorbic acid penetrating into guinea pig skin from a gel formulation varied with the pH of the formulation

Ascorbic acid (AA) has important physiological effects on the skin, including inhibition of melanogenesis, promotion of collagen biosynthesis and prevention of free radical formation.[1-3,5-9,13] Thus, AA is an excellent active ingredient for skin care.

The process of developing vehicle formulations is fundamental to the performance of final products.[11] A study by Kassem et al.[3] indicates that pH can influence the stability of ascorbic acid. These authors concluded that a pH of approximately 6.5 is ideal for maintaining ascorbic acid stability in aqueous solution. For any final product containing this type of active acid substance, the active's effectiveness depends upon the effect that the formulation's pH has on the cutaneous absorption of the active.

Smith[10] showed that topical treatment using formulations containing alpha hydroxy acids produces prolonged hydration when the formulation's pH is at high levels. At lower pH, however, the most noticeable effect with those formulations is peeling.[4]

These facts suggest that the pH of a formulation can provoke fundamental changes in the action of the final product.

At the pHs typically found in skin care products, AA is present in its ionized form. If it is not in the ionized form, then the pH value is typically higher than the pK_a value of AA, which is 4.2.[6] (The pK_a value, related to the dissociation constant K_a, is defined as the pH value when the ionized and un-ionized forms exist in the ratio of 1:1.) Having the pH value higher than the pK_a value can decrease the cutaneous absorption of ascorbic acid.

Presented at the XIV Congresso Latinoamericano e Ibérico de Quimicos Cosméticos & IFSCC International Conference, Santiago de Chile, May 23-26, 1999.

Formula 1. A nonionic gel

Ascorbic acid	2.0%
Hydroxyethylcellulose	2.0
Propylene glycol	5.0
Methyldibromo glutaronitrile (and) phenoxyethanol	0.2
Disodium EDTA	0.1
Sodium metabisulfite	0.5
Citrate buffer*	qs 100.0

*Buffer to pH of 2.8, 4.5 or 5.3

The focus of this study was to investigate the influence of a formulation's pH on the penetration of ascorbic acid into the skin.

Method

We used a nonionic gel formulation (Formula 1), buffered with citrate to pH 2.8, 4.5 or 5.3. All formulations contained 2% ascorbic acid.

The permeation of AA into the skin was studied by using a glass Franz diffusion cell model, where the receiver phase was phosphate buffer, pH 7.2. Two grams of each test formulation were applied to the abdomen skin of adult female guinea pigs (average weight 350 g). The receiver volume used for each cell was

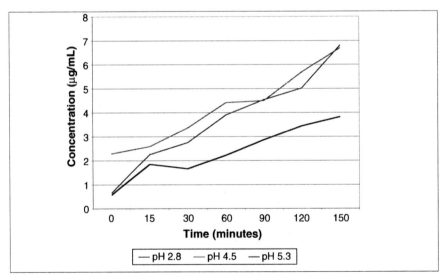

Figure 1. Effect of formulation pH on penetration of ascorbic acid into guinea pig skin in vitro over time (average value of five penetration measurements at each time point)

Table 1. Friedman test (n=5) statistical results from concentrations of ascorbic acid in the receiver solution

Formulation pH pairs compared	p value
2.8 and 4.5	non-significant
2.8 and 5.3	< 0.001
4.5 and 5.3	< 0.05

calibrated prior to use, and was constantly stirred with a Teflon-coated bar magnet. A uniform 37°C receiver temperature was maintained with a jacket connected to a circulating bath.[12]

Portions were obtained at 0, 30, 60, 90, 120 and 150 min. The amount of AA in the receiver phase was obtained by ultraviolet absorption analysis[a] at 254 nm. Five readings were taken at each time point. Data were analyzed statistically using the non-parametric Friedman test.

Results

Figure 1 shows the averaged results from this in vitro examination of AA penetration into guinea pig skin. Table 1 shows the results of the non-parametric Friedman test.

Discussion

In our experiments, the formulations were all the same, except for their pHs. The nonionic gel formulation was suggested for these experiments based on our previous studies in which we compared the cutaneous penetration in a nonionic gel, a nonionic gel cream and a nonionic o/w cream.

The cutaneous penetration of the active principle depends on certain characteristics of the vehicle. These characteristics include ideal pH and viscosity.

The level of AA penetration into the guinea pig skin increased throughout the time studied for all the test formulations. We observed that the amount of AA in the receiver solution and the velocity of its penetration varied according to the pH of the gel tested.

However, we observed that no statistically significant levels of penetration were obtained from formulations whose pH was between 2.8 and 4.5. We observed the smallest penetration into the skin (p< 0.05) when the formulation pH was 5.3 (Figure 1). These differences in the absorption levels are probably due to the nearness of formulation pH to the AA pK_a value.[6]

The more effective formulations in this study were those with pH 4.5 and 2.8; they were equal statistically, which agrees with results described in the technical

---
[a] Spectrophotometer DU 640 from Beckman Instruments, Beckman Coulter Inc., Fullerton, California USA

literature.[4] On the other hand, the literature cites cases in which an active acid substance was altered by the pH.[4] With AA, one wants the highest possible penetration into the skin because AA's effects occur in the dermis.

Thus, during the development of a skin care formulation, it is essential to define both the cutaneous penetration level required for the active substance and the influence that the vehicle may have on the efficacy of the active substance.

Conclusion

Under the experimental conditions described, the levels of ascorbic acid penetration into the skin varied according to the pH of the formulation used.

The gel formulations with pH 4.5 and 2.8 showed higher levels of absorption than the gel formulation with pH 5.3. A pH of 4.5 is compatible with the skin and favorable to its normal metabolism. Therefore, we recommend a pH of 4.5 for skin care products containing ascorbic acid as the major active substance.

—**Gisele Mara Silva and Patricia M. B. G. Maia Campos,** *Pharmaceutical Sciences of Ribeirão Preto, University of São Paulo, Brazil*

References

1. D Darr and I Fridovich, Free radicals in cutaneous biology, *J Invest Dermatol* 102(5) 670 (1994)
2. K Iozumi, GE Hogansom, R Pennela, MA Everett and B Fuller, Role of tyrosinase as the determinant of pigmentation in cultured human melanocytes, *J Invest Dermatol* 100(6) 806 (1993)
3. MA Kassem, AA Kassem, HO Ammar, Studies on the stability of injectable L-ascorbic acid solutions, *Pharm Act Helvet* 44 611 (1969)
4. PMBG Maia Campos and GR Leonardi, Emprego de alfa-hidrósi-ácidos em produtos cosméticos e dermatológicos, *Rev Cosm & Med Est* (3 trimestre) (3) 26-29 (1997).
5. CL Phillips, SB Combs and SR Pinnell, Effects of ascorbic acid on proliferation and collagen syntnesis in relation to the donor age of human dermal fibroblasts, *J Invest Dermatol* 103(2) 229 (1994)
6. SR Pinnell, Vitamina C tópica, *Rev Cosm Med Est* 3(4) 31 (1995)
7. G Prota, Regulatory mechanisms of melanogenesis: Beyond the tyrosinase concept, *J Invest Dermatol* 100(2) (Supp) 156-S (1993)
8. Y Shindo, E Witt and L Packer, Antioxidant defense mechanisms in murine epidermis and dermis and their responses to UL light, *J Invest Dermatol* 100(3) 261 (1993)
9. M Ponec, A Weerheim, J Kempenaar, A Mulder, GS Gooris, J Bouwstra and AM Mommaas, The formation of competent barrier lipids in reconstructed human epidermis requires the presence of vitamin C, *Soc Inv Derm* 109(3) 348-355 (1997)
10. WP Smith, Comparative effectiveness of !a-hydroxy acids on skin properties, *Int J Cosmet Sci* 18 75-83 (1996)
11. R Hermitte, Aged skin: Retinoids and Alpha Hydroxy Acids, *Cosmet Toil* 107(7) 63-67 (1992); see also, Pele envelhecida: retinóides e alfa-hidróxi ácidos, *Cosmet Toil* (Portuguese edition) 5(5) 55-58 (1993)
12. PMBG Maia Campos and GM Silva, Ascorbic acid and its derivatives in cosmetic formulations, Cosmet Toil 115(6) 59-62 (2000)
13. GM Silva and PMBG Maia Campos, Histopathological, morphometric and stereological studies of ascorbic acid and magnesium ascorbyl phosphate in a skin care formulation, Int J Cos Sci 22(3) 169-179 (2000)

O/w Emulsions Enriched with Vitamin E

Keywords: vitamin E, o/w emulsions, skin moisture, skin barrier

Effect of o/w emulsions with vitamin E on moisture content and barrier function was studied

Vitamin E has an important protective function for the entire organism. It is believed that the broad biological activities of vitamin E are due to its ability to inhibit lipid peroxidation and stabilize biological membranes.

When used in skin care products, vitamin E protects the skin from UV light, may reduce the appearance of facial lines and wrinkles and has a favorable effect on the skin moisture content. However, the effects of vitamin E largely depend on both the concentration and the type of the vehicle. In moisturizing cosmetic products, vitamin E acetate is usually used in the 2-10% range, although studies have shown that the most effective products contain 5% vitamin E acetate.[1-3]

O/w emulsions and lotions are commonly used as vehicles for vitamin E because the penetration and cutaneous absorption of the fat-soluble ingredient may be significantly improved by applying it in hydrophilic vehicles.[2]

Polymeric emulsifiers such as acrylates/C_{10}-C_{30} alkyl acrylate crosspolymer produce emulsions of o/w type with good aesthetic and tactile properties. Emulsions made with acrylates/C_{10}-C_{30} alkyl acrylate crosspolymer have a triggered release mechanism: the acrylic hydrophilic portion of the polymeric emulsifier hydrogel instantly collapses upon contact with salt content on the skin. This results in the rapid release of the oil phase and immediate formation of an occlusive layer that contains no residual surfactants and is not prone to removal through immersion in water.[4]

The main objective of our study was to formulate simple and stable o/w emulsions with vitamin E acetate (5%) using the novel polymeric emulsifier acrylates/C_{10}-C_{30} alkyl acrylate crosspolymer at a constant amount (25%) but with different types of oil phase. In addition, we performed dermatological studies to evaluate the effects of these emulsions on skin surface lipids, moisture content and skin barrier function.

Materials and Methods

The compositions of the test emulsions with and without vitamin E acetate (5% w/w) are presented in Table 1. The same table also shows the suppliers of the polymeric

Table 1. Compositions (% w/w) of the test emulsions using acrylates/C10-C30 alkyl acrylate crosspolymer as the polymeric emulsifier

	E1	P1	E2	P2	E3	P3
Acrylates/C10-C30 alkyl acrylate crosspolymer (Pemulen TR-1, BF Goodrich, Brecksville, OH USA)	0.3	0.3	0.3	0.3	0.3	0.3
Oleyl oleate (Cetiol, Henkel, Germany)	25.0	25.0	-	-	-	-
Decyl oleate (Tegosoft DO, Goldschmidt, Germany)	-	-	25.0	25.0	-	-
Octyl stearate (Tegosoft OS, Goldschmidt, Germany)	-	-	-	-	25.0	25.0
Vitamin E acetate (Hoffmann LaRoche, Austria)	5.0	-	5.0	-	5.0	-
Triethanolamine, 10% sol	qs	qs	qs	qs	qs	qs
Methylparaben (and) propylparaben (and) diazolidinyl urea (and) propylene glycol (Germaben II, Sutton Labs, NJ USA)	1.0	1.0	1.0	1.0	1.0	1.0
Water (aqua), purified, qs to	100.0	100.0	100.0	100.0	100.0	100.0

emulsifier and other ingredients. Oleyl oleate, triethanolamine and purified water were pharmacopoeia quality.[5]

Emulsions with vitamin E acetate were prepared using a direct method. The polymeric emulsifier was dispersed in the water phase (water and preservative). The lipophilic phase (emollient oil and vitamin E acetate) was slowly added under constant agitation, and the dispersion was neutralized with triethanolamine (10% w/w). Placebo emulsions were prepared in the same manner, except that the level of water phase was increased to compensate for the absence of vitamin E acetate.

Test samples were prepared at the room temperature and evaluated using physico-chemical and dermatological studies.

Physico-Chemical Studies

All physico-chemical measurements were performed 48 h after sample preparation. Appearance, tackiness and spreadability of the emulsion samples were evaluated using visual examination and sensory studies. Specific conductivity was measured instrumentally.[a] pH value was determined using a pH meter.[b]

Dermatological Studies

Subjects: Twenty healthy female volunteers (mean age 30±7), with no history or clinical signs of dermatological disease or dry and scaly skin, were enrolled in the studies of skin surface lipids content and hydration. Six female volunteers (mean age 40±5) were enrolled in the transepidermal water loss (TEWL) study.

---

[a] Conductometer CDM 230, Radiometer, Copenhagen, Denmark
[b] HI 8417, Hanna Instruments Inc., Woonsocket, Rhode Island, USA

Volunteers were instructed not to use moisturizers for at least seven days prior to the study. Informed consent was obtained for all participants, and the local ethics committee approved the study.

Short-term studies: After 20 min of adaptation to laboratory conditions, 2 mg/cm^2 of the emulsions were applied at precisely delineated test areas of the forearms.[6] To circumvent the effects of vehicle on the investigated parameters, the treatments were labeled as follows: P1, P2 and P3 were placebo treatments (emulsions without vitamin E acetate); E1, E2 and E3 were active treatments (vitamin E enriched emulsions); and treatment C was the control (no product applied).

The effects of the tested samples on skin surface lipids and moisture content as well as on skin barrier function were evaluated for 6 h. The skin surface lipids content, electrical capacitance and TEWL were measured before application of the emulsions and after 30, 60, 120, 240 and 360 min.

Long-term study: The study period was 30 days, with the volunteers applying test emulsions twice daily for 28 days. The treatments were labeled in the same manner as in the short-term studies. The participants were not allowed to use any other product on each forearm for the duration of the trial. Participants were instructed how to apply the test emulsions and how to fill out a checklist for daily recording of the treatment.

The skin hydration measurements were performed during the study on day 1 (baseline value), and on days 3, 7, 14, 21 and 28, as well as on day 30 (two days after the last treatment day). All measurements were obtained 12-16 h after last product application, as suggested in literature.[6]

Evaluation methods: Evaluation was carried out using the non invasive bioengineering techniques for measurement of electrical capacitance of the skin, TEWL and skin surface lipids.

The electrical capacitance of the skin is an indicator of the hydration level of the stratum corneum and was measured using a corneometer.[c] The probe was applied to the skin with a constant pressure for a recording time of 1 sec. Three consecutive measurements were performed for each subject, and the mean was reported as the capacitance value in relative corneometer units (RCU).

TEWL is the passive diffusion of water through the stratum corneum. It is an indicator of the integrity of the stratum corneum. TEWL was quantified instrumentally[d] following the guidelines of the Standardization Group of the European Society of Contact Dermatitis.[7] The probe was held in place perpendicular to the skin surface until a stable TEWL value was established. The average value of three recordings was taken as the TEWL value (mg/cm^2/h).

Skin surface lipids content was determined using a sebumeter[e], which consists of sebumeter cassette and main instrument housing. The sebumeter cassette consists of a measuring head covered with plastic tape of a certain opacity that becomes translucent in contact with surface lipids. The degree of transparency is proportional to the amount of skin surface lipids. For evaluation of strip

---

[c] Corneometer CM 900, Courage + Khazaka GmbH, Cologne, Germany
[d] Tewameter, Courage + Khazaka GmbH, Cologne, Germany
[e] Sebumeter SM 810, Courage + Khazaka GmbH, Cologne, Germany

transparency, the measuring head of the cassette was inserted into the aperture of the device where a photocell measured the transparency. The result was evaluated by microprocessor and transferred to a digital readout that was shown on the display in $\mu g/cm^2$ of skin.[6,7]

The relative humidity in the room where measurements took place was 60 ± 5%, and the ambient room temperature was 20 ± 2°C.

Statistics: Student t-test for paired data was used for comparative studies. A significance level of $p < 0.05$ was chosen.

Results

Physico-chemical characteristics: Table 2 shows the results of pH and specific conductivity studies of the test emulsions.

Short-term hydration: The short-term effects of o/w vitamin E acetate enriched emulsions and placebo emulsions (without vitamin E) on skin moisture content are shown in Figure 1.

Six hours following application, placebo emulsions had slightly increased skin moisture content of 1, 1.5 and 1.7% for emulsions P1, P2 and P3, respectively. Emulsions with vitamin E acetate increased skin capacitance 6.54, 9.04 and 10.45% for emulsions E1, E2 and E3, respectively ($p < 0.05$).

Placebo emulsions had slightly increased skin moisture content ($p > 0.05$) indicating that the occlusive effect of the oil phase is not a major factor in the hydrating potential of the test emulsions.[8]

Long-term hydration: The long-term study was carried out in order to explore the possibility of cumulative effects of vitamin E acetate on skin moisture content. Figure 2 shows the results in the form of percentage changes over time.

The mean readings for the vitamin E acetate-treated skin were significantly higher from day 3 through day 28 as well as on the treatment-free days 29 and 30 ($p < 0.05$). Four weeks after treatment with vitamin E enriched emulsions, skin moisture content increased 61.24, 63.87 and 70.17% in samples E1, E2 and E3, respectively. Two days after the final treatment (day 30), skin moisture content on

Table 2. pH and specific conductivity values of the emulsion samples

Sample	pH value	Specific Conductivity ($\mu S/cm^2$)
E1	5.2	276.6
E2	5.1	287.4
E3	5.05	302.3
P1	5.55	286.6
P2	5.76	294.6
P3	5.84	332.3

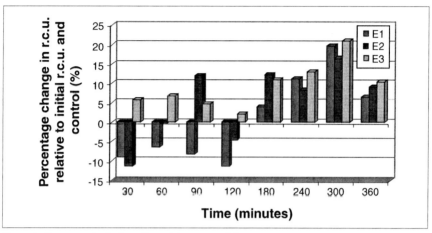

Figure 1. Short-term hydrating effect of the test emulsions

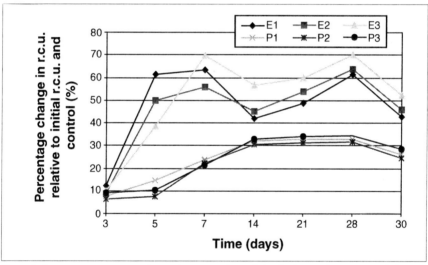

Figure 2. Long-term percentage change in mean values of skin hydration during twice-daily applications of vitamin E acetate emulsions and placebo emulsions

areas treated with vitamin E acetate emulsions E1, E2 and E3 was higher than the starting value by 42.89, 46.06 and 52.12%, respectively. Statistical analysis showed that the order of moisturizing potential of the test emulsions was E3>E2>E1 ($p < 0.05$).

TEWL: The effects of creams with vitamin E acetate and placebo emulsions (without vitamin E) on TEWL are shown in Figure 3.

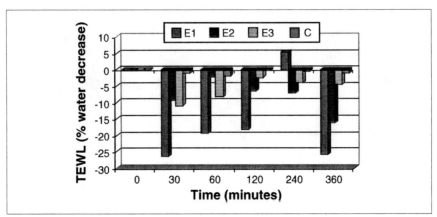

Figure 3. Effect of test emulsions on TEWL

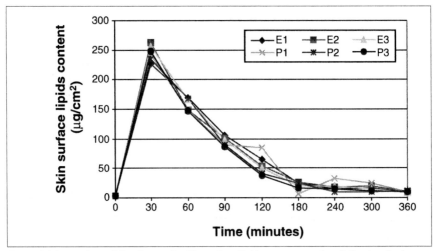

Figure 4. Mean values of the changes in skin surface lipid content over time following single application of the test emulsions

Six hours after application, placebo emulsions had no statistically significant effect on TEWL (p> 0.05) (data not shown). However, as Figure 3 shows, creams with vitamin E acetate decreased TEWL by 25.00, 16.26 and 4.64% for emulsions E1, E2 and E3, respectively (p< 0.05).

Skin surface lipids content: The effects of creams with vitamin E acetate and placebo emulsions (without vitamin E) on skin surface lipid content are shown in Figure 4.

During the trial, all samples significantly increased skin sebum content compared to baseline value (p< 0.05). Measurements of skin surface lipids showed a broad peak between 0 and 90 min followed by a gradual decline toward 0 during the remainder of the examination period. Six hours after treatment, skin surface lipids content was 7.86, 10.13 and 8.05 µg/cm^2 for emulsions E1, E2 and E3, respectively ($p< 0.05$).

We detected no statistically significant difference ($p> 0.05$) between samples on skin surface lipids content, which indicates that, in the case of the studied emulsions, the type of the oil phase is not the main factor affecting the change in the skin surface lipids content.

Discussion

Physico-chemical: Sensory and visual studies showed that all samples possess required aesthetics and spreadability. Values of specific conductivity (276.6-332.3) indicate that all emulsions were o/w type. The pH values of all test samples were mildly acidic (5.05-5.84), therefore corresponding to general requirements for topical preparations.

Short-term hydration: Shortly after the vitamin E acetate emulsions were applied (Figure 1), we saw an increase in the hydration of the horny layer followed by a plateau value and a slow decrease of the measured values toward baseline, according to the efficacy of the emulsions ($p< 0.05$).

The plateau value or slow decrease is the result of the effect of the lipids contained in the formulation and left on the skin surface after evaporation of water. Probably during the lipidization phase, emulsion lipids penetrated into the outer epidermis, with possible consequences for epidermal hydration.[9,10] In addition, during this same phase, it is probable that vitamin E acetate penetrated into the skin and increased skin moisture content due to its antioxidative and membrane-stabilizing effect.[1]

Long-term hydration: Long-term studies of moisturizing potential of vitamin E acetate emulsions indicated the cumulative effect of vitamin E acetate. Between the first and fourth weeks of treatment, we found a slight decrease in moisturizing potential of the emulsions with vitamin E acetate in the second and third weeks of treatment (Figure 2). The reason for these variations is unknown, but they might be due to the effect of environmental factors on the instrument or on the skin surface.[10] Placebo emulsions also increased skin moisture content during four weeks of treatment ($p< 0.05$).

The long-term hydration study did reveal a significant difference between the moisturizing potential of the placebo emulsions and that of the o/w vitamin E acetate enriched emulsions, whose great moisturizing potential is due to the cumulative effect of vitamin E acetate. Furthermore, two days after the end of treatment we observed a significant decrease in skin hydration, indicating that the amount of vitamin E acetate in the skin had been decreased.

TEWL: Hydration of the skin is known to affect its barrier function and thereby exerts an effect on the diffusion of water itself and on penetration of the other substances across the stratum corneum. However, the relationship between TEWL and skin moisture content is very complex. Some authors suggest that direct correlation between skin moisture content and TEWL can be established whereas others suggest that increased skin hydration is accompanied with no change in TEWL.[8,11]

In the present study, the increase in capacitance after short-term treatment with vitamin E acetate emulsions (E1, E2 and E3) was not accompanied by a simultaneous increase in TEWL, indicating that the skin was not hydrated to the extent that its water permeability was affected. Moreover, obtained results suggest that test emulsions had slightly decreased TEWL (Figure 3). This slight reduction of TEWL indicated that emulsion samples did not affect barrier function of the normal skin.

Skin surface lipids content: Measurements of skin surface lipids content showed that the largest increase was observed 30-90 min after sample application, followed by a decrease in the next few hours (Figure 4). This effect is probably due to the absorption of the applied oil phase by the stratum corneum.

Although there were no statistically significant differences between the samples regarding skin surface lipid content, the obtained data (Figure 3) suggests slower clearance of lipid components of emulsion E1 from the skin surface. In addition, emulsion E1 had the greatest affect on TEWL and the least effect on skin moisture content compared to the other emulsions. E1 was formulated using oleyl oleate as oil phase. Oleyl oleate is known to be an occlusive emollient, thus decreased TEWL is probably the consequence of the slow clearing of the nonvolatile emulsion lipid and vitamin E acetate from the skin surface.[12,13]

In emulsion E3, octyl stearate was used as an oil phase. Octyl stearate is known to have great compatibility with skin lipids. Thus, it seems probable that the hydrating potential of emulsion E3 was more likely due to penetration of the oil phase into the outer epidermis, than to its occlusive effect.[12,14]

Conclusion

The aim of our study was to formulate simple and stable o/w emulsions with vitamin E acetate (5%) using a novel polymeric emulsifier acrylates/C_{10}-C_{30} alkyl acrylate crosspolymer and an oil phase of constant amount but different types. Furthermore, we wanted to evaluate the effect of oil phase of an o/w emulsion with vitamin E acetate on skin moisture content and skin barrier function in dermatological studies.

Physico-chemical studies showed that all samples were o/w type, with good sensory properties and required pH (5.05-5.84).

The short-term and long-term studies revealed significant difference between the moisturizing potential of o/w vitamin E acetate enriched emulsions and placebo emulsions. However, placebo emulsions had no statistically significant effect on TEWL whereas emulsions with vitamin E acetate slightly decreased TEWL, indicating that vitamin E acetate-enriched emulsions did not change the barrier function of the normal skin. In addition, there were no statistically significant differences in skin surface lipid content between placebo and vitamin E acetate-enriched emulsions.

Results of our investigation showed that the type of oil phase used in the test emulsions had a great influence on their short-term and long-term moisturizing potential as well as on skin barrier function (p< 0.05). However, the type of oil phase had no influence on skin surface lipid content (p >0.05). Regarding the fact that all emulsions were formulated in the same manner and differed only in type of emollient oil, it is reasonable to assume that the observed difference is due to different affinity of vitamin E acetate to the vehicle and, thus, different rates of penetration of vitamin E acetate into the skin.[1-4,6] We can speculate that after penetration into the skin, vitamin E acetate has restored impaired barrier function and increased skin moisture content due to its membrane stabilizing and occlusive effect. To better understand the observed effects, we have started penetration studies of vitamin E acetate from the emulsion samples (manuscript in preparation).

—**Jelena Djordjevic, Gordana Vuleta, Jela Milic,** *Department of Pharmaceutical Technology and Cosmetology, University of Belgrade, Belgrade, Yugoslavia*
—**Hongbo Zhai, Howard Maibach,** *Department of Dermatology, University of California at San Francisco, San Francisco, California, USA*

References

1. M Rangarajan and J Zatz, Skin delivery of vitamin E, *Int J Cosm Sci* 50 249-279 (1999)
2. R Brigelius and M Traber, Vitamin E: function and metabolism, *FASEB* 13 1145-1153 (1999)
3. J Djordjevic and G Vuleta, Effect of formulations with different concentrations of vitamin E and emulsion type on sebum content and skin moisture, in *Proceedings of the Symposium on Lipid and Surfactant Dispersed Systems*, Moscow: APGI Publications (1999) 247-248
4. J Castner et al, Novel skin care formulations with optimized sensory parameters using acrylates/C10-30 alkyl acrylate crosspolymers as primary emulsifiers, in *In-Cosmetics Conference Proceedings*, Dusseldorf: H Ziolkowsky (1997) 247-248
5. *Pharmacopoeia Jugoslavica*, ed IV, Beograd: Savezni zavod za zastitu zdravlja (1984)
6. J Serup, Bioengineering and the skin: Standardization, *Clinics in Dermatology* 13 293-297 (1995)
7. H Elden, Protocols and strategies for biophysical testing, in *Clinical Safety and Efficacy Testing of Cosmetics*, W Waggoner, ed, New York: Marcel Dekker (1995) 93-107
8. W Hannon and H Maibach, Efficacy of moisturizers assessed through bioengin-eering techniques, in *Textbook of Cosmetic Dermatology*, R Baran, ed, London, UK: Martin Dunitz Ltd (1998) 529-535
9. M Loden and M Lindberg, Product testing: Testing of moisturizers (chapter 24), in *Bioengineering of the Skin: Water and the Stratum Corneum*, P Elsner, ed, Boca Raton: CRC Press (1997) 275-289
10. AO Barel et al, In vivo evaluation of the hydration state of the skin, in *Cosmetics: Controlled Efficacy Studies and Regulation*, P Elsner, ed, Berlin: Springer-Verlag (1999) 56-78
11. E Berardesca and HI Maibach, Stratum corneum water content and TEWL, in *Textbook of Cosmetic Dermatology*, R Baran, ed, London: Martin Dunitz Ltd (1998) 529-535
12. B Salka, Choosing emollients, *Cosmet Toil* 112(10) 101-106 (1997)
13. J Trivedi, S Krill and J Fort, Vitamin E as a human skin penetration enhancer, *Eur J Pharm Sci* 3 241-243 (1995)

14. T Dietz, Basic properties of cosmetic oils and their relevance to emulsion preparation, *SOFW* 125 2-7 (1999)

A Triply Stabilized System to Improve Retinol Stability

Keywords: wrinkles, retinol, encapsulation, liposome

Stability and skin penetration of retinol can be increased by triply encapsulating it

Retinol is an especially powerful ingredient to reduce wrinkles, but its stability depends on environmental factors such as solvent, temperature, oxygen and light. In this chapter we describe how its stability and skin penetration can be improved by using a triply stabilized system (TSS).

Retinol has a wide variety of biological functions such as immune reactions, epidermal differentiation, vision in vertebrates and stimulating embryonic growth and development. It is also a prime candidate for cancer prevention.[1-4]

Vitamin A is the generic name for a class of nutritionally active and unsaturated hydrocarbons. It is present in the animal kingdom as vitamin A_1 (retinol) and vitamin A_2 (3-dehydro-retinol), and in the vegetable kingdom as carotenoid (Figure 1).

Vitamin A_2 has approximately 40% of the effect of vitamin A_1, and both A_1 and A_2 exist in the ester form of fatty acid. Retinol contains at least one non-oxygenated beta-ionone ring with an attached isoprenoid side chain. Retinol that contains all trans double bonds in the isoprenoid side chain is the most bioactive form, however its efficaciousness declines over time because vitamin activity is decreased by isomerization, photochemical oxidation and thermal oxidation. Such degradation reactions can also reduce vitamin activity of stored and processed foods. In general, the stability of retinol and its relatives is slightly reduced in conditions of high humidity, low pH and high temperature.[5-6]

Retinol is a fat-soluble material abundant in fish and mammalian liver, milk fat and egg yolks. Due to its hydrophobic character, it is usually found in a complex with lipid droplets or micelles. Therefore if liposome technology could be applied, we could expect to protect retinol from reactions that degrade it.[7-9]

Liposomes are spherical closed vesicles of phospholipid bilayers with an entrapped aqueous phase. The lipid layers are mainly made up of phospholipids that have amphophilic character. In aqueous solution, they are arranged in bilayers, which form closed vesicles like artificial cells.

In the cosmetic area, liposomes are used to stabilize unstable materials against exterior conditions, maximize their efficacy and enhance skin absorption because of the phospholipid's great affinity for skin. The stability and delivery of liposome-incorporated retinol have been studied in several articles. However, the stability of retinol in liposomes has not been sufficiently studied.[10-11] *(Editor's note: The accompanying Editor's Note lists some other techniques that have been suggested for improving the stability of retinol in cosmetic formulations-see page 485.)*

The Triply Stabilized System

In this study reported here, the stability and skin penetration of retinol are improved by a triply stabilized system (TSS). According to this system:

1. A porous silica is prepared and retinol is adsorbed in the pores.
2. The porous silica and any remaining retinol are encapsulated in solid lipid nanoparticles (SLN) to improve retinol's stability against light and heat.
3. The SLN are then organized into a multi-lamellar matrix of skin lipids (SLM) to improve retinol's skin penetration effect (Figure 2).

Preparation of the TSS

Adsorbing retinol into silica pores: Porous silica was prepared using the sol-gel method (see sidebar). One of the most important reasons for using porous silica is to give thermal and chemical stability that are not easily obtained from organic materials. Then retinol was entrapped into the porous silica by dispersion to give thermal and chemical stability and induce the diffusion of the active ingredient during long periods, giving an extended-release property.

Creating the SLN liposome system: Using a system of non-phospholipid vesicles (NPV) and a high-pressure homogenizer[a], the free retinol and the retinol that was adsorbed in silica pores were encapsulated to give double stabilized retinol. We call this system the Solid Lipid Nanoparticles (SLN) liposome system because it contains solid particles in its core. We also refer to this as the "primary liposome." Primary liposomes are non-phospholipid vesicles that contain free retinol and retinol entrapped in silica. Primary liposomes are prepared by steps 1 and 2 of the TSS. In contrast, we will also refer to a general liposome system of non-phospholipid vesicles that contain only retinol liquid in their core.

Synthesizing Skin Lipid Matrix and TSS: The Skin Lipid Matrix (SLM) is a high viscous compound consisting of 20% hydrogenated phosphatidyl choline, 20% caprylic (and) capric triglyceride, 10% propylene glycol and 50% ceramide 3 and other ingredients. We mixed and heated all the ingredients together until they dissolved. Then we used high-pressure homogenizer to get the SLM. This matrix also contains membranes in the form of vesicles and multi-lamellar sheets and can be used as an emulsifier.

After obtaining the SLM, we slowly stirred the primary liposome and the SLM with the homogenizer to make the TSS. In other words, we combined primary liposomes (steps 1 and 2) and SLM (step 3) to get the TSS.

---

[a] High-pressure Homogenizer, Model M110F, Microfluidics, Newton, Massachusetts, USA

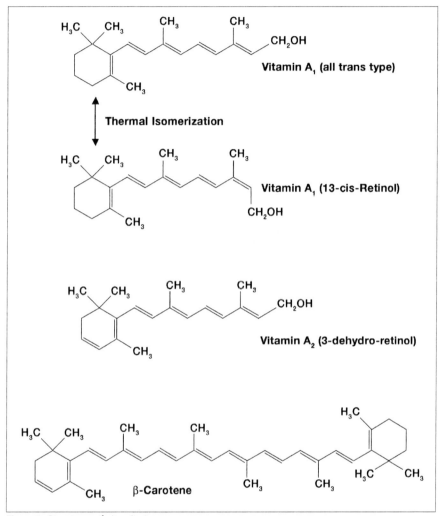

Figure 1. Structures of vitamin A

As with the SLN, the SLM also plays an important role in stability. So two materials, in this case, act simultaneously to give stability to retinol.

Materials and Measurement Equipment

Materials: In this experiment, we used retinol[b] and two liposomes[c] from commercial sources. This retinol–a viscous yellow oil that crystallizes at low temperatures–is approximately 50% solution of retinol in polysorbate 20, stabilized with BHT and BHA. Other materials were also from commercial sources. We used purified water that had been passed through an anion-cation exchange resin column. The TSS was prepared according to the steps described earlier.

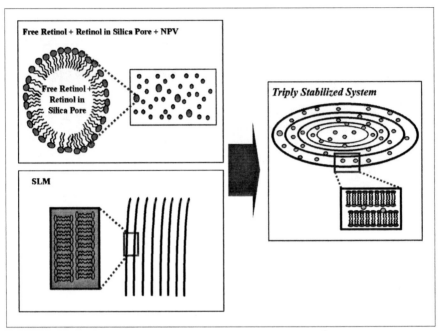

Figure 2. Schematic diagram of the Triply Stabilized System

Analysis equipment: We used transmission electron microscopy (TEM) and freeze-fracture scanning electron microscopy (FF-SEM) to observe liposome formation. The particle sizes of liposomes were measured by using a laser light scattering system[d] and the evaluation of color change was performed using a chromameter[e]. We performed quantitative analysis of retinol by using HPLC[f] under the following conditions: UV spectrophotometer (325 nm) detector; C18 column (3.9 X 150 mm); 1.0 mL/min flow rate; methanol:water (90:10) solution as a mobile phase.

Method

Color stability against light and heat: The color stability after one month was measured for each liposome, for the general liposome and the TSS, under storage conditions of 25°C without light, 40°C under normal lighting conditions and at room temperature under normal lighting conditions.

Survivability of retinol: Under the conditions of 25°C without light, 40°C and light exposure for one month, we carried out quantitative analysis on variation of retinol content.

Skin penetration effect: We prepared two different types of creams: one contained 2% of general liposome and the other contained 2% of the TSS.

[b] Retinol 50C (1.58 Million I.U./g), BASF, Ludwigshafen, Germany.
[c] Lipoid S 100-3 and Lipoid S 75-3, Lipoid GmbH, Ludwigshafen, Germany
[d] Zetasizer 3000H, Malvern Instruments Ltd, Malvern, Worcestershire, UK
[e] Color JS 555, Color Techno System Corporation, Tokyo, Japan
[f] Model 510, Waters Corporation, Milford, Massachusetts, USA

> **Editor's Note: Some techniques suggested for improving the stability of retinol in cosmetic formulations**
>
> The Retinol 50 C used in the Triply Stabilized System from H&A PharmaChem is one of two retinol active ingredients available from BASF. In this note the editors report on some alternative approaches to stabilizing retinol in cosmetic formulations.
>
> RetiSTAR Stabilized Retinol from BASF is an oily dispersion containing retinol, tocopherol and sodium ascorbate in caprylic/capric triglycerides. The ratio and absolute concentration of these three ingredients is reportedly well balanced to achieve maximum stability (at least 12 months) of retinol in cosmetic emulsions without the need for manufacturing or packaging under inert gas. (RetiSTAR is a registered trademark of BASF Aktiengesellschaft, Ludwigshafen, Germany.)
>
> A Liu et al patent (US Pat 5,976,555) assigned to Johnson & Johnson from 1996 describes topical o/w emulsions containing retinol and other retinoids and possessing good physical and chemical stability. The compositions contain an emulsifier system, a co-emulsifier, an oil phase and the retinoid. A stabilizing system may contain a chelating agent or antioxidants or both. The compositions are claimed to retain at least about 70% of the retinoids after 13 weeks storage at 40°C.
>
> A search through the archives of *Cosmetics & Toiletries* magazine disclosed the following recent articles touching on retinol and its stability:
> - Don Orth et al, Stability and skin persistence of topical products, *Cosmet Toil* 113(10) 51-63 (1998)
> - Song et al, Polyethoxylated retinamide as an anti-wrinkle agent, *Cosmet Toil* 114(6) 53-58 (1999)
> - H Zi, Retinyl palmitate at 5% in a cream: Its stability, efficacy and effect, *Cosmet Toil* 114(3) 61-70 (1999)
> - A Jentzsch, Defending against photoaging: A new perspective for retinol, *Cosmet Toil* 117(10) 58 (2002)

After applying the creams to the skin of a hairless mouse, we examined the retinol penetration into the skin after 12 h. The experiment was carried out in Frantz-type diffusion cells. Also HPLC was used for quantitative analysis of extracted retinol.

Results and Discussion

Identifying formation of liposomes: We confirmed the primary liposome, SLM and TSS by using TEM and FF-SEM. As shown in Figure 3, primary liposomes had a spherical shape, as expected. SLM had multi-lamella sheets. And the TSS had a mixture of primary liposome and SLM.

The Sol-Gel Method

The sol-gel method–a simple and cheap process–involves the phase transformation of colloidal suspension (sol) into a continuous liquid phase (gel) by gelation of the sol to form a network; then this gel is going to be densified by thermal annealing. The precursors for preparing these suspensions consist of metal or metalloid elements surrounded by various reactive ligands. Metal alkoxides are mainly used because they easily react with water. The preferred metal alkoxides are alkoxysilanes such as tetramethoxysilane (TMOS) and tetraethoxysilane (TEOS). Three reactions are generally used to describe the sol-gel method: hydrolysis, alcohol condensation and water condensation.

Particle size distribution: A laser light scattering system was used to measure particle size distribution. The sizes of primary liposomes ranged from 20 – 150 nm; the mean size was 70 nm (Figure 4). Particles distributed in the TSS were 20 – 1,000 nm; the mean size was 210 nm. Broad distribution of TSS is attributed to mixing of primary liposomes and SLM.

Color stability against light and heat: In the case of exposure to light, the TSS was approximately twice as stable as the general liposome (Figure 5). For 40°C and 25°C without light, it was more stable by factors of 1.5 and almost 3, respectively. Considered as a whole, the TSS was more stable compared to the general liposome. We believe the reason for stability in our system is that retinol is triply stabilized; that is, it is stabilized in porous silica, NPV and SLM.

Survivability of retinol: Figure 6a shows the content change under light exposure. In the general liposome system, after 30 days, retinol content remaining was about 35%. In our system, its content was about 80%. Under storage at 40°C (Figure 6b), retinol content in the general liposome system was decreased to about 55% and about 90% in the TSS. When stored at 25°C without light exposure (Figure 6c), the general liposome system retained about 90% of its retinol, while the TSS stored similarly retained about 95% of its retinol. The enhanced survivability is again attributed to the triple stabilizing actions of the TSS.

Skin penetration effect: We prepared two different types of cream containing 2% of general liposome and 2% of TSS. We had already determined the initial concentration of retinol by HPLC analysis.

We applied the creams onto the skin of a hairless mouse. The retinol that had penetrated into the skin was extracted after 12 h. Then we calculated the penetration ratio in percent as [(the concentration of retinol penetrated into the skin after 12 h) / (the initial concentration of retinol)] x 100.

Figure 7 shows that the amount of retinol that penetrated the mouse skin after 12 h was 60% greater from TSS than from the general liposome. If we consider only the particle size of the liposome, and if we assume the general liposome is the same size as the primary liposome because they both have the same composition

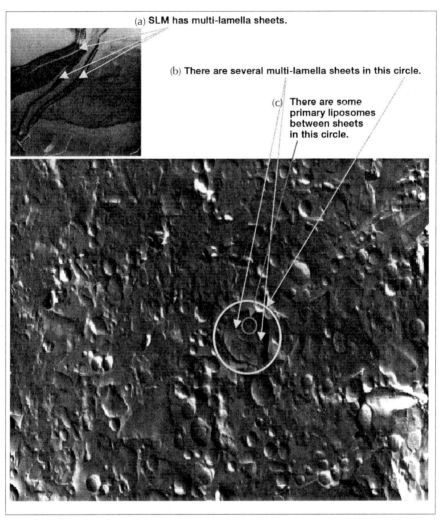

Figure 3. Images of stages of the Triply Stabilized System
a = TEM image of primary liposomes
b = FF-SEM image of SLM showing multi-lamella sheets
c = FF-SEM image of the TSS showing some primary liposomes between sheets

except for the presence of porous silica, we would expect the general liposome to have a better penetration rate than TSS (which has a larger particle size). But we discovered that with TSS, the penetrating materials are actually primary liposomes, not the whole TSS, and these primary liposomes are similar to general liposomes in size. That is, our SLM had multi-lamellar structure. When TSS is in contact with skin–especially the stratum corneum–only the primary liposomes in the outermost layer of TSS are released and pass into the skin.

Because the lamellar layers of SLM are composed of ingredients that are similar to skin components, they have good skin affinity. Therefore, layers will

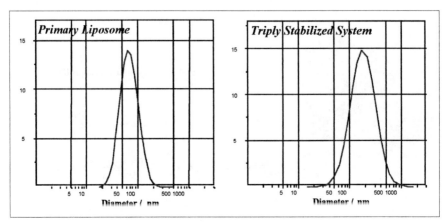

Figure 4. Particle size distribution

be fused with skin. Primary liposomes in a new outermost layer are released and pass through. This process repeats many times. So unlike the retinol in a general liposome that has only one or just a few bylayers, retinol in the TSS is not released all at once. That means retinol is slowly and steadily released from the outermost lamellar layer to the inmost lamella. Therefore, penetration efficiency is enhanced due to SLM. This fact is confirmed by Figure 7, but we think we need more data (such as retinol concentration penetrated into the skin vs. time) to more fully explain this phenomenon.

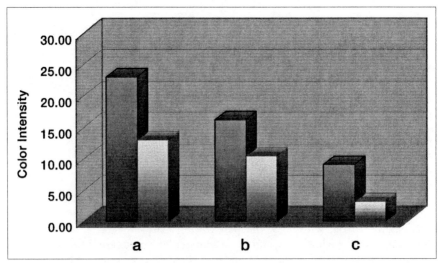

Figure 5. Color stability against light and heat for general liposome (dark gray) and TSS (light gray)
a = Normal lighting and room temperature
b = Normal lighting and 40°C
c = No light and 25°C

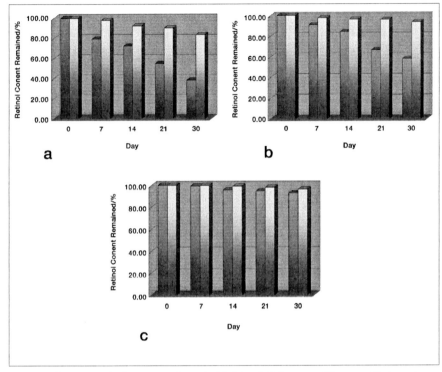

Figure 6. Change in retinol content at day 0, 7, 14, 21 and 30 for general liposome (dark gray) and TSS (light gray)
a = Normal lighting and room temperature
b = Normal lighting and 40°C
c = No light and 25°C

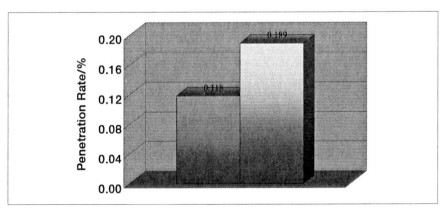

Figure 7. Skin penetration rate (%) of general liposome (dark gray) (0.118%) and TSS (light gray) (0.189%)

Conclusions

The skin adsorption and bio-availability of retinol can be improved by putting the retinol in a general liposome (a non-phospholipid vesicle), but the general liposome is relatively unstable.[12] In order to improve its stability, we used TSS, a system that triply encapsulated the retinol. TSS was prepared by combining primary liposomes (a non-phospholipid vesicle containing both free retinol and retinol adsorbed in the pores of silica) and a matrix of skin lipids. Primary liposome was 20 – 150 nm in size; its mean diameter was 70 nm. In the TSS, particle size was 20 – 1,000 nm; mean diameter was 210 nm.

Our system was more stable than general liposomes. According to chromameter data, the color stability of the TSS was 1.5 – 3 times greater than the color stability of general liposome systems. We also confirmed through HPLC analysis that retinol in our system was more long lasting. The TSS also improved skin penetration of retinol. Finally, The TSS delivers these benefits at low cost (25-80% cheaper than other retinol products commercially available in the Korean market) and with a high content (approximately 4.5%) of pure retinol.

—Hong Geun Ji and Jung Sik Choi, *H&A PharmaChem, Bucheon-si, Korea*

References

1. B Idson, Vitamins and the skin, *Cosmet Toil* 108(12) 79-94 (1993)
2. CM Lee et al, Review of animal models in carotenoid research, *J Nutr* 129 2271-2277 (1999)
3. CC Willhite, Structure-activity relationships of retinoids in developmental toxicology. II. Influence of the polyene chain of the vitamin A molecule, *Toxicol Appl Pharmacol* 83 563-575 (1986)
4. H Oikarinen et al, Modulation of procollagen gene expression by retinoids, *J Clin Invest* 75 1545-1553 (1985)
5. HB Sauberlich, Bioavailability of vitamins, *Prog Food Nutr Sci* 9 1-33 (1985)
6. SC Lee et al, Stabilization of retinol through incorporation into liposomes, *J Biochem Molecular Biol* 35 358-363 (2002)
7. CJ Kirby, Controlled delivery of functional food ingredients: Opportunities for liposomes in the food industry; in *Liposome Technology II*, G Gregoriadis, ed, Boca Raton: CRC Press (1984) 215-232
8. HH Kim and IC Baianu, Novel liposome microencapsulation techniques for food applications, *Trends Food Sci Technol* 2 55-61 (1991)
9. G Fex and G Johannesson, Retinol transfer across and between phopholipid bilayer membranes, *Biochimet Biophy Acta* 944 249-255 (1988)
10. AM Young and G Gregoriadis, Photolysis of retinal in liposomes and its protection with tocopherol and oxybenzone, *Photochem & Photobiol* 63 344-352 (1996)
11. AK Singh and J Das, Liposome encapsulated vitamin A compounds exhibit greater stability and diminished toxicity, *Biophys Chem* 73 155-162 (1998)
12. IF Uchegbu et al, Non-ionic surfactant based vesicles (niosomes) in drug delivery, *Intl J Pharmaceutics* 172 33-70 (1998)

Delivering Antiaging Actives

Keywords: moisturizers, nanoparticles, microcapsules

A discussion of award-winning products with interesting new active ingredients and delivery systems

It takes two to tango; in the battle against aging, cosmetic chemists are dancing with both new ingredients and new ways to deliver them. This point was illustrated in April 2005 when the Cosmetic Executive Women (CEW) announced the 2005 Beauty Award winners. The 3,400 CEW members vote annually on awards for outstanding product innovation in 25 categories of the beauty industry.

In the category "Moisturizer $20 and Over" there were five finalists, each judged to be innovative in terms of its patented technology or R&D breakthrough. As described on CEW's Web site[1] and summarized in Table 1, these finalists illustrate innovation in both ingredients and delivery systems. In this chapter we'll look at these five moisturizers and see how their innovations fit into related trends in anti-aging products. (Just to avoid further suspense, CEW voters chose Lancôme's Aqua Fusion Continuous Infusing Moisturizer as the winner in the "Moisturizer $20 and Over" category.)

Actives and Efficacy

Among the ingredients cited by CEW in the high-end moisturizer finalists are the following:

- Phytosan in Avon's Anew Retroactive+ Repair Face Cream is a trademarked product of Chemisches Laboratorium Dr. Kurt Richter GmbH (Berlin). Phytosan contains proteins, glycoproteins and polysaccharides isolated from soybeans. In vitro and in vivo tests show it protects against ATP decrease after UV irradiation, stimulates protein biosynthesis, reduces skin redness after UV irradiation and reduces the depth of wrinkles.[2]
- Sodium RNA and cat's claw plant extract support skin's defense system, removing damaging substances before they cause harm, according to CEW's description of the Clinique moisturizer. In addition, a new grade of antioxidants is claimed to provide immediate and sustained protection against environmental damage. Sodium RNA is used to aid skin repair, according to one supplier.[3]
- Criste Marine (*Crithmum maritimum*) is a botanical found on rock formations alongside the Mediterranean Sea. This unusual flower is claimed to

Table 1. Finalists for the 2005 Cosmetic Executive Women's beauty award in the category "Moisturizer $20 and Over"

Product (and manufacturer)	Patented technology or R&D breakthrough
Anew Retroactive+Repair Face Cream (Avon)	Rejuvi-cell technology Phytosan
Superdefense Triple Action Moisturizer SPF 25 (Clinique)	Sodium RNA Cat's claw plant extract New grade of antioxidants
Future Perfect Anti-Wrinkle Radiance Moisturizers (Estée Lauder)	Cell Vector technology
Cryste Marine Cream (Kiehl's)	Cryste Marine, a botanical
Aqua Fusion Continuous Infusing Moisturizer (Lancome)	Patented water complex

boost cell renewal for a supple skin effect. Fine lines and wrinkles are visibly softened.[4]

- The patented water complex in the Lancôme Aqua Fusion moisturizer is claimed to be almost identical to skin's own water. It contains 16 essential minerals, such as calcium, manganese and zinc, along with amino acids, organic acids and sugars, all of which are found in young, hydrated, healthy skin.[1]

The array of antioxidants, growth factors and other ingredients used as antiaging actives is broad and broadening. Table 2 lists some of those ingredients along with their antiaging benefits.

Formulators expect suppliers to demonstrate proof of efficacy, and suppliers are complying. For example, Bio-Botanica standardizes its botanical products against marker compounds inherent to the species extracted. Its laboratory uses a variety of analytical instruments, including LC/MS, HPLC, GC/MS and a recently installed ICP-MS (inductively coupled plasma mass spectrophotometer), because the cosmetics and pharmaceutical industries are demanding an advanced level of analysis.[5]

"In the botanical segment, the industry has changed markedly in the last five years, and currently most leading personal care companies demand substantiated actives (i.e., the content of the extract is guaranteed) for use in cosmeceuticals where a performance label claim is to be made," according to Gillian Morris, manager for chemicals practice at Kline & Company, a market research and consulting firm based in Little Falls, New Jersey.

Kline's recently published study, "Competitive Intelligence in Specialty Actives and Active Delivery Systems for Cosmetics and Toiletries 2004: U.S. and Western

Table 2. Antiaging benefits claimed for active ingredients obtained from vitamins (V), acids (A), enzymes (E), growth factors (G), peptides (P), yeast (Y) or botanical extracts (B)[5]

Type	Active	Skin care antiaging function
V	Vitamin A (Retinol)	Antioxidant to inhibit lipid peroxidation Stimulate collagen synthesis
A	Alpha hydroxy acid (glycolic, lactic, malic)	Reduce blotchiness Increase epidermal shedding Stimulate production of collagen Increase the presence of hyaluronic acid in dermis and epidermis
A	Alpha-lipoic acid	Break down in the cell to dihydrofolic acid, an antioxidant Assist in the recycling of antioxidants such as vitamins C and E Anti-inflammatory
E	Coenzyme Q-10	Antioxidant Protect cells from UV-induced oxidative damage
E	2,3-dimethoxy-5-methyl-6-(10-hydroxydecyl)-1,4-benzoquinone*	Antioxidant (synthetic analog to coenzyme Q-10)
G	N(6)-furfuryladenine (kinetin)	Slows certain changes associated with cellular aging Antioxidant
P	Copper peptides	Increase collagen synthesis Stimulate angiogenesis to firm the skin Stimulate glycosaminoglycan synthesis to firm the skin
P	Lys-Thr-Thr-Lys-Ser	Stimulate collagen synthesis
P	Acetyl hexapeptide-3 (Argireline, Lipotec)	Temporarily paralyze muscles affecting crow's feet Temporarily paralyze muscles affecting wrinkle lines
P	Hexapeptide (Minoxinol LS 9736, Inhibit facial muscle contraction Cognis)	Inhibit free radical oxidation
P	Hexapeptide (Peptamide 6, Arch)	Firm the skin Deplete the effect of lines
Y	Ozone-stressed yeast lysate (Biodynes O3, Arch)	Protects skin from ozone and environmental pollution
B	Rhodiola extract (NAB Rhodiola Extract, Arch)	Helps skin adapt to changes in thermal stress
B	Mushroom extract (Kombuchka, Croda)	Lipo-filling Smoothes skin by adding volume through fat cell differentiation Smoothes skin by adding volume through fat cell deposition
B	Centipeda cunninghamii extract (Phytoplenolin, Bio-Botanica)	Anti-inflammatory Renews cells Screens sunlight

*Idebenone, Allergan

Europe," pegs U.S. and Western European consumption of specialty active ingredients—compounds that provide a demonstrable therapeutic benefit to the skin or hair—and the systems used to deliver these ingredients, at more than US$500 million at the manufacturer level and rising steadily.

"Specialty actives and delivery systems represent the only growth area in the overall U.S. and European cosmetics and toiletries market for the short term," Morris said. "Suppliers of specialty actives and delivery systems—particularly nanoparticles, peptides and substantiated botanicals—have to come up with new performance claims and innovations to stay competitive."

Delivery Systems

As categorized by Kline & Company, the most important categories of delivery systems (aside from skin patches) are nanoparticles, nanoemulsions, microcapsules and millicapsules, in order of increasing size from less than 100 nm to approximately 4 mm.[6] Nanoparticles and microcapsules (as well as skin patches) are used in antiaging skin care products.

Nanoparticles include various types of delivery systems that are broadly grouped on the basis of size and ease of penetration. The most commonly used base/starting material for nanoparticle formation is lecithin or fractionated phospholipids. Nanoparticles can be subdivided on the basis of the encapsulating membrane structure into liposomes and nanoemulsions/nanosomes/nanotopes.[6]

Nanoparticles are used to deliver such actives as tocopherol, vitamin E acetate, retinol, retinyl palmitate, ascorbyl palmitate, magnesium ascorbyl phosphate, sodium ascorbyl phosphate, lactic acid and glycolic acid, among others. Initially used in prestige brands, these delivery systems have now transitioned across all market sectors, including mass, direct and specialty markets (see sidebar on trends in professional antiaging products).

"We already have excellent active ingredients, so now we're finding better ways to get them into the skin," said Emma Palfreyman in a *New York Times* article from January 2005.[7] Palfreyman is a senior scientist for the Olay division of Procter & Gamble.

P&G's concept is to use the existing anti-wrinkle ingredients, but add tiny mineral spheres to Olay's latest body lotion and night cream as a way to get those ingredients into the skin. The mineral spheres are microscopic particles (polyethylene and aluminum starch octenyl succinate) that act like mini-ball bearings, gliding on the surface of the skin to make the product feel soft and silky.

"The particle system allows us to use high levels of active—such as glycerin and niacinamide—in the lotion and the product still feels pleasant to use. Women will be more likely to use the product regularly and also rub the product in for longer. Thus more active gets delivered to the skin where it can work to moisturize and improve dryness," said Judith Russell at P&G's UK corporate press office.

Nanosomes are used at L'Oréal to get pure vitamin E into the skin. They are used particularly to transport lipophilic ingredients. Given that the interstices of the outer layer of skin measure about 100 nm, these tiny, rigid-walled, intercellular "vesicles" with a diameter of roughly 20-50 nm alter the skin's affinities, penetrate it more easily and then release their active ingredients. Nanosomes can distribute the vitamin throughout the outer layers of the skin, providing significant amounts of vitamin E to help block the attack of free radicals, according to L'Oréal's Web site.[8]

> **Trends in the Professional Skin Care Market**
>
> While much of the growth in antiaging product sales has been attributed to traditional retail brands, suppliers and marketers also need to take note of a burgeoning segment: professional skin care brands.
>
> Professional skin care products—those sold in spas and salons, retail stores and doctor's offices—have outpaced the general facial treatments segment, which has been expanding at about 6.5% a year over the last five years. Since 1998, professional skin care products have been expanding at a robust average annual rate of 12.3%, according to Kline & Company's recent study titled "U.S. Professional Skin Care Market 2004."
>
> "With over $675 million in sales in 2003, the professional products market could prove attractive for marketers of traditional cosmetic and toiletry products," says Carrie Bonner, industry manager for Kline's Consumer Products Practice. "It might not become a core channel of distribution for a large marketer, but it could be an attractive alternative channel to drive growth and build sales, especially since many traditional outlets have become stagnant."
>
> Source: Kline & Company[11]

L'Oréal's 20 years of experience in the field of liposomes enabled its research and development group to acquire wide-ranging experience with vectors. A vector is an agent capable of transporting active ingredients to the sites where they are most useful, in as specific and long-lasting a way as possible. A good vector should protect the active ingredients during their "journey" through the skin, hone in on its target and release the ingredients on arrival.

Among the high-end moisturizer finalists considered by CEW, only one reported an innovative delivery system,[a] It is the "Cell Vector" technology from Estée Lauder. It claims to target areas of skin deficient in moisture and radiance, and increase skin's antioxidant concentration. Future Perfect Antiwrinkle Moisturizer, launched earlier in 2005, uses this technology in a formulation that includes refined vitamin C, muru muru butter and coffee seed extract, among other ingredients.

What are cell vectors? Lauder's Web site describes them as interactive delivery agents specialized into three key types: Antiaging, Radiance and Hydra Cell Vectors.[9] Each reportedly contains several ingredients.

- **Antiaging.** Vectorized Vitamin C, Siegesbeckia, high-potency antioxidants like NDGA and proven anti-irritants help boost and maintain skin's natural production of collagen and elastin, neutralize free radicals and calm irritation.
- **Radiance.** A unique "recharge cocktail" of NADH, AMP and creatine complements the natural power source ATP to increase skin's energy levels, and restore its youthful look, vitality and inner light.
- **Hydration.** (Sorry, Lauder gives no details on this Vector.)

---

[a] Details on Avon's Rejuvi-cell technology are scant, except that it reportedly is patent pending, works deep within the skin and improves cell communication.

Another patented delivery technology is Novasomes, developed for pharmaceutical applications at Novamax Inc. in Malvern, Pennsylvania. Drugs or other materials can be encapsulated in Novasomes for delivery into the body topically or orally. Novasomes are made from a variety of readily available chemicals called amphiphiles, using the company's patented manufacturing processes. Amphiphiles include fatty alcohols and acids, ethoxylated fatty alcohols and acids, glycol esters of fatty acids, glycerol fatty acid mono and diesters, ethoxylated glycerol fatty acid esters, glyceryl ethers, fatty acid diethanolamides and dimethyl amides, fatty acyl sarcosinates, alkyds and phospholipids.

IGI in Buena, New Jersey, uses the technology to make nano-sized Novasome capsules that reportedly can deeply penetrate skin and do not degrade while on the shelf. Chief executive Frank Gerardi said top customers include Johnson & Johnson for its Neutrogena skin care line, and Estée Lauder for several product lines including Renutriv, Resilience and others.[10]

Summary

By employing novel delivery systems like nanoparticles, companies like L'Oréal and Estée Lauder and their competitors have not only introduced several new product lines, they've also been able to breathe new life into existing ones. This has led to double-digit growth rates for skin care products—the only segment to show such excitement for the short term in an otherwise mature cosmetics and toiletries market, according to Kline's Gillian Morris.

"Skin care marketers have always focused a lot of attention on finding better and more powerful active ingredients, but they're also developing new and more effective ways to deliver the actives they're already using," Morris says. "These new delivery systems can provide more targeted results to the application site and can really differentiate a marketer's product from the rest of the field."

—**Bud Brewster,** *Allured Publishing Corporation, Carol Stream Illinois USA*

References

1. Who knows beauty best? A press release from the Cosmetic Executive Women (Apr 1, 2005) http://cew.org/eventa/beauty/ba2005/ba05finalistmoisture/ (visited Apr 30, 2005)
2. CLR products, Chemisches Laboratorium Dr. Kurt Richter GmbH (undated product brochure)
3. http://www.connock.co.uk/product_specials.htm (visited Apr 30, 2005)
4. http://www.kiehls.com/Default.aspx?p=580&sc=580&st=&cc=134 (visited Apr 30, 2005)
5. C Boswell, Cosmeceutical actives build on success, a focus report in *Chemical Market Reporter* (Oct 18, 2004) p FR10
6. GS Morris, Forever Young?... Specialty Actives and Active Delivery Systems for Anti-Aging Skin Care Applications, an undated presentation to the New York Society of Cosmetic Chemists, http://www.nyscc.org/news/archive/tech0402.htm (visited Apr 5, 2005)
7. CH Deutsch, Cosmetics break the skin barrier, *The New York Times* (Jan 8, 2005) www.nytimes.com/2005/01/08/business/08skin.html
8. www.lorealusa.com/research/nanosomes.aspx (visited Apr 5, 2005)
9. http://www.esteelauder.co.uk/templates/products/sp_nonshaded.tmpl?CATECORY_ID=CAT2724&PRODUCT_ID=PROD62524 (visited Apr 30, 2005)
10. http://www.askigi.com/current_customers.htm (visited Apr 5, 2005)
11. Innovative specialty actives and delivery systems spur growth in skin care (Nov 1, 2004) http://www.kilnegroup.com/6_20041101.htm (visited Apr 5, 2005)

Index

symbols
β-glucan 73, 123

a
AHAs... 245
anatomy.. 25
anti-acne 293
antiaging................................. 93, 293
anti-inflammation 293
anti-inflammatory 123, 181, 189
anti-inflammatory agents................. 3
anti-irritant.................................. 123
antioxidant 93, 111, 277, 373
antioxidants........................... 287, 331
anti-pollution 293
Artèmia extract 133
ascorbic acid...........................271, 467
ascorbyl 2-phosphate................... 277

b
barrier function 433, 441
barrier repair 313
basement membrane.................... 401
beratinocytes............................... 433
botanical blend 161

c
cell communication 313
Centella asiatica................... 103, 189
ceramides..................................... 433
cherry tree leaves 111
clinical trial grading 455
collagen 45, 167, 173, 301
collagen III 359
collagen degradation 161
collagen growth 161

connective tissue 103
copper peptides............................ 411
curcuminoids 181
cytokines 359

d
d-α-tocopherol 339
desensitization 301
diet supplement................... 373, 391
DNA... 133
DNA damage................................. 61
DNA repair.................................... 61

e
elastase 3
elasticity.......................................391
encapsulation 481
Engelhardtia chrysolepis................. 3
essential fatty acid 441
estrogren 173
eye 25
extracellular matrix.............. 229. 401

f
ferulic acid 155
fibroblast activity 359
free radicals 61, 245, 287, 373

g
galactomannan............................... 73
genistein...................................... 147
ginseng.. 167
green tea 93

h
hair care 49, 293

hair growth................................. 173
heat shock proteins...................... 213
homeostasis.................................. 441
honeysuckle flower.......................... 3
hyaluronan................... 221, 229, 245
hyaluronic acid 45, 221, 229, 245
hyaluronidases.............................. 245
hydration 391
hydroquinone 255

i

inflammation................. 229, 313, 441
integrins.. 35
isoflavone (soy)..................... 147, 173

k

keratinocyte differentiation........... 95
keratinocytes................................ 359
kojic acid..................................... 255

l

laminin 401
light imaging techniques.............. 455
lipid.. 441
lipid/protein synthesis 95
liposome.............................. 359, 481
long-term trial.............................. 449

m

Macrocystis pyrifera..................... 263
madecassoside 189
maple tree leaves.......................... 111
matrix................................... 17, 267
matrix metallo proteinases 331
melanin 155, 383
melatonin..................................... 287
metalloproteinases................ 17, 263
metalloproteinase inhibitors..........267
microcapsules 491

minimal erythemal dose (MED) 17
moisturization..................... 301, 441
moisturizers 491
moisturizing................................. 339

n

nanoparticles................................ 491
nerve growth factor 417
neuroprotection............................ 417

o

oat fractions 49
oligosaccharide 293
o/w emulsions 471

p

penetration 467
peptide... 401
percutaneous absorption.............. 323
pH 467
photoaging 17, 277, 331
photographic evaluation............... 455
Phyllanthus emblica..................... 197
pigmentation.......................... 25, 255
polyethoxylated retinamide.......... 349
postmenopausa/skin 147
potassium ayeloyl diglycinate....... 383
product claim support 449
profilometry................................. 449
protein oxidation.......................... 313
phytoplankton.............................. 313

q

Quebracho extract......................... 61

r

Retin-A.. 45
retinol 331, 349, 481

s

sallowness 391
Schizophyllum commune............ 123
seaweed................................ 301
sebum................................. 383
Secale cereale (rye) seed extract...213
skin barrier................................ 471
skin barrier function....................... 95
skin firmness.................. 35, 103, 161
skin homeostasis 313
skin hydration 95, 271
skin lightening 155
skin moisture 471
skin penetration 271
skin protection................................ 95
skin repair...................................... 95
skin restructuring 95
skin sagging.................................. 161
skin whitening.......................267, 383
slimming 301
soft focused/scattered reflection 427
sphericals 427
stress proteins............................... 213
swelling.. 25
synthetic dermis 229

t

tannin 197
terminoloside................................. 189
tocopheryl acetate UV.................. 277

u

ursolic acid................................... 433
UV 35, 49, 123, 133, 181, 197
       229, 245, 267, 287, 313, 331, 417
UVA.. 3
UVB 17

v

vitamin A palmitate 323
vitamin E 339, 471

w

Waltheria indica........................... 155
wound healing 229, 411
wrinkles....... 3, 25, 111, 167, 263, 301
     331, 349, 391, 427, 455, 481

x

xyloglucan 73

For Further Reading...

If you find this book useful, you may be interested in other books from

Global Information Leader

- Preservatives for Cosmetics 2nd Edition 2006
- The Chemistry & Manufacture of Cosmetics
    - Volume I – Basic Science
    - Volume II – Formulating
    - Volume III – Ingredients (2 book set)
- Beginning Cosmetic Chemistry
- Cosmeceuticals: Active Skin Treatment
- Fragrance Applications: A Survival Guide
- Hair Care
- Personal Care Formulas – Revised
- Silicones for Personal Care
- Surfactants: Strategic Personal Care Ingredients
- Physiology of the Skin II
- Asian Botanicals

COMING in 2006:
- Formulating for Sun
- Skin Care Handbook
- Encyclopedia of UV Filters
- Patent Peace of Mind

For more information or to order products, please visit our web site,
www.Allured.com/bookstore or e-mail us at Books@Allured.com
Fax: 630-653-2192

We welcome your comments and ideas for book topics.